Introduction to Applied and Surface Chemis

T0261129

GEORGIOS M. KONTOGEORGIS AND SØREN KIIL
Department of Chemical and Biochemical Engineering,
Technical University of Denmark,
Denmark

WILEY

Library of Congress Cataloging-in-Publication Data

Names: Kontogeorgis, Georgios M., author. | Kiil, Søren, author.
Title: Introduction to applied colloid and surface chemistry / Georgios M.
   Kontogeorgis and Søren Kiil.
Description: Chichester, UK ; Hoboken, NJ : John Wiley & Sons, 2016. |
   Includes bibliographical references and index.
Identifiers: LCCN 2015045696 | ISBN 9781118881187 (pbk.) | ISBN 9781118881217 (ePDF) | ISBN 9781118881200 (ePUB)
Subjects: LCSH: Surface chemistry. | Colloids.
Classification: LCC QD506 .K645 2016 | DDC 541/.33–dc23
LC record available at http://lccn.loc.gov/2015045696

A catalogue record for this book is available from the British Library.

Cover image: Rudisill, PLAINVIEW, Wragg, Margouillatphotos/Getty

Set in 10/12pt Times by SPi Global, Pondicherry, India

1   2016

Contents

Introduction to Applied Colloid and Surface Chemistry

9 Characterization Methods of Colloids – Part II: Optical Properties (Scattering, Spectroscopy and Microscopy)

10 Colloid Stability – Part I: The Major Players (van der Waals and Electrical Forces)

Preface

Colloid and surface chemistry is a subject of immense importance and has implications both to our everyday life and to numerous industrial sectors from paints and materials to medicine and biotechnology.

When we observe nature, we are impressed by mosquitos and other small insects that can walk on water but are drawn into the water when detergents (soaps) are added in their neighbourhood. We are fascinated by the spherical shape of water and even more by the mercury droplets that can roll around without wetting anything. We know that for the same reasons we should use plastic raincoats when it is raining. We are also impressed by some of natural wonders like the "delta" created by rivers when they meet the sea and the non-sticky wings and leaves of butterflies, lotus and some other insects and plants. We are also fascinated by the blue colour of the sky and the red colour of the sunset.

When we are at home we are constantly surrounded by questions related to colloids or interfaces. We would like to know how detergents really clean. Why can we not just use water? Can we use the detergents at room temperature? Why do we often clean at high temperatures? Why do so many products have an expiration date (shelf-life) of a few days or weeks? Why can't milk last for ever? And why this "milky" colour that milk has? Is it the same thing with the well-known drink Ouzo? Why does Ouzo's colour change from transparent to cloudy when we add water? And what about salt? Why does it have such large effect on foods and on our blood pressure? Why do we so often use eggs for making sauces?

Those who have visited the famous VASA museum in Stockholm are impressed by the enormous efforts made in preserving this ship, which sank 400 years ago, after it was taken out of the sea. Why was a solution of poly(ethylene glycol) spayed on the ship?

Of course, similar problems occur in industries that focus on the development and manufacturing of a wide range of products ranging from paints, high-tech materials, detergents to pharmaceuticals and foods. In addition, here it is not just curiosity that drives the questions! Paint industries wish to manufacture improved coatings that can be applied to many different surfaces but they should also be environmentally friendly, e.g. should be less based on organic solvents and if possible exclusively on water. Food companies are interested in developing healthy, tasty but also long-lasting food products that appeal to both the environmental authorities and the consumer. Detergent and enzyme companies have worked in recent years, sometimes together, to develop improved cleaning formulations containing both surfactants and enzymes that can clean much better than before, working on more persistent stains, at lower temperatures and amounts, to the benefit of both the environment and our pocket! Cosmetics is also big business. Many of creams, lotions and other personal care products are complex emulsions and the companies involved are interested in optimizing their performance in all respects and even connecting consumer's reactions to products characteristics.

Some companies, often inspired by nature's incredible powers as seen in some plants and insects, are interested in designing surface treatment methods that can result in self-cleaning surfaces or self-ironing clothes; surfaces that do not need detergents, clothes that do not need ironing!

There are many more other questions and applications! Why can we get more oil from underground by injecting surfactants? Can we deliver drugs for special cases in better and controlled ways?

All of the above and actually much more have their explanation and understanding in the principles and methods of colloid and surface chemistry. Such a course is truly valuable to chemists, chemical engineers, biologists, material and food scientists and many more. It is both a multi- and interdisciplinary topic, and as a course it must serve diverse needs and requirements, depending on the profile of the students. This makes it an exciting topic to teach but also a very difficult one. Unfortunately, as Woods and Wasan (1996) showed in their survey among American universities, a relatively small number of universities teach the course at all! This is a problem, as several universities try to "press" concepts of colloid and surface chemistry (especially the surface tension, capillarity, contact angle and a few more) into general physical chemistry courses. This is no good! This is no way to teach colloids and interfaces. This exciting topic is a science by itself and deserves at least one full undergraduate course and of course suitable books that can be used as textbooks.

This brings us to the second major challenge, which is to have a suitable book for teaching a course to undergraduate students of a (technical) university. More than ten years ago, we were asked to teach a 5 ECTS theory course on colloids and interfaces at the Technical University of Denmark (DTU). The time allocated for a typical 5 ECTS point (ECTS = European Credit System) course at DTU is one four-hour block a week during a 13 week semester, followed by an examination. This course is part of the international program of our university, typically at the start of M.Sc. studies (7th–8th semester) and can be followed by students of different M.Sc. programs (Advanced and Applied Chemistry, Chemical and Biochemical Engineering, Petroleum Engineering). B.Sc. students towards the end of their studies can also follow the course. In Denmark, students submit a written anonymous evaluation of the course at the end of semester, providing feedback on the teaching methods and course content, including course material (books used).

Our experience from 12 years of teaching the course is that we found it particularly difficult to choose a suitable textbook that could fulfil the course requirements and be appealing to different audiences and the increasing number of students. This may appear to be a "harsh comment". First of all, there are many specialized books in different areas of colloid and surface chemistry, e.g. Jonsson *et al.* (2001) on surfactants and Israelachvili (2001) on surface forces. These and other excellent books are of interest to researchers and also to students as supplementary material but are not suitable –and we do not think they were meant by these authors to be – as a stand-alone textbook for a general colloid and surface chemistry course. Then, there are some books, for example Hunter (1993) and Barnes and Gentle (2005), that focus either only on colloid or on surface science. There are, nevertheless, several books that, more or less, target to cover a large part of a standard colloid and surface chemistry curriculum. Examples are those written by Shaw (1992), Goodwin (2004), Hamley (2000), Myers (1991) and Pashley and Karaman (2004). Some of these could and are indeed used as textbooks for colloid and surface chemistry courses. Each of these books has naturally their own strengths and weaknesses. We have reported our impressions and those of our students on the textbooks which we have used in the course over the years in a previous publication (Kontogeorgis and Vigild, 2009) where we also discuss other aspects of teaching colloids and interfaces.

What we can state, somewhat generally, is that unlike other disciplines of chemical engineering (which we know well, both of us being chemical engineers) we lack in colloid and surface chemistry what we could call "classical style" textbooks and with an applied flavour. In other fields of chemical engineering, e.g. unit operations, thermodynamics, reaction engineering and process control, there are textbooks with a clear structure and worked out examples along with theory and numerous exercises for class or homework practice. We could not generalize that "all is well done" in the textbooks for so many different disciplines, but we do see in many of the classical textbooks for other disciplines many common features that are useful to teachers and of course also to students. As the structure in several of these disciplines and their textbooks is also rather established, things appear to be presented in a more or less smooth and well-structured way.

We felt that many of these elements were clearly missing from existing textbooks in colloid and surface chemistry and this book makes an attempt to cover this gap. Whether we succeed we cannot say a priori but the positive comments and feedback from the students to whom parts of this material has been exposed in draft form over the years is a positive sign.

Thus, this book follows the course we have taught ourselves and we hope that the content and style may be appealing to others as well, both colleagues and students. We would like to present the main elements of this book in two respects (i) content and (ii) style.

First of all, the book is divided approximately equally into a surface and a colloid chemistry part although the division is approximate due to the extensive interconnection of the two areas. The first two chapters are introductory, illustrating some applications of colloids and interfaces and also the underlying – for both colloids and interfaces – role of intermolecular and interparticle/intersurface forces. The next two chapters present the concepts of surface and interfacial tension as well as the "fundamental" general laws of colloid and surface science; the Young equation for the contact angle, the Young–Laplace and Kelvin equations for pressure difference and vapor pressure over curved surfaces, the Harkins spreading coefficient and the Gibbs equation for adsorption. We also present in some detail in Chapter 3 various estimation methods for surface and interfacial tensions, with special focus on theories using concepts from intermolecular forces. These estimation methods will be used later (Chapter 6) in applications related to wetting and adhesion. We hope that, already after Chapter 4, it will have become clear that colloid and surface chemistry have a few "general laws" and many concepts and theories, which should be used with caution. Another major result from these early chapters should be the appreciation of the Gibbs adsorption equation, as one of the most useful tools in colloid and surface chemistry. This is an equation that can link adsorption theories, two-dimensional equations of state (surface pressure–area equations) and surface tension/concentration equations. These will be fully appreciated later, in Chapter 7, where adsorption is discussed in detail, as well as in Chapter 14, which discusses multicomponent adsorption theories.

After Chapters 1–4 comes the discussion of surfactants (Chapter 5), solid surfaces with wetting and adhesion (Chapter 6) and adsorption (Chapter 7). In the surfactant chapter, we emphasize the structure–property relationships as quantified via the critical packing parameter (CPP) and the various factors affecting micellization and the values of the critical micelle concentration (CMC). The complexity of solid surfaces is discussed in Chapter 6. Wetting and adhesion phenomena are analysed with the Young equation and the theories presented in Chapters 3 and 4 but several practical aspects of adhesion are discussed as well. Chapter 7 provides a unified discussion of the adsorption at various interfaces. Thus, the similarities and differences (both in terms of physics and equations) of the adsorptions at various interfaces: gas–liquid, liquid–liquid, liquid–solid, solid–gas are shown. We discuss how information from one type of interface, e.g. solid–gas, can be used in analysing data in liquid–solid/liquid interfaces. Finally, the adsorption of surfactants and polymers which is crucial, e.g. in the steric stabilization of colloids, is presented in some detail. Quantitative tools like CPP have also a role here.

Chapter 8 starts the presentation of colloidal properties. Chapters 8 and 9 discuss the kinetic, optical and rheological properties of colloids and we illustrate how measurements on these properties can yield important information for the colloidal particles especially their molecular weight and shape.

Chapters 10 and 11 are devoted entirely to aspects of colloid stability. First, the essential concepts of the electrical and van der Waals forces between colloid particles are presented with special emphasis on the concepts of the zeta potential, double-layer thickness and Hamaker constants. Then, the DLVO theory for colloidal stability is presented. This is a major tool in colloid chemistry and we discuss how stability is affected by manipulating the parameters of by the classical DLVO theory. Chapter 11 closes with a presentation of kinetics of colloid aggregation and structure of aggregates. Chapters 12 and 13 are about emulsions and foams, respectively – two important categories of colloid systems where DLVO and other principles of colloid and surface science are applied. In this case, DLVO is often not sufficient. Steric forces and solvation effects are not covered by the classical DLVO and their role in colloid stability is also discussed in Chapter 12.

Chapter 14 presents three theories that can be used for describing multicomponent adsorption. This is a very important topic, which unfortunately is only very briefly touched upon in most colloid and surface chemistry books. It is also a rather advanced topic which may be omitted in a regular course. The same can be said for Chapter 15, which presents in some more detail, compared to previous chapters, theories for interfacial tension and their strengths and weaknesses are discussed.

Finally, we close with a concluding chapter and some remarks on research aspects in colloid and surface chemistry.

In terms of style, we have attempted to provide a book for those students/engineers/scientists interested in a first course about colloids and interfaces. We have decided to cover the most important topics in a generic form, rather than emphasizing specific applications, e.g. of relevance to food or pharmaceuticals. Nevertheless, many applications are illustrated via the exercises and selected case studies. Each chapter starts with an introduction and ends with a conclusion, when needed with links to what will be seen next. The basic equations are presented in the main text but to avoid "focusing on the trees and losing the forest" the derivations of some equations are, when we feel necessary, added as appendices in the corresponding chapters. Some of these derivations, e.g. those in Chapters 4, 10 and 14, require advanced knowledge of thermodynamics but our aim is to have the main principles and applications of colloids and interfaces understood also by those students who do not have extensive background in thermodynamics or electrochemistry. We have included several worked-out examples dispersed in the various chapters and many exercises at the end (with answers in the book website). A full solution manual is available to instructors from the authors. We have deliberately selected a large variety of problem types, which illustrate different aspects of the theory but also a variety of calculation techniques. Thus, the problems vary from simple calculations – demonstrations of the general laws of the applicability of theories – to derivations, problems from industrial case studies and a range of combined/review problems which are presented at the end of the book.

We have decided, largely because we had to draw a line with respect to book size and purpose at some point, to limit this book on theoretical aspects without extensive discussions of experimental equipment and techniques used in colloid and surface chemistry. Experiments are extremely important for colloids and interfaces and fortunately certain key properties can be readily measured. Even though we do not discuss them extensively we believe, however, that it is important for the student to know which properties can indeed be measured and which cannot – and for which theories are very important. A comprehensive table for this is included in Chapter 1.

We mentioned that we tried to "write a book with a purpose", to balance theory and applications, to present the basic principles of colloid and surface chemistry in an easy to understand way, suitable for students and beginners in the field and with the applied flavour in mind. But we are not aware if we have succeeded. We certainly tried and it has been enjoyable for us to write a book in this way and on this very exciting topic.

If part of this excitement is passed to our readers, students, colleagues and engineers, we have certainly succeeded.

We wish to thank many colleagues for their contribution to this book. We are particularly grateful to Professor Martin E. Vigild who participated together with us in the teaching of the course on "colloid and surface chemistry" over many years and his influence and input is to be found in many places in this book.

Finally, we would like to thank The Hempel Foundation for financial support of this book project. This has enabled us to hire one of our former students, Emil Kasper Bjørn, to help us prepare many of the figures in the book.

Georgios M. Kontogeorgis and Søren Kiil
DTU Chemical Engineering
Copenhagen, Denmark, February 2015

References

G.T. Barnes and I.R. Gentle, 2005. Interfacial Science – An Introduction. Oxford University Press.
J. Goodwin, 2004. Colloids and Interfaces with Surfactants and Polymers. An Introduction. John Wiley & Sons, Inc. Chichester.

I. Hamley, 2000. Introduction to Soft Matter. Polymers, Colloids, Amphiphiles and Liquid Crystals. John Wiley & Sons, Inc. Chichester.

P. C. Hiemenz, and R. Rajagopalan, 1997. Principles of Colloid and Surface Science, 3rd edn, Marcel Dekker, New York.

R. J. Hunter, 1993. Introduction to Modern Colloid Science. Oxford Science.

J. Israelachvilli, 2011, Intermolecular and Surface Forces, 3rd edn, Academic Press.

B. Jonsson, B. Lindman, K. Holmberg, B. Kronberg, 2001. Surfactants and Polymers in Aqueous Solution. John Wiley & Sons, Inc. Chichester.

G.M. Kontogeorgis and M.E. Vigild, 2009. Challenges in teaching "Colloid and Surface Chemistry". A Danish Experience. *Chem. Eng. Educ.*, 43(2), 1.

D. Myers, 1991. Surfaces, Interfaces, and Colloids. Principles and Applications. VCH, Weinheim.

R. M. Pashley and M. E. Karaman: Applied Colloid and Surface Chemistry, John Wiley & Sons Ltd, Chichester, 2004.

D. Shaw, 1992. Introduction to Colloid & Surface Chemistry. 4th edn, Butterworth-Heinemann.

D.R. Woods and D.T. Wasan, 1996. Teaching colloid and surface phenomena. *Chem. Eng. Educ.*, 190–197.

Useful Constants

Acceleration of gravity, $g = 9.8066$ m·s^{-2}

Avogadro Number, $N_A = 6.022 \times 10^{23}$ mol^{-1}

Boltzmann's constant, $k_B = 1.381 \times 10^{-23}$ J K^{-1}

Relative permittivity (or dielectric constant) of water at 20 °C = 80.2

Relative permittivity (or dielectric constant) of water at 25 °C = 78.5

Dielectric permittivity of vacuum, $\varepsilon_o = 8.854 \times 10^{-12}$ C^2·J^{-1}·m^{-1}

Dipole moment unit, 1 $D(ebye) = 3.336 \times 10^{-30}$ C·m

Electronic (elementary) charge, $e = 1.602 \times 10^{-19}$ C

Ideal gas constant, $R_{ig} = 8.314$ J·K^{-1}·mol^{-1} $(= N_A k_B)$

Ideal gas volume, $V = 22414$ cm^3·mol$^{-1} = 2.2414 \times 10^{-2}$ m^3 mol$^{-1} = 22.414$ L·mol^{-1} (at s.t.p, 0 °C, 1 atm)

"Natural" kinetic energy, $k_B T = 4.12 \times 10^{-21}$ J (298 K)

Planck's constant, $h = 6.626 \times 10^{-34}$ J·s

Viscosity of water at 20 °C, 10^{-3} N·s·m$^{-2} = 10^{-3}$ kg m^{-1}·s^{-1}

Viscosity of water at 25 °C, 8.9×10^{-4} N·s·m$^{-2} = 8.9 \times 10^{-4}$ kg m^{-1} s^{-1}

Important unit relationships

N = J m^{-1} = kg m s^{-2}

J = N m = kg m^2 s^{-2}

J = kg m^2 s^{-2}

Pa = N m^{-2}

V = J C^{-1}

Symbols and Some Basic Abbreviations

Latin

A	surface area, m^2
Λ, A_{123}, A_{121}, A_{11}, A_{22}	Hamaker constants, J
A_{eff}	effective Hamaker constant, J
ΔA	change in surface area, m^2
AB	acid-base (interactions, concept)
A_0	area occupied by a gas molecule, m^2
A_{spec}	specific surface area, typically in m^2/g
a	energy parameter in two- or three-dimensional equations of state
a	acceleration, m/s^2
a_0	area of the head of a surfactant molecule, m^2
B	parameter in the Langmuir equation
B_1	first Virial coefficient, m^3/kg
B_2	second Virial coefficient, $m^6/(kg \cdot mol)$
b	co-volume parameter in two- or three-dimensional equations of state
C,c	molar concentration (often in mol/L or mol/m^3) or concentration (in general)
C	parameter in the BET equation
CCC	critical coagulation concentration, mol/L
CFT	critical flocculation temperature, K
CMC	critical micelle concentration, mol/L
CPP	critical packing parameter
d,d_p	(particle) diameter (and d can also be distance), m
D	diffusion coefficient, m^2/s
D_o	diffusion coefficient of equivalent unsolvated spheres, m^2/s
E	electric field strength, V/M
E	elasticity, J/m^2 or mN/m

E	entry coefficient, J/m^2 or mN/m
e	electronic (unit) charge, C
EO	ethylene oxide
F	foam number
F	force, N
F_V	viscous force, N
f	friction(al) coefficient, kg/s
f_0	friction(al) coefficient of unsolvated sphere, kg/s
f_{RPM}	rounds per minute (centrifuge), min^{-1}
g	acceleration of gravity, m/s^2
h_c	critical rupture thickness, m
H	distance between two particles or surfaces or films, m
HB	hydrogen bonds/hydrogen bonding
HLB	Hydrophilic-lipophilic balance
HSP	Hansen solubility parameter $(cal/cm^3)^{1/2}$
I	ionization potential, J
I	ionic strength, mol/m^3
IEP	isoelectric point
K_L	equilibrium adsorption constant, m^3/mol
K_{ow}	octanol-water partition coefficient
k	parameter in the Langmuir equation
k_B	Boltzmann constant, J/K
k_2^{o}	rate constant, $m^3/(numbers \cdot s)$
l	parameter in the Hansen/Beerbower equation
l_c	length of a surfactant molecule, m
LA	Lewis acid
LB	Lewis base
M	molar mass (molecular weight), kg/mol
MW	molecular weight, kg/mol
m	mass (of colloid particles), kg
n_i^{σ}	number of molecules at surface, numbers/m^2

n	refractive index
n	molar amount, mol
n	number of particles per volume, m^{-3}
n_o	initial number of particles per volume, m^{-3}
N_A	Avogadro number, 6.0225×10^{23} molecules/mol=mol^{-1}
N_{agg}	Aggregation (or aggregate) number of a micellar structure
[P]	Parachor
P	(vapour) pressure, Pa
P^{sat}, P_o	equilibrium vapour pressure over a flat surface ("ordinary" vapour pressure), Pa
PZC	point of zero charge
PIT	phase inversion temperature, K
R_{ig}	ideal gas constant, 8.314 J/(mol·K)
R	particle of drop radius, m
R	Hansen radius of solubility $(cal/cm^3)^{1/2}$
R_o	initial radius of bubble, m
R_g	radius of gyration, m
R_f	roughness factor
R_f	horizontal film length, m
R	radius of curved surface (spherical particle, droplet, bubble), m
Re_p	particle Reynolds number
r	intermolecular distance, m
S	Harkins spreading coefficient, N/m or mN/m
S_{eq}	solubility of gas in liquid, mol/(m³·Pa)
s	solubility, mol/L
s	sedimentation coefficient, s
SDS	sodium dodecyl sulphate
T	temperature, K
t	time, s
t	plate thickness, m
$t_{1/2}$	half life, s
T_{br}	T/T_b (T_b = boiling temperature)
V	molar volume, m³/mol
V	volume per g of solid, m³/g
V	potential energy (F=−dV/dH), J or J/m²
V_A	attractive potential energy, J or J/m²
V_R	repulsive potential energy, J or J/m²
V_m	maximum volume occupied by a gas (in adsorption in a solid), cm³/g
V_{max}	maximum value of potential energy, J or J/m²

V_g	gas volume at standard T & P conditions (=22414 cm³/mol)
v_{av}	average drainage velocity, m/s
V_F	volume of foam, m³
V_L	volume of liquid, m³
W	work, J
W	stability ratio
x	mole fraction
x	distance (e.g. in centrifuge), m
$\bar{x}$	Brownian end-to-the-end distance, m
u	velocity (of a colloid particle), m/s
Q	quadrupole moment
w	weight fraction
vdW	van der Waals (forces)
z, z_i	ionic valency (including sign)

Greek

α_0	electronic polarizability, $C \cdot m^2 \cdot V^{-1}$
β	parameter in the Zisman equation
γ	surface or interfacial tension, N/m or J/m²
$\gamma \infty$	infinite dilution activity coefficient
γ_o	water (solvent) surface tension, N/m or J/m²
Γ_i	adsorption of compound (i), mol/g
Γ_{max}	maximum adsorption, mol/g
δ	solubility parameter $(cal/cm^3)^{1/2}$
δ	adsorbed layer thickness, m
δ	film (lamella) thickness, m
ΔP	pressure difference across a curved surface / capillary pressure, Pa
ΔG	Gibbs energy change (and of micellization), J/mol
Δh	height change (osmotic pressure), m
ΔH	Enthalpy change (and of micellization), J/mol
ΔH^{vap}	Enthalpy of vaporization, J/mol
ΔS	Entropy change (and of micellization), J/mol
ε_0	permittivity of free space (vacuum), $8.854 \cdot 10^{-12}$ $C^2 \cdot J^{-1} \cdot m^{-1}$
ε	relative permittivity (dielectric constant)
ζ	zeta potential, V
η	viscosity (of dispersion medium), kg/(m·s)
η_o	viscosity of particle-free medium, kg/(m·s)
η_r	relative viscosity

η	number of particles per unit volume, m^{-3}
η_0	start number of particles per unit volume, m^{-3}
ϑ	contact angle
θ	theta temperature, K
κ^{-1}	Debye length (double-layer thickness), m
μ	electrophoretic mobility, $m^2 \cdot V^{-1} \cdot s^{-1}$
μ	dipole moment, $C \cdot m$
π	surface pressure ($=\gamma_w - \gamma$), N/m or mN/m
π_{sv}	spreading pressure ($=\gamma_s - \gamma_{sv}$), N/m or mN/m
π, Π	osmotic pressure, Pa
ρ	(molar) density (mol/m^3) or number density
ρ_L	density of liquid, kg/m^3
σ	surface
σ	shear stress, N/m^2
σ_o	charge density, C/m^2
φ	correction parameter in the Girifalco-Good equation
φ_G	volume fraction of gas
φ_L	volume fraction of liquid
ψ_0	surface potential, V
ω	angular acceleration, s^{-1}

Superscripts and subscripts

A	attraction/attractive
AB	acid/base interactions
adh/A	adhesion
ads	adsorption
coh	cohesion
crit	critical (in critical surface tension, different from critical point
c	critical
d	dispersion
dw	dirt-water
ds	dirt-solid
eff	effective
exp	experimental
EO	ethoxylate group
i	gas, solid or liquid in expressions for surface or interfacial tensions
ind	induction
j	gas, solid or liquid in expressions for surface or interfacial tensions
ij	gas/liquid, liquid/liquid, liquid/solid or solid/solid in expressions for interfacial tension
g	gas
h	hydrogen bonding
H_g	mercury
l,L	liquid
lg	liquid-gas
LW	London/van der Waals
m	mixture
m	metallic bonding/forces
max	maximum
mix	mixing
o	oil (in the "broader" sense used in colloid and surface science)
OA	oil-air interface
OW	oil-water interface
[P]	parachor
p	polar or particle
R	repulsion/repulsive
r	reduced (e.g $T_r = T/T_c$)
sat	saturated
s	solid
s	steric
sd	solid-dirt
spec	specific (forces/contribution)
sl	solid-liquid interface
surf	surfactant
spec	specific (non-dispersion) effects e.g. due to polar, hydrogen bonding, metallic,...
sw	solid-water
sv	solid-vapor (with vapor coming from liquid)
theor	theoretical
w	water
WA	water-air interface
+	acid effects (van Oss-Good theory)
−	base effects (van Oss-Good theory)
1	particle or droplet
2	medium

About the Companion Web Site

This book is accompanied by a companion website

www.wiley.com/go/kontogeorgis/colloid

This website includes:

- PowerPoint slides of all figures from the book for downloading
- Solutions to problems

1

Introduction to Colloid and Surface Chemistry

1.1 What are the colloids and interfaces? Why are they important? Why do we study them together?

Colloid and surface chemistry is a core subject of physical chemistry. It is a highly interdisciplinary subject, of interest to diverse fields of science and engineering (pharmaceuticals, food, cosmetics, detergents, medicine and biology, up to materials and microelectronics, just to mention a few). Being challenging to teach, it is often either incorporated or presented very briefly in general physical chemistry courses or, even worse, completely neglected (Panayiotou, 1998).

Colloidal systems have a minimum of two components. Colloidal dispersions are systems of particles or droplets with the "right dimensions" (the dispersed phase), which are dispersed in a medium (gas, liquid or solid). The medium is called the continuous phase, which is usually in excess. But which are the "right dimensions"? The particles or droplets have

dimensions (or one key dimension) between (typically) 1 nm and 1 μm and their special properties arise from the large surfaces due to precisely these dimensions (Figure 1.1).

However, sometimes even larger particles, with diameters up to 10 or even up to 50 micrometre (μm), e.g. in emulsions, or very small particles as small as 5×10^{-10} m can present colloidal character. Thus, despite the above definition, it is sometimes stated that "If it looks like and if it acts like a colloid, it is a colloid".

Colloids are characterized by their many interesting properties (e.g. kinetic or optical) as well as by observing their stability over time.

The characteristic properties of colloidal systems are due to the size of the particles or droplets (i.e. the dispersed phase), and not to any special nature of the particles. However, their name is attributed to Thomas Graham (Figure 1.2), who was studying glue-like (gelatinous or gum-like polymeric) solutions (from the Greek word for glue which is "*colla*").

Introduction to Applied Colloid and Surface Chemistry, First Edition. Georgios M. Kontogeorgis and Søren Kiil.
© 2016 John Wiley & Sons, Ltd. Published 2016 by John Wiley & Sons, Ltd.
Companion website: www.wiley.com/go/kontogeorgis/colloid

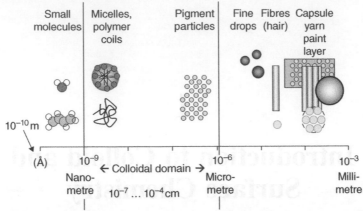

Figure 1.1   *Scales in colloid and surface science. Typically, colloidal particles have one key dimension between 1 nm and 1 μm (micrometre). Adapted from Wesselingh et al. (2007), with permission from John Wiley & Sons, Ltd*

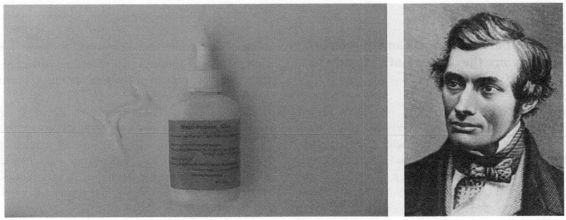

Figure 1.2   *Thomas Graham (1805–1869), the pioneer in the study of colloidal systems, used the term "colloids" derived from the Greek word for glue ("colla"). He thought that their special properties were due to the nature of the compounds involved. Later, it was realized that the size of particles (of the "dispersed phase", as we call it) is solely responsible for the special properties of colloidal systems. (Right) T. Graham, H407/0106. Courtesy of Science Photo Library*

Many colloidal systems like milk are easily identified by their colour, or more precisely their non-transparent appearance (Figure 1.3). The optical properties of colloids are very important, also in their characterization and study of their stability – as discussed in later chapters.

Colloidal particles (or droplets) are not always spherical. They can have various shapes (e.g. spherical and rod- or disk-like), as shown in Figure 1.4. Proteinic and polymeric molecules are usually large enough to be defined as colloid particles. Moreover, their shape may be somewhat affected by solvation (hydration) phenomena, where solvent molecules become "attached" to them and influence their final properties. Solutions of proteins and polymers may be stable and they are classified as lyophilic colloids. Many colloidal particles (e.g. Au or AgI) are (near) spherical, but others are not. For example, proteins are often ellipsoids, while many polymers are random coils.

1.1.1   Colloids and interfaces

What about surfaces and interfaces? Colloidal systems are composed of small particles dispersed in a medium. The fact that these particles have such small dimensions is the reason that a huge surface (interfacial) area is created. Their high interfacial area is the reason why colloidal systems have special properties and also why we study colloids and interfaces together. As shown in Figure 1.5, the surfaces or interfaces are sometimes

considered to be "simply" the "dividing lines" between two different phases, although they are not really lines; they *do* have a certain thickness of a few Å (of the order of molecular diameters).

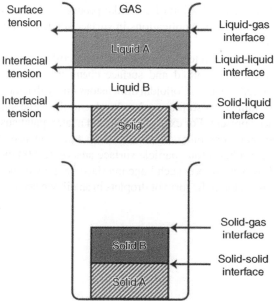

Figure 1.5 *Surfaces and interfaces involving solids, liquids, and gases. An interface has a thickness of a few ångstrøm (1 Å = 10^{-10} m)*

Figure 1.3 *A non-colloidal (water) and a colloidal liquid system (milk)*

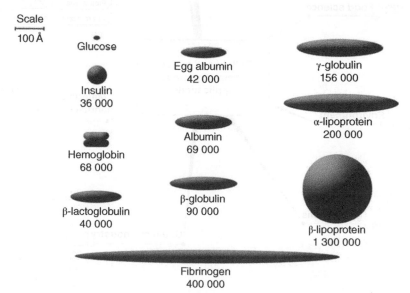

Figure 1.4 *Different shapes of colloid particles with molecular weights provided in g mol^{-1}. Pr J. L. Onclev. Harvard Medical School*

We often use the term "surfaces" if one of the phases is a gas and the term "interface" between liquid–liquid, liquid–solid and solid–solid phases. All of these interfaces are important in colloid and surface science, in the understanding, manufacturing or in the application of colloidal products. However, there are many applications in surface science which are not directly related to colloids.

The huge interface associated with colloids is the reason why colloid and surface chemistry are often studied together. Colloidal dimensions imply that there are numerous surface molecules due to the large surfaces present. For example, 1 litre of a latex paint suspension containing 50% solids with a particle size of 0.2 μm has a total particle surface area of 15 000 m². However, to form such huge interfaces, e.g. by dispersing water in the form of droplets in an oil, we need "to

do a lot of work". This work remains in the system and thus the dispersed phase is *not* in the lowest energy condition. There is a natural tendency for droplets to coalesce and for particles to aggregate. To maintain the material in the colloidal state, we need to manipulate the various forces between particles/droplets and achieve stability. Colloidal stability is one of the most important topics in colloid chemistry.

1.2   Applications

Colloids and interfaces are present and of importance in many (everyday) products and processes, ranging from food, milk and pharmaceuticals to cleaning agents, and paints or glues (Figure 1.6). These are

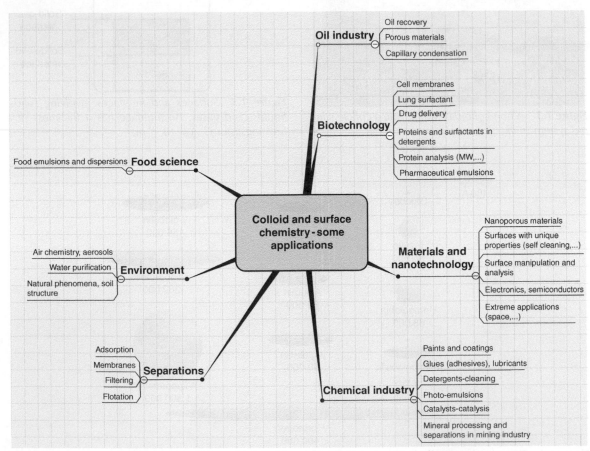

Figure 1.6   *Selected applications of colloid and surface chemistry*

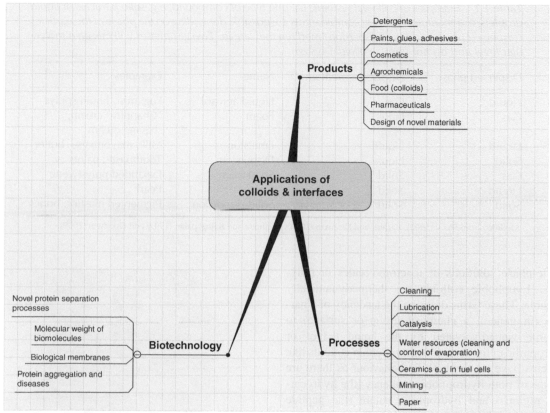

Figure 1.7 *A few applications of colloids and interfaces related to various types of products and processes*

some examples of what we call "structured products". Most of these products are colloidal systems, e.g. milk (liquid emulsion) or paint (emulsions or dispersions). The production and/or use of many colloidal-based products involve knowledge of surface science, e.g. the adhesion of glues and paints or cleaning with detergents. Most of these everyday "consumer" products are rather complex in the sense that they contain many components, e.g. polymers, solids, surfactants, and water or other solvents. As already mentioned, colloids and interfaces are linked and they are best studied together. Figure 1.7 shows some interrelations.

1.3 Three ways of classifying the colloids

Colloids (or colloidal dispersions) can be classified according to the state of the dispersed phase and

the dispersion medium (gas, liquid, solid), see Table 1.1, or according to their stability. The most well-known colloids are emulsions (both phases are liquids), dispersions (solid particles in a liquid medium), foams (gases in liquids), liquids in solids (gels) and aerosols (liquids or solids in a gas).

The common colloidal dispersions (e.g. food or paint) are thermodynamically unstable, while association colloids (surfactants) and polymer/protein solutions are thermodynamically stable. In addition, there can be multiple or complex colloids which are combinations of the above, e.g. dispersion, emulsion, surfactants and/or polymers in a continuous phase. Finally, network colloids, also called gels, are sometimes considered to be a separate category.

Lyophobic (i.e. solvent hating) colloids are those in which the dispersoid (dispersed object) constitutes a distinct phase, while lyophilic colloids refer to

Table 1.1   *Examples of colloidal systems, i.e. one type of compound, e.g. solid particles or liquid droplets, in a medium. Different combinations are possible depending on the phase of the particles (dispersed phase) and the (dispersion) medium they are in. Two gas phases will mix on a molecular level and do not form a colloidal system.*

Dispersed phase	Dispersion medium	Name	Examples
Liquid	Gas	Liquid aerosol	Fog, mist, liquid sprays
Gas	Liquid	Foam	"Chantilly" cream, shaving cream
Liquid	Liquid	Emulsion	Milk, mayonnaise, butter
Solid	Liquid	Dispersion	Toothpaste, paints
Gas	Solid	Solid foam	Expanded polystyrene
Liquid	Solid	Gel	Pearl
Solid	Solid	Solid dispersion	Pigmented plastics, bones

Modified from Shaw (1992), Pashley and Karaman (2004), Hiemenz and Rajagopalan (1997) and Goodwin (2009).

single-phase solutions of macromolecules or polymers. Lyophobic colloids are thermodynamically unstable. These terms describe the tendency of a particle (in general a chemical group or surface) to become wetted/solvated by the liquid (=lyo- or hydrophilic in the case of water). Certain colloids like proteins have an amphiphilic behaviour as there are groups of both hydrophobic tendency (the hydrocarbon regions) and hydrophilic nature (the peptide linkages and the amino and carboxyl groups).

The terms hydrophobic and hydrophilic can also be used for surfaces. Both surfaces and colloid particles can "change" character from hydrophilic to hydrophobic and vice versa. For example, clean glass surfaces are hydrophilic but they can be made hydrophobic by a coating of wax, as discussed by Pashley and Karaman (2004). In addition, the hydrophobic (hydrocarbon) droplets in an oil-in-water emulsion can be made hydrophilic by the addition of protein to the emulsion – the protein molecules adsorb onto the droplet surfaces.

Unstable colloids can be kinetically stable (i.e. stable over a limited time period). The stability of colloids, one of their most important characteristics, is discussed in Chapters 10 and 11.

1.4   How to prepare colloid systems

There are various ways to "trick" particle formations and create a colloidal system. The most important ones are the "aggregation" of molecules or ions and

Figure 1.8   *Ouzo, an example of a colloidal system. The reduced transparency upon addition of water is due to the reduction of anise oil solubility in alcohol*

"grinding" or "milling" methods, typically in a mill/stirrer with the application of shear stress and adding some dispersants, e.g. surfactants. Other methods are based on the precipitation or the reduction of the solubility of a substance in a solvent such as in the case of the well-known Greek drink Ouzo (Figure 1.8), whose opaque colour when water is added is due to the reduction of the alcohol content. In Ouzo's standard state (conventional alcohol content) the drink is colourless because the anise oil fully dissolves in the alcohol. But as soon as the alcohol content is reduced (by adding water), the essential oils transform into white crystals, which you cannot see through (like in milk, another classical colloidal system). The same phenomenon occurs when it is stored in a refrigerator. But Ouzo resumes its former state as soon as it is placed at room temperature.

Typically, the colloids need, after their preparation, to be purified e.g. to remove the electrolytes that destabilize them and there are many techniques for doing that. Among the most popular ones are the dialysis, the ultrafiltration, the size exclusion chromatography (SEC) and the gel permeation chromatography (GPC). The basic separation principle is the size difference between the colloid and the other substances that need to be removed.

1.5 Key properties of colloids

Colloidal systems are special and exciting in many ways. They have very interesting kinetic, rheological and optical properties (Chapters 8 and 9) which are important for their characterization (determination of molecular weight and shape) and application. But their most important feature is possibly the large surface area, and this is why these systems are often unstable (or metastable). The stability of colloids involves the relative balance between the attractive van der Waals and the repulsive forces; the latter are often due to the electrical charge that most colloid particles have. The van der Waals attractive forces in colloids are much stronger than those between molecules and lead to aggregation (instability), but there are (fortunately) also repulsive electrical forces, when the particles are charged, which "help stability". There are other types of repulsive forces, e.g. steric, solvation. Manipulating colloidal stability implies knowing how we can change or influence the various forces, especially the van der Waals attractive and the electrical and steric repulsive forces.

We emphasize, thus, from the start that almost all lyophobic colloids are in reality metastable systems. When we use the term "stable" colloids throughout this book, we imply a kinetically stable colloid at some arbitrary length of time (which can be, for example, two days or two years! depending on the application).

1.6 Concluding remarks

Colloids and interfaces are present and important in many (everyday) products and processes, ranging from food, milk and pharmaceuticals to cleaning agents and paints or glues. They are intimately linked and are best studied together. Colloids have many important, exciting properties of which stability is possibly the most important. Some properties of colloids and interfaces can be measured while others cannot and are obtained best via theories/models. An overview of what can be measured and what cannot in colloid and surface science is given in Appendix 1.1.

Colloids can be classified according to the phase (gas, liquid, solid) of the dispersed phase and the dispersion medium or according to their stability. Colloidal dispersions are thermodynamically unstable, while association colloids (surfactants) and polymer/protein solutions are stable. The former are often called lyophobic (hydrophobic if the dispersion medium is water) and the latter lyophilic (hydrophilic) colloids. These terms can be also used for surfaces.

Crucial in the study of both colloids and interfaces is knowledge of the forces between molecules and particles or surfaces and this is discussed next. While, as explained, a strict division is not possible, Chapters 3–7 discuss characteristics and properties of interfaces (surface and interfacial tensions, fundamental laws in interfacial phenomena, wetting & adhesion, surfactants and adsorption), while Chapters 8–13 present the kinetic, rheological and optical properties of colloids, as well as their stability and also a separate discussion of two important colloid categories, emulsions and foams.

Appendix 1.1

Table A1 *Overview of what can be measured and what can be calculated in the area of colloid and surface chemistry*

Property	Can we measure it? (How?)	Can we estimate it? (How?)	Comments – applications
Surface tension of pure liquids and liquid solutions	Yes (Du Nouy, pendant drop, Wilhelmy plate, capillary rise)	Yes (parachor, solubility parameters, corresponding states)	Wetting, adhesion, lubrication
Interfacial tension of liquid–liquid interfaces	Yes (Du Nouy)	Yes (many methods, e.g. Fowkes, Hansen, Girifalco–Good)	Surfactants
Surface tension of solids		Yes (Zisman plot; extrapolation from liquid data, solubility parameters, parachor)	Wetting and adhesion
Interfacial tension of solid–liquid and solid–solid interfaces		Yes (many methods, e.g. Fowkes, Hansen, van Oss–Good)	Wetting, adhesion, characterization and modification of surfaces… (paints, glues…)
Contact angle between liquid and solid	Yes (many goniometers and other methods)	Yes (combination of Young equation with a theory for solid–liquid interfaces)	Wetting, adhesion, characterization and modification of surfaces…
Critical micelle concentration of surfactants	Yes (change of surface tension or other properties with concentration)		Detergency
Surface or zeta potential of particles	Yes (micro-electrophoresis)		Stability of colloidal dispersions
Adsorption of gases/liquids on solids	Yes (many methods)	Yes (many theories, e.g. Langmuir, Brunauer–Emmett–Teller (BET), Freudlich)	Stability, surface analysis
Topography of a surface	Yes (AFM, STM)		Surface analysis and modification
HLB (hydrophilic–lipophilic balance)		Yes (group contribution methods, solubility parameters)	Design of emulsions including stability of emulsions and determining the emulsion type
Work of adhesion	Yes (JKR, AFM)	Yes (the ideal one is via Young–Dupre and similar equations)	Adhesion, detergency

Table A1   (*continued*)

Property	Can we measure it? (How?)	Can we estimate it? (How?)	Comments – applications
Interparticle forces and colloid stability	Yes (surface force apparatus, AFM and other methods for stability, e.g. Turbiscan)	Yes (DLVO theory)	Stability of all types of colloids (paints, food colloids…)
Molecular weight of polymers and proteins	Yes (many methods, e.g. ultracentrifuge and osmotic pressure)		Characterization of high molecular weight molecules
Creaming and sedimentation of suspensions and emulsions	Yes (Turbiscan)	Yes (Stokes equation for dilute dispersions)	Stability of colloids
Critical coagulation concentration	Yes (series of experiments adding salts in colloidal dispersions until coagulation occurs)	Yes (DLVO theory)	Stability of colloids
Determining emulsion type		Yes (HLB and Bancroft rule)	Emulsion design

Problems

Problem 1.1: Colloids in everyday life and colloid types

Give at least three examples of products with colloids from your everyday life, and describe for each the nature and function of the components. You can use Table 1.1 for inspiration.

Describe the following colloidal categories, by naming the state/form and function of their components: dispersion, aerosol, micelle, emulsion, foam, paste, gel, latex. Place the examples you gave above into one of these categories.

Problem 1.2: Which ones are colloids?

You have the following three ingredients available:

- oil (a liquid)
- polymer (a low molecular hydrophobic liquid)
- water (a liquid)

You are allowed to mix/blend two ingredients together into one sample. How many of the three resulting binary blends could be classified as colloids? For each sample state why/why not. In addition, give details about what the criterion could be to judge whether or not the blend is a colloid. For the samples you have classified as colloids you should give the correct term for the type of colloid the sample belongs to.

Problem 1.3

Try to characterize an interface in terms of thickness, molecular arrangement and importance for product properties.

Problem 1.4

Try to mix 100 ml of (preferably unskimmed) milk with 100 ml of vinegar. What do you observe? Explain what you think has happened (compare this with the Ouzo example in Figure 1.8).

Problem 1.5

What particle shapes are possible for colloids? Give some examples.

References

J. Goodwin, (2009) Colloids and Interfaces with Surfactants and Polymers: An Introduction, John Wiley & Sons, Ltd, Chichester.

P.C. Hiemenz, R. Rajagopalan, (1997) Principles of Colloid and Surface Chemistry, Marcel Dekker, Inc., New York.

K. Panayiotou, (1998) Interfacial Phenomena and Colloid Systems (In Greek). Zitis Publications, Thessaloniki, Greece, 2nd edn.

R.M. Pashley, M.E. Karaman, (2004) Applied Colloid and Surface Chemistry, John Wiley & Sons, Ltd, Chichester.

D.J. Shaw, (1992) Colloid & Surface Chemistry, 4th edn, Butterworth and Heinemann, Oxford.

J.A. Wesselingh, S. Kiil, M.E. Vigild, (2007) Design and Development of Biological, Chemical, Food and Pharmaceutical Products, John Wiley & Sons, Ltd, Chichester.

2

Intermolecular and Interparticle Forces

2.1 Introduction – Why and which forces are of importance in colloid and surface chemistry?

The role of intermolecular and interparticle forces is crucial in colloid and surface science and for this reason we devote this early chapter to a short description of these forces, especially those aspects that are important in practical applications. For an excellent and more detailed discussion of forces, as applied to colloid and surface science, the reader is referred to the book of Israelachvili (1985, 2011).

A useful way to represent intermolecular (and interparticle) forces is via the potential energy–distance function, $V(r)$, which is related to the intermolecular force:

$$F(r) = -\frac{dV(r)}{dr} \qquad (2.1)$$

As a convention, a negative sign indicates attractive forces and a positive sign indicates repulsive forces.

The total energy of molecules includes their kinetic energy, depending on the temperature, and the potential energy, depending on their positions and forces.

First of all, a few words on the importance and practical implications of intermolecular and interparticle forces related to colloids and interfaces. In surface science, we see the intermolecular forces in the discussion of surface and interfacial forces that are directly connected to forces between molecules. Actually, as we will see in Chapter 3, the "surface component" theories of interfacial tension take into account exactly this connection to some of the most important forces, the dispersion, polar and hydrogen bonding ones. Of course, the high surface tension of water with all its implications (e.g. insects walking on water and spherical droplets) is due to the extensive hydrogen bonds of water and to the associated hydrophobic phenomenon. The latter has a crucial role also in micelle formation in surfactant solutions. The nature and value of interfacial tension in liquid–liquid interfaces is connected to the extent of miscibility in these systems, which is in itself linked to the

Introduction to Applied Colloid and Surface Chemistry, First Edition. Georgios M. Kontogeorgis and Søren Kiil.
© 2016 John Wiley & Sons, Ltd. Published 2016 by John Wiley & Sons, Ltd.
Companion website: www.wiley.com/go/kontogeorgis/colloid

interactions between molecules. These are also connected to intermolecular forces. The highest interfacial tensions are seen for mixtures containing water with very hydrophobic compounds (hydrocarbons and even more fluorocarbons). When enhanced interactions, e.g. for water–aromatic hydrocarbons, result in higher miscibility, then the liquid–liquid interfacial tension is also higher. Of course, the intermolecular forces, which are forces between molecules due to their nature and position (potential energy), depend on temperature, directly and indirectly as they are competing with the kinetic energy of molecules (which is proportional to $k_B T$).

As surface/interfacial energies are connected to the intermolecular forces, so also are the related phenomena of wetting and adhesion. In fact, we will see in Chapter 6 that the most widely used theory of adhesion is explained by thermodynamic forces at the interfaces. Enhanced hydrogen bonding forces (or in general acid–base interactions) are often very useful for achieving good adhesion.

The forces between particles and/or surfaces are as important as those between molecules. Actually, the van der Waals forces (typically attractive) between particles or surfaces are of much longer range than the corresponding forces between the molecules, while the electrostatic forces (typically repulsive) between particles/surfaces are also long range but in many cases with a different distance dependency. These two types of forces play a crucial role and often entirely determine the stability of colloidal particles.

Thermodynamic aspects are very important in colloid and surface science. They have been reviewed in several published articles, e.g. in *Current Opinion in Colloid & Interface Science* (Aveyard, 2001; Texter, 2000; Lynch, 2001). This chapter offers a brief introduction to the most important concepts, especially those related to intermolecular and interparticle forces.

2.2 Two important long-range forces between molecules

We first discuss two very important long-range forces, the Coulombic and those originating from hydrogen bonds. For Coulombic forces, the expression for the potential energy is:

$$V_{12} = \frac{q_1 q_2}{(4\pi\varepsilon_0\varepsilon)r} = \frac{(z_1 z_2)e^2}{(4\pi\varepsilon_0\varepsilon)r} \qquad (2.2)$$

where ε_0 is the dielectric permittivity of a vacuum (8.854×10^{-12} C^2 J^{-1} m^{-1}), q_i is the electric charge (C), z_i is the ionic valency, ε is the relative permittivity or dielectric constant (dimensionless), r is the distance between charged molecules, and e is the electronic charge. The relative permittivity is dimensionless because it is the ratio of the permittivity of the material to that of vacuum.

The relative permittivity is a measure of the extent of reduction of electric fields by a given medium and consequently of the reduced strengths of electrostatic interactions in a (polar) medium. A high relative permittivity means that charges are easily maintained in the medium. The relative permittivity of water is about 80, a high value that allows ion dissociation to occur, whereas in air and in non-polar liquids, e.g. hexane or benzene (ε of about 2), we would expect no dissociation. For this reason, most solid dispersions are studied in aqueous media. The permanent dipoles in the water molecules stabilize the charges on ions.

Relative permittivity values are useful in many respects. For example, if we plot the solubilities of sodium chloride and glycine in solvents of different relative permittivities at 25 °C a linear relationship is obtained. Hydrogen bonding solvents are effective solvents and high relative permittivity values imply high solubilities. For example, the solubility of NaCl in acetone is 4×10^{-7} kg NaCl per kg solvent while that of glycine in acetone is 2×10^{-6} kg NaCl per kg solvent.

The very strong Coulombic forces partially explain the difficulties associated with constructing suitable theories for electrolyte solutions (Prausnitz, Lichtenthaler and de Azevedo, 1999). They also form the basis of the electrostatic repulsive forces between colloidal particles, which are discussed in Chapter 10, as the colloids are typically charged in aqueous media.

The Coulombic forces are in the range 100–600 kJ mol^{-1}, and are much stronger than the van der Waals forces (often <1 kJ mol^{-1}) and even the "quasi-chemicals" hydrogen bonds (10–40 kJ mol^{-1}).

However, as the distance increases and especially in media with high relative permittivities, e.g. water ($\varepsilon = 80$), the Coulombic forces can decrease substantially. This is illustrated in Example 2.1 and thus we understand why ions dissociate in water but not in non-polar media.

Unfortunately, we do not know a precise expression for the potential energy due to hydrogen bonding but it is considered that the distance dependency is around r^{-2}, i.e. it is much longer range than the van der Waals forces between molecules and similar to Coulombic forces, though weaker. We can nevertheless present the most qualitative characteristics of the hydrogen bonds and related enhanced interactions.

First of all, hydrogen bonds, i.e. those bonds typically occurring between H and F, O or N atoms, belong to a broader family of enhanced interactions under the name Lewis acid–Lewis base (LA–LB) interactions. Hydrogen bonds explain the dimerization of organic acids in the vapour phase, the oligomeric structure of alcohols and phenols in liquids and the three-dimensional network of water, while LA–LB interactions in general explain the weak complexes between a polar self-associating compound and an aromatic or olefinic hydrocarbon, e.g. water or alcohols with benzene. All of these interactions are extremely important in the interpretation of interfacial phenomena and wetting/adhesion.

Example 2.1. *Coulombic forces and colloids. Why is water the typical medium in colloid chemistry?*
Consider two ions Cl^- and Na^+ in contact ($r = 0.276$ nm) and also at a distance of $r = 56$ nm.

i. What is the inter-ion potential energy at both distances and both in vacuum and in water? (At 25 °C when the relative permittivity of water is equal to $\varepsilon = 78.41$)

ii. Calculate the potential energies of question (i) in both J (per molecule) and kJ mol^{-1} and as fractions of $k_B T$ (at room temperature) where k_B is the Boltzmann constant. How do these energies compare to the kinetic energy ($= 1.5 k_B T$) and to the hydrogen-bonding energy (about 20 kJ mol^{-1})?

Comment on the results.

Solution
Using the Coulomb's equation we have:

$$V_{12} = \frac{q_1 q_2}{(4\pi\varepsilon_0\varepsilon)r} = \frac{(z_1 z_2)e^2}{(4\pi\varepsilon_0\varepsilon)r}$$

For NaCl, $z_1 = +1$, $z_2 = -1$
For vacuum:
At $r = 0.276$ nm $= 0.276 \times 10^{-9}$ m, we get $V_{12} = -8.36 \times 10^{-19}$ J or -503 kJ $mol^{-1} = 203 k_B T = 135 \times$ Kinetic energy (the values of Boltzmann constants and Avogadro numbers are required in these calculations).
At $r = 56$ nm, $V_{12} = -0.0414 \times 10^{-19}$ J or -2.49 kJ mol^{-1}, which is about equal to $k_B T$.
For water:
At $r = 0.276$ nm, $V_{12} = -0.106 \times 10^{-19}$ J or -6.38 kJ mol^{-1}, about $2.6 k_B T$.
At $r = 56$ nm, $V_{12} = -0.0005279 \times 10^{-19}$ J or -0.03 kJ mol^{-1}, about $0.0127 k_B T$.
We note that the forces are very high at contact (actually much higher than the kinetic energy) and thus they dominate in the case of vacuum. However, they reduce dramatically with distance or when the salt is in a polar medium like water and this explains why NaCl dissociates easily to ions in water.
(The kinetic energy is $1.5 k_B T$ (per molecule) $= 1.5 R_{ig} T$ (per mole) and at 25 °C, it is equal to: 6.17×10^{-21} J per molecule or 37.156×10^2 J mol^{-1} or 3.7 kJ mol^{-1})

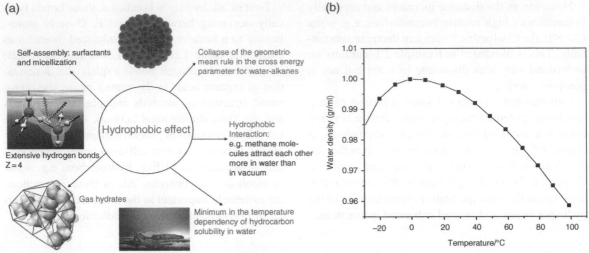

Figure 2.1 (a) Some implications of the hydrophobic effect, one of the most unique properties of water molecules; (b) water density as a function of temperature illustrating the maximum at 4 °C. Adapted from Kontogeorgis and Folas (2010), with permission from John Wiley & Sons, Ltd

Acid–base interactions can be understood and to some extent quantified with various parameters, e.g. the so-called solvatochromic (or Kamlet–Taft) parameters (Kamlet *et al.*, 1983). Self-associating compounds have both an acid and a base parameter.

While the van der Waals forces are always present between molecules (see Section 2.3), when hydrogen bonding is present it often completely dominates the properties of molecules and their mixtures. For example, the boiling temperature of ethanol is 79 °C, while for the isomeric dimethyl ether it is only –25 °C. Ethanol and dimethyl ether have more or less the same polarity (1.7 and 1.3 Debye, respectively).

While much stronger than the van der Waals forces, hydrogen bonds are still weaker (8–40 kJ mol^{-1}) than the ordinary chemical (covalent) bonds (150–900 kJ mol^{-1}, i.e. 100–300k_BT).

Let us now return to water, a liquid of immense importance in colloid and surface science. The very strong hydrogen bonds of water explain many of its special properties, first of all its liquid nature. Water has a boiling temperature, T_b, of 100 °C, with a molecular weight equal to 18 g mol^{-1}; compare this to the gas methane with T_b = –167 °C and a molecular weight of 16 g mol^{-1}. Hydrogen bonding is responsible for many other properties, including the maximum of density at 4 °C, its high relative permittivity and

surface tension and especially the hydrophobic effect (Figure 2.1).

According to the hydrophobic effect, water molecules in liquid water are connected with strong extensive hydrogen bonds and they are "forced" into even more structured cavities (so that hydrogen bonding is restored) if "foreign" non-polar molecules (e.g. alkanes and fluorocarbons) are added to the solution. Thus, water molecules "like themselves" too much and wish to stick to each other and "be far away from enemies" especially the very non-polar molecules. For this reason, water molecules alone are often considered to have 3–3.5 hydrogen bonds/molecules, and this increases to about four hydrogen bonds per molecule in water–alkane systems. This higher degree of local order in water/non-polar mixtures compared to pure liquid water explains the entropy decrease shown in Table 2.1. It is more this loss of entropy rather than enthalpy changes that leads to the unfavourable positive Gibbs energy change associated with the non-mixing of hydrocarbons and other similar molecules in water.

Related to the hydrophobic effect is also the so-called hydrophobic interaction, a term describing the strong attraction between non-polar (hydrophobic) molecules and surfaces in water. This attraction is often stronger than the interaction in the free space

Table 2.1 *Change in standard molar Gibbs energy, enthalpy and entropy (all in kJ mol^{-1}) for the transfer of hydrocarbons from pure liquids into water at 25 °C (Prausnitz, Lichtenthaler and de Azevedo, 1999; Gill and Wadso, 1976). Notice the large negative entropy changes due to the hydrophobic effect. In the case of n-butane, the entropy decrease amounts to 85% of the Gibbs energy of solubilization, while for other hydrocarbons the entropic contribution is even larger*

Hydrocarbon	ΔG (kJ mol^{-1})	ΔH (kJ mol^{-1})	$T\Delta S$ (kJ mol^{-1})
Ethane	16.3	−10.5	−26.8
Propane	20.5	−7.1	−27.6
n-Butane	24.7	−3.3	−28.0
n-Pentane	28.6	−2.0	−30.61
n-Hexane	32.4	0	−32.4
Cyclohexane	28.1	−0.1	−28.2
Benzene	19.2	+2.1	−17.1
Toluene	22.6	+1.7	−20.9
Ethylbenzene	26.2	+2.02	−24.2

Reprinted from Kontogeorgis and Folas (2010), with permission from John Wiley & Sons, Ltd.

and it naturally cannot be explained via the van der Waals forces, which would predict the opposite effect (lower attraction of molecules in a medium compared to water)! Moreover, also associated with the hydrophobic effect is the density–temperature profile which explains why ice floats over water. Without this, life might be impossible in the sea. Water is a molecule of immense importance in colloid and surface science and at the same time, an entirely unique molecule that has created heated debates in the literature. A short account of these discussions is presented in Appendix 2.1.

2.3 The van der Waals forces

2.3.1 Van der Waals forces between molecules

The van der Waals forces between molecules are rather short range; the potential energy decreases with the inverse of the sixth power of the distance of molecules. The van der Waals forces between two molecules 1 and 2 are given by the general expression (for the potential):

$$V_{12} = -\frac{C}{r^6} = -\frac{C^{disp} + C^p + C^{ind}}{r^6} \qquad (2.3)$$

where the individual contributions are due to dispersion (*disp*), polar (*p*) and induction (*ind*) forces and the potential are given by the following equations:

$$V_{12} = -\frac{C^p}{r^6} = -\frac{1}{3 k_B T (4\pi\varepsilon_0)^2 r^6} \mu_1^2 \mu_2^2 \qquad (2.4a)$$

$$V_{12} = -\frac{C^{ind}}{r^6} = -\frac{\alpha_{01}\mu_2^2 + \alpha_{02}\mu_1^2}{(4\pi\varepsilon_0)^2 r^6} \qquad (2.4b)$$

$$V_{12} = -\frac{C^{disp}}{r^6} = -\frac{3}{2} \frac{\alpha_{01}\alpha_{02}}{(4\pi\varepsilon_0)^2 r^6} \left[\frac{I_1 I_2}{I_1 + I_2}\right] \qquad (2.4c)$$

In these equations, k_B is the Boltzmann's constant (1.38×10^{-23} J K^{-1}), T is the temperature, I is the first ionization potential (J), α_{0i} is the electronic polarizability (C^2 m^2 J^{-1}) and μ is the dipole moment ($\mu = ql$) often given in Debye (1 Debye = 3.336×10^{-30} Cm). In the dipole moment equation q is the electric charge (C), and l is the distance between the positive and negative charge within a given molecule (m). The electronic polarizability is defined as the ease with which the electrons of molecules are displaced by an electric field, e.g. that created by an ion or a polar molecule. The polarizability is expressed in m^2 V^{-1} or C^2 m^2 J^{-1}. In volume units, polarizability is expressed as $\alpha_0/(4\pi\varepsilon_0)$. Notice that only dipolar–dipolar (polar) forces are temperature-dependent. This has a relevance, for example, when looking at the temperature dependencies of emulsions (Chapter 12). Dispersion forces are always present.

There are also weaker (shorter range) forces than the van der Waals ones, e.g. when quadrupoles are present (CO_2 is a typical quadrupole), but these are of little importance in colloid and surface science and will not be discussed further here (for more details on these weaker forces see Prausnitz, Lichtenthaler and de Azevedo, 1999; Kontogeorgis and Folas, 2010).

When using the Debye and Keesom expressions (induction and polar forces) for interactions in a medium other than vacuum or air, the permittivity of vacuum could be simply multiplied by the relative permittivity (dielectric constant). However, this cannot be done for the London forces, which are due to fluctuating dipoles. In the general case, the van der Waals forces between *molecules* 1 and 2 in a solvent-medium 3 can be expressed (in terms of the potential) using the electrostatic quantum field theory of McLachlan (1963) (see Israelachvili, 2011, for a detailed discussion):

$$V = V_{\nu=0} + V_{\nu>0} = -\frac{3k_B T}{r^6} R_1^3 R_2^3 \left[\frac{\varepsilon_1 - \varepsilon_3}{\varepsilon_1 + 2\varepsilon_3}\right]\left[\frac{\varepsilon_2 - \varepsilon_3}{\varepsilon_2 + 2\varepsilon_3}\right]$$

$$-\frac{\sqrt{3}h\nu_e R_1^3 R_2^3}{2r^6} \frac{(n_1^2 - n_3^2)(n_2^2 - n_3^2)}{\sqrt{(n_1^2 + 2n_3^2)(n_2^2 + 2n_3^2)}\left[\sqrt{(n_1^2 + 2n_3^2)} + \sqrt{n_2^2 + 2n_3^2}\right]}$$

$$(2.5)$$

$V_{\nu=0}$ = zero-frequency contribution due to polar/induction forces (Keesom/Debye);

$V_{\nu>0}$ = finite frequency contributions due to dispersion (London) forces;

R_i = radius of particle i;

ε_i = static (zero frequency) relative permittivity of the three media;

ν_e = main electronic absorption frequency in the UV region (about 3×10^{15} Hz), assumed to be the same for all three media;

n = refractive index in the visible region (this parameter is a measure of the speed of light in vacuum compared to that in the substance and $n = 1$ for vacuum/air);

k_B = Boltzmann's constant;

h = Planck's constant;

T = absolute temperature.

Repulsive forces due to overlapping electron clouds (Born repulsion) are much less well-understood and are typically represented with a potential energy expression ($V^{rep} = C/r^m$) similar to that of the van der Waals forces but with a higher exponent than 6 (m is typically between 8 and 16).

2.3.2 Forces between particles and surfaces

In colloid and surface literature it is customary to use H for the distance instead of r and this nomenclature is

used here. While, for molecular interactions, details of the shape and size of molecules do not enter into the expressions for the van der Waals forces, the opposite is the case for the forces between particles or surfaces. In this case, we need to account for the size/shape of the particles and this leads to a long list of expressions depending on the shape/size of the particles as well as their proximity. Some of the most used expressions are shown in Table 2.2. The expressions are derived upon integrating the van der Waals forces between molecules and considering also the different geometries of particles/surfaces. Notice that while for two atoms or small molecules the interaction energy has no explicit size dependence, as seen in Equations 2.4a–2.4c, the potential energy between particles increases linearly with their radius. Consequently, when should we use Equations 2.4 and when those of Table 2.2, i.e. when is a molecule "large enough" to be considered a particle? According to Israelachvili (2011), "molecules" with diameters larger than 0.5 nm should be treated as particles and the van der Waals forces should be estimated with the expressions shown in Table 2.2, otherwise the strength of the interaction will be underestimated.

In the derivation of the equations shown in Table 2.2, it is assumed that the intermolecular distance dependency is that shown in Equations 2.3 and 2.4 (i.e. the one used in the Lennard-Jones potential).

The key property in these calculations is the so-called Hamaker constant (A), which is directly linked to the C parameter of Equations 2.3 and 2.4 via the so-called microscopic (London) approach:

$$A = \pi^2 C \rho^2 \qquad (2.6)$$

where ρ is the number density (molecules per volume) and C is the London coefficient as given in Equation 2.4. In this way, the Hamaker constant is estimated based on the atomic polarizabilities and densities of the materials involved. When A is estimated from Equation 2.6 and used in the equations of Table 2.2 then we obtain the expression for the interparticle potentials when the medium is vacuum or air. However, in another medium (other than vacuum or air), as is usually the case in colloid

Table 2.2 *Van der Waals interaction (potential) energies between particles/surfaces; V_A is the potential energy (in J or J m^{-2} for the interaction between two surfaces), H is the interparticle/intersurface distance and R is the radius (for spherical particles); A is the Hamaker constant (see Equations 2.6–2.8 and Hamaker, 1937) and, depending on the application, is evaluated under conditions of either vacuum/air or a dielectric (i.e. a liquid medium, in which case an effective Hamaker constant must be used). C is defined in Equation 2.6 and ρ is the number density (molecules/volume)*

Geometry	Expression for the potential energy V_A (H)
Surface-molecule	$V_A = -\dfrac{\pi C \rho}{6 H^3}$
Two equal-sized spheres ($H \ll R$)	$V_A = -\dfrac{AR}{12H}$
Two equal-sized spheres, valid at all distances	$V_A = -\dfrac{AR}{12H}\left[\dfrac{4R}{(H+4R)} + \dfrac{4RH}{(2R+H)^2} + \dfrac{2H}{R}\ln\left(\dfrac{H(H+4R)}{(2R+H)^2}\right)\right]$
Two unequal-size spheres (radii R_1 and R_2) ($H \ll R_1$, R_2)	$V_A = -\dfrac{A}{6H}\left(\dfrac{R_1 R_2}{R_1 + R_2}\right)$
Two spherical particles of unequal/equal radii, valid at all separations. In the equation, use $r = R_1 + R_2 + H$.	$V_A = -\dfrac{A}{6}\left[\dfrac{2R_1 R_2}{r^2 - (R_1 + R_2)^2} + \dfrac{2R_1 R_2}{r^2 - (R_1 - R_2)^2} + \ln\left\{\dfrac{r^2 - (R_1 + R_2)^2}{r^2 - (R_1 - R_2)^2}\right\}\right]$
Two flat plates/surfaces of "infinite" thickness (per unit area, J m^{-2})	$V_A = -\dfrac{A}{12\pi H^2}$
Two flat plates/surfaces per unit area of surface (J m^{-2}) (finite thickness t)	$V_A = -\dfrac{A}{12\pi}\left[\dfrac{1}{H^2} + \dfrac{1}{(H+2t)^2} - \dfrac{2}{(H+t)^2}\right]$
Two parallel cylinders with radii R_1 and R_2 ($H \ll R_1$, R_2) (per unit length, J m^{-1})	$V_A = -\dfrac{A}{12\sqrt{2}H^{3/2}}\left(\dfrac{R_1 R_2}{R_1 + R_2}\right)^{1/2}$
Cylinder–cylinder (crossed at 90°) interactions (radii R_1, R_2, respectively)	$V_A = -\dfrac{A}{6H}\sqrt{R_1 R_2}$
Sphere or macromolecule of radius R near a flat surface ($H \ll R$)	$V_A = -\dfrac{AR}{6H}$
Cylinder of radius R near and parallel with a flat surface (per unit length, J m^{-1}) ($H < R$)	$V_A = -\dfrac{A\sqrt{R}}{12\sqrt{2}H^{3/2}}$

science (where often water is the medium), we should use the so-called effective Hamaker constant. The effective Hamaker constant can be estimated either from combining rules or via the more rigorous macroscopic Lifshitz approach. They are both presented next.

According to the combining rules approach, the Hamaker constant in the case of two different types of particles (1 and 3) in a medium 2 is given by the equation:

$$A_{123} = \left(\sqrt{A_{11}} - \sqrt{A_{22}}\right)\left(\sqrt{A_{33}} - \sqrt{A_{22}}\right) \quad (2.7a)$$

This expression is simplified in the case of particles of the same type (1) in a medium (2):

$$A_{121} = \left(\sqrt{A_{11}} - \sqrt{A_{22}}\right)^2 \quad (2.7b)$$

When the Hamaker constant is estimated using Equation 2.6, the total interaction is assumed to be the sum of the interactions between all interparticle atom pairs and is assumed to centre around a single oscillation frequency. The influence of neighbouring atoms on the interaction of a given pair of atoms is ignored. Thus, the vdW interaction energies calculated in accordance to the microscopic approach are likely to be in error but the error involved is not likely to be so great as to change the general conclusions concerning colloid stability.

According to the more accurate macroscopic (Lifshitz) approach, the Hamaker constant for particles 1 and 2 in a medium 3 is given as (Israelachvili, 2011):

$$A_{123} = A_{\nu=0} + A_{\nu>0} \approx \frac{3k_B T}{4} \left[\frac{\varepsilon_1 - \varepsilon_3}{\varepsilon_1 + \varepsilon_3} \right] \left[\frac{\varepsilon_2 - \varepsilon_3}{\varepsilon_2 + \varepsilon_3} \right]$$

$$+ \frac{3h\nu_e}{8\sqrt{2}} \frac{\left(n_1^2 - n_3^2\right)\left(n_2^2 - n_3^2\right)}{\sqrt{\left(n_1^2 + n_3^2\right)\left(n_2^2 + n_3^2\right)}\left[\sqrt{\left(n_1^2 + n_3^2\right)} + \sqrt{n_2^2 + n_3^2}\right]}$$

$$(2.8)$$

The symbols are the same as in Equation 2.5.

As Figure 2.2 shows, the Hamaker constant is a smooth function of the refractive index for a large number of compounds, which include organics (hydrocarbons, acetone, ethanol), polymers, water, ceramics (mica, SiO_2) and diverse particles like alumina, zirconia and diamond. There are only a few exceptions to this smooth trend. On the other hand, an equally smooth trend of Hamaker constant with the relative permittivity is not seen. Water with a relative permittivity of 80.2 at 20 °C has almost the same Hamaker constant (3.7×10^{-20} J) as pentane (3.8×10^{-20} J) with a relative permittivity of only 1.84. Other polar compounds (e.g. acetone and ethanol) also have quite low Hamaker constants (around 4.1×10^{-20} J while their relative permittivities are about 21–26). Nonetheless, excluding water and other polar compounds, there is an approximate increasing trend of the Hamaker constant with the relative permittivity, similar to the trend we see in Figure 2.2 with the refractive index.

The Lifshitz theory is more complicated than the London/Hamaker approach, but the full theory also accounts for retardation effects of the induced dipole interactions (time taken for the dipoles to re-orient). These effects are ignored in the London theory, but they are often not very important. In the Lifshitz theory, the interacting particles and the intervening medium are treated as continuous phases. The calculations are complex and require the availability of bulk optical/dielectric properties of the interacting materials over a sufficiently wide frequency range. In practice, such data are obtained using, for example, dielectric relaxation spectroscopy, where measurements of the dielectric permittivity are conducted by exposing a sample material to a wide range of electric field frequencies. The so-called static relative permittivity is obtained by using a zero frequency field (direct current) over sufficient time to let the material fully relax. Dielectric relaxation spectroscopy equipment is typically designed to use a potential of 1 V, but some machines can go up to 1 kV.

In some few cases, particle–particle forces can be measured and, thus, a few experimental values are available. They agree well with the theoretical calculations of the Hamaker constant via the London or Lifshitz theories (Israelachvili, 1985).

Examples 2.2 and 2.3 show calculations of the Hamaker constant and of the van der Waals forces for some colloidal systems.

Hamaker constant

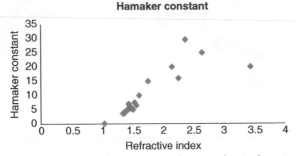

Figure 2.2 *Hamaker constant (for two identical particles interacting in air or vacuum, A_{11}) against the refractive index for 24 compounds (based on the data from Israelachvili, 1985). In the figure, the Hamaker constant is given in 10^{-20} J. Silicon (Si) with refractive index of 3.44 and zirconia particles with a refractive index of 2.15 both have a Hamaker constant of about 20 × 10^{-20} J. Diamond, with a refractive index of 2.375, has the highest Hamaker constant of these 24 compounds (29.6×10^{-20} J). The refractive index of a substance is dimensionless and defined as the ratio of the speed of light in vacuum to the speed of light in the substance*

Example 2.2. Van der Waals forces in colloidal systems.

1. Derive for the Lifshitz theory, starting from the Hamaker equation for two different particles in a medium, Equation 2.8, the following equation for the Hamaker constant of particles type (1) in a medium (2):

$$A_{121} = \frac{3k_BT}{4}\left[\frac{\varepsilon_1-\varepsilon_2}{\varepsilon_1+\varepsilon_2}\right]^2 + \frac{3h\nu_e}{16\sqrt{2}}\frac{\left(n_1^2-n_2^2\right)^2}{\left(n_1^2+n_2^2\right)^{3/2}}$$

Explain all the variables and the universal constants involved in this equation. How is this equation simplified if the particles are in vacuum (i.e. the medium is a vacuum)?

The Hamaker constants of polystyrene (PS), alumina and zirconia particles are 6.1–7.9, 14 and 17, all in 10^{-20} J. The Hamaker constant for water is equal to 3.7×10^{-20} J.

2. What is the value of the Hamaker constant of each of the following particles in water: polystyrene (PS), alumina and zirconia?

3. Teflon has a Hamaker constant equal to 3.8×10^{-20} J. Estimate the Hamaker constant of Teflon in water and compare the result to the experimental value (0.3×10^{-20} J). Comment on the result. How can we achieve better agreement?

Solution

1. Following the macroscopic (Lifshitz) approach, the Hamaker constant for particles 1 and 2 in a medium 3, is given by Equation 2.8:

$$A_{123} = \frac{3k_BT}{4}\left[\frac{\varepsilon_1-\varepsilon_3}{\varepsilon_1+\varepsilon_3}\right]\left[\frac{\varepsilon_2-\varepsilon_3}{\varepsilon_2+\varepsilon_3}\right] + \frac{3h\nu_e}{8\sqrt{2}}\frac{\left(n_1^2-n_3^2\right)\left(n_2^2-n_3^2\right)}{\sqrt{\left(n_1^2+n_3^2\right)\left(n_2^2+n_3^2\right)}\left[\sqrt{\left(n_1^2+n_3^2\right)}+\sqrt{n_2^2+n_3^2}\right]}$$

Setting $1 = 2$ (identical particles) and 2 instead of 3 for the medium, and after some algebra, we arrive at the requested equation.

The variables involved are the static relative permittivity (ε), the refractive index (n) and the main electronic absorption frequency in the UV region, ν_e. The universal constants are h = Planck's constant = 6.63×10^{-34} s and k_B is the Boltzmann constant; T is the temperature (in K).

If the medium is a vacuum, then both relative permittivity and refractive index are equal to one.

2. We choose the average Hamaker constant value of PS (7×10^{-20} J). The remaining Hamaker values are provided in the assignment. The effective Hamaker constant is calculated using these values and the Hamaker constant value given for water (3.7×10^{-20} J) and the combining rule:

$$A_{121} = \left(\sqrt{A_{11}} - \sqrt{A_{22}}\right)^2$$

For PS in water we obtain $A_{121} = 0.52 \times 10^{-20}$ J, for alumina $A_{121} = 3.31 \times 10^{-20}$ J and for zirconia $A_{121} = 10.7 \times 10^{-20}$ J.

3. Using the combining rule approach, we obtain $A_{121} = 0.000653 \times 10^{-20}$ J, which is clearly far from the experimental value (0.3×10^{-20} J). We could obtain better agreement by using the Lifshitz theory, if relative permittivity and refractive index data are available.

Example 2.3.   Estimation of Hamaker constants with various methods.
The following information is provided:

  i.   Hamaker constant of water $= 3.7 \times 10^{-20}$ J;
 ii.   Hamaker constant of gold particles $= 30–50 \times 10^{-20}$ J (Israelachvili, 1985, based on exact solutions);
iii.   the relative permittivity and refractive index for octane are equal to 1.95 and 1.387, respectively.

Estimate using this information the following Hamaker constants:

1.   octane particles in water $(0.41 \times 10^{-20}$ J);
2.   water–air in octane $(0.53 \times 10^{-20}$ J);
3.   octane–air in water $(-0.2 \times 10^{-20}$ J);
4.   gold particles in water $(40 \times 10^{-20}$ J).

Compare the results to the experimental values which are given in parenthesis. Comment on the results. In which cases do we have attractive and when do we have repulsive van der Waals forces?

Solution
The Hamaker constant of octane is not given but can be calculated from the Lifshitz theory and more specifically from Equation 2.8 in the form shown in Example 2.2 where in addition we assume that the "medium" is vacuum or air with a relative permittivity and a refractive index equal to 1. Thus, the final equation to estimate the Hamaker constant of pure octane is:

$$A_{11} = \frac{3k_B T}{4}\left[\frac{\varepsilon_1 - 1}{\varepsilon_1 + 1}\right]^2 + \frac{3h\nu_e}{16\sqrt{2}}\frac{(n_1^2 - 1)^2}{(n_1^2 + 1)^{3/2}}$$

Then, substituting the given information for the relative permittivity and refractive index and the other constants (see under Equation 2.5), we find $A_{11} = 4.5 \times 10^{-20}$ J at 300 K (Israelachvili, 1985).

For octane particles in water we can estimate the effective Hamaker constant using Equation 2.7b and obtain:

$$A_{121} = \left(\sqrt{A_{11}} - \sqrt{A_{22}}\right)^2 = \left(\sqrt{4.5 \times 10^{-20}\text{J}} - \sqrt{3.7 \times 10^{-20}\text{J}}\right)^2 = 0.039 \times 10^{-20} \text{ J}$$

which indicated attractive forces but of much smaller magnitude compared to the experimental value. If instead, we would have used the Lifshitz theory (Equation 2.8) for which we also need the relative permittivity and refractive index for water as well (80, 1.333), we would have estimated a value of 0.36×10^{-20} J, which is much close to the experimental (exact solution) value.

For dissimilar particles we can use either the general combining rule, Equation 2.7a or the Lifshitz theory, Equation 2.8. Using the combining rule we get:

water–air in octane:

$$A_{123} = \left(\sqrt{A_{11}} - \sqrt{A_{22}}\right)\left(\sqrt{A_{33}} - \sqrt{A_{22}}\right) = \left(\sqrt{3.7 \times 10^{-20}} - \sqrt{4.5 \times 10^{-20}}\right)\left(0 - \sqrt{4.5 \times 10^{-20}}\right) = 0.42 \times 10^{-20} \text{ J}$$

octane–air in water:

$$A_{123} = \left(\sqrt{A_{11}} - \sqrt{A_{22}}\right)\left(\sqrt{A_{33}} - \sqrt{A_{22}}\right) = \left(\sqrt{4.5 \times 10^{-20} \text{J}} - \sqrt{3.7 \times 10^{-20} \text{J}}\right)\left(0 - \sqrt{3.7 \times 10^{-20} \text{J}}\right)$$
$$= -0.38 \times 10^{-20} \text{ J}$$

In the first case (medium is octane) we have attractive vdW forces but in the second case we have repulsive vdW forces as the Hamaker constant is negative. Moreover, in the first case (non-aqueous medium), the agreement is rather satisfactory (21% deviation), while for the octane–air in water system the agreement is poor (90% in absolute deviation). Again we can obtain much better agreement with the experimental data if we use the Lifshitz theory (which results in 0.51×10^{-20} J and -0.24×10^{-20} J, respectively).

Finally, for the system with gold particles we have (using the average value for the Hamaker constant of gold particles):

$$A_{121} = \left(\sqrt{A_{11}} - \sqrt{A_{22}}\right)^2 = \left(\sqrt{40 \times 10^{-20} \text{J}} - \sqrt{3.7 \times 10^{-20} \text{J}}\right)^2 = 19.4 \times 10^{-20} \text{ J}$$

This value is in gross disagreement with the experimental value. This is an expected result for many reasons. First, we have already seen that the combining rule method breaks down for aqueous medium systems. Moreover, for these metal solutions, the Hamaker constant does not appear to decrease very much with the medium, and thus has more or less the same value in air and in water. The combining rule always predicts a decrease of the Hamaker constant (compared to air or vacuum) and thus this trend cannot be predicted. Unfortunately, as we lack experimental relative permittivity/refractive index data for metals (no meaning) the Lifshitz theory in the form of Equation 2.8 cannot be used in this case.

2.3.3 Importance of the van der Waals forces

The van der Waals forces are important in the study of both surface and colloid phenomena. Of the vdW forces, only the dispersion forces are universal, while polar and induction forces (as well as other forces) are sometimes called "specific" forces, i.e. they are only present when the substances involved have some specific characteristics, e.g. electric charges, presence of dipoles or hydrogen bonds.

Of the three van der Waals forces, only the polar ones depend directly on temperature (Equation 2.4a). Moreover, for polar forces, the potential energy depends on the fourth power of the dipole moment for pure polar fluids and is thus quite important for highly polar molecules (having a dipole moment above 1 Debye).

The dispersion forces are semi-additive (see Figure 2.3 for *n*-alkanes; also Goodwin, 2004) and thus the van der Waals forces, especially the dispersion

ones, dominate often in the case of colloidal particles and surfaces. This is because the distance dependency is much more pronounced (i.e. much longer range forces) than in the case of the van der Waals forces between molecules, as can be seen in Table 2.2. The vdW forces for particles are computed by summing the attractions between all interparticle pairs.

A comparative evaluation of the van der Waals forces for molecules of same type is shown in Table 2.3. It can be seen that the van der Waals forces are important in all cases and they cannot be ignored, not even for highly polar and hydrogen bonding molecules. For example, the dispersion contribution of water is 15% of the total van der Waals forces at 0 °C and 24% at 298 K. The dispersion forces account for over 80% for water–methane and 60% for ammonia–ammonia interactions.

The van der Waals forces (either between molecules or between particles) are always attractive when the

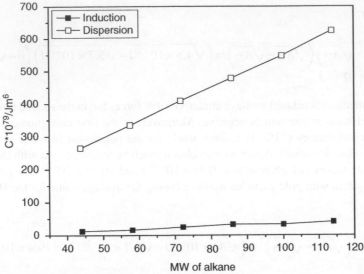

Figure 2.3 *Dispersion and induction contributions for* n-*alkanes (C as calculated in Equations 2.4a–2.4c). This illustrates the importance and additivity of dispersion forces. Reprinted from Kontogeorgis and Folas (2010), with permission from John Wiley & Sons, Ltd*

Table 2.3 *Comparison of intermolecular forces between two identical molecules. The C-values of the van der Waals forces* $(V = -C/r^6)$ *for identical molecules are given at 0 °C; C-values are expressed in* 10^{-79} *J m*^{6}

Molecule	Dipole moment (D)	C-dipole	C-induction	C-dispersion
Cyclohexane	0	0	0	1560
Methane	0	0	0	106
Methanol	1.7	148	18.7	135.4
Water	1.84	203	10.8	38.1
Acetone	2.87	1200	104	486

Kontogeorgis and Folas (2010). Reproduced with permission from John Wiley & Sons, Ltd. With data from Prausnitz et al. (1999) and from Tassios (1993).

molecules are in vacuum or air or between identical molecules or particles but they can be repulsive between different molecules or different particles if they are in a third medium (if the refractive index of the medium has a value intermediate to that of the two molecules or particles, see Equations 2.5 and 2.8). Repulsive vdW forces have important applications, e.g. predicting immiscibility in polymer blends (Problem 2.7) and engulfing. The vdW forces decrease because of an intervening medium (see Equation 2.7b).

The most important applications of vdW forces between particles or surfaces are in the understanding of colloid stability via the well-known DLVO theory or in adhesion studies, as discussed in Chapters 6, 10 and 14 (see e.g. Myers, 1991, 1999 and Israelachvili, 1985).

In most cases, the second (non-zero frequency) term in Equations 2.5 or 2.8 (including the refractive indexes), which is due to the London forces, dominates the Hamaker constant value and thus the forces between particles or surfaces, but for highly polar molecules, e.g. water, the first term in Equations 2.5 or 2.8 (with the relative permittivities) can be significant. This is investigated in the following example.

Example 2.4. Relative importance of polar and dispersion forces for water/alkanes.
The refractive index of methane is 1.30, while that of water is 1.33. The relative permittivity of water is about 80 and that of alkanes about 2.

1. Show that for two methane molecules in water, the major contribution of the van der Waals forces comes from the polar (zero-frequency) term and that Equation 2.5 for the potential reduces approximately to:

$$V = V_{\nu=0} + V_{\nu>0} \approx V_{\nu=0} \approx -\frac{k_B T R_1^3}{r^6}$$

2. Experimental measurements on hydrophobic interactions, e.g. alkane (here methane) attractions in water, have been reported to be about 10 kJ mol^{-1} at 25 °C. How does this value compare to the one calculated from the previous equation? Comment on the results.

Solution
Substituting the values of refractive indices and relative permittivities of methane and water in Equation 2.5 we obtain for the potential:

$$V = V_{\nu=0} + V_{\nu>0}$$

$$= -\frac{3k_B T}{r^6} R_1^3 R_2^3 \left[\frac{\varepsilon_1 - \varepsilon_3}{\varepsilon_1 + 2\varepsilon_3}\right] \left[\frac{\varepsilon_2 - \varepsilon_3}{\varepsilon_2 + 2\varepsilon_3}\right] - \frac{\sqrt{3}h\nu_e R_1^3 R_2^3}{2r^6} \frac{(n_1^2 - n_3^2)(n_2^2 - n_3^2)}{\sqrt{(n_1^2 + 2n_3^2)(n_2^2 + 2n_3^2)}\left[\sqrt{(n_1^2 + 2n_3^2)} + \sqrt{n_2^2 + 2n_3^2}\right]}$$

Then the second term of the right-hand side of the equation essentially disappears and the whole value of the potential energy reduces approximately to:

$$V = V_{\nu=0} + V_{\nu>0} \approx V_{\nu=0} \approx -\frac{k_B T R_1^3}{r^6}$$

which is purely entropic, as is well-known for the "hydrophobic interaction" between small hydrocarbon molecules in water (although the measured values suggest an interaction stronger than that expected from the above equation).

According to Israelachvili (1985), for two small molecules we would expect a free energy of dimerization of water of the order 0.04 kJ mol^{-1} at 25 °C, a value too small to induce immiscibility. This value is at least 100 times larger than the reported experimental values. The lack of agreement may be related to the breakdown of the simple model of the excess polarizability for highly polar molecules.

Finally, we should mention that the van der Waals forces provide an explanation for the origin of the geometric mean rule for the molecular energy parameters, e.g. as used in thermodynamic models or also as we will see later (Chapter 3) for the dispersion part of surface tension. Starting from the London expression for the intermolecular potential, it can be easily shown that the attractive cross intermolecular potential between two unlike non-polar molecules 1 and 2 is given by the equation:

$$V_{12} = \sqrt{V_1 V_2}\left(\frac{2\sqrt{I_1 I_2}}{I_1 + I_2}\right) \qquad (2.9)$$

where I_i is the ionization potential of component i.

For the potential we will use the attractive part of the Mie potential function (which for $m = 12$, $n = 6$ is the Lennard-Jones potential):

$$V_{12} = \frac{m}{m-n}\left(\frac{m}{n}\right)^{\frac{n}{m-n}}\varepsilon_{12}\left[\left(\frac{\sigma_{12}}{r}\right)^m - \left(\frac{\sigma_{12}}{r}\right)^n\right] \quad (2.10)$$

where ε_{12} is the molecular cross-energy parameter, σ_{12} is the molecular cross diameter and r is the distance between the molecules.

Using plausible assumptions for the ionization potential (as shown, for example, by Coutinho, Vlamos and Kontogeorgis, 2000):

$$1 \propto \frac{1}{\sigma^3} \Rightarrow \frac{2\sqrt{I_1 I_2}}{I_1 + I_2} \cong \left(\frac{\sqrt{b_1 b_2}}{b_{12}}\right)^{-1} \quad (2.11)$$

and the relationship between "microscopic" (ε, σ) and "macroscopic" properties-parameter:

$$(T_c, V_c, a, b): \quad \begin{array}{c} \varepsilon \propto \dfrac{a}{b} \propto T_c \\[2mm] \sigma^3 \propto b \propto V_c \end{array} \quad (2.12)$$

where T_c, V_c, a, b are, respectively, the critical temperature, critical volume and the energy and co-volume parameters of cubic equations of state, then:

$$a_{12} = \sqrt{a_1 a_2}\left(\frac{\sqrt{b_1 b_2}}{b_{12}}\right)^{\frac{n}{3}-2} \quad (2.13)$$

or:

$$a_{12} = \sqrt{a_1 a_2} \quad \text{for } n = 6$$

The latter equation provides a justification for the geometric mean rule often used in component theories for the interfacial tension, as seen in Chapter 3. Strictly speaking, this rule is valid only for dispersion/van der Waals forces and should be expected to break down for hydrogen bonding molecules and related interactions, e.g. for aqueous systems. This is illustrated in Example 2.5.

Example 2.5.   *Validity range of the geometric mean rule.*
As discussed in Section 2.3, the van der Waals forces are given in terms of the potential by the general equation:

$$V = -\frac{C}{r^6}$$

where C is calculated from the dispersion, polar and induction forces.

Calculate, using the data given below, the C-parameters of the van der Waals forces and the percentage (%) contribution of dispersion, induction and polar contributions at 293 K for:

i.   two methane molecules
ii. two Ne-molecules
iii. two water molecules
iv. Ne–methane
v.   methane–water

Compare the "experimental" interactions between two unlike molecules (C_{12}) parameters to those computed by the geometric mean rule ($C_{12} = \sqrt{C_1 C_2}$). What do you observe? When needed, calculate the interaction parameter-correction to the geometric mean rule (k_{ij}) so that experimental and calculated van der Waals forces match:

$$C_{12} = \sqrt{C_1 C_2}\left(1 - k_{ij}\right)$$

Discuss briefly the results.

Data needed: Reduced electronic polarizabilities: methane (2.6×10^{-30}), Ne (0.39×10^{-30}), water (1.48×10^{-30}) – all values in m^3. Ionization potential: methane (12.6), Ne (21.6), water (12.6) – all values in eV ($= 1.602 \times 10^{-19}$ J). Only water has a dipole moment (1.85 D).

Solution

Using the polarizability and dipole moment values given in the problem and the expressions for the "three" vdW potentials (induction, polar and dispersion), Equations 2.4a–2.4c, we can calculate the C-values. These values, together with the C_{12} values, those estimated from the geometric mean rule and the interaction parameters are presented in the table below at 293 K. The percentage dispersion energy contribution to the total contribution is also given in parentheses. All values are given in 10^{-79} J m^6.

System	$C_{induction}$	C_{dipole}	$C_{dispersion}$ (% total)	C_{11}/C_{12}	C_{12} (GM rule) ($= \sqrt{C_1 C_2}$)	Interaction parameter k_{12} $C_{12} = \sqrt{C_1 C_2}(1 - k_{12})$
Ne–Ne	0	0	4 (100)	4		
CH$_4$–CH$_4$	0	0	102 (100)	102		
H$_2$O–H$_2$O	10	96	33 (24)	139		
Ne–CH$_4$	0	0	19 (100)	19	20.2	0.06
CH$_4$–H$_2$O	9	0	58 (87)	67	119.1	0.4374

We note that in the case of the non-polar system neon–methane, which only has dispersion forces, the geometric mean rule offers an excellent approximation and, thus, the interaction parameter k_{12} correction to this rule is very close to zero (0.06).

On the other hand, for aqueous systems, the geometric mean rule deteriorates and large correction parameters are needed, as shown here for the water–methane system. This is because of the hydrophobic effect, in general and, more specifically, the strong water–water interactions which are much more important than the water–methane interactions. Moreover, as indicated above, the geometric mean rule is "strictly" valid – derived for molecules having only dispersion (London) forces.

2.4 Concluding remarks

Intermolecular forces play a crucial role in colloid and surface chemistry as well as in physical chemistry and thermodynamics in general. Coulombic forces (of importance to electrolyte solutions and colloid particles) are very strong and long-range, while the van der Waals (vdW) forces are far weaker and substantially shorter range but are often equally important, especially for describing the attractive forces of colloidal particles.

Of the three types of van der Waals (vdW) forces, the polar and dispersion ones are the most important; with the latter being universal and the former especially strong for highly polar molecules, e.g. those having dipole moments above 1 Debye. Quadrupolar forces are less important and more short-range than the vdW ones, but can be of importance at low temperatures for strongly quadrupolar molecules, e.g. CO_2. The quadrupolar forces are less important in colloid and surface science.

Quasi-chemical forces especially hydrogen bonding ones are very important and often dominate in molecules such as water, alcohols, organic acids, amines, glycols and many biomolecules and polymers. They are extremely important, especially in surface science for understanding adhesion and micellization phenomena. The hydrophobic effect in water and solvating interactions are attributed to the hydrogen bonding or in general the Lewis acid–Lewis base interactions. Micellization is also explained by the hydrophobic phenomenon.

For macromolecules and, in general, for particles/droplets in the colloid domain, the vdW forces are much longer range than the forces between molecules. But both intermolecular and interparticle or interfacial forces depend on the intervening medium, which can be quantified via the relative permittivities or the Hamaker constants. The vdW forces are typically attractive, but because of the presence of an intervening medium they can be repulsive in some systems that consist of at least two different types of molecules or particles.

Intermolecular potential functions can be used to represent intermolecular forces and are often used directly in thermodynamic models. The Lennard-Jones potential includes both repulsive and attractive forces. The distance dependency of the attractive part of the Lennard-Jones potential is the same as for the vdW forces.

Input from the intermolecular forces is useful in numerous direct or indirect applications in colloid and surface science. For example, the geometric mean rule typically used for the combining rule of the energy parameter in equations of state and in theories for the interfacial tension has its origin in the geometric mean rule for the intermolecular potential, as derived from the dispersion (London) forces. Concepts from intermolecular forces will be used in, for example, theories for interfacial tensions and in the study of colloid stability.

Appendix 2.1 A note on the uniqueness of the water molecule and some of the recent debates on water structure and peculiar properties

Water is the most important substance in the world; it covers two-thirds of the Earth and our own cells include two thirds water by volume. Nevertheless, we know so little about it. In the words of Philip Ball, for many years a consultant of *Nature*: "No one really understands water. It's still a mystery" (Ball, 2008). Water has over 50 exceptional or "anomalous" properties; their magnitude or trends with temperature, pressure and composition (for mixtures) do not follow what we know from other compounds. No other liquid behaves this way. Among the most exciting properties are the maximum of density at 4 °C, see Figure 2.1 (ice floats, lakes freeze from top to bottom), the high values of heat capacity (thus stabilizing Earth's climate), relative permittivity (leading to dissociation of ions) and surface tension (resulting in, for example, small insects walking on water) and the maxima and minima of many thermodynamic properties as a function of temperature, e.g. the minimum hydrocarbon solubility in water at room temperature (related to the hydrophobic effect) and the speed of sound.

What is the true reason for all these questions? Hydrogen bonding is the evident explanation and especially the hydrogen bonding structure and its changes with temperature, additives and other conditions. The important question is, however, whether we are certain as to which is the dominant hydrogen bonding structure of water. Figures 2.1 and 2.4 illustrate some of the problems.

Numerous theories for water structure are all up for debate. It is unclear (Figures 2.1 and 2.4a) whether liquid water maintains the tetrahedral structure (as we know it from ice) or whether it should best be described by a two-state model, where most molecules are in the form of rings or chains (Wernet, 2004; Zubavicus and Grunze, 2004) and the literature is full of heated discussions (Ball, 2003, 2008). Neither molecular simulation (Israelachvili, 2011; Etherington *et al.*, 1998) nor advanced experimental methods (Kontogeorgis *et al.*, 2010; Thøgersen *et al.*, 2008) provide the full answer. Direct spectroscopic measurements could provide quantitative information on the degree of hydrogen bonding of water but they are not in good agreement with each other (Figure 2.4b) and the data can be interpreted in different ways (Kontogeorgis *et al.*, 2010; Kontogeorgis and Folas, 2010). Novel statistical-thermodynamic based theories are promising but, using an unclear picture of water, it is not surprising that they cannot explain many of the anomalous properties of pure water and aqueous solutions (Kontogeorgis and

(a)

(b)

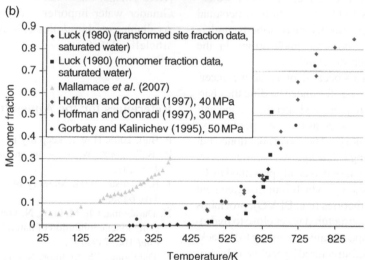

Figure 2.4   *A discussion of hydrogen bonds in water: (a) A new theory claiming that water chains and rings constitute the dominant structure for water instead of 3D-networks (Zubavicus and Grunze, 2004). Courtesy of Hirohito Ogasawara, A. Nilsson. (b) Monomer fraction of water as function of temperature from different experimental techniques and sources. From Liang* et al. *(in press)*

Folas, 2010). While much research has focused on "pure" water, its tendency to absorb almost everything and the role of impurities is notorious and often used in the explanation or throwing out of "discoveries" such as polywater (a polymer-like water) and water memory where water appears to "remember" the effect of molecules it has been in contact with (Benveniste, 1994).

The complexity increases when the effects of salts (Frosch, Bilde and Nielsen, 2010), biomolecules or solid surfaces on water structure are also considered. For example, experiments and simulation indicate that only a few nanometres of water are tightly bound to a solid, while friction and other studies indicate 100 nm thick layers (I. Fabricius, personal communication). Nobody understands why and similar questions arise due to the apparent low elasticity of water close to solids.

Finally, let us touch upon water memory (Benveniste, 1994) which is almost a "taboo" subject in water science. A similar issue appears in gas hydrate research. Gas hydrates are solid structures formed from water and small gas molecules like methane under conditions of high pressure and low temperature. For example natural gas at 200 bar forms solid hydrates with water at 25 °C. It is widely accepted (Daraboina, Malmos and von Solms, 2013a, b; Mazloum *et al.*, 2011; Lee, Wu and Englezos, 2007) that it is easier (faster) to make hydrates from water that has previously been in hydrate form rather than directly from fresh water. In fact this is used as an experimental technique to run through experiments faster (this phenomenon is termed "memory water" by researchers in this field; N. von Solms, personal communication). Why this happens nobody understands. What are the possible implications in the understanding of water structure?

In addition, or perhaps because of the above uncertainties, "peculiar" theories for understanding the characteristics of water were presented after the second half of the twentieth century, such as the concept of "polywater" and "water memory". They were abandoned but were they entirely wrong? New light has been shed with a potentially pioneering theory recently presented by G. Pollack (2013). This theory, which is in disagreement with classical colloid theories (e.g. DLVO), proclaims the existence of highly structured water close to hydrophilic surfaces. More specifically, according to Pollack, water forms an extended structured layer close to hydrophilic surfaces. This layer excludes everything (exclusion zone or EZ) and is negatively charged. Beyond that there is a positive layer (hydronium ions), thus water appears to behave as battery due to charge separation. According to Pollack, his "exclusion zone" theory can explain many water anomalies.

Both in relation to this theory and claims by others questions arise: Does water have "unusual powers" such as the possibility of storing charge, acting as a battery and even remembering its contact with other molecules, storing and transferring information? While some of these questions surely border on science fiction, devices claiming the above have been presented (http://www.granderwater.co.uk) but are

not understood. Methods to produce fuels from sun and water have also been reported (Muhich *et al.*, 2013).

Pollack's theory gives new insight and it remains to be seen whether it can be verified by other researchers and whether it will explain some of water's anomalous properties or at least if it will become an accepted way of viewing the behaviour of water near (hydrophilic) surfaces. We believe that real progress can be achieved when water scientists from the "conventional side" can meet and discuss with those claiming novel or peculiar theories (Wilson, 2009; Benveniste, 1994; Radin *et al.*, 2006; Walach *et al.*, 2005). A debate is absolutely needed involving scientists (and even practitioners) from both "fronts" (e.g. DR2 Tema: Kan vand huske? [DR2 Theme: Can water remember?] with P. Westh, Grander water importer and DR journalists, 2007. Available at http://presse.dr.dk/presse/Article.asp?articleID=24659).

References for the Appendix 2.1

P. Ball, 2003. How to keep dry in water. *Nature*, 423, 25.

P. Ball, 2008. Water – an enduring mystery. *Nature*, 452, 291.

J. Benveniste, 1994. Memory of water revisited. *Nature*, 370, 6488.

N. Daraboina, Ch. Malmos, N. von Solms, 2013a. Synergistic kinetic inhibition of natural gas hydrate formation, *Fuel* 108, 749–757.

N. Daraboina, Ch. Malmos, N. von Solms, 2013b. Investigation of kinetic hydrate inhibition using a high pressure micro differential scanning calorimeter, *Energy & Fuels* 27, 5779–5786.

J.R. Errington, G.C. Boulougouris, I.G. Economou, A.Z. Panagiotopoulos, D.N. Theodorou, 1998. Molecular Simulation of Phase Equilibria for Water – Methane and Water - Ethane Mixtures, *J. Phys. Chem. B*, 102(44), 8865–8873.

M. Frosch, M. Bilde, O.F. Nielsen, 2010. From Water Clustering to Osmotic Coefficients. *J. Phys. Chem. A*, 114, 11933.

J.N. Israelachvili, 2011. Intermolecular and Surface Forces. 3rd edn, Academic Press.

G.M. Kontogeorgis, G.K. Folas, 2010. Thermodynamic Models for Industrial Applications – from Classical and Advanced Mixing Rules to Association Theories, John Wiley & Sons, Inc.

G.M. Kontogeorgis, I. Tsivintzelis, N. von Solms, A. Grenner, D. Bogh, M. Frost, A. Knage-Rasmussen, I.G. Economou, 2010. Use of monomer fraction data in the parametrization of association theories. *Fluid Phase Equilibria*, 296(2), 219–229.

J.D. Lee, H. Wu, P. Englezos, 2007. Cationic starches as gas hydrate kinetic inhibitors, *Chemical Engineering Science*, 62, 6548–6555.

X. Liang, B. Maribo-Mogensen, I. Tsivintzelis, G.M. Kontogeorgis, 2016. A comment on water's structure using monomer fraction data and theories. *Fluid Phase Equilibria*, 407, 2–6.

S. Mazloum, J. Yang, A. Chapoy, B. Tohidi, 2011. A New Method for Hydrate Early Warning System Based on Hydrate Memory, Proceedings of the 7th International Conference on Gas Hydrates (ICGH 2011), Edinburgh, Scotland, United Kingdom, July 17–21. Available at http://www.pet.hw.ac.uk/icgh7/papers/icgh2011Final 00341.pdf (accessed 10 November 2015).

C.L. Muhich *et al.*, 2013. Efficient generation of H2 by splitting water with an isothermal redox cycle. *Science*, 341, 540.

G.H. Pollack, 2013. The Fourth Phase of Water. Beyond Solid, Liquid, and Vapor. Ebner & Sons Publishers, Seattle WA.

D. Radin, G. Hayssen, M. Emoto, T. Kizu, 2006. Double-Blind test of the effects of distant intention on water crystal formation. *Explore*, 2(5), 408–411.

J. Thøgersen, S.K. Jensen, C. Petersen, S.R. Keiding, 2008. Reorientation of hydroxide ions in water. *Chem. Phys. Lett.*, 466, 1.

H. Walach, W.B. Jonas, J. Ives, R. van Wijk, O. Weingartner, 2005. Research on homeopathy: state of the art. *J. Alternative Complementary Res.*, 11(5), 813–829.

Ph. Wernet *et al.*, 2004. *Science*, 304, 995.

E.K. Wilson, 2009. Watering down science? *Chemical & Engineering News*, 32.

Y. Zubavicus, M. Grunze, 2004. New insights into the structure of water with ultrafast probes. *Science*, 304, 974–976.

Problems

Problem 2.1: Multiple choice questions on intermolecular and interparticle forces

Fill in Table 2.4. Only one answer is correct for each question.

Table 2.4 *Multiple choice questions*

---

1. Intermolecular forces are *not* present…
a. ☐ In ideal solutions
b. ☐ In ideal gases
c. ☐ In the liquid solution polyethylene/hexane
d. ☐ When we have mixtures with alkanes

2. In which of the following molecules are hydrogen bonds likely to exist?
a. ☐ *n*-Pentane
b. ☐ Carbon dioxide
c. ☐ Ethanol
d. ☐ Polystyrene

3. All three types of intermolecular "van der Waals forces"…
a. ☐ Depend on the dipole moment
b. ☐ Depend on temperature
c. ☐ Have the same dependency on intermolecular distance
d. ☐ Are not important in thermodynamic calculations

4. "Intramolecular association" is a special form of hydrogen bonding (HB) indicating…
a. ☐ … HB between two different molecules of same type
b. ☐ … HB between two different molecules of different types
c. ☐ … HB between two atoms inside the same molecule
d. ☐ … HB between water and another molecule

(continued overleaf)

Table 2.4   *(continued)*

5. Ethanol and dimethyl ether are isomers but ethanol's boiling point is 79 °C, while ether's boiling point is only −25 °C. This large difference is due to…
a. ☐   The difference in polarity between the two compounds
b. ☐   The hydrogen bonding character of ethanol
c. ☐   The difference in the quadrupole moments of the two compounds
d. ☐   The difference in molecular weight between the two compounds

6. Hydrogen bonding is *not* at all related to only *one* of the following phenomena. Which one?
a. ☐   The DNA double-helix
b. ☐   The floating of ice on water
c. ☐   The strong deviations from Raoult's law in a polyethylene–hexane mixture
d. ☐   The formation of dimers of organic acids in the vapour phase

7. Which of the following mixtures is *not* likely to exhibit hydrogen bonding interactions?
a. ☐   Methanol–water
b. ☐   Propanol–hexane
c. ☐   Polyethylene–octane
d. ☐   Acetic acid–water

8. In which of the following mixtures are "cross-association" (solvation) phenomena expected due to hydrogen bonding interactions?
a. ☐   Methanol–hexane
b. ☐   Water–octane
c. ☐   Chloroform–acetone
d. ☐   Acetone–octane

9. The intermolecular potential of an ideal gas is…
a. ☐   Given by the Lennard-Jones equation
b. ☐   Zero
c. ☐   Approximated by the square-well equation
d. ☐   Given by a complex function that is not known

10. At high temperatures …
a. ☐   Dipole forces become more pronounced
b. ☐   Hydrogen-bonding interactions become weaker
c. ☐   Dispersion interactions increase
d. ☐   There are no intermolecular forces

11. Dispersion forces…
a. ☐   Depend on the dipole moments of the compounds
b. ☐   Are always smaller than "dipole–dipole" forces
c. ☐   Are often very significant
d. ☐   Depend on temperature

12. Hydrogen bonds are often called "quasi-chemical" (forces) because they…
a. ☐   Are equally strong as the "chemical" (covalent) bonds
b. ☐   Have a strength close to that of chemical bonds
c. ☐   Are being taught in chemistry courses
d. ☐   Are bonds between chemicals

Table 2.4   (continued)

13. Coulombic and hydrogen bonding forces have some common features...
a.   They are both rather weak forces
b.   They are both long-range forces
c.   They are both less important than the van der Waals forces between molecules
d.   They are both important in determining stability for colloids in organic solvents

14. The van der Waals forces between molecules and between particles...
a.   Have the same dependency with respect to distance
b.   Depend on the shape of the involved molecules or particles
c.   Both depend on the relative permittivities and refractive indices
d.   Are not important in colloid and surface science

15. The van der Waals forces between particles...
a.   Are always positive
b.   Are always negative
c.   Do not depend on the shape of particles
d.   Can be positive or negative

16. For estimation of the Hamaker constant, the more reliable method is...
a.   Equation 2.7a
b.   Equation 2.7b
c.   Equation 2.8
d.   It is impossible to estimate it

17. The Hamaker constants of polystyrene (PS) and silver particles are 6.6 and 40 (both in $\times 10^{-20}$ J). These values imply that:
a.   The vdW forces between two PS particles are stronger than between two silver particles
b.   The vdW forces between two silver particles are stronger than between two PS particles
c.   The vdW forces cannot be assessed based on this information
d.   The vdW forces are not important for such values of the Hamaker constant

18. Only one of the following statements about the vdW forces between particles is wrong. Which one?
a.   They are always attractive in air or vacuum
b.   They are always attractive between two identical particles
c.   They increase because of an intervening medium
d.   They can be repulsive between different particles in a medium

19. The rutile (TiO_2) particles have in vacuum or air a Hamaker constant equal to 43. The corresponding Hamaker constant for rutile particles in water is 3.7. The following is true:
a.   The rutile Hamaker constant value in water is the same as its value in vacuum
b.   The rutile Hamaker constant in water is about 22
c.   The rutile Hamaker constant in water is about 35
d.   The rutile Hamaker constant in water cannot be estimated from the given information
(All Hamaker constant values are given in 10^{-20} J).

20. The Hamaker constant of pentane in air is 3.8, while the Hamaker constant of pentane in water is 0.28. The results mean:
a.   There is small difference in the vdW forces of pentane in air and in water
b.   The vdW forces of pentane are stronger in water than in air
c.   Nothing special about the vdW forces between pentane molecules
d.   The vdW forces are stronger in air than in water

Problem 2.2: London forces from different expressions

Consider two identical molecules interacting in free space. Compare the London contribution from Lifshitz theory (Equation 2.5) to the expression for the dispersion forces (Equation 2.4c). What do you observe? Are the two expressions identical?

Problem 2.3: Range of Hamaker constants

The Hamaker constants of various pure materials/fluids are given below (all in $\times 10^{-20}$ J): water: 3.7, pentane: 3.8, *n*-hexadecane: 5.1, CCl4: 5.5, acetone: 4.1, ethanol: 4.2, Teflon (PTFE): 3.8.

Comment on the values in connection to the size and forces present in these molecules. Which are the dominant factors that determine the Hamaker constant values?

Problem 2.4: The origin of the geometric mean rule

Starting from the van der Waals (London) forces, prove first Equation 2.9 and then prove Equation 2.13. What are the most important assumptions in the calculations and what is the significance of the final result for colloid and surface science?

Problem 2.5: Van der Waals forces in water

a. Calculate for two water molecules the C-parameters of the van der Waals forces (Equations 2.3 and 2.4) and the percentage (%) contribution of dispersion, induction and polar contributions at two temperatures (273 and 373 K). What do you observe? How can the Hamaker constant of water be estimated based on the C-values?
b. Repeat question (a) for two methane molecules.
c. Repeat question (a) for the water–methane interaction. Could the interaction between water and methane be estimated using the geometric mean rule ($C_{12} = \sqrt{C_1 C_2}$)? What do you observe? Discuss briefly the results.

Data needed: Reduced electronic polarizabilities: methane (2.6×10^{-30}), water (1.48×10^{-30}) – all values in m³. Ionization potential: methane (12.6), water (12.6) – all values in eV (=1.602×10^{-19} J). Only water has a dipole moment (1.85 D).

Problem 2.6: Effective Hamaker constants

Consider the following three systems:

System 1: pentane–pentane in water
System 2: mica–mica in water
System 3: water–air in pentane

The Hamaker constants of the pure materials/fluids are known (in $\times 10^{-20}$ J): water: 3.7, pentane: 3.8, mica: 10

a. Calculate using the combining relations the effective Hamaker constants in all three cases and compare the results to the experimental (rigorously computed) Hamaker constant values, which are, respectively, 0.34, 2.0 and 0.11, all in $\times 10^{-20}$ J. Comment on the results.
b. For which systems do you expect attractive and for which repulsive van der Waals forces? What is the physical meaning of the repulsive van der Waals forces?

Problem 2.7: Predicting miscibility in polymer blends (data from Oss et al. (1980))

In the case of unlike particles, effective Hamaker constants can be positive or negative, and the latter implies repulsive van der Waals forces which contribute to stability. One application of such repulsive vdW forces is in predicting solubility for blend-solvent mixtures, as discussed by van Oss *et al.* (1980). Using the Hamaker constant values for some polymers and solvents, fill in Table 2.5 and compare

Table 2.5 *Blend-solvent miscibility*

Polymer(1)–solvent(2)–Polymer(3)	Hamaker constant	Visual observation
PIB–THF–cellulose acetate		S
PIB–benzene–PMMA		S
PIB–cyclohexanone–PS		S
PIB–benzene–PS		S
PS–chlorobenzene–PMMA		M
Cellulose acetate–MEK–PS		M
PVC–THF–cellulose acetate		M
PMMA–MEK–cellulose acetate		M

S = separation, *M* = miscible

the conclusions using the Hamaker constants with the experimental (visual) observation.

Hint: Consult the article of van Oss *et al.* (1980) for values and estimation methods of the Hamaker constants.

References

R. Aveyard, Curr. Opin. Colloid Interface Sci., 2001, 6, 338.

J.A.P. Coutinho, P.M. Vlamos, G.M. Kontogeorgis. *Ind. Eng. Chem. Res.*, 2000, 39(8), 3076.

S.J. Gill and I. Wadso, Proc. Natl. Acad. Sci. USA, 1976, 73, 2955.

J. Goodwin, 2004. Colloids and Interfaces with Surfactants and Polymers. An Introduction, John Wiley & Sons Ltd, Chichester.

H.C. Hamaker, 1937, Physica, 4, 1058.

J. Israelachvili, 1985. Intermolecular and Surface Forces, 1st edn, Academic Press.

J. Israelachvili, 2011, Intermolecular and Surface Forces, 3rd edn, Academic Press.

M.J. Kamlet, J.M. Abboud, M.H. Abraham, R.W. Taft, J. Org. Chem., 1983, 48, 2877.

G.M. Kontogeorgis, G.K. Folas, 2010. Thermodynamic Models for Industrial Applications. From Classical and Advanced Mixing Rules to Association Theories. John Wiley & Sons Ltd, Chichester.

M. Lynch, Curr. Opin. Colloid Interface Sci., 2001, 6, 402.

E. McCafferty, J. Adhesion Sci. Technol., 2002, 16(3), 239.

A.D. McLachlan, Proc. Roy. Soc. London, 1963, Ser. A 271, 387; 274, 80.

D. Myers, 1991 & 1999. Surfaces, Interfaces, and Colloids. Principles and Applications. VCH, Weinheim.

J.M. Prausnitz, R.N. Lichtenthaler, E.G. de Azevedo, 1999. Molecular Thermodynamics of Fluid-Phase Equilibria (3rd edn). Prentice Hall International.

D.P. Tassios, 1993. Applied Chemical Engineering Thermodynamics. Springer-Verlag.

J. Texter, Curr. Opin Colloid Interface Sci., 2000, 5, 70.

C.J. van Oss, D.R. Absolom, A.W. Neumann, *Colloids and Interfaces*, 1980, 1, 45.

3

Surface and Interfacial Tensions – Principles and Estimation Methods

3.1 Introduction

All types of surfaces can be characterized by surface and interfacial tensions, which are key concepts in surface chemistry. They are of relevance to many products and processes in colloid science, e.g. they are very important in understanding wetting, spreading, lubrication and adhesion phenomena. Surface and interfacial tensions can be directly measured for liquid–gas and liquid–liquid interfaces but they are equally important for all interfaces (solid–gas, solid–liquid, solid–solid). Estimation methods have been developed for calculating surface and interfacial tensions when these are not available experimentally and/or when solid surfaces are present.

3.2 Concept of surface tension – applications

Surface tension (or surface energy) is defined as the work, W, required in order to isothermally increase the area of a surface, A, usually in air by unit amount (dW/dA). On a molecular level, the surface tension describes the unequal distribution of forces at a (liquid–air) interface compared to the forces in the bulk of the liquid (Figure 3.1). Because of this, molecules at the gas–liquid interface experience unbalanced attractive forces and there is a tendency for the liquids to conform to the shape that corresponds to the minimum surface (i.e. spherical).

The non-balance of forces at the interface between two immiscible liquids is represented by the liquid–liquid interfacial tension (or interfacial energy). This interfacial tension usually (but not always) lies between the individual surface tensions of the two liquids. For fully miscible liquids, like water and methanol or water and ethanol, the interfacial tension is zero.

The above picture implies a static state of affairs. However, the actual state is that of great turbulence on the molecular scale as a result of two-way traffic between the bulk of the liquid and the surface, and between the surface and the vapour phase (for a gas–liquid surface). The average lifetime of a molecule at the surface of a liquid is about 10^{-6} s.

We sometimes use the term surface tension for liquids or solids in contact with air (or vacuum) and the term interfacial tension when we have two condensed phases in contact. e.g. liquid–liquid or liquid–solid. Nowadays, both terms are used, with the term interfacial tension or energy being the more rigorous one.

Typical units of surface and interfacial tensions are mN m^{-1} or the equivalent dyn cm^{-1} (or mJ m^{-2}). Typical values of surface tension are, for liquids, 20–50 mN m^{-1}, though they are higher for water (72 mN m^{-1} at 20 °C) and other hydrogen bonding liquids like glycerol and can be much higher for *clean*

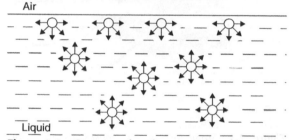

Figure 3.1 *Concept of surface tension and its origin due to the inequality of surface forces compared to the forces in the bulk liquid. Shaw (1992). Reproduced with permission from Elsevier*

solids (Table 3.1). Ethers and Teflon are among materials (fluids) with the lowest surface tension. Fluorocarbons in general have very low surface tensions, typically lower than alkanes and other hydrocarbons. Ionic and metallic surfaces have, on the other hand, high surface energies (in the case of clean solids).

As illustrated in Table 3.1, many solids have typically higher surface tensions than most liquids. In particular, *clean* metals, glass, ceramics and metal oxides have very high surface tensions, while polymers and fabrics are low energy surfaces.

The surface energies of solids are, however, not as well defined as those of liquids, as solid surfaces are usually inhomogeneous, they may change with time and they can be contaminated or rough. As a result, their surface energy varies with position. Moreover, the surface energies of solids cannot be (easily) measured, while those of liquids can be measured with many techniques.

Nevertheless, surface tensions are related to wetting phenomena. Roughly, a liquid will wet a solid if its surface tension is lower than that of the solid ($\gamma_l < \gamma_s$). "High energy" surfaces like *clean* metals, glass, ceramics and metal oxides are much easier to wet than plastic (polymer) surfaces. Wetting becomes more difficult when there are adsorbed liquids or contaminations as the surface energy is reduced. Solid

Table 3.1 *Surface tension values for typical liquids and solids (in mN m^{-1}). The solid surface tension values are "ideal" (production under vacuum) and will be much lower under normal laboratory conditions or after exposure in air. Except otherwise indicated, the values are at 20–25 °C*

Liquids	Surface tension	Solids	Surface tension
Mercury	485	Glass	3000
Water	73	Steel	2000
Glycerol	63–66	Tungsten	>1000
Epoxy glue	40	Aluminium	1000
Polyurethane	35	Gold (1200 °C)	1120
Aromatic hydrocarbons	28–30	Copper (1130 °C)	1100
Octanol	27.5	NaCl	110
CCl$_4$	26.8	Tree	70
Hexane	18	Ethylene glycol	48.4
Perfluoro-octane	12	Polystyrene	40
He (–270 °C)	0.24	Polyethylene	31
Oxygen (–198 °C)	17	Teflon	20

surfaces, as well as the related wetting and adhesion phenomena, will be discussed in Chapter 6.

There are many applications of surface tension that we can recognize immediately. Owing to their (own) surface tension, liquids tend to contract to a sphere to minimise their surface area as the sphere is the shape which corresponds to the minimum surface area. The spherical shape of rain (Figure 3.2) and especially of mercury droplets, the fact that small insects walk on water (Figure 3.3) and the fact that water rises in capillaries and in porous materials (soil, etc.) are manifestations of the high surface tension of water or mercury. High surface tensions such as that of water can be greatly decreased (e.g. from 72 to less than 30 mN m^{-1}) by the addition of surfactants, which may have a major impact on aquatic organisms (Figure 3.4).

Figure 3.3 *Small insects, e.g. mosquitos and other aquatic organisms, seem to "stand on" or walk on water. This is also largely attributed to the high surface tension of water which appears to give rise to an "elastic skin" on the surface. Markus Gayda, licensed under the Creative Commons Attribution-Share Alike 3.0 Generic license*

Figure 3.2 *The high surface tension of water (72 mN m^{-1}), which is largely due to the strong hydrogen bonds, is responsible for the spherical shape of rain droplets. Mightyhansa, licensed under the Creative Commons Attribution-Share Alike 3.0 Generic license*

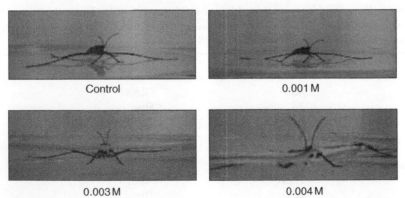

Figure 3.4 *Effect of surfactants on water striders. Surfactants (surface-active compounds), e.g. sodium dodecyl sulfate (SDS), can reduce the surface tension of water from 72 to 30 mN m^{-1} or less. Thus, the mosquito dies as it can no longer rely on water's high surface tension. In this sense, surfactants, which are contained in all detergents and many other products, can, in some cases, affect the aquatic environment*

(a)

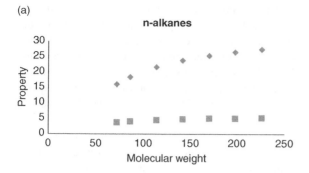

(b)

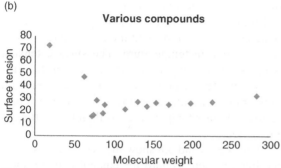

Figure 3.5 *(a) Surface tension (rhombuses) and Hamaker constants (squares) of alkanes from n-pentane to n-hexadecane as a function of molecular weight (g mol^{-1}). The surface tension values are given in mN m^{-1} and the Hamaker constants are in ×10^{-20} J. (b) Surface tension of alkanes and other compounds (shown in Table 3.1) as a function of molecular weight (g mol^{-1}). The surface tension values are given in mN m^{-1}. The small compounds with the high surface tension are water (73 mN m^{-1}) and MEG (monoethylene glycol) (48.4 mN m^{-1})*

Surface tensions are dominated by dispersion forces. Figure 3.5a shows the variation of surface tensions and Hamaker constants of *n*-alkanes with molecular weight. The dispersion forces become increasingly more pronounced as the molecular weight increases. Heavy alcohols and other similar molecules have only slightly higher surface tensions than their "corresponding" alkanes (having similar molecular weights). For some polar/hydrogen bonding molecules, like water or glycerol, the effect of polar/hydrogen bonding forces is significant (Figure 3.5b). In the case of water, the dispersion

forces account for about 30% of the total surface tension, as discussed in Example 3.3. For water, 21.8 mN m^{-1} of the total surface tension value (72.8 mN m^{-1}) is attributed to the dispersion effects.

Indeed, dispersion effects are important in surface tension and small changes in molecular weight often have small effects in the surface tension value. Consider for example the alcohol series: methanol (32.1), ethanol (46.1), propanol (60.1), isobutanol (74.12), pentanol (88.1) and octanol (130). The molecular weights, given in parentheses, clearly increase across the series. There is little variation, however, in the respective surface tensions: 22.5, 22.3, 23.7, 22.9, 25.2 and 27.5 mN m^{-1}. Similar conclusions are seen for other organic families and with the exception of very polar compounds the surface tensions are, in many cases, in the range 20–40 mN m^{-1}.

It is important to have an engineer's feeling for surface tension values. Dr Jeff Smith gave the following answer in the column "Ask the experts" published in the journal *Chemical Engineering Progress*. The question posed was: "Surface tension is a property for which I don't have a feel of what is a reasonable value. Is there a rule-of-thumb and/or a temperature-dependence rule for surface tension?"

The answer was as follows:

For most organic liquids, the surface tension at 25 °C is usually in the range of 15–40 mN m^{-1}, although glycols, certain nitrogen-compounds and a limited number of other compounds have values in the 40s. As for the temperature dependence, surface tension should decrease (and almost linearly) with increasing temperature, going to zero at the critical temperature. In one documented case at my company, one of the major commercial process simulators (which shall remain nameless) estimated at 50 °C surface tension for a seven-carbon glycol equal to 26.2 mN m^{-1}, which, of course, is off by about three orders of magnitude. Once it was spotted, the error was quickly circumvented by choosing an alternative estimation method within the simulator, which resulted in a much more reasonable value of 45 mN m^{-1}".

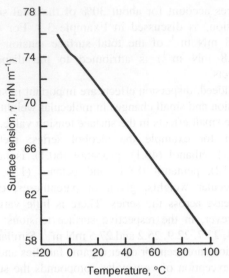

Figure 3.6 *Decrease in water surface tension with temperature. This dependency can be described by the Ramsay–Shields or the Eotvos equations (T_c is the critical temperature):*

$$\gamma = k\left(\rho_l - \rho_g\right)^{2/3}(T_c - T - 6)$$

$$\gamma = k\rho_l^{2/3}(T_c - T)$$

Reprinted from Atkins (2009), with permission from Oxford University Press

This example shows how "very basic information" such as knowledge of the order of magnitude of surface tension can be very useful. Even when using commercial process simulators we must be rather careful. Estimation methods are useful when we lack data, but common sense is equally important. Estimation methods are discussed in Section 3.4.

As intermolecular forces, especially the polar and hydrogen bonding ones, decrease at higher temperatures so does the surface tension, as Figures 3.6 and 3.7a illustrate for water. The caption in Figure 3.6 also gives some empirical equations which are used to represent the dependency of surface tension on temperature. The surface tension decreases with temperature and is zero at the critical point. The decrease of surface tension with temperature partially explains why cleaning is preferred at higher temperatures as then the wetting is facilitated.

Figures 3.7b and 3.8 show the decrease of surface tension with temperature for ethanol and for some alkanes, respectively. Similar near-linear trends are observed in all cases. The calculations in Figures 3.7 and 3.8 were made with a thermodynamic model (CPA equation of state) coupled with the gradient theory (Queimada *et al.*, 2005).

(a)

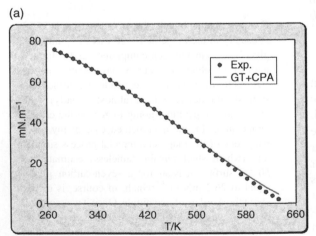

(b)

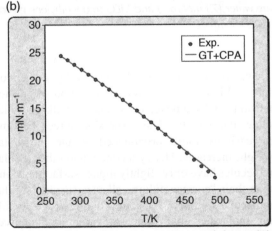

Figure 3.7 *Water (a) and ethanol (b) surface tension versus temperature. The points are experimental data, while the lines are calculations made using the gradient theory (GT) combined with the CPA equation of state. Adapted from Queimada et al. (2005), with permission from Elsevier*

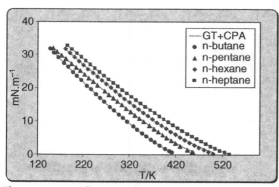

Figure 3.8 *n-Alkane surface tension versus temperature. The points are experimental data, while the lines are calculations made using the gradient theory (GT) combined with the CPA equation of state. Adapted from Queimada et al. (2005), with permission from Elsevier*

3.3 Interfacial tensions, work of adhesion and spreading

3.3.1 Interfacial tensions

Interfacial tensions are important for all interfaces (liquid–liquid, solid–liquid, solid–solid) but can be measured directly only for liquid–liquid interfaces.

Liquid–liquid interfacial tensions exist for immiscible liquid–liquid systems, e.g. water or glycols with hydrocarbons and water–alcohols. In most cases, the interfacial tension value is between the surface tensions of the two liquids involved, e.g. the value of 51 mN m^{-1} reported for water–hexane is between the 18 and 72 mN m^{-1} for hexane and water, respectively. The lower, compared to hexane–water, value for the interfacial tension of benzene–water (35 mN m^{-1}) is due to the higher solubility of benzene compared to hexane in water (Figure 3.9). This is due to the weak complexes formed between aromatics and water which exist because of the so-called "π"-electrons of the aromatic rings. For this reason, the benzene–water interface is much smaller than the hexane–water one and this is why water–benzene has a much lower interfacial tension than the "more insoluble" water–hexane.

Owing to adsorption of alcohols at the water interface, alcohol–water interfacial tensions are much lower than the surface tensions of the individual liquids. Table 3.2 gives some typical values for liquid–liquid interfacial tensions. In the case of water–alcohol systems, we notice that the higher the molecular weight of the alcohol the lower its solubility in (miscibility with) water (Figure 3.10); this is due to the higher liquid–liquid interface, which in turn results in higher interfacial tension against water. Of course, for all alcohols that are smaller than butanol and are completely miscible in water the interfacial tensions are zero. The large reduction of water surface tension due to the presence of alcohols is due to the fact that the hydroxyl group of alcohols faces the water phase, a phenomenon we will see again when discussing surfactants.

The connection between miscibility and the interfacial tension values can be also observed for the highly immiscible water/perfluorocarbons (both perfluoroalkanes and perfluoro-aromatics). As seen in Figure 3.11, perfluorocarbons are much less soluble in water compared to hydrocarbons, which explains the higher interfacial tension for water/perfluorocarbons.

Actually, perfluorocarbons behave in many ways quite differently from hydrocarbons, as already established in the literature. For example, their behaviour in the context of thermodynamic models differs significantly from all other compounds, including hydrocarbons (Figure 3.12). Moreover, in the context of the regular solution/solubility parameter theory (see below), solutions containing fluorocarbons show deviations not shared by other mixtures of non-polar compounds, which constitutes the main area of applicability of regular solution/solubility parameter theory (Prausnitz, Lichtenthaler and de Azevedo, 1999).

Finally, as illustrated in Figure 3.13, the link between thermodynamics (miscibility) and the magnitude of the interfacial tension is not limited to interfacial tensions of aqueous mixtures but is also valid for other immiscible mixtures such as glycol–hydrocarbons. The higher solubilities of aromatic hydrocarbons with MEG (compared to alkanes–MEG) are due to Lewis acid–base interactions.

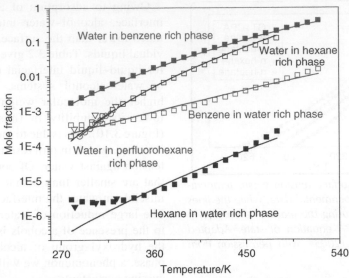

Figure 3.9 *Liquid–liquid equilibrium data (points) for water–hexane, water–benzene and water–perfluorohexane. Notice the much higher solubilities in water of the aromatic benzene compared to hexane, which explains the higher interfacial tensions for the latter system (51 mN m^{-1} for water–hexane versus 35 mN m^{-1} for water–benzene). The lines are modelling results using a successful thermodynamic model (the CPA equation of state; more information on this model is given by Kontogeorgis and Folas, 2010)*

Table 3.2 *Liquid–liquid interfacial tensions (and liquid surface tensions) for some liquids. All values are given at room temperature (20 °C). Differences in miscibility in, for example, aqueous solutions with hydrocarbons, fluorocarbons or alcohols can be elucidated from the values of the interfacial tensions. The higher the interfacial tensions, the lower the miscibility. For alcohols smaller than butanol, the interfacial tensions with water are zero, as such alcohols are completely miscible in water. Fluorocarbons have among the lowest surface tensions due to their very weak van der Waals forces*

System	Interfacial tension (mN m^{-1})	Liquid	Surface tension (mN m^{-1})
Water–hexane	51.1	Water	72.8
Water–octane	50.8	Octane	21.8
Water–tetradecane	52.0	Tetradecane	25.6
Water–benzene	35.0	*n*-Hexane	18.4
Water–C$_6$F$_{14}$	57.2	Benzene	28.9
Water–CHCl$_3$	28–31.6	CHCl$_3$	27.2
Water–CCl$_4$	45.1	CCl$_4$	26.8
Water–diethyl ether	11.0	Diethyl ether	17.0
MEG-hexane	16.0	C$_6$F$_{14}$	11.5
Water–butanol	1.8	MEG	47.7
Water–pentanol	6.0	Pentanol	25.2
Water–octanol	8.5	Octanol	27.5
Water–cyclohexanol	4.0	Cyclohexanol	32.7
Water–oleic acid	15.7	Oleic acid	32.5
Water–aniline	5.8	Aniline	42.9
Water–bromoform	40.8	Bromoform	41.5
Water–mercury	415	Mercury	485

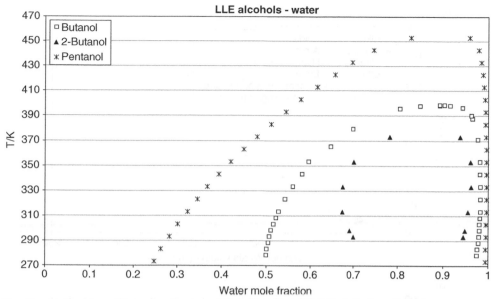

Figure 3.10 Liquid–liquid equilibria data for three alcohol–water systems. The immiscibility increases from butanol to pentanol and so does the alcohol–water interfacial tension (see Table 3.2; water–butanol: 1.8 mN m⁻¹; water–pentanol: 6 mN m⁻¹). Reprinted from Kontogeorgis and Folas (2010), with permission from John Wiley & Sons

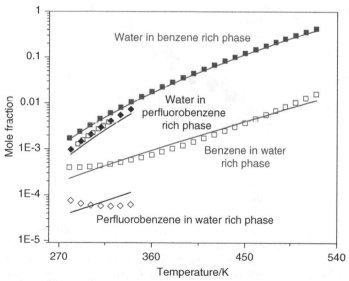

Figure 3.11 Liquid–liquid equilibrium data (points) for water–benzene and water–perfluorobenzene. Notice the much lower solubilities in water of perfluorobenzene compared to benzene, which explains the higher interfacial tensions (57.2 mN m⁻¹ for perfluorobenzene–water versus 35 mN m⁻¹ for benzene–water). Fluorocarbons are extremely hydrophobic compounds having very weak van der Waals forces. As in Figure 3.9, the lines are modelling results using the CPA equation of state. Reprinted from Tsivintzelis and Kontogeorgis (2012), with permission from American Chemical Society

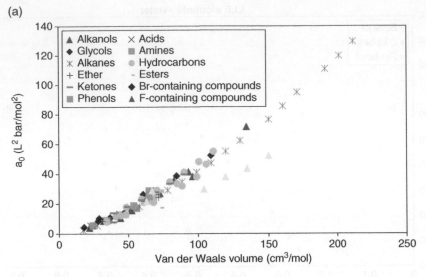

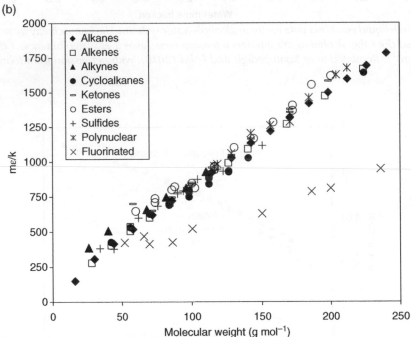

Figure 3.12 *Energy parameter of a wide range of molecules for two thermodynamic models (CPA – (a) and sPC-SAFT – (b)) against a size parameter (van der Waals volume or molecular weight). Notice that, in both cases, the trends for the fluorinated compounds (mostly fluorocarbons) differ significantly from all other families of compounds. (a) Reproduced with permission from Anders Schlaikjer (2014, MSc. Thesis) and (b) Reprinted from Tihic et al. (2006), with permission from Elsevier*

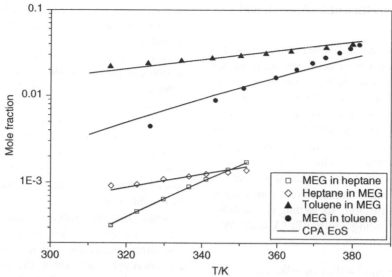

Figure 3.13 *Liquid–liquid equilibrium data (points) for MEG–heptane and MEG–toluene. The solubilities are much lower in the case of the aliphatic alkanes, which explains the much higher interfacial tensions for MEG–heptane (about 17 mN m⁻¹) compared to MEG–toluene (about 7 mN m⁻¹). As in Figures 3.9 and 3.11 the lines are modelling results using the CPA equation of state. Reprinted from Kontogeorgis and Folas (2010), with permission from John Wiley & Sons, Ltd*

3.3.2 Work of adhesion and cohesion

An important property connected to interfacial tensions is the thermodynamic or ideal work of adhesion, which is defined as the work needed to separate an interface into two separate surfaces (in contact with air).

The expression for this thermodynamic or ideal work of adhesion, see also Figure 3.14, is given by the Dupre equation:

$$W_{adh} = \gamma_i + \gamma_j - \gamma_{ij} \qquad (3.1)$$

where the phases i and j are indicated as A and B in Figure 3.14.

The work of adhesion is defined in all cases where we have two immiscible interfaces, e.g. liquid–liquid, solid–liquid and solid–solid.

Similarly, we can define the so-called work of cohesion for a single surface (work needed to break the surface), W_{coh}, which is equal to two times the value of its surface tension ($W_{coh} = 2\gamma$).

In reality, the actual contact surface area between two different materials is (often much) less than the

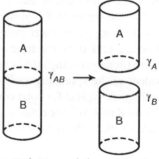

Figure 3.14 *Definition of the work of adhesion–see also Equation 3.1*

geometric area. For this reason adhesion is not produced by just pressing two solids together and in most cases an "adhesive" (glue) is needed to increase the contact area (Pashley and Karaman, 2004). The adhesive material will increase the area significantly and thus the contact between the solids, and in this way strong van der Waals and hydrogen bonding forces can be applied that can hold the materials together. Adhesion is discussed in more detail in Chapter 6.

3.3.3   Spreading coefficient in liquid–liquid interfaces

Interfacial tensions are also used to calculate the spreading for different interfaces. Spreading of one liquid in another immiscible liquid (or solid) can be expressed via the Harkins spreading coefficient.

When a liquid, termed generally as an "oil" (O), is placed on water (W) it is possible for it to either stay as a lenz on the surface (non-spreading) or to spread as a thick film (complete spreading). It is also possible to spread as a monolayer film (partial spreading).

When a drop of liquid is placed on another liquid with which it is immiscible, its wetting is quantified by the Harkins spreading coefficient (A = air, W = water, O = oil):

$$S = \gamma_{WA} - (\gamma_{OA} + \gamma_{OW}) = W_{adh,OW} - W_{coh,O} \quad (3.2)$$

According to the Harkins definition of the spreading coefficient, it is the difference between the original energy (surface energy of the water–air interface, WA) and the final energy, in the case of an oil–water (OW) and an oil–air (OA) interface.

The Harkins spreading coefficient is a tool for determining whether one liquid will spread upon another, for example, "oil" on water. If S is positive, then the two interfaces have together lower energy than the water–air interface and thus we have spreading – and the opposite (non-spreading) if S is negative.

Positive values of the spreading coefficient which indicate spreading are typical for polar compounds in water, e.g. $S = 36.8$ mN m^{-1} for octanol in water. On the other hand, $S = 0.2$ mN m^{-1} for octane in water,

indicating minimal spreading, and $S = -9.3$ mN m^{-1} for hexadecane in water, indicating that it does not spread at all in water.

Spreading increases with solubility, but mutual saturation can also affect the results, e.g. for benzene–water $S = 8.9$ mN m^{-1} initially and $S = -1.4$ mN m^{-1} (non-spreading) after mutual saturation. Thus, while initially S is positive and benzene appears to spread on water, after mutual saturation the surface tension of (especially) air–water changes and S becomes negative. Then the spreading stops. Similar observations can be made for hexanol and other water immiscible systems.

Notice that the Harkins coefficient is related to the works of adhesion and cohesion: positive spreading coefficient and thus spreading ($S > 0$) are obtained only when the work of adhesion between oil and water is higher than the work of cohesion of the oil, or, in other words, when oil "likes" (adheres to) water more than it likes itself (coheres to itself) (see Problem 3.11). Similar concepts are true for solid–liquid interfaces, as discussed in Chapter 6.

The spreading of a liquid on water is a measure of its affinity to water. Thus, unsurprisingly, the spreading coefficient is, as illustrated in Figure 3.15, directly linked to the solubility in water. The higher the water solubility of a compound the higher its spreading in water is. Thus, while hydrocarbons have spreading coefficients in water ranging from –3 to 10 (Baker *et al.*, 1967), many ketones and esters have coefficients in the range 25–46 and alcohols/ethers in the range 15–50 mN m^{-1}.

Example 3.1.   *Calculation of the initial spreading coefficient.*
Tables 3.1 and 3.2 present surface tension values for many liquids as well as interfacial tensions for many water-containing interfaces. Calculate the initial spreading coefficient of aniline, oleic acid, butanol and bromoform. Which of the liquids will spread on water and which will not? Comment on the results.

Solution
Using Equation 3.2 together with the values for the surface and interfacial tensions available in Tables 3.1 and 3.2, we obtain the following results (all in mN m^{-1}):

Aniline:

$$S = \gamma_{WA} - (\gamma_{OA} + \gamma_{OW}) = 72.8 - (42.9 + 5.8) = 24.1$$

Oleic acid:

$$S = \gamma_{WA} - (\gamma_{OA} + \gamma_{OW}) = 72.8 - (32.5 + 15.7) = 24.6$$

Butanol:

$$S = \gamma_{WA} - (\gamma_{OA} + \gamma_{OW}) = 72.8 - (24.6 + 1.8) = 46.4$$

Bromoform:

$$S = \gamma_{WA} - (\gamma_{OA} + \gamma_{OW}) = 72.8 - (41.5 + 40.8) = -9.5$$

We observe that the first three liquids will initially spread on water, while bromoform will not. Butanol will exhibit the strongest initial spreading. We notice that these large differences in spreading are due to the significant variation of water–liquid interfacial tension, which varies here between 1.8 and 40.8 mN m^{-1}. The actual chemical surface tension varies much less (25–42 mN m^{-1}). The water–liquid interfacial tension is also an indication of the miscibility of the chemical in water, with the lower interfacial tension values indicating higher miscibility. Thus, it is not surprising (see also Figure 3.15) that the lower the interfacial tension (against water) is, and thus the higher the water miscibility, the higher is the value of the initial spreading coefficient.

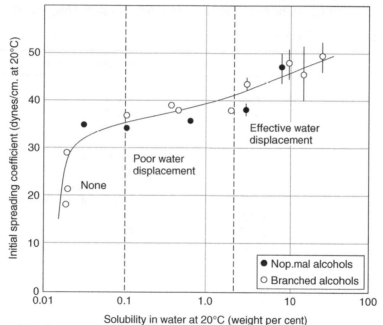

Figure 3.15 *Relationship of water solubility to the initial spreading coefficient and the water displacing ability of simple alcohols. The vertical lines through points in the region of strong displacement are proportional in length to the areas of a 2-mm water film displaced by the respective alcohols. Reprinted from Baker et al. (1967), with permission from American Chemical Society*

3.4 Measurement and estimation methods for surface tensions

There are many methods for measuring surface and interfacial tensions of liquids and liquid–liquid interfaces. The most common ones are based on the Du Nouy Tensionmeter, the Wilhelmy plate, measuring the capillary rise of liquids or the Lying Drop and Pendant Drop methods. These and other methods are described in several textbooks (Pashley and Karaman, 2004; Shaw, 1992; Hiemenz and Rajagopalan, 1997; Goodwin, 2004). They will not be further discussed here.

When not available, liquid surface tensions (pure compounds and solutions) can be estimated using predictive methods like those based on parachors and group contributions, solubility parameters (including Hansen solubility parameters) and corresponding states. Alternatively, they can be estimated from thermodynamic models like UNIFAC and SAFT combined with the gradient or the density functional theories. For a review of the latter, see Kontogeorgis and Folas (2010). Some of the most important direct methods for estimating surface tension are briefly described here.

3.4.1 The parachor method

The parachor method of McLeod–Sudgen (1923; see, for example, Sudgen, 1930) is one of the most widely

used methods for estimating the surface tensions of liquids and it can be used for both pure liquids and mixtures.

For liquids, the surface tension is estimated as:

$$\gamma^{1/4} = [P]\left(\rho_l - \rho_g\right) \tag{3.3a}$$

where the liquid and gas molar densities (ρ_l, ρ_g) are given in mol cm^{-3} and the surface tensions in mN m^{-1}.

The Parachor is estimated from group contributions:

$$[P] = \sum_i n_i [P]_i \tag{3.3b}$$

with group parameters made available by many researchers, e.g. by Quayle (1953) (Table 3.3). There

Table 3.3 *Parachor group contributions, [P]$_i$*

Carbon–hydrogen		R-[-CO-]-R′ (ketone)	
C	9.0	R + R′ = 2	51.3
H	15.5	R + R′ = 2	49.0
CH$_3$	55.5	R + R′ = 2	47.5
CH$_2$ in (CH$_2$)$_n$		R + R′ = 2	46.3
$n < 12$	40.0	R + R′ = 2	45.3
$n > 12$	40.3	R + R′ = 2	44.1
		-CHO	66.0
Alkyl groups:		O (not noted above)	20.0
1-Methylethyl	133.3	N (not noted above)	17.5
1-Methylpropyl	171.9	S	49.1
1-Methylbutyl	211.7	P	40.5
2-Methylpropyl	173.3	F	26.1
1-Etyhylpropyl	209.5	Cl	55.2
1,1-Dimethylethyl	170.4	Br	68.0
1,1-Dimethylpropyl	207.5	I	90.3
1,2-Dimethylpropyl	207.9		
1,1,2-Trimethylpropyl	243.5	Ethylenic bonds:	
C$_6$H$_5$	189.6	Terminal	19.1
		2,3-position	17.7
Special groups:		3,4-position	16.3
–COO–	63.8		
–COOH	73.8	Triple bond	40.6
–OH	29.8		
–NH$_2$	42.5	Ring Closure	
–O–	20.0	Three-membered	12.0
–NO$_2$	74.0	Four-membered	6.0
–NO$_3$ (nitrate)	93.0	Five-membered	3.0
–CO(NH$_2$)	91.7	Six-membered	0.8

have been several updates of this table, e.g. by Knotts *et al.* (2001), but this version is sufficient for our purposes.

For example, the parachor of ethanol is :

$$[P] = \sum_{i=1}^{3} n_i [P]_i = 1 \bullet [P]_{CH_3} + 1 \bullet [P]_{CH_2} + 1 \bullet [P]_{OH}$$
$$= 55.5 + 40 + 29.8 = 125.3$$

A useful approximation of Equation 3.3a for liquids (ignoring the gas density) is:

$$\gamma^{1/4} = [P]\rho_l \Rightarrow \gamma = \left(\frac{[P]}{V}\right)^4 \qquad (3.4)$$

where *V* is the molar volume in cm^3 mol^{-1}.

For liquid mixtures, the surface tension is not always equal to the linear sum of the surface tensions of the individual liquids, i.e. in general:

$$\gamma \neq \sum_i x_i \gamma_i \qquad (3.5)$$

The difference between the liquid surface tension and the above linear sum is called the "surface excess" and is usually negative, but can be positive in the case of strong intermolecular interactions.

The McLeod–Sudgen method can be extended to mixtures as:

$$\gamma_m^{1/4} = \sum_{i=1}^{n} [P]_i (x_i \rho_{lm} - y_i \rho_{gm}) \qquad (3.6)$$

where the surface tension is in mN m^{-1}, when the liquid and vapour mixture densities are given in mol cm^{-3}, and x_i and y_i are the mole fractions of the compounds in the liquid and vapour phases, respectively.

As in the case of pure liquids, at low pressures the gas term can be ignored. Then, Equation 3.6 in combination with Equation 3.3b can be written as:

$$\gamma_m^{1/4} = \sum_{i=1}^{n} [P]_i x_i \rho_{lm} - \rho_{lm} \sum_{i=1}^{n} \frac{x_i \gamma_i^{1/4}}{\rho_{li}} \qquad (3.7)$$

Another often used approximation is:

$$\gamma_m^r = \sum_i x_i \gamma_i^r \qquad (3.8)$$

where *r* is close to unity for many hydrocarbon mixtures, with values typically between −1 and −3.

The McLeod method for mixtures has an accuracy of 5–10%, but it can be less than that if the parachors are obtained from experimental data (Reid, Prausnitz and Poling, 1987).

Equation 3.7 yields the correct limit, i.e. the surface tension of the mixture is that of the pure liquid when the concentration of the compound tends to unity.

Example 3.2. *Estimation of surface tension using the Parachor method.*

Estimate, using the parachor method of McLeod–Sugden, the surface tension of liquid isobutyric acid at 333 K. The liquid density is 0.912 g cm^{-3}, the molecular weight is 88.107 g mol^{-1} and it can be assumed that the liquid density is much higher than the vapour density.

How does the calculated value compare to the experimental value (21.36 mN m^{-1})?
The structure of isobutyric acid is CH$_3$-CH(CH$_3$)-COOH.

Solution

The units in the parachor equation are mol cm^{-3} for the molar density and mN m^{-1} for the surface tension.

We first need the density value in mol cm^{-3}, which is 0.912/88.107 = 0.01035 mol cm^{-3}.

Based on the structure of isobutyric acid, we can estimate the parachor from the group contribution tables (Table 3.3):

$$[P] = \sum_{i=1}^{3} n_i [P]_i = 1 \bullet [P]_{(CH_3)_2 CH} + 1 \bullet [P]_{COOH} = 1 \bullet 133.3 + 1 \bullet 73.8 = 207.1$$

Since the gas density effect can be ignored, the parachor equation can be written as:

$$\gamma^{1/4} = [P]\rho_l \Rightarrow \gamma = \left(\frac{[P]}{V}\right)^4 = (207.1 \cdot 0.01035)^4 = 21.1\,\text{mN}\,\text{m}^{-1}$$

This is rather close to the experimental value; the percentage deviation is equal to [(experimental − calculated)/experimental × 100] about 1.2%.

Comment

An alternative way to estimate the parachor of the acid is to consider it as the sum of the parachor values of C, H, CH_3 (two times) and COOH. This gives a similar value for the parachor as above (209.3), thus resulting in a similar value for the surface tension (22.02 mN m^{-1}).

3.4.2   Other methods

There are many more methods for estimating the surface tension. Two useful methods are those based on the corresponding-states principle (CSP) and on solubility parameters.

The CSP method has been proposed by Brock–Bird–Miller (Reid, Prausnitz and Poling, 1987) and the surface tension (mN m^{-1}) is given by the equation:

$$\gamma = P_c^{2/3} T_c^{1/3} Q (1 - T_r)^{11/9}$$

$$Q = 0.1196 \left[1 + \frac{T_{br}\ln(P_c/1.01325)}{1 - T_{br}} \right] - 0.279 \quad (3.9)$$

where T_r is the reduced temperature ($=T/T_c$), the critical pressure P_c is in bar, the critical temperature T_c is in K and T_{br} is the reduced boiling point temperature (=boiling point, T_b, divided by the critical temperature, both in K).

According to Reid, Prausnitz and Poling (1987) the corresponding states method performs satisfactorily for non-polar and polar liquids, but not for hydrogen bonding ones, and the average deviation is less than 5%. On the other hand, the parachor method has a deviation of 5–10% but is more widely applicable, including to hydrogen bonding liquids, e.g. alcohols and acids.

Finally, we mention a method based on solubility parameters. As both solubility parameters and the surface tensions reflect the strength and nature of the intermolecular forces of molecules it is anticipated that they can be connected.

The solubility parameter introduced by Hildebrand (Hildebrand and Scott, 1964) is a widely used concept in thermodynamics and physical chemistry in general. Several handbooks and reference books provide extensive lists of solubility parameters of numerous chemicals. The solubility parameter is defined as:

$$\delta = \sqrt{E^{coh}} = \sqrt{\frac{\Delta E^{VAP}}{V}} = \sqrt{\frac{\Delta H^{VAP} - RT}{V}} \quad (3.10)$$

where V is the liquid molar volume and ΔE^{VAP} is the change in the internal energy of vaporization, equal to the heat of vaporization minus RT (the product of gas constant and temperature). This quantity has been traditionally named the cohesive energy (E_{coh}).

Traditionally, solubility parameters are given in (cal cm^{-3})$^{1/2}$ = Hild(ebrands), in honour of the founder of the regular solution theory, Joel Hildebrand. The solubility parameters were originally defined in conjunction with the regular solution theory. Now, they are more commonly listed in (MPa)$^{1/2}$ (= (J cm^{-3})$^{1/2}$). In the absence of experimental data, the solubility parameters of both solvents and polymers can be estimated via group contribution methods, as discussed by van Krevelen and Hoftyzer (1972).

The Danish scientist Charles Hansen (Hansen, 2000) proposed an extension of the solubility parameter concept that is particularly suitable for solubility assessments for strongly polar and hydrogen bonding fluids. He identified three contributions to the cohesive energy and thus to the solubility parameter, one stemming from non-polar (dispersion or van der Waals forces, d), one from (permanent) polar (p) and one from hydrogen bonding forces (h). He suggested that these three effects contribute additively to the cohesive energy density (i.e. the ratio of the cohesive energy to the volume):

$$E_{coh} = E_d + E_p + E_h \qquad (3.11)$$

Thus, the total solubility parameter is estimated from the equation:

$$\delta = \sqrt{\delta_d^2 + \delta_p^2 + \delta_h^2} \qquad (3.12)$$

The solubility parameter may be thought of as a vector in a three-dimensional d–p–h space. The above equation provides the magnitude of this vector. Each solvent and each polymer can be characterized by the three "solubility parameter increments", δ_d, δ_p, δ_h due to dispersion, polar and hydrogen bonding forces, respectively. These are often called Hansen solubility parameters (HSPs).

Hansen has presented extensive tables with the solubility parameter increments (d,p,h). In his book (2000), he has collected an extensive compilation of such values. When not available, van Krevelen and Hoftyzer (1972) have proposed group contribution methods for estimating the three Hansen solubility parameters. More recent group contributions methods are also available for estimating the Hansen solubility parameters (e.g. Modarresi *et al.*, 2008; Stefanis and Panayiotou, 2008 and 2012). Some of these methods provide accurate estimations of the HSP, as they contain both first- and higher-order group contributions. Recently, it also has been shown that Hansen solubility parameters can be estimated from modern equations of state having separate contributions from dispersion, polar and/or hydrogen bonding effects. One example is the hydrogen bonding lattice-fluid theory pioneered by Panayiotou (see, for example, Panayiotou, 2003).

Hansen observed that when the solubility parameter increments of the solvents and polymers are plotted in three-dimensional plots, then the "good" solvents lie approximately within a sphere of radius R (with the centre being the polymer). This is expressed via the equation:

$$\sqrt{4(\delta_{d1} - \delta_{d2})^2 + (\delta_{p1} - \delta_{p2})^2 + (\delta_{h1} - \delta_{h2})^2} \le R$$
$$(3.13)$$

where 1 denotes the solvents and 2 the polymer. The quantity under the square root is the distance between the solvent and the polymer. Hansen found empirically that a universal value of 4 should be added as a factor in the dispersion term to approximately attain the shape of a sphere.

The Hansen method is very useful. It has found widespread use particularly in the paint and coating industry, where the choice of solvents to meet economical, ecological and safety constraints is of critical importance. It can explain cases in which polymer and solvent solubility parameters are almost perfectly matched, and yet the polymer will not dissolve. The Hansen method can also predict cases where two non-solvents can be mixed to form a solvent.

Another application of the Hansen solubility parameters is in surface tension determination. The method was proposed by Beerbower and Hansen–Skaarup (Hansen, 2000) and can be expressed as:

$$\gamma = 0.0715 V^{1/3} \left[\delta_d^2 + l \left(\delta_p^2 + \delta_h^2 \right) \right] \qquad (3.14)$$

In Equation 3.14 the solubility parameters are given in $(cal\ cm^{-3})^{1/2}$, the molar volume V in cm^3 mol^{-1}, and then the surface tension is expressed in $mN\ m^{-1}$. Values of the molar volumes and Hansen solubility parameters in these units are provided in Appendix 3.1; l is an adjustable parameter.

There are variations of Equation 3.14 for specific families of compounds, i.e. using different relationships for the adjustable parameters l (which can be also related to the Hansen solubility parameters). On average l is equal to 0.8 for many homologous series of compounds, but it should be obtained

Charles Hansen

Courtesy of Dr C. Hansen. http://hansen-solubility. com/CharlesHansen.html

Charles Hansen is a Danish scientist widely known for the so-called Hansen solubility parameters (HSP). Charles Hansen began his work with solvents in 1962, and almost immediately began producing new and groundbreaking results. Since then, his Hansen solubility parameters have been extensively used and proven valuable to various industries, including coatings, adhesives, plastics, protective clothing and environmental protection. They allow correlations and systematic comparisons previously not possible, such as polymer solubility, swelling and permeation, surface wetting and dewetting, the solubility of organic salts, and many biological applications. Their seemingly universal ability to predict molecular affinities has been generally accepted as being semi-empirical. Charles Hansen published a book summarizing his many years of experience in the field of solubility parameters (Hansen, 2000 and subsequent editions). For a profile of this great scientist and an interesting interview see Meyn (2002).

from experimental surface tension data for precise calculations.

Equation 3.14 can be used (when l is known) to determine the dispersion, polar and hydrogen bonding contributions of the surface tension:

$$\gamma^d = 0.0715V^{1/3}\delta_d^2$$

$$\gamma^p = 0.0715V^{1/3}l\delta_p^2 \qquad (3.15)$$

$$\gamma^h = 0.0715V^{1/3}l\delta_h^2$$

Naturally $\gamma = \gamma^d + \gamma^p + \gamma^h$.

Equation 3.15 is useful, as discussed in the next section, for estimating the interfacial tension using the Hansen method.

As a concluding remark, let us consider the often complex aqueous solutions. In such cases there are many more methods that can be used, like those proposed by Szyszkowski (dilute aqueous solutions) and Tamura–Kurata–Odani (concentrated binary aqueous solutions). These have been presented by Reid, Prausnitz and Poling (1987).

3.5 Measurement and estimation methods for interfacial tensions

Liquid–liquid interfacial tensions can be measured by some of the same methods as for liquids. However, for other interfaces (solid–liquid, solid–solid) of importance in wetting and adhesion studies as well as for surface modification such direct measurements are not possible. Nonetheless, interfacial tension values for all interfaces are important in practical applications and thus development of estimation methods for interfacial tensions has become a very active research field.

There are many theories for estimating interfacial tensions. They roughly belong to two families: "surface component theories" and "direct theories", with the former being the most widely used. In the "surface component theories", the link between surface tensions and intermolecular forces is exploited. The surface tension is divided into different "components" (contributions) due to the various intermolecular forces. Among the various intermolecular forces, the dispersion forces are universal (and the only ones present in alkanes), while all the other forces are "specific", i.e. they only exist for specific molecules. Examples of "specific" contributions are those due to polar and hydrogen bonding forces, e.g. for glycols and water, for which the surface tensions are high, 66 and 72.8 mN m^{-1}, respectively, at 25 °C. Equally important "specific" forces are the so-called metallic ones (for mercury and other metals). For example, due to the "metallic bonding" (mobile electrons shared by atoms of a metal), the surface tension of mercury (Hg) is very high (485 mN m^{-1}).

3.5.1 "Direct" theories (Girifalco–Good and Neumann)

In "direct theories", the interfacial tension is expressed directly as a function of the surface tensions in some universal way, although some adjustable parameters are often needed.

In the early method of Girifalco and Good (1957, 1960), the interfacial tension is given as:

$$\gamma_{ij} = \gamma_i + \gamma_j - 2\varphi\sqrt{\gamma_i\gamma_j} \qquad (3.16)$$

The correction parameter φ is provided in tables (Appendix 3.2) for liquid–liquid interfaces and can be calculated via the following equation using the volumes for solid–liquid interfaces:

$$\varphi = \frac{4(V_s V_l)^{1/3}}{\left(V_s^{1/3} + V_l^{1/3}\right)^2} \qquad (3.17)$$

where the volumes are expressed in cm^3 mol^{-1} (V_s is the molar volume of the solid and V_l is the molar volume of the liquid).

In the more modern and widely used theory of Neumann and co-workers (e.g. Kwok and Neumann, 1998, 1999 and 2000), the interfacial tension is expressed as:

$$\gamma_{ij} = \gamma_i + \gamma_j - 2\sqrt{\gamma_i\gamma_j}\,e^{-\beta(\gamma_i - \gamma_j)^2} \qquad (3.18)$$

It is based on observations like those presented for example in Figure 3.16. The parameter β depends on the surface, but is almost universal, $\beta = 0.0001247$ (m^2 mJ^{-1})^2. The major characteristic of the Neumann equation is that the solid–liquid

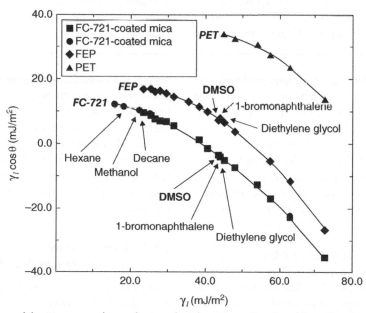

Figure 3.16 *Validation of the Neumann theory for interfacial tensions. Reprinted from Kwok and Neumann (1999), with permission from Elsevier*

Professor Robert J. Good (1920–2010)

Courtesy of the School of Engineering and Applied Sciences at State University of New York at Buffalo.

Professor Robert J. Good (MSc University at California at Berkley, PhD University of Michigan) served as a professor for 26 years in the Department of Chemical Engineering at State University of New York at Buffalo. He also worked as a chemist and consultant for Dow, Monsanto and other corporations. He was a rare combination of an academic scientist with a long career in industry.

He has been involved in both theoretical and experimental – applied research in surface and molecular thermodynamics, wetting, adhesion, microemulsions and many other applications in colloid and surface science.

He has received several awards for his work in surface and colloid chemistry, including the Jacob F. Schoellkopf Award in 1979, and the Kendall Award from the American Chemical Society in 1976, the latter "for his pioneering investigations of atomic and molecular interactions across interfaces, and the application of the resulting, new concepts to scientific, technological, and biological problems."

He wrote about 160 research papers, including those about the well-known theories for interfacial tensions discussed in this book, such as the earlier Good–Girifalco theory and the more modern acid–base theory of Carel van Oss and M. K. Chaudhury. The first papers from these two theories have so far (2014) received about 1800 citations.

interfacial tension depends, in a universal way, on the solid and liquid surface tensions only (and not on their force-components, as in the "surface component" theories). The equation-of-state theory of Neumann has been criticized for not accounting for strong chemical effects, e.g. hydrogen bonding. Even its thermodynamic background and derivation are disputed. It is, however, still in use and often compared against the other recent major theory, that of van Oss–Good (van Oss *et al.*, 1987, see Section 3.5.3).

Neumann and co-workers have published numerous articles supporting their approach. They have showed that the product of the liquid surface tension with the cosine of the contact angle is a universal function of the surface tension of the liquid for a specific solid. They presented curves like those shown in Figure 3.16 for several fluoro-surfaces, polymers and numerous additional surfaces, e.g. co-polymeric and polymeric (acrylates, methyl acrylates) surfaces. Such universal trends are impressive and seem to validate, at least for some surfaces, Neumann's theory that the interfacial tension should be a function of *only* the solid and liquid surface tensions and *not* of their components (dispersion, specific, etc.). The Neumann theory is discussed more extensively in Chapter 15.

3.5.2 Early "surface component" theories (Fowkes, Owens–Wendt, Hansen/Skaarup)

Early theories of interfacial tensions are based on the extension of the Fowkes equation. These theories are those often cited in classical colloid and surface chemistry textbook literature (e.g. Myers, 1991,

1999; Shaw, 1992; Hiemenz and Rajagopalan, 1997; Goodwin, 2004).

In the Fowkes theory, the surface tension is divided into dispersion (d) and specific (spec) components:

$$\gamma = \gamma_i^d + \gamma_i^{spec} \tag{3.19}$$

The interfacial tension is given as:

$$\gamma_{ij} = \gamma_i + \gamma_j - 2\sqrt{\gamma_i^d \gamma_j^d} \tag{3.20}$$

Notice that, unlike the Girifalco–Good equation, the cross term includes a contribution only from the dispersion forces. Thus, the Fowkes equation is based on the fundamental assumption that the cross-interaction term across the interface (work of adhesion) is due to dispersion forces alone.

Hydrocarbons, especially alkanes, only have dispersion forces (d), but other molecules may have other forces as well, which are termed specific (spec), e.g. polar, metallic, m (e.g. for Hg), hydrogen bonding, etc. As examples, for *n*-hexane (*n*-C$_6$), mercury (Hg) and *n*-octanol (*n*-C$_8$OH) the surface tension is divided, respectively, as:

$$\gamma_{n\text{-}C_6} = \gamma_{n\text{-}C_6}^d$$

$$\gamma_{Hg} = \gamma_{Hg}^d + \gamma_{Hg}^m \tag{3.21}$$

$$\gamma_{n\text{-}C_8OH} = \gamma_{n\text{-}C_8OH}^d + \gamma_{n\text{-}C_8OH}^{spec}$$

In the case of mercury (Hg) the specific contribution includes only the metallic part, while in the case of alcohols, the specific contribution includes both polar and hydrogen bonding contribution. Examples 3.3 and 3.4 illustrate some applications of the Fowkes equation.

Example 3.3. *Use of the Fowkes equation for liquid–liquid interfaces.*
The liquid–liquid interfacial tension of *n*-hexane (o) and water (w) is 51.1 mN m^{-1}. Estimate based on the Fowkes equation the dispersion and specific surface tensions of water. The surface tensions of hexane and water are 18.4 and 72.8 mN m^{-1}, respectively, at room temperature.

Then, estimate the interfacial tension of perfluorohexane (11.5 mN m^{-1}) with water and compare the result with the experimental value (Table 3.1). Comment on the results.

Finally, estimate the interfacial tension between mercury (485 mN m^{-1}) and water and compare the result with the experimental value (426 mN m^{-1}). The experimental interfacial tension between mercury and octane (21.8 mN m^{-1}) is 375 mN m^{-1}.

Solution
For *n*-hexane, which is a hydrocarbon, the dispersion part of the surface tension is equal to the total surface tension (18.4 mN m^{-1}).

Thus, using the Fowkes equation for the interfacial tension of water/hexane we obtain:

$$\gamma_{ow} = \gamma_o + \gamma_w - 2\sqrt{\gamma_o^d \gamma_w^d} \Rightarrow$$
$$51.1 = 18.4 + 72.8 - 2\sqrt{(18.4)\gamma_w^d} \Rightarrow$$
$$\gamma_w^d = 21.8\,\text{mN m}^{-1}$$

The specific part of the surface tension of water is 72.8 – 21.8 = 51.0 mN m^{-1}.

The specific value includes both polar and hydrogen bonding effects. Noticeably, the "dispersion" value indicates that approximately 30% of the surface tension value of water is due to dispersion forces, which agrees well with theoretical considerations and the molecular theories mentioned in Chapter 2. (The contribution of dispersion to water's potential energy is around 24% at 298 K.) This illustrates the importance of the van der Waals forces even for strongly hydrogen bonding compounds such as water. The dispersion surface tension value of water from the Fowkes theory can be considered to be widely accepted and is therefore used in further calculations.

The interfacial tension of perfluorohexane–water is then calculated with the Fowkes equation using the dispersion value of water previously estimated:

$$\gamma_{ow} = \gamma_o + \gamma_w - 2\sqrt{\gamma_o^d \gamma_w^d} \Rightarrow$$
$$\gamma_{ow} = 11.5 + 72.8 - 2\sqrt{(11.5)(21.8)} \Rightarrow$$
$$\gamma_{ow} = 52.633\,\mathrm{mN\,m^{-1}}$$

We assume that perfluorohexane only has dispersion contributions (non-polar molecule). We can see that the Fowkes equation indeed predicts higher interfacial tension than for water–hexane, in qualitative agreement with the experimental data. Quantitatively, the agreement is not good as there is about 8% deviation from the experimental value. The Fowkes equation underestimates the interfacial tension significantly and predicts a behaviour much closer to water–alkanes than the data indicate.

Unlike the previous cases, mercury cannot be considered to have solely dispersion forces. A significant contribution to the surface tension should come from metallic forces: $\gamma_{Hg} = \gamma_{Hg}^d + \gamma_{Hg}^m$. Thus, first we will use the mercury–octane interfacial tension data together with the Fowkes equation to estimate the dispersion part of the surface tension of mercury:

$$\gamma_{HgC_8} = \gamma_{Hg} + \gamma_{C_8} - 2\sqrt{\gamma_{Hg}^d \gamma_{C_8}^d} \Rightarrow$$
$$375 = 485 + 21.8 - 2\sqrt{\gamma_{Hg}^d\,21.8} \Rightarrow$$
$$\gamma_{Hg}^d = 199.21\,\mathrm{mN\ m^{-1}}$$

Then, we apply again the Fowkes equation, this time to the mercury–water interface using the previously estimated dispersion values for the surface tensions of mercury and water:

$$\gamma_{Hgw} = \gamma_{Hg} + \gamma_w - 2\sqrt{\gamma_{Hg}^d \gamma_w^d} \Rightarrow$$
$$\gamma_{Hgw} = 485 + 72.8 - 2\sqrt{(199.21)(21.8)} \Rightarrow$$
$$\gamma_{Hgw} = 426\,\mathrm{mN\,m^{-1}}$$

This result is in excellent agreement with the experimental value. This and other similar calculations have given many researchers the feeling that the Fowkes equation is a very successful theory with wide applicability. This is, unfortunately, generally not the case.

The Fowkes theory, like other approaches discussed later, has strengths and weaknesses. Success stories have been reported, e.g. for mercury–water and alkanes, or interfaces between water or glycol and *n*-alkanes (see Example 3.3 and Problems 3.5 and 3.6). For other interfaces, e.g. water–aromatics, the Fowkes equation exhibits problems. See also the next example for some more capabilities and limitations of the Fowkes equation.

Example 3.4. Fowkes equation for various water-based liquid–liquid interfaces.
The dispersion contribution to the surface tension of water estimated in Example 3.3 can generally be considered correct in the context of the Fowkes equation. Using this value, estimate with the Fowkes equation the interfacial tension of *n*-tetradecane (25.6)–water and diethyl ether (17)–water mixtures and compare them to the experimental values which are, respectively, 52 and 11. All surface tension values are given at mN m^{-1} and are at the same (room) temperature. Comment on the results.

Solution

Both *n*-tetradecane and diethyl ether are non-polar fluids and we will assume that the whole surface tension contribution is due to dispersion forces.

For tetradecane–water we have:

$$\gamma_{ow} = \gamma_o + \gamma_w - 2\sqrt{\gamma_o^d \gamma_w^d} \Rightarrow$$
$$\gamma_{ow} = 25.6 + 72.8 - 2\sqrt{(25.6)(21.8)} \Rightarrow$$
$$\gamma_{ow} = 51.15 \, \text{mNm}^{-1}$$

This is in good agreement to the experimental value (1.6% deviation).

For diethyl ether–water we have:

$$\gamma_{ow} = \gamma_o + \gamma_w - 2\sqrt{\gamma_o^d \gamma_w^d} \Rightarrow$$
$$\gamma_{ow} = 17 + 72.8 - 2\sqrt{(17)(21.8)} \Rightarrow$$
$$\gamma_{ow} = 51.298 \, \text{mNm}^{-1}$$

Clearly, the Fowkes equation cannot describe the interfacial tension for the diethyl ether–water system. Essentially, the Fowkes equation cannot distinguish between diethyl ether and hydrocarbons and predicts an interfacial tension similar to the one exhibited for water–alkanes. This is certainly not correct as diethyl ether is much more miscible with water than alkanes, and therefore it has a much lower interfacial tension value in mixtures with water.

We wish to investigate whether this problem is due to the assumption we adopted for diethyl ether that all contributions to the surface tension are due to dispersion forces. If we use the Fowkes equation again assuming that the interfacial tension is known and back-calculate the dispersion value of diethyl ether's surface tension that would fit the experimental value we obtain:

$$\gamma_{ow} = \gamma_o + \gamma_w - 2\sqrt{\gamma_o^d \gamma_w^d} \Rightarrow$$
$$11 = 17 + 72.8 - 2\sqrt{\gamma_o^d (21.8)} \Rightarrow$$
$$\gamma_o^d = 71.21 \, \text{mNm}^{-1}$$

Of course this value, which is almost equal to the surface tension of water (!) and higher than the total surface tension of diethyl ether, has no physical meaning. Thus, the Fowkes equation clearly cannot be improved in this way and better theories are needed.

The success of the Fowkes equation, but also the need for better theories, has been illustrated in Examples 3.3 and 3.4. Extensions of the Fowkes equation have been proposed, which account explicitly for polar and hydrogen bonding effects in the expression for the interfacial tension using geometric-mean rules for all terms. Two such well-known theories are the Owens–Wendt theory, which is often used for polymer surfaces, and the Hansen/Skaarup (van Krevelen and Hoftyzer, 1972; Hansen, 2000) model. Both models are presented below. The Hansen/Skaarup theory is sometimes called the Hansen/Beerbower theory as all three scientists have contributed to its development.

It can be seen that these theories resemble the Fowkes equation but one or two additional cross terms are added to account for the "specific" interactions (a combined "specific" terms is used for Owens–Wendt while both polar and hydrogen bonding terms are used in the Hansen equation). The relevant equations for the surface and interfacial tensions for these two theories are given in Equations 3.22–3.24:

Owens–Wendt:

$$\gamma = \gamma^d + \gamma^{spec} \qquad (3.22a)$$

$$\gamma_{ij} = \gamma_i + \gamma_j - 2\sqrt{\gamma_i^d \gamma_j^d} - 2\sqrt{\gamma_i^{spec} \gamma_j^{spec}}$$

$$= \left(\sqrt{\gamma_i^d} - \sqrt{\gamma_j^d}\right)^2 + \left(\sqrt{\gamma_i^{spec}} - \sqrt{\gamma_j^{spec}}\right)^2 \qquad (3.22b)$$

Hansen/Beerbower (or Hansen/Skaarup):

$$\gamma_{ij} = \left(\sqrt{\gamma_i^d} - \sqrt{\gamma_j^d}\right)^2 + \left(\sqrt{\gamma_i^p} - \sqrt{\gamma_j^p}\right)^2 + \left(\sqrt{\gamma_i^h} - \sqrt{\gamma_j^h}\right)^2 \qquad (3.23)$$

or alternatively:

$$\gamma_{ij} = \gamma_i + \gamma_j - 2\sqrt{\gamma_i^d \gamma_j^d} - 2\sqrt{\gamma_i^p \gamma_j^p} - 2\sqrt{\gamma_i^h \gamma_j^h} \qquad (3.24)$$

For the Hansen equation the individual surface tensions (dispersion, polar and hydrogen bonding parts; d,p,h) are estimated based on the total value of surface tension and the Hansen solubility parameters using Equations 3.15. The *l*-parameter can be estimated first via Equation 3.14 using an experimental value of the surface tension. In this way, "reasonably correct" values for the three "surface components", i.e. the dispersion (d), polar (p) and hydrogen bonding (h) parts of the surface tensions ($\gamma^d, \gamma^p, \gamma^h$) are obtained.

Notice that in the Hansen/Beerbower theory we have a separate treatment of polar (p) and hydrogen bonding (h) effects.

Example 3.5.   Application of the Hansen–Beerbower equation for the surface tension.
Estimate using the Hansen–Beerbower method the surface tension components of ethyl acetate (surface tension equal to 23.9 mN m^{-1}). Then, using plausible assumptions estimate the surface tension of *n*-butyl acetate.

Solution
We need to calculate the surface tension of *n*-butyl acetate. First, we will use the information for ethyl acetate to calculate the parameter *l*-value from the Hansen–Beerbower equation.

The (total) surface tension of ethyl acetate is given (23.9 mN m^{-1}). According to the Hansen equation provided in the problem, we have:

$$\gamma = 0.0715 V^{1/3} \delta_d^2 + 0.0715 l V^{1/3} \delta_p^2 + 0.0715 l V^{1/3} \delta_h^2 = 23.9$$

or:

$$\gamma = 0.0715 \times 98.5^{1/3} 7.7^2 + 0.0715 l \times 98.5^{1/3} 2.6^2 + 0.0715 l \times 98.5^{1/3} 3.5^2 = 23.9$$

The values of volume and (Hansen) solubility parameters are obtained from the tables provided in Appendix 3.1.

From the above equation, we calculate the *l*-value that best fits the data. This is *l* = 0.6891.

Thus, the dispersion, polar and hydrogen bonding parts of the surface tensions as calculated from Equation 3.15 are 19.5749, 1.53797 and 2.787 mN m^{-1}.

Using this value and the Beerbower equation, we can estimate the surface tension of *n*-butyl acetate (in mN m^{-1}):

$$\gamma = 0.0715 \times 132.5^{1/3} 7.7^2 + 0.0715 l \times 132.5^{1/3} 1.8^2 + 0.0715 l \times 132.5^{1/3} 3.1^2 = 24.8$$

It is reasonable to assume that the same *l*-value can be used for both acetates, as it depends largely on the homologous series used.

3.5.3 Acid–base theory of van Oss–Good (van Oss *et al.*, 1987) – possibly the best theory to-date

Despite the success of the classical "surface components theories", described above, in many practical situations such as for polymer surfaces, they are gradually being abandoned. This may be attributed to the doubtful use of the geometric-mean rule for the polar and especially for the hydrogen bonding interactions. As discussed in Chapter 2, this geometric mean rule is rigorously valid only for the dispersion forces.

One of the most successful and widely used recent methods is the van Oss *et al.* (1987) theory or acid–base theory, hereafter called van Oss–Good. In this theory the hydrogen bonding (in general "Lewis acid–Lewis base" interactions) are expressed via asymmetric combining rules (for a review of the theory see Good, 1992). More specifically, according to this theory, the surface tension is given as the sum of the van der Waals forces (LW) and an asymmetric acid–base term, which accounts for hydrogen bonding and other Lewis acid (electron acceptor +)/Lewis base (electron donor –) interactions. A Fowkes-type geometric-mean term is used for the LW forces. The equations for the surface and interfacial tensions are:

$$\gamma = \gamma^{LW} + \gamma^{AB} = \gamma^{LW} + 2\sqrt{\gamma^+ \gamma^-} \qquad (3.25)$$

$$\gamma_{ij} = \left(\sqrt{\gamma_i^{LW}} - \sqrt{\gamma_j^{LW}}\right)^2 + 2\left(\sqrt{\gamma_i^+} - \sqrt{\gamma_j^+}\right)\left(\sqrt{\gamma_i^-} - \sqrt{\gamma_j^-}\right)$$

$$(3.26a)$$

or alternatively:

$$\gamma_{ij} = \gamma_i + \gamma_j - 2\sqrt{\gamma_i^{LW} \gamma_j^{LW}} - 2\sqrt{\gamma_i^+ \gamma_j^-} - 2\sqrt{\gamma_i^- \gamma_j^+}$$

$$(3.26b)$$

where LW are the London/van der Waals forces (dispersion, induction and polar) and AB are the acid(+)/base(–) forces. Liquids or solids containing only LW terms are characterized as "apolar", those containing only an acid or base component "monopolar" and those having both are "bipolar". Table 3.4 shows some typical values for liquids often used in the analysis of solid surfaces. These are original values from the work of van Oss *et al.* (1987). Della Volpe and Siboni (1997, 2000) have provided revised parameters, shown in Table 3.5, where the acid and base contributions to water's surface tension are not assumed to be equal. Instead the acid contribution is much higher than the base one, in agreement to the Kamlet–Taft solvatochromic parameters, where the acid parameter is also higher than the base one (the acid/base ratio is 4.33 in the Della Volpe–Siboni (2000) parameters and 6.5 in the case of the Kamlet–Taft parameters as well as in the Della Volpe–Siboni 1997 parameters). More is discussed in Chapter 15.

A major observation from the van Oss–Good theory is that, unlike other theories, the interfacial tension may be positive or negative. Note that the LW term is given by a geometric-mean Fowkes equation, but this is not the case for the Lewis AB asymmetric term.

Following the success of the van Oss–Good theory and other investigations, many leading scientists have

Table 3.4 *Parameters of the van Oss–Good theory for some classical test liquids (in mN m⁻¹). Notice the high values of the base contribution to the surface tension in all cases*

Liquid	γ	γ^{LW}	γ^{AB}	γ^+	γ^-
cis-Decalin	33.3	33.2	0	0	0
α-Bromonaphthalene	44.4	43.5	0	0	0
Diiodomethane (methylene iodide)	50.8	50.8	0	0	0
Water	72.8	21.8	51.0	25.5	25.5
Glycerol	64.0	34.0	30.0	3.92	57.4
Formamide	58.0	39.0	19.0	2.28	39.6
Ethylene glycol	48.0	29.0	19.0	1.92	47.0
Dimethyl sulfoxide	44.0	36.0	8.0	0.5	32.0

Table 3.5   *Recent parameters of the van Oss–Good theory for some classical test liquids (in mN m⁻¹) as proposed by Della Volpe and Siboni (1997, 2000). The water acid and base parameters are not assumed to be equal*

Liquid	γ	γ^{LW}	γ^{AB}	γ^+	γ^-
Year 2000 values					
Diidomethane	50.8	50.8	0	0	0
Water	72.8	26.2	46.6	48.5	11.2
Glycerol	63.5	35.0	28.5	27.8	7.33
Formamide	58.1	35.5	22.6	11.3	11.3
Ethylene glycol	48	33.9	14.1	0.966	51.6
Year 1997 values					
Diidomethane (methylene iodide)	50.8	50.8	0	0	0
Water	72.8	21.8	51	65.0	10.0
Glycerol	64	34.4	29.6	16.9	12.9
Formamide	58	35.6	22.4	1.95	65.7
Ethylene glycol	48.0	31.4	16.4	1.58	42.5

abandoned the Owens–Wendt theory (and other traditional theories) that "assume" geometric mean rules for the "asymmetric polar (and hydrogen bonding) interactions". For example, the Owens–Wendt theory erroneously predicts that acetone and ethanol are as immiscible in water as benzene is.

Gilbert Newton Lewis (1875–1946)

Kasha (1945). Reproduced with permission from College of Chemistry, UC Berkeley.

Discussion of Lewis acid–Lewis base interactions would be incomplete without saying a few words about Gilbert Lewis. This great and highly influential American scientist received his PhD from Harvard in 1899 and in 1905 he established physical chemistry at MIT. He became Dean of Chemistry at the University of California, Berkeley (in 1912). He conducted research in very many areas of physical chemistry ranging from thermodynamics (fugacity, activity, etc.), acid–bases, covalent bonds and spectroscopy. The many followers of his work included Nobel Prize winner Irving Langmuir. Actually, many consider it very surprising that Lewis never received the Nobel Prize himself. He had been nominated numerous times for the Nobel Prize between 1922 and 1940 and there are many explanations as to why he never got it. An interesting account of the events is presented by Gussman (2009).

He died as a result of a laboratory accident (hydrogen cyanide vapour leak).

3.5.4 Discussion

Many theories for estimating the interfacial tensions have been presented in Sections 3.5.1–3.5.3. The equations for the surface and interfacial tensions as well as for the work of adhesion are summarized in Table 3.6. Notice that the work of adhesion corresponds to the "cross term" of the interfacial tension expression (under the square roots), which reflects different contributions of intermolecular forces, according to the various theories (either the total surface tensions in Girifalco–Good and Neumann, only those contributions due to dispersion forces in Fowkes, due to both dispersion and specific forces in Owens–Wendt, separately dispersion, polar and hydrogen bonding ones in Hansen/Beerbower, or the van der Waals and asymmetric acid/base effects in van Oss *et al.*).

The subscripts *i* and *j* indicate the various surfaces: gas, liquid, solids – and in principle all combinations are possible, i.e. all of these theories can be used for all interfaces, but there are exceptions.

The theories discussed above have been applied extensively and it is now clear that all of them have strengths and weaknesses. We have mentioned that even some of the simplest theories like that of Fowkes are successful for some interfaces but do have problems for others. The theories that are most useful are those based on a – as much as possible – complete understanding of the many intermolecular forces present. Thus, the Hansen/Beerbower and especially the van Oss–Good are possibly the best tools to-date.

Even the van Oss–Good theory has been criticized, e.g. due to the very high basic values, but despite that it has found wide applicability in describing interfacial phenomena (interactions) involving polymers, paints, proteins and other complex systems (like polymer surface characterization, CMC determination of surfactants, protein adsorption, cell adhesion, enzyme–substrate interactions).

For liquid–liquid interfaces we often have experimental data. Thus, the most important reason for developing and using theories for interfacial tension is to apply them to solid–liquid and solid–solid interfaces, which cannot be easily accessed experimentally, but which are very important in many practical applications, e.g. for understanding adhesion and wetting phenomena.

Figure 3.17 illustrates how solid-containing interfaces and solids in general are analysed ("characterized") using theories for interfacial tension. The terms "analysis" or "characterization" in this context mean

Table 3.6 Theories for interfacial tension and expressions for the work of adhesion

Theory	γ	γ_{ij}	W_{adh}
Girifalco–Good	—	$\gamma_i + \gamma_j - 2\varphi\sqrt{\gamma_i\gamma_j}$	$2\varphi\sqrt{\gamma_i\gamma_j}$
Neumann	—	$\gamma_i + \gamma_j - 2\sqrt{\gamma_i\gamma_j}\,e^{-\beta(\gamma_i-\gamma_j)^2}$	$2\sqrt{\gamma_i\gamma_j}\,e^{-\beta(\gamma_i-\gamma_j)^2}$
Fowkes	$\gamma = \gamma^d + \gamma^{spec}$	$\gamma_i + \gamma_j - 2\sqrt{\gamma_i^d\gamma_j^d}$	$2\sqrt{\gamma_i^d\gamma_j^d}$
Owens–Wendt	$\gamma = \gamma^d + \gamma^{spec}$	$\gamma_i + \gamma_j - 2\sqrt{\gamma_i^d\gamma_j^d} - 2\sqrt{\gamma_i^{spec}\gamma_j^{spec}} =$ $\left(\sqrt{\gamma_i^d}-\sqrt{\gamma_j^d}\right)^2 + \left(\sqrt{\gamma_i^{spec}}-\sqrt{\gamma_j^{spec}}\right)^2$	$2\sqrt{\gamma_s^d\gamma_i^d} + 2\sqrt{\gamma_s^{spec}\gamma_i^{spec}}$
Hansen	$\gamma = \gamma^d + \gamma^p + \gamma^h$	$\gamma_i + \gamma_j - 2\sqrt{\gamma_i^d\gamma_j^d} - 2\sqrt{\gamma_i^p\gamma_j^p} - 2\sqrt{\gamma_i^h\gamma_j^h}$	$2\sqrt{\gamma_i^d\gamma_s^d} + 2\sqrt{\gamma_i^p\gamma_s^p} + 2\sqrt{\gamma_i^h\gamma_s^h}$
Van Oss *et al.*	$\gamma^{LW} + \gamma^{AB} = \gamma^{LW} + 2\sqrt{\gamma^+\gamma^-}$	$\left(\sqrt{\gamma_i^{LW}}-\sqrt{\gamma_j^{LW}}\right)^2 + 2\left(\sqrt{\gamma_i^+}-\sqrt{\gamma_j^+}\right)\left(\sqrt{\gamma_i^-}-\sqrt{\gamma_j^-}\right) =$ $\gamma_i + \gamma_j - 2\sqrt{\gamma_i^{LW}\gamma_j^{LW}} - 2\sqrt{\gamma_i^+\gamma_j^-} - 2\sqrt{\gamma_i^-\gamma_j^+}$	$2\sqrt{\gamma_i^{LW}\gamma_j^{LW}} + 2\sqrt{\gamma_i^+\gamma_j^-} + 2\sqrt{\gamma_i^-\gamma_j^+}$

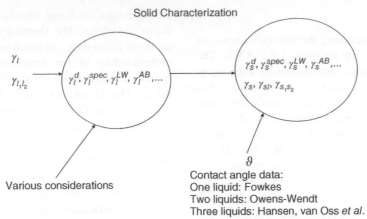

Figure 3.17 *Procedure for characterizing a solid surface. The work of adhesion and the characterization/profile of the solid surface can be obtained as well as ideas about what contaminants may be present, for surface modification and adhesion. Data from studies of liquid–liquid interfaces may be required*

determining the solid surface tension and its "components" (dispersion, specific, etc.), which assist in determining which impurities are present on the surface and which modifications are required for improving, for example, wetting or adhesion for a certain application.

Such analysis is based on the theories presented in this chapter, the concept of the contact angle and the associated Young equation discussed in Chapter 4. The analysis of solid interfaces and its application in understanding wetting and adhesion will be illustrated in Chapter 6, after the concept of contact angle is presented in Chapter 4 and surfactants in Chapter 5. Theories for interfacial tension will be discussed in more detail in Chapter 15.

3.6   Summary

Surface and interfacial tensions are key concepts in surface chemistry. They can be directly measured only for liquid–gas and liquid–liquid interfaces but they *do* exist and are equally important for all interfaces (solid–gas, solid–liquid, solid–solid).

When not available, liquid surface tensions (pure compounds and solutions) can be estimated using predictive methods like those based on parachors and group contributions, solubility parameters (including Hansen parameters) and corresponding states.

Surface/interfacial tensions are linked to intermolecular forces. This link is much used in estimating liquid–liquid (and other) interfacial tensions via different theories. There are many theories for

estimating interfacial tensions and they can be divided into two families; "surface component theories" and "direct theories", of which the former are the most well-known. In principle, all theories can be used for all interfaces, but there are exceptions.

All theories have strengths and weaknesses and it may be difficult to decide a priori which theory is the best. Perhaps for this reason, several textbooks (Shaw, 1992; Hamley, 2000; Goodwin, 2004; Israelachvilli, 1985) present only the very simple theories of Fowkes and Girifalco–Good, possibly for illustration purposes, while others (Pashley and Karaman, 2004) do not present any theories at all for estimating the interfacial tension(!).

Interestingly, none of the above textbooks discuss modern theories nor do they present comparisons between theories or mention in any detail the limitations of the Fowkes approach. We believe that, despite limitations, the acid–base theory of van Oss–Good is the most widely used method today and is therefore recommended for practical applications.

All theories, in combination with contact angle data and information from liquid interfaces, will be used later in the book (Chapter 6) for estimating the interfacial tensions of solid-interfaces (solid–liquid, solid–solid) and for characterizing solid surfaces and thus for understanding important phenomena such as wetting, lubrication and adhesion. Chapter 15 offers a more detailed presentation of theories for estimating the interfacial tension as well as some comparisons between them.

Appendix 3.1 Hansen solubility parameters (HSP) for selected solvents

To be used in Equations 3.14 and 3.15. HSP in $(cal\ cm^3)^{-1/2}$, V in $cm^3\ mol^{-1}$.

Table 3.7 *Solubility parameters of various liquids at 25 °C. For a complete list, see Hansen*

Class	Code	Name	Molar Volume	Parameters		
				δ_d	δ_p	δ_h
Paraffin	1	n-Butane	101.4	6.9	0.0	0.0
Hydrocarbons	2	n-Pentane	116.2	7.1	0.0	0.0
	3	i-Pentane	117.4	6.7	0.0	0.0
	4	n-Hexane	131.6	7.3	0.0	0.0
	5	n-Heptane	147.4	7.5	0.0	0.0
	6	n-Octane	163.5	7.6	0.0	0.0
	7	i-Octane	166.1	7.0	0.0	0.0
	8	n-Nonane	179.7	7.7	0.0	0.0
	9	n-Decane	195.9	7.7	0.0	0.0
	10	n-Dodecane	228.6	7.8	0.0	0.0
	11	n-Hexadecane	294.1	8.0	0.0	0.0
	12	n-Eicosane	359.8	8.1	0.0	0.0
	13	Cyclohexane	108.7	8.2	0.0	0.1
	14	Methyl Cyclohexane	128.3	7.8	0.0	0.5
	14.1	cis-Decalin	156.9	9.2	0.0	0.0
	14.2	trans-Decalin	159.9	8.8	0.0	0.0
Aromatic	15	Benzene	89.4	9.0	0.0	1.0
Hydrocarbons	16	Toluene	106.8	8.8	0.7	1.0
	16.1	Naphthalene	111.5	9.4	1.0	2.9
	17	Styrene	115.6	9.1	0.5	2.0
	18	o-Xylene	121.2	8.7	0.5	1.5
	19	Ethyl Benzene	123.1	8.7	0.3	0.7
	19.1	l-Methyl Naphthalene	138.8	10.1	0.4	2.3
	20	Hesitylene	139.8	8.8	0.0	0.3
	21	Tetralin	136.0	9.6	1.0	1.4
	21.1	Biphenyl	154.1	10.5	0.5	1.0
	22	p-Diethyl Benzene	156.9	8.8	0.0	0.3
Halocarbons	23	Methyl Chloride	55.4	7.5	3.0	1.9
	24	Methylene Dichloride	63.9	8.9	3.1	3.0
	24.1	Chloro Bromo Methane	65.0	8.5	2.8	1.7
	25	Chloro Difluoro Methane	72.9	6.0	3.1	2.8
	26	Dichloro Fluoro Methane	75.4	7.7	1.5	2.8
	27	Ethyl Bromide	76.9	8.1	3.9	2.5
	27.1	1,1 Dichloro Ethylene	79.0	8.3	2.3	1.6
	28	Ethylene Dichloride (1,2 Dichloro Ethane)	79.4	9.3	3.3	2.0
	28.1	Methylene Diiodide[a]	80.5	8.7	1.9	2.7
	29	Chloroform	80.7	8.7	1.5	2.8
	29.1	1,1 Dichloro Ethane	84.8	8.1	4.0	0.2

(continued overleaf)

***Table* 3.7**   (continued)

Class	Code	Name	Molar Volume	δ_d	δ_p	δ_h
	29.2	Ethylene Dibromide	87.0	9.6	3.3	5.9
	30	Bromoform	87.5	10.5	2.0	3.0
	31	n-Propyl Chloride	88.1	7.8	3.8	1.0
	32	Trichloro Ethylene	90.2	8.8	1.5	2.6
	33	Dichloro Difluoro Methane	92.3	6.0	1.0	0.0
	34	Trichloro Fluoro Methane	92.8	7.5	1.0	0.0
	35	Bromo Trifluoro Methane	97.0	4.7	1.2	0.0
	36	Carbon Tetrachloride	97.1	8.7	0.0	0.3
	37	1.1.1 Trichloro Ethane	100.4	8.3	2.1	1.0
	38	Tetrachloro Ethylene	101.1	9.3	3.2	1.4
	39	Chloro Benzene	102.1	9.3	2.1	1.0
	39.1	n-butylchlorid	104.9	8.0	2.7	1.0
	39.2	Tetrachloro Ethane	105.2d	9.2	2.5	4.6
	40	Bromo Benzene	105.3	10.0	2.7	2.0
	41	o-Dichloro Benzene	112.8	9.5	3.1	1.6
	42	Benzyl Chloride	115.0	9.2	3.5	1.3
	42.1	Tetrabromo Ethane[a]	116.8	11.1	2.5	4.0
	43	Dichloro Tetrafluoro Ethane[a]	117.0	6.2	0.9	0.0
	44	Trichloro Trifluoro Ethane	119.2	7.2	0.8	0.0
	45	Cyclohexyl Chloride	121.3	8.5	2.7	1.0
	46	l-Bromo Naphthalene	140.0	9.9	1.5	2.0
	47	Trichloro Binhenyl[a]	187.0	9.4	2.6	2.0
	48	Perfluoro Methyl Cyclohexane	196	6.1	0.0	0.0
	49	Perfluoro Dimethyl Cyclohexan	220	6.1	0.0	0.0
	50	Perfluoron-Heptane	226	5.9	0.0	0.0
Ethers	51	Furan	72.5	8.7	0.9	2.6
	51.1	Epichlorhydrin	79.9	9.3	5.0	1.8
	51.2	Tetrahydrofuran	81.7	8.2	2.4	3.6
	51.3	1.4 Dioxane[a]	85.7	9.3	2.4	2.1
	51.4	Methylal[a]	88.8	7.4	0.9	4.2
	52	Diethyl Ether	104.8	7.0	1.4	3.0
	53	2.2 Dichloro Diethyl Ether	117.6	9.2	4.4	2.8
	53.1	Anisole[a]	119.1	8.7	2.0	1.8
	53.2	Di-(2-Methoxy Ethyl) Ether	142.0	7.7	3.0	4.5
	53.3	Di-(Chloro-i-Propyl) Ether	146.0	9.3	4.0	2.5
	53.4	Dibenzyl Ether[a]	192.7	8.5	1.8	3.0
	54	bis-(m-Phenoxy Phenyl) Ether	373	9.6	1.5	2.5
	55	Acetone	74.0	7.6	5.1	3.4
	56	Methyl Ethyl Ketone	90.1	7.8	4.4	2.5
	57	Cyclohexanone	104.0	8.7	3.1	2.5
	58	Diethyl Ketone	106.4	7.7	3.7	2.3
	58.1	Mesityl Oxide	115.6	8.0	3.5	2.5
	59	Acetophenone	117.4	9.6	4.2	2.5
	60	Methyl i-Butyl Ketone	125.8	7.5	3.0	2.0

Table 3.7 (continued)

Class	Code	Name	Molar Volume	Parameters		
				δ_d	δ_p	δ_h
	61	Methyl i-Amyl Ketone	142.8	7.8	2.8	2.0
	61.1	Isophorone	150.5	8.1	4.0	3.6
	62	Di i-Butyl Ketone	177.1	7.8	1.8	2.0
Aldehydes	63	Acetaldehyd[a]	57.1	7.2	3.9	5.5
	63.1	Furfaraldehyde	83.2	9.1	7.3	2.5
	64	Butyraldehyde	88.5	7.2	2.6	3.4
	65	Benzaldehyde	101.5	9.5	3.6	2.6
Esters	66	Ethylene Carbonate	66.0	9.3	10.7	2.5
	66.1	Y-Butyrolactone	76.8	9.3	8.1	3.6
	66.2	Methyl Acetate	79.7	7.6	3.5	3.7
	67	Ethyl Formate	80.2	7.6	4.1	4.1
	67.1	Propylene Carbonate	85.0	9.0	9.6	2.0
Esters (ctd.)	68	Ethyl Chloroformate	95.6	7.6	4.9	3.3
	69	Ethyl Acetate	98.5	7.7	2.6	3.5
	69.1	Trimethyl Phosphate	99.9	8.2	7.8	5.0
	70	Diethyl Carbonate	121	8.1	1.5	3.0
	71	Diethyl Sulfate	131.5	7.7	7.2	3.5
	72	n-Butyl Acetate	132.5	7.7	1.8	3.1
	72.1	i-Butyl Acetate	133.5	7.4	1.8	3.1
	72.2	2-Ethoxyethyl Acetate	136.2	7.8	2.3	5.2
	73	i-Amyl Acetate	148.8	7.5	1.5	3.4
	73.1	i-Butyl i-Butyrate	163	7.4	1.4	2.9
	74	Dimethyl Phthalate	163	9.1	5.3	2.4
	75	Ethyl Cinnamate	166.8	9.0	4.0	2.0
	75.1	Triethyl Phosphate	171.0	8.2	5.6	4.5
	76	Diethyl Phthalate	198	8.1	3.5	4.8
	76.1	Dibutyl Phtalate	266	8.7	4.7	2.0
	76.2	Butyl Benzyl Phthalate	306	9.3	5.5	1.5
	77	Tricresyl Phosphate	316	9.3	6.0	2.2
	78	Tri n-Butyl Phosphate	345	8.0	4.0	3.0
	79	i-Propyl Palmitate	330	7.0	1.9	1.8
	79.1	Dibutyl Sebacate	339	6.8	2.2	2.0
	79.2	Methyl Oleate[c]	340	7.1	1.9	1.8
	79.3	Dioctyle Phthalate	377	8.1	3.4	1.5
	80	Butyl Stearate	382	7.1	1.8	1.7
Nitrogen Compounds	81	Acetonitrile	52.6	7.3	9.0	3.7
	81.1	Acrylonitrile	67.1	8.0	8.5	3.3
	82	Propionitrile	70.9	7.5	7.0	2.7
	83	Butyronitrile	87.0	7.5	6.1	2.5
	84	Benzonitrile	102.6	8.5	4.4	2.1
	85	Nitromethane	54.3	7.7	9.2	3.0
	86	Nitroethane	71.5	7.8	7.6	2.7
	87	2-Nitropropane	86.9	7.9	5.9	2.5

(continued overleaf)

Table 3.7   (continued)

Class	Code	Name	Molar Volume	Parameters		
				δ_d	δ_p	δ_h
	88	Nitrobenzene	102.7	9.8	4.2	2.0
	89	Ethanolamine	60.2	8.4	7.6	10.4
	89.1	Ethylene Diamine	67.3	8.1	4.3	8.3
Nitrogen Compounds	89.2	1,1-Dimethyl Hydrazine[a]	76.0	7.5	2.9	5.4
(ctd.)	89.3	2-Pyrrolidone	76.4	9.5	8.5	5.5
	90	Pyridine	80.9	9.0	4.4	3.5
	91	n-Propylamine	83.0	8.3	2.4	4.2
	92	Morpholine	87.1	9.2	2.4	4.5
	93	Aniline	91.5	9.5	2.5	5.0
	93.1	n-Methyl 2-Pyrrolidone	96.5	8.8	6.0	3.5
	94	n-Butylamine	99.0	7.9	2.2	3.9
	95	Diethylamine	103.2	7.3	1.1	3.0
	95.1	Diethylene Triamine	108.0	8.2	6.5	7.0
	96	Cyclohexylamine	115.2	8.5	1.5	3.2
	96.1	Quinoline	118.0	9.5	3.4	3.7
	97	Di n-Propylamine	136.9	7.5	0.7	2.0
	98	Formamide	39.8	8.4	12.8	9.3
	99	Dimethylformamide	77.0	8.2	6.7	5.0
	99.1	Dimethyl Acetamide	92.5	8.2	5.6	5.0
	99.2	Tetramethyl Urea	120.4	8.2	3.5	4.5
	99.3	Hexamethyl Phosphorsyreamide[a]	175.7	9.5	2.7	4.3
Sulfur Compounds	100	Carbon Disulfide	60.0	10.0	0.0	0.3
	101	Dimethyl Sulfoxide	71.3	8.2	8.0	5.0
	101.1	Ethyl Morcaptan[a]	74.3	7.7	3.2	3.5
	102	Dimethyl Sulfone[b]	75	9.3	9.5	6.0
	103	Diethyl sulfide	108.2	8.1	1.5	2.0
Acid Halides and	104	Acetyl Chloride	71.0	7.7	5.2	1.9
Anhydrides	104.1	Succinio Anhydride[b]	66.8	9.1	9.4	8.1
	105	Acetic Anhydride	94.5	7.8	5.7	5.0
Mono-Hydric Alcohols	120	Methanol	40.7	7.4	6.0	10.9
	121	Ethanol	58.5	7.7	4.3	9.5
	121.1	Hydracrylonitrile	68.3	8.4	9.2	8.6
	121.2	Allyl Alcohol[a]	68.4	7.9	5.3	8.2
	122	1-Propanol	75.2	7.8	3.5	8.5
Monohydric Alcohols	123	2-Propanol	76.8	7.7	3.0	8.0
(ctd.)	123.1	3-Clorpropanol	84.2	8.6	2.8	7.2
	124	Furfuryl Alcohol	86.5	8.5	3.7	7.4
	125	1-Butanol	91.5	7.8	2.8	7.7
	126	2-Butanol	92.0	7.7	2.8	7.1
	126.1	i-Butanol	92.8	7.4	2.8	7.8
	126.2	Benzyl Alcohol	103.6	9.0	4.0	7.0
	127	Cyclohexanol	106.0	8.5	2.0	6.6
	128	1-Pentanol	109.0	7.8	2.2	6.8
	129	2-Ethyl Butanol	123.2	7.7	2.1	6.6

Table 3.7 (continued)

Class	Code	Name	Molar Volume	Parameters		
				δ_d	δ_p	δ_h
	129.1	Diacetone Alcohol	124.2	7.7	4.0	5.3
	130	Ethyl Lactate	115	7.8	3.7	6.1
	130.1	Butyl Lactate	149	7.7	3.2	5.0
	131	2-Methoxy Ethanol	79.1	7.9	4.5	8.0
	132	2-Ethoxy Ethanol	97.8	7.9	4.5	7.0
	132.1	2-(2-Methoxy Ethoxy) Ethanol	118.0	7.9	3.8	6.2
	132.2	2-(2-Ethoxy Ethoxy) Ethanol	130.9	7.9	4.5	6.0
	133	Butoxy Ethanol	132	7.8	2.5	6.0
	133.1	2-Ethyl Hexanol	157.0	7.8	1.6	5.8
	134	1-Octanol	157.7	8.3	1.6	5.8
	134.1	2-(2-Butoxy Ethoxy) Ethanol	170.6	7.8	3.4	5.2
	135	1-Decanol	190	8.6	1.3	4.9
	136	i-Tridecyl Alcohol[c]	242	7.0	1.5	4.4
	136.1	i-Nonyl Phenoxy Ethanol[c]	275	8.2	5.0	4.1
	137	Oleyl Alcohol[c]	316	7.0	1.3	3.9
	137.1	Oleyl-Triethylene Glycol Ether[c]	418.5	6.5	1.5	4.1
Acids	140	Formic Acid	37.8 mono	7.9	6.2	10.0
	141	Acetic Acid	57.1 mono dim.	8.1	5.5	8.5
				7.2	3.0	5.0
	141.1	Benzoic Acid[b]	100	8.9	3.4	4.8
	142	n-Butyric Acid	110 mono	8.1	4.7	7.3
	142.1	n-Octoic Acid[a]	159 mono	8.2	3.5	6.0
Acida (ctd.)	143	Oleic Acid	320	7.0	1.5	2.7
	143.1	Stearic Acid[b]	326	8.0	1.6	2.7
Phenols	144	Phenol	87.5	8.8	2.9	7.3
	144.1	Resorcinol[b]	87.5	8.8	4.1	10.3
	145	m-Cresol	104.7	8.8	2.5	6.3
	145.1	Guiacol[a]	109.5	8.8	4.0	6.5
	146	Methyl Salicylate	129	7.8	3.9	6.0
	147	i-Nonyl Phenol	231	8.1	2.0	4.5
Water	148	Water[a]	18.0	7.6	8.1	20.6
Polyhydric Alcohols	149	Ethylene Glycol	55.8	8.3	5.4	12.7
	150	Glycerol	73.3	8.5	5.9	14.3
	150.1	Propylene Glycol	73.6	8.2	4.6	11.4
	150.2	1.3 Butanediol	89.9	8.1	4.9	10.5
	151	Diethylene Glycol	95.3	7.9	7.2	10.0
	152	Triethylene Glycol	114.0	7.8	6.1	9.1
	153	Hexylene Glycol	123	7.7	4.1	8.7
	154	Dipropylene Glycol[a]	131.3	7.8	8.0	9.0

n – Values uncertain.

b – The "δ" in solubility parameters (all of them i.e., both d,p,h) cannot be seen in the whole table.

c – Impure commercial product of this nominal formula.

Appendix 3.2 The "φ" parameter of the Girifalco–Good equation (Equation 3.16) for liquid–liquid interfaces. Data from Girifalco and Good (1957, 1960)

Table 3.8 *Mercury–non-metallic liquid interfaces at 20 °C (except for water, heptane and benzene which are at 25 °C)*

Liquid	φ	Liquid	φ
n-Hexane	0.62	Methanol	0.56
p-xylene			
bromobenzene			
n-Heptane	0.61	Ethanol	0.53
n-nonane			
o-xylene			
toluene			
n-Octane	0.60	n-Propanol	0.57
n-propylbenzene			
n-butylbenzene			
diisoamylamine			
m-Xylene	0.63	n-Butanol	0.58
isooctane		isoamyl alcohol	0.58
diethyl ether		n-butyl acetate	
Benzene	0.59	n-Hexanol	0.59
chlorobenzene			
nitrobenzene			
		2-Octanol	0.69
Ethyl mercaptan	0.88	n-Octanol	0.70
Nitroethane	0.82	Isobutanol	0.75
Iodomethane	0.84	Acetic acid	0.76
		valeric acid	
Iodoethane	0.79	Oleic acid	0.75
1,2-Dichloroethane	0.76	Aniline	0.66
Dichloromethane	0.70	1,2-Dibromoethane	0.59
Chloroform	0.64	Water	0.32
carbon tetrachloride			

Table 3.9 *Water–organic liquid interfaces at 25 °C*

Liquid	φ	Liquid	φ
Isobutanol	1.15	Diethyl ether	1.12
n-Butanol	1.13	Diisopropyl ether	1.01
2-octanol			
Isoamyl alcohol	1.11	Di-n-propylamine	1.17
n-Pentanol	1.09	Di-n-butylamine	1.06
n-Hexanol	1.06	Aniline	0.98
n-heptanol			
Cyclohexanol	1.04	Butyronitrile	1.00
n-Octanol	1.03	Isovaleronitrile	0.97
Isovaleric acid	1.11	Nitromethane	0.97

Table 3.9 (continued)

Liquid	φ	Liquid	φ
Heptanoic acid	1.04	Nitrobenzene	0.81
Octanoic acid	1.03	o-Nitrotoluene	0.79
Oleic acid	0.92	m-Nitrotoluene	0.97
Methyl n-hexyl ketone	1.08	Ethyl mercaptan	0.85
Methyl n-propyl ketone	1.07	Phenyl isothiocyanate	0.68
Methyl n-butyl ketone	1.03	Carbon disulfide	0.58
Mono-chloroacetone	1.00		
Methyl amyl ketone	0.99	Ethyl acetate	1.08
Ethyl n-propyl ketone	0.98	Ethyl carbonate	0.99
n-Heptaldehyde	0.97	n-Butyl acetate	0.97
1,1-Dichloroacetone	0.96	Ethyl n-butyrate	0.97
Benzaldehyde	0.90	Ethyl isovalerate	0.94
		Isoamyl n-butyrate	0.88
n-Pentane	0.58	Ethyl n-nonylate	0.85
n-Hexane–n-decane	0.55	Ethyl n-octanoate	0.84
Cyclohexane decalin	0.55		
n-Tetradecane	0.53	Chlorobenzene	0.70
		Bromobenzene	0.69
Benzene	0.72	o-Chloronaphthalene	0.67
Toluene	0.71	Iodobenzene	0.66
Ethylbenzene	0.69	o-Bromotoluene	0.66
		o-bromonaphthalene	
Mesitylene	0.67	Isobutyl chloride	0.88
Chloroform	0.76	Carbon tetrachloride	0.69

Problems

Problem 3.1: Estimation of the surface tension using the corresponding states method

Estimate, using the corresponding states method, the surface tension of liquid ethyl mercaptan at 303 K. The critical temperature is 499 K, the boiling point temperature is 308.2 K and the critical pressure is 54.9 bar. Compare the result to the experimental value $(22.68 \text{ mN m}^{-1})$ and the estimation using the parachor method (which results in a deviation of 9.1%).

Problem 3.2: Estimation of the surface tension of liquid mixtures using the parachor method (data from Hammick and Andrew (1929))

Estimate, using the parachor method for mixtures (both the "full" and the approximate method), the surface tension of a liquid mixture having a density of 0.7996 g cm^{-3} and containing diethyl ether (42.3%) and benzene. The temperature is 298 K. Compare the two methods to each other and to the experimental value which is 21.81 mN m^{-1}.

The structure of diethyl ether: CH_3CH_2-O-CH_2CH_3.

The following data is available for the two pure substances (Hammick and Andrew, 1929).

Property	Benzene	Diethyl ether
Density $(g \text{ cm}^{-3})$	0.8722	0.7069
Surface tension (mN m^{-1})	28.23	16.47
Molecular weight $(g \text{ mol}^{-1})$	78.114	74.123
T_c (K)	562.2	466.7
P_c (bar)	48.9	36.4
T_b (K)	353.2	307.6

Problem 3.3: Adhesion and spreading for liquid–liquid interfaces

At 20 °C the surface tensions of water, perfluorohexane and *n*-octane are, respectively, 72.8, 11.5 and 21.8 mN m^{-1}, while the interfacial tension of the *n*-octane–water interface is 50.8 mN m^{-1}.

Calculate:

a. the work of adhesion between *n*-octane and water;
b. the work of cohesion for (1) *n*-octane and (2) water;
c. the initial spreading coefficient of *n*-octane on water. Will octane initially spread on water?
d. the dispersion component of the surface tension of water, assuming the validity of the Fowkes equation;
e. the interfacial tension between water and perfluorohexane using the Fowkes equation and the result to question (d). How does your result compare to the experimental value (which is equal to 57.2 mN m^{-1})? Discuss briefly the results.

Problem 3.4: The dispersion component of the surface tension of water using the Fowkes equation

Estimate the dispersive and specific components of the surface tension of water using the Fowkes equation and the experimental value for the liquid–liquid interfacial tension of water–cyclohexane (50.2 mN m^{-1}). The surface tension of water is 72.8 and of cyclohexane is 25.5, all values are in mN m^{-1} at 20 °C. How do the results compare to the value calculated in Example 3.3?

Problem 3.5: Interfacial tensions for liquid–liquid systems with the Fowkes equation

The interfacial tension of mercury with benzene is (at 20 °C) $\gamma_{HgB} = 357$ mN m^{-1}. Using the values given in the table below for the surface tensions of mercury (Hg), benzene (b) and water (w), estimate using the Fowkes equation for the liquid–liquid interfacial tension:

1. The dispersion part of the surface tension of mercury at 20 °C (γ_{Hg}^{d}).

2. The interfacial tension of the mercury–water system (γ_{Hgw}). Compare the result with the experimental value, which is in the range 415–426 mN m^{-1} at 20 °C.
3. The interfacial tension of water/benzene and compare it to the experimental value (35 mN m^{-1}).

Explain the assumptions required and discuss briefly the results. If, in some cases, the results using the Fowkes equation are not satisfactory, which other method would you recommend using?

Compound	Surface tension (at 20 °C) (mN m^{-1})
Mercury	485
Benzene	28.9
Water	72.8
n-Pentane	16.8
n-Octane	21.8

Problem 3.6: Liquid–liquid interfacial tensions with the Fowkes equation

1. Glycol–alkane interfaces

The structure of monoethylene glycol (MEG) is: OH-CH$_2$-CH$_2$-OH. MEG has a surface tension equal to 47.7 mN m^{-1}.

The following table shows experimental data for the surface tension of two hydrocarbons and their liquid–liquid interfacial tensions with MEG.

Alkane	Surface tension (mN m^{-1})	Interfacial tension against MEG (mN m^{-1})
Cyclohexane	25.5	14
n-Hexane	18.4	16

Based on these data, can you conclude whether the Fowkes equation can be applied to glycol–alkane interfaces? Justify your answer with calculations.

How will the MEG-benzene interfacial tension compare to the MEG-hexane one? Justify briefly your answer.

2. Halogenated hydrocarbon–water interfaces

Perform the same exercise as part 1 for the CHCl$_3$–water and CCl$_4$–water interfaces using the data of the table below. Use the value for the dispersion surface tension of water from example 3.3 (21.8 mN m^{-1}).

Compound	Surface tension (mN m^{-1})	Interfacial tension against water (mN m^{-1})
Chloroform (CHCl$_3$)	27.1	28
Carbon tetrachloride (CCl$_4$)	26.9	45

Problem 3.7: Interfacial tensions for hydrocarbon/water systems with the Hansen–Beerbower equation

Two sets of the dispersion, polar and hydrogen bonding solubility parameters have been reported for water: the ones given in the Hansen Tables (Appendix 3.1) (7.6, 8.1, 20.6) and the set (10.8, 14.3, 15.6). All values are given in (cal cm^{-3})$^{1/2}$. The experimental surface tensions of pure liquids are known (table in Problem 3.5).

1. Calculate, using the Hansen–Beerbower equation, the dispersion and specific (polar and hydrogen bonding) parts of the surface tension of water based on both sets of water solubility parameters. Which set of water solubility parameters results in values for the dispersion and specific surface tensions for water closer to those estimated from the Fowkes theory?
2. Estimate, using the Hansen–Beerbower equation, the dispersion, polar and hydrogen bonding parts of the surface tension of pentane, octane and benzene Comment on the results.
3. Estimate, using the Hansen–Beerbower equation for the interfacial tension, the interfacial tensions for water–octane and water–benzene. Which water solubility parameter set gives better results (closer to the experimental values)? The experimental interfacial tensions are 50.8 and 35.0 mN m^{-1}, respectively.

Problem 3.8: Interfacial tension for "complex" liquid–liquid interfaces with various models

Many theories have been proposed for the estimation of interfacial tensions, many more than just the Fowkes, Girifalco–Good and Hansen–Beerbower. These theories can be tested against experimental data for liquid–liquid interfaces but testing is more difficult for solid–liquid interfaces (where the interfacial tension cannot be measured directly).

In this problem we will examine aqueous mixtures with organic compounds like aniline and alcohols and will select the best among a number of these theories.

The first application is the water–aniline system. The surface tension of aniline is 42.9 mN m^{-1} and the ratio of dispersion to specific surface tension is, for the same liquid, 1.294.

1. Calculate the water/aniline interfacial tension with the Fowkes equation and with the following two versions of the harmonic mean equation:

$$\gamma_{ij} = \gamma_i + \gamma_j - \frac{4\gamma_i^d \gamma_j^d}{\gamma_i^d + \gamma_j^d} \qquad \text{(HARM-1)}$$

$$\gamma_{ij} = \gamma_i + \gamma_j - \frac{4\gamma_i^d \gamma_j^d}{\gamma_i^d + \gamma_j^d} - \frac{4\gamma_i^{spec} \gamma_j^{spec}}{\gamma_i^{spec} + \gamma_j^{spec}} \qquad \text{(HARM-2)}$$

Compare the results to the experimental value, which is 5.8 mN m^{-1}? What do you observe?

2. Calculate the water/aniline interfacial tension with the Owens–Wendt expression, which is similar to Fowkes but includes an extra term for the specific forces:

$$\gamma_{ij} = \gamma_i + \gamma_j - 2\sqrt{\gamma_i^d \gamma_j^d} - 2\sqrt{\gamma_i^{spec} \gamma_j^{spec}}$$

How does the Girifalco–Good model perform for this system?

Which of the four models compares best with the experimental data?

3. The second application involves the immiscible water–heavy alcohols (i.e. heavier than propanol). The surface tensions of butanol, hexanol, heptanol and octanol are 24.6, 25.8, 25.8 and

27.5 mN m^{-1}, respectively. Which of the above models (considered in questions 1 and 2) would you employ for estimating the interfacial tensions of these alcohols with water? The experimental values are: 1.8 for butanol/water, 6.8 for hexanol/water, 7.7 for heptanol/water and 8.5 for octanol/water (all values in mN m^{-1}). How does the best model you have chosen perform compared to the experimental data?

Problem 3.9: Interfacial tensions and spreading of hexanol–water

The surface tension of n-hexanol is 24.8 mN m^{-1} (at 20 °C). Estimate, with a method of your own choice, the interfacial tension of the water–n-hexanol liquid–liquid interface and compare it with the experimental value (6.8 mN m^{-1}).

Then describe what happens over a period of time (at the start and after some time) when a small amount of n-hexanol is dropped onto a clean water surface at 20 °C. The surface tension for a saturated solution of hexanol in water is 28.5 mN m^{-1}.

Problem 3.10: The diversity of surface tension values

The surface tension of liquid metals reaches rather high values. For Na it is (at 403 K) equal to 198 mN m^{-1} and for Ag it is (at 1373 K) 878.5 mN m^{-1}. In contrast, many liquids like hydrocarbons and alcohols have surface tensions of 18–30 mN m^{-1}. How can you explain why there is such a large difference between the values of the surface tensions for these fluids? Why is the surface tension of water higher (72.1 mN m^{-1} at 298 K) than that of most other liquids? Mention three consequences or applications of the high surface tension of water.

Problem 3.11: Surface tension, interfacial tensions and spreading

1. Explain why water has a high surface tension (equal to 72.8 mN m^{-1} at room temperature) and mention two other liquids that also have high surface tensions. Explain the reason for these high values.

2. The high surface tension of water can result in wetting problems. Mention two ways that can be used to decrease the surface tension of water and consequently facilitate the wetting.

3. Write the Fowkes equation for the interfacial tension and derive the expression for the work of adhesion from the Fowkes equation.

4. According to the Fowkes equation, the dispersion contribution to the surface tension of water is 21.8 mN m^{-1}. Comment on this result. Is this value reasonable?

5. The interfacial tension of water with hexane (18 mN m^{-1}), benzene (28.9 mN m^{-1}) and perfluorohexane (11.5 mN m^{-1}) is 51, 35 and 57.2 mN m^{-1}, respectively. In parenthesis are the surface tension values of the pure liquids. All values are given at room temperature. Explain in molecular terms these values of interfacial tensions.

6. Starting from the definition of the Harkins spreading coefficient S, show that the spreading coefficient can be equivalently given as the difference between the work of adhesion of oil (o) and water (w) and the work of cohesion of oil (o):

$$S = W_{adh,ow} - W_{coh,o}$$

Problem 3.12: Parachors and surface tension data for uncovering a compound's structure

We know that a certain compound has a structure of the general type $(C_3H_8O)_x$ and molecular weight equal to 60 g mol^{-1}. From measurements of its density and surface tension at the same temperature, we were able to determine the experimental parachor value to be equal to 167.5. What is the exact molecular structure of the compound?

Problem 3.13: Surface and interfacial tensions and intermolecular forces

Which answer in each of the following questions is correct?

1. The high surface tension value of water is largely due to its:
   a. polarity
   b. high dispersion forces

c. hydrogen bonding

d. metallic effects.

2. Which of the following systems are *not* colloids?
   a. a latex composed of polymer molecules dissolved in water;
   b. sugar in water;
   c. milk from the local farm;
   d. micelles of a non-ionic surfactant in an aqueous solution.

3. Wetting is *most* difficult for:
   a. clean metals
   b. plastics
   c. ceramics
   d. clean glasses.

4. The Fowkes, Girifalco–Good and Hansen are all theories for estimating the interfacial tension of solid–liquid (and other) interfaces. Which statement is *always* correct?
   a. All of them have the same accuracy.
   b. None of them is really good.
   c. Girifalco–Good is the best theory because it does not include separation of the surface-tension into dispersion, specific, etc. parts.
   d. All of them are typically useful when combined with Young's equation for the contact angle for characterizing a solid surface.

5. The dispersion value of the surface tension of water (21.8 mN m^{-1}) estimated from the Fowkes equation is widely accepted because:
   a. Fowkes is the best theory for estimating interfacial tensions.
   b. It is very close to the experimental value.
   c. It represents a reasonable percentage of dispersion forces in agreement with molecular theories.
   d. Other theories give the same value.

6. The measured interfacial tension of benzene–water is only 35 mN m^{-1}, which is much lower than that of a hexane/water system (51 mN m^{-1}) because:
   a. benzene has a lower molecular weight than hexane;
   b. there must be some mistake in the measurements;
   c. of no special reason;

d. there is (an acid–base type) attraction between benzene and water molecules.

7. Based on the following data for surface and interfacial tensions:

System	Interfacial tension (mN m^{-1})	Liquid	Surface tension (mN m^{-1})
Water–hexane	51	Water	72.8
Water–benzene	35	*n*-hexane	18
Water–C$_6$F$_{14}$	57.2	Benzene	28.9
MEG-hexane	16	C$_6$F$_{14}$	11.5
Water–butanol	1.8	MEG	47.7

You can conclude that:
   a. hydrocarbons are the most hydrophobic compounds;
   b. fluorocarbons are the most hydrophobic compounds;
   c. alcohols are fully miscible with water;
   d. hexane is more soluble in water than in MEG.

8. Determine the correct type of dominant intermolecular forces for the following compounds based on the reported values of surface tensions:

Liquid	Surface tension (mN m^{-1})	Dominant types of intermolecular forces
Mercury	485	
Water	72.8	
Pentanol	25.2	
Benzene	29	
MEG	47.7	
Perfluoro-octane	12	
Hexane	18.4	

Problem 3.14: Component theories and intermolecular forces (open project)

Chapter 2 presents the framework of intermolecular forces and how the contributions of dispersion, polar and induction forces can be estimated for different molecules using information from polarizabilities, dipole moments and ionization potentials. Chapter 3 presents different theories for interfacial tensions, e.g. Fowkes, Owens–Wendt and Hansen–Beerbower, which divide the surface tension into contributions from dispersion and other forces. We have seen that the dispersion contribution of water estimated from Fowkes is in good agreement with the theory of intermolecular forces. Can the same be concluded for other molecules? Select a few molecules of your own choice (hydrocarbons, alcohols, glycols, esters, ethers, etc.) and determine both the contribution of intermolecular (van der Waals) forces and the surface components of some of the theories for estimating the surface tension. Comment on the results as well as the conclusions extracted from the analysis of the relative contributions of Hansen solubility parameters.

References

P. W. Atkins, 2009. Physical Chemistry, 9th ed. Oxford University Press.

H.R. Baker, P.B. Leach, C.R. Singleterry, W.A. Zisman, Ind. Eng. Chem., 1967, 59(6), 29.

G.T. Barnes and I.R. Gentle, 2005. Interfacial Science – An Introduction. Oxford University Press.

C. Della Volpe, S. Siboni, J. Colloid Interface Sci., 1997, 195, 121.

C. Della Volpe, S. Siboni, J. Adhesion Sci. Technol., 2000, 14, 235.

L.A. Girifalco, R.J. Good, J. Phys. Chem., 1957, 61, 904.

L.A. Girifalco, R.J. Good, J. Phys. Chem., 1960, 64, 561.

J. Goodwin, 2004. Colloids and Interfaces with Surfactants and Polymers. An Introduction, John Wiley & Sons, Ltd, Chichester.

R.J. Good, J. Adhesion Sci. Technol., 1992, 6(12), 1269.

Gussman, N. (2009) *Chem. Eng. Prog.*, p. 56.

I. Hamley, 2000. Introduction to Soft Matter. Polymers, Colloids, Amphiphiles and Liquid Crystals, John Wiley & Sons, Ltd, Chichester.

Hammick, D.L. and Andrew, L.W. (1929) *J. Am. Chem. Soc.*, 51, 754.

C.M. Hansen, 2000. Hansen solubility parameters. A User's Handbook, CRC Press, Boca Raton, FL.

P.C. Hiemenz, and R. Rajagopalan, 1997. Principles of Colloid and Surface Science, 3rd edn, Marcel Dekker, New York.

J.R. Hildebrand and R.L. Scott, 1964. The Solubility of Non-Electrolytes, Dover, New York.

J. Israelachvilli, 1985. Intermolecular and Surface Forces, Academic Press.

T.A. Knotts, W.V. Wilding, J.L. Oscarson, R.L. Rowley, J. Chem. Eng. Data, 2001, 46, 1007.

G.M. Kontogeorgis, G.K. Folas, 2010. Thermodynamic Models for industrial Applications. From Classical and Advanced Mixing Rules to Association Theories, John Wiley & Sons, Ltd, Chichester.

D.Y. Kwok, A.W. Neumann, Colloids Surfaces, A: Physicochem. Eng. Aspects Sci., 2000, 161, 31 and 49.

D.Y. Kwok, A.W. Neumann, Prog. Colloid Polym. Sci., 1998, 109, 170.

D.Y. Kwok, A.W. Neumann, Adv. Colloid Interface Sci., 1999, 81, 167.

E. McCafferty, J. Adhesion Sci. Technol., 2002, 16(3), 239.

K. Meyn, 2002 Charles Medom Hansen: A life with solubility parameters. Dansk Kemi, 83 (3), page 10 (in Danish).

H. Modarresi, E. Conte, J. Abildskov, R. Gani, P. Crafts, Ind. Eng. Chem. Res. 2008, 47, 5234.

D. Myers, 1991 & 1999. Surfaces, Interfaces, and Colloids. Principles and Applications. VCH, Weinheim.

J. Owens, Appl. Polymer Sci., 1970, 14, 1725.

C.G. Panayiotou, 2003, "Hydrogen Bonding in Solutions: The Equation-of-State Approach" in *Handbook of Surface and Colloid Chemistry*, 2nd edn, ed. K.S. Birdi. CRC Press, Boca Raton, FL, ch 2.

R. M. Pashley, M.E. Karaman, 2004, Applied Colloid and Surface Chemistry, John Wiley & Sons, Ltd, Chichester.

J.M. Prausnitz, R.N. Lichtenthaler, E.G. de Azevedo, 1999. Molecular Thermodynamics of Fluid-Phase Equilibria, 3rd edn, Prentice Hall International.

O.R. Quayle, Chem. Rev., 1953, 53, 439.

A.J. Queimada, Ch. Miqueu, I.M. Marrucho, G.M. Kontogeorgis, J.A.P. Coutinho, 2005. Fluid Phase Equilibria, 228-229: 479.

R. C. Reid, J.M. Prausnitz, B.E. Poling, 1987. The Properties of Gases and Liquids. 4th edition, McGraw-Hill International.

D. Shaw, 1992. Introduction to Colloid & Surface Chemistry. 4th edn, Butterworth-Heinemann, Oxford.

E. Stefanis, C. Panayiotou, 2008. Int. J. Thermophys., 29, 568.

E. Stefanis, C. Panayiotou, 2012. Int. J. Pharmaceutics., 426, 29.

S. Sudgen, 1930. The Parachor and Valency, Routledge, London.

A. Tihic, G.M. Kontogeorgis, N. von Solms, M.L. Michelsen, *Fluid Phase Equilibria*, 2006, 248, 29.

I. Tsivintzelis, G.M. Kontogeorgis, G.M, 2012. *Ind. Eng. Chem. Res.*, 51(41): 13496-13517.

D.W. van Krevelen and P.J. Hoftyzer, 1972. Properties of Polymers. Correlations with Chemical Structure. Elsevier.

C.J. van Oss, M.K. Chaudhury, R.J. Good, Adv. Colloid Interface Sci., 1987, 28, 35.

W.A. Zisman, 1964, in *Contact Angle, Wettability and Adhesion*, Advances in Chemistry Series, vol. 43, American Chemical Society, Washington D.C., 1.

Fundamental Equations in Colloid and Surface Science

4.1 Introduction

Surface and colloid chemistry have only a few fundamental equations that are (almost) always true and they control many practical phenomena. All these general equations were developed more than a century ago. One of these fundamental equations is the Young equation for the contact angle, which is presented next. The other equations are the Young–Laplace, Kelvin and the famous Gibbs adsorption equation. They are also discussed in this chapter.

4.2 The Young equation of contact angle

The component theories, which are presented in Chapter 3, are useful for estimating the interfacial tensions of solid-interfaces (solid–liquid, solid–solid) and for characterizing solid surfaces using the experimental data for the few properties that can be measured (liquid–gas surface tension, liquid–liquid interfacial tension, contact angle). Important equations in this context are the Young equation

for contact angle and the work of adhesion (Equations 4.1–4.3).

Solid surfaces are much more complex than liquids because of roughness, chemical heterogeneity, impurities and other phenomena and the contact angle plays an important role in understanding them. As Figure 4.1 shows, the contact angle is the angle between the liquid and the solid, a property that can be measured and is, together with the theories for interfacial tension, widely used for characterizing solid surfaces.

4.2.1 Contact angle, spreading pressure and work of adhesion for solid–liquid interfaces

When a drop of liquid is placed in contact with a solid, we have *three* interfaces: the solid/liquid, solid/vapour and liquid/vapour interfaces. Each of these has its own interfacial energy. For a drop that partially wets a solid, the total interfacial energy is minimal when the horizontal components of the interfacial tensions are in equilibrium. At this point, the *contact angle* between liquid and solid has a value that is determined by the

Introduction to Applied Colloid and Surface Chemistry, First Edition. Georgios M. Kontogeorgis and Søren Kiil.
© 2016 John Wiley & Sons, Ltd. Published 2016 by John Wiley & Sons, Ltd.
Companion website: www.wiley.com/go/kontogeorgis/colloid

(a)

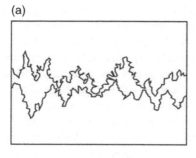

(b)

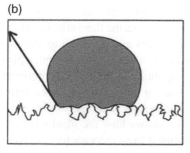

(c)

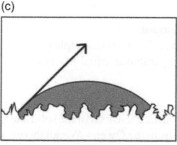

Figure 4.1 *Concept of the contact angle. Values below 90° (c) indicate partial wetting of the solid by the liquid (e.g. water on clean glass), whereas a value equal to 0° indicates excellent (complete) wetting. Values above 90° (b) indicate partial non-wetting, e.g. water on solid mercury, which becomes "worse" as the contact angle increases, up to 180°. This maximum value of contact angle indicates complete non-wetting and is rather rare (many liquids on mercury or on some other metals approach this value). Measurements and interpretations of contact angle data can be complicated by solid roughness (a)*

three interfacial energies. The relation is known as the Young equation (Equation 4.1).

Thus, the Young equation is derived as a force balance across the x-axis ($\gamma_l \cos \vartheta + \gamma_{sl} = \gamma_s$) and provides the relationship between the contact angle (ϑ) and surface tensions of solid and liquid (s, l) and the liquid–solid interfacial tension (sl):

$$\cos \vartheta = (\gamma_s - \gamma_{sl})/\gamma_l \qquad (4.1)$$

Equation 4.1 is correct if the spreading pressure is zero, which is almost always true for low energy surfaces like polymers. The spreading pressure ($\pi_{sv} = \gamma_s - \gamma_{sv}$) is the difference between the solid surface tension (in contact with vacuum or air) and the solid surface tension in presence with vapour coming from the liquid drop.

Thus, in the Young equation, in the general case we should use γ_{sv} which is the surface tension of the solid in contact with the vapour which comes from the liquid drop. In the general case, it is not identical to the surface tension of the pure solid γ_s (i.e. the surface tension in vacuum or in contact with its own vapour). This is due to some adsorption from the vapour of the liquid to the solid, which decreases the solid surface tension. In these cases, the solid–vapour surface tension should be corrected for adsorption phenomena (spreading pressure), but this is a rather complex procedure.

In general, this spreading pressure is small for moderately large values of the contact angle and for low energy surfaces like polymers, but it may be significant as the contact angle approaches zero or for high energy surfaces like metals and metal oxides. In most applications, we will for simplicity ignore the spreading pressure and assume that the solid–vapour interfacial tension in Young's equation is indeed simply the surface tension of the pure solid, i.e. $\gamma_{sv} = \gamma_s$. Spreading pressure phenomena are discussed further in Chapter 15.

Due to the adsorption from environment, i.e. the presence of spreading pressure, materials are in general stronger in vacuum (space) where the high surface tension is not being reduced by adsorbed water, gases, etc. that may exist in ordinary environments (Pashley and Karaman, 2004).

Contact angle values equal to zero indicate full wetting and values below 90° partial wetting, e.g. water on clean glass, while values above 90° correspond to partial non-wetting, e.g. water on solid mercury and at the extreme and rare case of 180° we have complete non-wetting (many liquids on mercury or on some other metals approach this value).

According to Equation 4.1, liquids with high surface tensions, e.g. mercury (485 mN m^{-1}) and water

(72 mN m^{-1}), often exhibit poor wetting with many solids (high contact angle). Water wets only some solids while mercury wets only a few metals. On the other hand, hydrocarbons with (rather low) surface tensions, typically 20–30 mN m^{-1}, can wet many surfaces.

We now address again the work of adhesion that we defined in its general form in Chapter 3. The general equation (Equation 3.1) is applied to all interfaces. In the case of solid–liquid interfaces, the work of adhesion is given by the Young–Dupre equation:

$$W_{adh} = W_{sl} = \gamma_s + \gamma_l - \gamma_{sl} = \gamma_l (1 + \cos \vartheta) \qquad (4.2)$$

Thus, the Young–Dupre equation is *not* as general as the definition equation (Equation 3.1). This is because the spreading pressure term is neglected and due to the uncertainties associated with the contact angle (e.g. hysteresis, roughness and heterogeneity). Still, and although the Young–Dupre equation is a useful tool as all properties can be assessed experimentally, the difficulties in interpretations are significant. Thus, the Young–Dupre equation is sometimes criticized. Importantly, the "maximum" value of the work adhesion predicted by Equation 4.2 (when the contact angle is zero) is equal to two times the surface tension of the liquid. The work of adhesion (even the theoretical one) can indeed be higher than this value due to all of the above assumptions. Thus, if appropriate theories for the interfacial tensions are available, the general equation (Equation 3.1) may actually be more useful for estimating the theoretical work of adhesion.

Equation 4.2 gives the theoretical work of adhesion for ideal interfaces and due only to surface phenomena. Adhesion is a much more complex phenomenon with many applications (e.g. paints, polymer technology and cleaning via surfactant-based formulations). We discuss wetting and adhesion in Chapter 6. Example 4.1 shows how the ideal work of adhesion can be estimated from theories for interfacial tension.

Example 4.1. Work of adhesion from theories for interfacial tension.
A very simple theory for the interfacial tension that performs satisfactorily in some simple cases is given by a modified form of the Girifalco–Good equation (using a correction parameter equal to one):

$$\gamma_{sl} = \left(\sqrt{\gamma_s} - \sqrt{\gamma_l} \right)^2$$

Derive an expression for the work of adhesion as a function of the surface tensions of the solid and the liquid.
Then derive the expression for the solid–liquid work of adhesion from the Owens–Wendt theory. Compare the results from the two theories. Which one is expected to perform best?

Solution
Expanding the given expression for the interfacial solid–liquid tension, we obtain:

$$\gamma_{sl} = \left(\sqrt{\gamma_s} - \sqrt{\gamma_l} \right)^2 = \gamma_s + \gamma_l - 2\sqrt{\gamma_s \gamma_l}$$

Consequently, based on the definition of the work of adhesion (Equation 4.2), it can be seen that the work of adhesion for the solid–liquid interface is equal to the last term of the above equation (the one with the square root), without the negative sign, i.e.:

$$W_{adh} = 2\sqrt{\gamma_s \gamma_l}$$

Similarly for the Owens–Wendt theory (see Chapter 3), the work of adhesion is the cross term again without the negative sign:

$$W_{adh} = 2\sqrt{\gamma_s^d \gamma_l^d} + 2\sqrt{\gamma_s^{spec} \gamma_l^{spec}}$$

We can see that the Owens–Wendt contains two terms from the contribution of dispersion and "specific" forces, while in the first theory the work of adhesion depends only on the values of the solid and liquid surface tensions. The two theories discussed here belong to the two different theory families (component and direct) which we have discussed in Chapter 3. We cannot a priori, without data or other information, determine which of the two theories will provide more correct values for the work of adhesion.

Notice, however, that in all cases for these theories the work of adhesion is equal to the intermediate "cross" term of the corresponding equations for the interfacial tension.

4.2.2 Validity of the Young equation

The simplicity of Young's equation (Equation 4.1) is only apparent and somewhat deceiving as only two of the four properties can be measured directly, the contact angle and the liquid surface tension (ϑ, γ_l). Actually, there are only a few cases where all four properties of the Young equation have been independently measured (Israelachvili, 1985). For example, for

a droplet of 8×10^{-4} M HTAB solution on a mono-layer-covered mica surface the measured surface tension of the solid is 27 ± 2 mN m^{-1}, the surface tension of the liquid is 40 mN m^{-1} and the interfacial surface tension for the solid–liquid interface is 12 ± 2 mN m^{-1}. The measured contact angle is 64°, which is in very good agreement with the value calculated from the Young equation.

Thomas Young (1773–1829)

Reproduced from http://commons.wikimedia.org/wiki/File:Thomas_Young_(scientist).jpg

Thomas Young was a brilliant English physician and physicist. By the age of 14 it was said that he was acquainted with Latin, Greek, French, Italian, Hebrew, Arabic and Persian. So great was his knowledge that he was called Phenomena Young by his fellow students at Cambridge. He studied medicine in London, Edinburgh and Göttingen and set up medical practice in London. His initial interest was in sense perception, and he was the first to realize that the eye focuses by changing the shape of the lens. He discovered the cause of astigmatism, and was the initiator, with Helmholtz, of the three colour theory of perception, believing that the eye constructed its sense of colour using only three receptors, for red, green and blue. In 1801 he was appointed Professor of Physics at Cambridge University. His famous double-slit experiment established that light was a wave motion, although this conclusion was strongly opposed by contemporary scientists who believed that Newton, who had proposed that light was corpuscular in nature, could not possibly be wrong. However, Young's work was soon confirmed by the French scientists Fresnel and Arago. He proposed that light was a *transverse* wave motion (as opposed to longitudinal) whose wavelength determined the colour. Since it was thought that all wave motions had to be supported in a material medium, light waves were presumed to travel through a so-called *aether*, which was supposed to fill the entire universe. He became very interested in Egyptology, and his studies of the Rosetta stone, discovered on one of Napoleon's expeditions in 1814, contributed greatly to the subsequent deciphering of the ancient Egyptian hieroglyphic writing. He carried out work in surface tension, elasticity (Young's modulus, a measure of the rigidity of materials, is named after him) and gave one of the first scientific definitions of energy.

4.2.3   Complexity of solid surfaces and effects on contact angle

Equation 4.1 is valid for ideal "smooth" solid surfaces. Solid–liquid and solid–solid interfaces are very important for understanding wetting, lubrication, adhesion and other phenomena and for characterizing solid surfaces, as discussed further in Chapter 6.

However, we underline here that solid surfaces are complex in many ways; they are often heterogeneous (chemically or geometrically) and rough, they easily become dirty or contaminated (adsorbed liquids or solids, e.g. dust), and their properties change with time due to chemical or physical adsorption.

Especially on high energy surfaces, part of the vapour of the liquid will adsorb on the solid. This lowers the energy of the solid–vapour interface. On a rough solid, the actual interfacial area is larger than the one that we see: this causes the effective energies of the solid interfaces to be larger than those of a flat surface. If there are contaminants (dirt) in the system, these will adsorb on the solid and lower the effective interfacial energies of the solid. As a result, simple calculations using "pure" interfacial energies are often inaccurate. Even so, they can be quite useful for a first understanding.

Let us discuss some of the above-mentioned factors in some detail.

Clean metals, metal oxides, glass and ceramics are high energy surfaces, while polymers and fabrics are low energy surfaces. However, for the high energy surfaces, cleaning of the surface is of paramount importance. Even the thinnest monolayer of adsorbed oil/grease can dominate the surface properties. For example, "clean tin cans" have been reported with surface tensions as low as 31 mN m^{-1} (Bentley and Turner, 1998).

Aging phenomena are also crucial. Even carefully prepared surfaces under vacuum rapidly adsorb (physically or chemically) material from the environment, e.g. water, hydrocarbons, dirt, etc. For example, Mica prepared under vacuum has a surface energy of 4500 mN m^{-1} but the surface energy decreases to about 300 mN m^{-1} under normal laboratory conditions, where water can easily adsorb (Israelaschvili, 1985). Moreover, the way the solid surface is prepared (crystallized in water or air) may have a large effect on its surface properties.

The contact angle is indeed the key property that we can measure related to solid interfaces, but the Young equation (Equation 4.1) applies in principle only to smooth surfaces. Thus, much care must be exercised when using contact angles from practical experiments in relation with the Young equation. One of the most important reasons for problems is the so-called contact angle hysteresis (i.e. the difference between the advancing and receding contact angles). For that and other reasons, the experimentally determined contact angle is not always the same as the theoretical value that is related to the various interfacial tensions in the Young equation.

The so-called contact angle hysteresis can have more than one cause:

- surfaces are not (almost never) absolutely smooth, i.e. they have some roughness (perhaps the most important reason);
- equilibrium is not reached;
- surfaces can be contaminated (adsorbed liquids or even solid particles, e.g. dust);
- surfaces may undergo some changes during contact with the test liquid;
- the spreading pressure may be significant for low values of contact angle.

For rough surfaces, we need to account for the roughness effect, e.g. via the so-called roughness factor, R_f, introduced by Wenzel (1936), which can be obtained, for example, by atomic force microscopy and other techniques.

Its definition and the Young equation for contact angle corrected for roughness are:

$$R_f = \frac{\cos \vartheta^{exp(rough)}}{\cos \vartheta^{theor(smooth)}} \tag{4.3}$$

$$R_f(\gamma_s - \gamma_{sl}) = \gamma_l \cos \vartheta^{exp}$$

As can be seen from Equation 4.3, the roughness factor R_f is defined as the ratio of the cosine of the experimental (exp) contact angle (for a rough surface) over the cosine of the theoretical (theor) contact angle (for the hypothetical smooth surface where ideally the Young equation is applied). The roughness factor is always above unity, and takes the value of 1 in the ideal case of a completely smooth surface. Thus, if

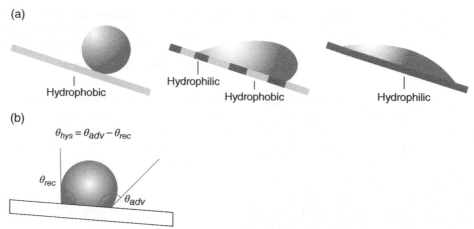

Figure 4.2 *Advancing and receding contact angles. (a) How a droplet is affected by the hydrophilic and hydrophobic parts of a solid surface. (b) Illustration of the definition of the advancing and receding contact angles as well as of the contact angle hysteresis. Holmgren and Persson (2008) with data from Li et al. (2007). Reproduced with permission from J. M. Holmgren and J. M. Persson*

the contact angle for a smooth surface is less than 90° (partial wetting), the roughness helps wetting (the contact angle becomes even smaller), while if we have partial non-wetting (contact angle above 90°) the roughness contributes to further non-wetting.

Thus, employing Wenzel's equation, the Young equation can be revised using the experimental contact angle. In this case, the roughness factor must be known. Calculation of this factor from the topography of the surface is possible but rather controversial.

Let us look closer now at the well-established phenomenon of "contact angle hysteresis". As mentioned, if we have contamination, roughness, significant spreading pressure, heterogeneity or even when equilibrium is not reached we can experience the phenomenon called "contact angle hysteresis". This means that the contact angle depends on how the liquid is applied, i.e. whether it is advancing on a dry or receding from a wet surface. The surface has spots with a high effective energy, and spots with a low effective energy. A liquid advancing along the surface will be retarded by spots with a low interfacial energy. This is illustrated in Figure 4.2.

A liquid retreating (receding) from the surface will be kept back by spots with a high interfacial energy. As a result, the advancing and retreating (or receding) contact angles differ. The advancing value is always the largest (it is not rare for it to be 20° or 30° greater than the receding angle, and sometimes even higher;

Pashley and Karaman, 2004). Drops with different advancing and retreating (receding) contact angles are in a metastable state: they are not in true equilibrium. This effect can be seen with drops of rain on a window pane. It is exactly this difference between the advancing and the receding contact angles that is called contact angle hysteresis.

4.3 Young–Laplace equation for the pressure difference across a curved surface

The Young–Laplace equation gives the pressure difference (inside–outside) across a curved surface. For spherical droplets, the equation is written as (see Appendix 4.1 for the derivation):

$$\Delta P = P_1 - P_2 = \frac{2\gamma}{R} \qquad (4.4)$$

The curvature is very important. Water droplets with radius of 10 μm show a pressure difference of 0.15 bar, which is greatly increased to 15 bar for a radius of 100 nm and 1500 bar if the radius is at the lower limit of colloids (1 nm).

There are other forms for other curved surfaces, e.g. for air bubbles the 2-factor in Equation 4.4 should be substituted by 4, while for cylinders, the 2 should be substituted by unity, i.e. for cylinders:

$$\Delta P = P_1 - P_2 = \frac{\gamma}{R}$$

Note also the convention that the radius of curvature is measured in the liquid phase and is thus positive for a liquid drop but negative for a gas bubble. This means – see also later the Kelvin equation – that the vapour pressure of a liquid in a small drop is higher than for a flat surface but is lower in a bubble compared to a flat surface.

The Young–Laplace equation gives the pressure difference across a curved surface and has many applications, e.g. in the understanding/control of liquid rise in capillaries, capillary forces in soil pores and nucleation and growth of aerosols in atmosphere. The importance of curvature is evident if we think, for example, that a bubble with radius equal to 0.1 mm in champagne results in a pressure difference of 1.5 kPa, enough to sustain a column of water of 15 cm height.

The most important application of the Young–Laplace equation is possibly the derivation of the Kelvin equation. The Kelvin equation gives the vapour pressure of a curved surface (droplet, bubble), P, compared to that of a flat surface, P^{sat}. The vapour pressure (P) is higher than that of a flat surface for droplets but lower above a liquid in a capillary. The Kelvin equation is discussed next.

4.4   Kelvin equation for the vapour pressure, P, of a droplet (curved surface) over the "ordinary" vapour pressure P^{sat} for a flat surface

The Kelvin equation is derived from the Young–Laplace equation and the principles of phase equilibria. It gives the vapour pressure, P, of a droplet (curved surface) over the "ordinary" vapour pressure (P^{sat}) for a flat surface (see Appendix 4.2 for the derivation):

$$\ln\left(\frac{P}{P^{sat}}\right) = \frac{V^l}{R_{ig}T}\frac{2\gamma}{R} = \frac{M}{\rho_l R_{ig}T}\frac{2\gamma}{R} \qquad (4.5)$$

According to Equation 4.5 the ratio of droplet vapour pressure (P) divided by that over a flat surface (P^{sat}) is proportional to the liquid volume and surface tension and inversely proportional to the temperature and droplet radius, R. Thus, we cannot ignore the effects of surface tension and radius. The vapour pressure of a droplet, P, is greater than that over a flat surface, P^{sat}.

The Kelvin equation is valid down to radii as low as 2.5 nm for many organic liquids. For water and mercury, the Kelvin effect has been experimentally verified down to dimensions as small as 30 Å (Barnes and Gentle, 2005).

An application example is presented in Example 4.2.

Example 4.2.   Application of the Kelvin equation to water droplets.
For water at 20 °C, the vapour pressure is 17.5 mmHg, the surface tension is 72 mN m^{-1} and the liquid volume is equal to 18×10^{-6} m^3 mol^{-1}. Estimate the effect of Kelvin equation on the vapour pressure when the water is in the form of liquid droplets having radii within the colloidal domain (choose four radii values in this region).

Solution
We will use the Kelvin equation as given in Equation 4.5 and we choose four radii in the colloidal domain. The results are summarized in the table below.

R (mm)	10^{-3}	10^{-4}	10^{-5}	10^{-6}
P/P^{sat}	1.001	1.011	1.114	2.95

We note that the effect of radius on vapour pressure is very significant at low radii values. It can possibly be ignored at the micrometre range but not at the nanometre range.

4.4.1 Applications of the Kelvin equation

The Kelvin equation has numerous applications, e.g. in the stability of colloids (Ostwald ripening, see below), supersaturation of vapours, atmospheric chemistry (fog and rain droplets in the atmosphere), condensation in capillaries, foam stability, enhanced oil recovery and in explaining nucleation phenomena (homo- and heterogeneous). The Kelvin (as well as the Gibbs equations, see Equation 4.7a) are also valid for solids/solid–liquid surfaces, and they can be used for estimating the surface tensions of solids. We discuss hereafter several applications of the Kelvin equation.

4.4.1.1 Capillary rise

An important application of the Kelvin equation, of great importance to oil applications and environmental engineering (processes in soils), is the presence of liquids in capillaries. As there is a vapour pressure decrease outside the concave surface (negative curvature), we have condensation of the liquids inside the cracks, a phenomenon called capillary condensation. In this case the vapour pressure is reduced relative to that of a flat surface. Liquids that wet the solid will therefore condense into the pores at pressures below the equilibrium vapour pressure corresponding to a plane surface.

When the radius is above 1000 nm, the normal vapour pressure is not affected, whereas for radii below 100 nm, and especially below 10 nm, there is an appreciable increase of the vapour pressure compared to the vapour pressure of flat surface. For liquids of high molar volume and surface tension, e.g. mercury, the effect is even more significant.

In conclusion, liquids in capillaries condense and evaporate "harder" than in the open space.

4.4.1.2 Condensation

The Kelvin equation points out the difficulties associated with the formation of a new phase. Suppose we start with a vapour phase and compress it to the saturation pressure, P_0. Formation of the liquid phase requires molecules coming together and coalescing. It is unlikely that a very large number of molecules will come together at once to form a drop. Even the formation of a small drop with $r = 100$ nm involves the simultaneous aggregation of 1.4×10^8 molecules! What is likely to happen is that a few molecules coalesce, starting with a pair, to form very tiny droplets. The Kelvin equation shows that the equilibrium vapour pressure for such droplets is several times larger than that for the bulk liquid. Hence the droplets are likely to evaporate rather than come together to form the bulk liquid. We can understand from this analysis that the pressure has to be considerably larger than the equilibrium vapour pressure of the bulk liquid (P_0) before condensation can occur. Thus, we say that the vapours are in this case "supersaturated".

In practice, condensation usually starts on dust particles or on surface sites on the walls of a vessel (heterogeneous nucleation). The presence of these sites for nucleation prevents supersaturation of the vapour phase. Often we add other "foreign matter" (particles) to by-pass this barrier of supersaturation.

The Kelvin equation also shows why liquids become superheated before boiling. Consider a bubble inside the bulk of a liquid. The pressure inside the bubble must be greater than that of the bulk liquid. The smaller the bubble is, the greater is the pressure. Hence there is a barrier for the formation of small bubbles, which leads to the superheating of liquids. Similar behaviour is observed with the freezing of liquids and the crystallization of solutes from saturated solutions. For example, extremely pure water may be cooled to −40 °C before it begins to solidify.

4.4.1.3 Ostwald ripening

The so-called "Ostwald ripening" is a major destabilizing mechanism in colloidal systems and at the same time is one of the most exciting applications of the Kelvin equation. In this context the Kelvin equation can be used to describe phenomena related to the stability of emulsions and other colloidal dispersions. For liquids, very small bubbles, droplets and particles tend to evaporate (due to their higher vapour pressure) or dissolve faster than the bigger ones and this phenomenon destabilizes a dispersion, see also later for suspensions. Thus, the large drops grow at the expense of the smaller ones. This is because material evaporates from the smaller droplets (that have higher vapour pressures) and condenses on the larger drops. Thus, Ostwald ripening explains why big droplets seem to "eat" the small ones, or in other words why big droplets grow while the small droplets become smaller or disappear.

Ostwald ripening is equally important for suspensions (dispersions of solid particles in a liquid medium) and it can be explained via the form of the Kelvin equation shown below:

$$\ln\left(\frac{s(r)}{s(r \to \infty)}\right) = \frac{V_s}{R_{ig}T} \frac{2\gamma_{sl}}{R} \qquad (4.6)$$

This equation describes the change in solubility of a solid particle in a liquid as the particle size is reduced. According to this equation, the solubility increases as the size of the particle is reduced in the same manner as the vapour pressure of the liquid in a drop increases as the size is reduced. This means that larger particles grow by the transference of material from the smaller particles, the phenomenon we called Ostwald ripening. As most suspensions are polydispersed, the Kelvin equation is a crucial indicator of their stability. The smaller particles will have a greater solubility and will tend to dissolve, and the larger particles will tend to grow at their expense. In a suspension of a highly insoluble substance, such as silver iodide (AgI) hydrosol, this phenomenon will be of little consequence, since both large and small particles have extremely little tendency to dissolve. In a suspension of more soluble material, such as calcium carbonate ($CaCO_3$) hydrosol, however, Ostwald ripening occurs to such an extent that it is not possible to prepare a long-lived dispersion with particles of colloidal dimensions unless a stabilizing agent, such a gelatin or a surfactant, is incorporated. Ostwald ripening and the stabilization mechanisms of colloids, in general, will be discussed further in Chapters 10–13 (colloid stability, emulsions and foams).

4.5 The Gibbs adsorption equation

We continue with the presentation of the Gibbs adsorption equation and our first discussion about adsorption. It is no exaggeration to state that much in colloid and surface science is about adsorption phenomena, e.g. in (steric) stabilization of colloids, detergent science, surface modification and numerous separations. Adsorption is thus a fundamental, crucial phenomenon in colloid and surface science. Before proceeding and discussing all the various applications of adsorption, we need to present the Gibbs equation

for adsorption. The Gibbs adsorption equation is one of the most important and fundamental equations in colloid and surface science.

An interface in real systems is an area of small but non-zero thickness (of about 10 molecular diameters). The adsorption, as defined in Equation 4.7a, is the number of moles at an interface per area and has units of concentration (molar) per area. The number of moles at the interface and thus the adsorption itself can be either positive or negative (for compounds avoiding the interface).

Using some algebra and concepts from thermodynamics (see Appendix 4.3), the Gibbs adsorption equation can be derived:

$$\Gamma_i = \frac{n_i^\sigma}{A} = -\frac{C_i}{R_{ig}T}\frac{d\gamma}{dC_i} = -\frac{1}{R_{ig}T}\frac{d\gamma}{d\ln C_i} \qquad (4.7a)$$

Equation 4.7a provides the relationship between adsorption (which cannot be easily measured directly) and the change of surface tension with concentration, $d\gamma/dC_i$ (which is rather easily obtained from experiments). This equation is derived for the case of a single adsorbing solute, e.g. a non-ionic surfactant. For ionic surfactants (Chapter 5) we have two species that adsorb at the interface and we should use "$2R_{ig}T$" instead of "$R_{ig}T$" in the denominator:

$$\Gamma_i = \frac{n_i^\sigma}{A} = -\frac{C_i}{2R_{ig}T}\frac{d\gamma}{dC_i} = -\frac{1}{2R_{ig}T}\frac{d\gamma}{d\ln C_i} \qquad (4.7b)$$

There are a couple of interesting points: First, we cannot determine "absolute adsorptions" using the Gibbs equation, only relative ones, typically of the solute with respect to the solvent, e.g. of a surfactant assuming no solvent present at the interface. In addition, when the liquid phase is non-ideal, we should use activities (activity coefficient × concentration) instead of the concentration in Equation 4.7a, but typically we assume ideal dilute solutions and thus only Equation 4.7a,b will be used in our applications.

There are many methods that have been used for experimental verification of the Gibbs adsorption equation. McBain, using very thin films (0.05 mm) from surfactant solutions, measured the concentration of the interfacial phase. This and other methods have provided excellent agreement with the Gibbs adsorption equation (Shaw, 1992).

Amphiphilic molecules, especially surfactants (discussed in Chapter 5) but also polar molecules such as alcohols, acids and others, tend to stick on the water–air interface and thus decrease the surface tension of water. This decrease is higher the higher the chain length of the alcohol (polar molecule), which is also true for surfactants. This effect can be quantified by the so-called Traube's rule. For some of these molecules, and especially at low concentrations, we observe that the surface tension decreases linearly with concentration. The area occupied by an adsorbed molecule, e.g. a surfactant molecule, can be calculated once the adsorption is known (in nm^2 when the adsorption Γ is given in mol m^{-2}):

$$A = \frac{10^{18}}{N_A \Gamma} \tag{4.8}$$

4.6 Applications of the Gibbs equation (adsorption, monolayers, molecular weight of proteins)

Adsorption is a central topic in the science of colloids and interfaces with applications in the stability of colloidal dispersions, adsorption of polymers, surfactants and proteins, cleaning, catalysis, surface modification, separations (e.g. adsorption on activated carbon and drying) and biotechnology. The Gibbs equation for the adsorption is the basic tool in the field and it can be applied, in principle, to all interfaces. We can estimate the adsorption from surface tension–concentration data, thus connecting these important concepts. At low concentrations, e.g. dilute aqueous alcohol or surfactant solutions, surface tension decreases linearly with concentration ($\gamma = \gamma_0 - bC$) and the two-dimensional ideal gas law is derived (see Example 4.3):

$$\pi A = n_i^\sigma R_{ig} T = k_B T \tag{4.9}$$

where the surface (sometimes also called spreading) pressure is defined as: $\pi = \gamma_0 - \gamma$, i.e. the difference between the surface tension of water (solvent) and that of the solution.

Equation 4.9 is the two-dimensional analogue of the well-known equation of state for ideal gases ($PV = nR_{ig}T$) and is the mathematical description of what is called "ideal gas films" (Section 4.7).

Example 4.3. Ideal gas law in surface science.
In some cases, e.g. at low concentrations, the surface tension of aqueous surfactant solutions, γ, decreases linearly with concentration, C:

$$\gamma = \gamma_0 - bC \tag{4.3.1}$$

where γ_0 is the surface tension of pure water and b is a constant depending on the system.
Show that, in this case, the surface pressure π is given by the so-called two-dimensional ideal gas law equation:

$$\pi A = n_i^\sigma R_{ig} T \tag{4.3.2}$$

where A is the surface area, R_{ig} is the ideal gas law constant and T is the temperature.

Solution
We will use the definition of the surface pressure:

$$\pi = \gamma_0 - \gamma \tag{4.3.3}$$

and the Gibbs adsorption equation, as in Equation 4.7a. Thus, we have:

$$\Gamma = \frac{n_i^\sigma}{A} = -\frac{C_i}{R_{ig}T}\frac{d\gamma_i}{dC_i} = \frac{-C_i}{R_{ig}T}(-b) = \frac{bC}{R_{ig}T} = \frac{\pi}{R_{ig}T} \Rightarrow$$

$$\pi A = n_i^\sigma R_{ig} T \tag{4.3.2}$$

The slope is $-b$ as obtained from Equation 4.3.1. Equation 4.3.2 is often called the ideal gas law in surface science and is, in fact, the equivalent in two dimensions of the well-known (three-dimensional) ideal gas law ($PV = nR_{ig}T$).

Equation 4.9 can be used, for example, to estimate the molecular weight of proteins (Shaw, 1992). Proteins adsorb and denaturate at high-energy air–water and oil–water interfaces because unfolding allows the polypeptide chains to be oriented with most of their hydrophilic groups in the water phase and most of the hydrophobic groups away from the water phase. The techniques used for studying spread monolayers of insoluble material can be used in the study of protein films. Some unfolding of the polypeptide chains takes place at the air–water surface and even further at oil–water interfaces. At low compressions, up to about 1 mN m^{-1}, protein films tend to be gaseous, thus permitting relative molecular weight determination. To realize ideal gaseous behaviour, experimental surface pressure data must be extrapolated to zero surface pressure.

Thus, the molecular weight of proteins, M, from surface pressure data assuming ideal gas film behaviour at low pressures can be calculated from Equation 4.9 written as:

$$\lim_{\pi \to 0}(\pi A) = \frac{R_{ig}T}{M} \qquad (4.10)$$

The area A in Equation 4.10 is expressed in m^2 g^{-1}.

The relative molecular masses of several spread proteins have been determined from surface-pressure measurements. In many cases they compare well with the relative molecular masses in bulk solution, although differences have been also reported, and they indicate surface dissociation of the protein molecules into sub-units. One application is shown in Example 4.4.

Example 4.4.   *Molecular weight of proteins from surface pressure data.*
Protein films tend to be gaseous at low concentrations, thus permitting relative molecular weight determination (see also Equation 4.10). Based on this methodology, estimate the molecular weight of the protein valinomycin from the following measurements of the surface pressure for various areas at 25 °C:

Surface pressure (mN m^{-1})	1.0	1.5	2.0	2.5
area A (m^2 mg^{-1})	3.8	3.33	2.66	2.60

Solution
We will calculate for every experimental point the product of surface pressure with the area and plot it against the surface pressure. The plot shown in this solution is almost linear. We extrapolate at zero surface pressure and then we can estimate the molecular weight of the protein:

$$\lim_{\pi \to 0}(\pi A) = 2.15 = \frac{R_{ig}T}{M} \Rightarrow M = \frac{R_{ig}T}{\lim_{\pi \to 0}(\pi A)} = \frac{8.314 \times 298}{2.15} = 1152 \, \text{g mol}^{-1}$$

The value 2.15 (mN m mg^{-1}) is obtained from the extrapolation mentioned above.
Of course the calculation depends somewhat on the accuracy of the extrapolation. For example, if for a limiting value equal to 2.2 (mN m mg^{-1}), a molecular weight about 1126 g mol^{-1} is obtained.

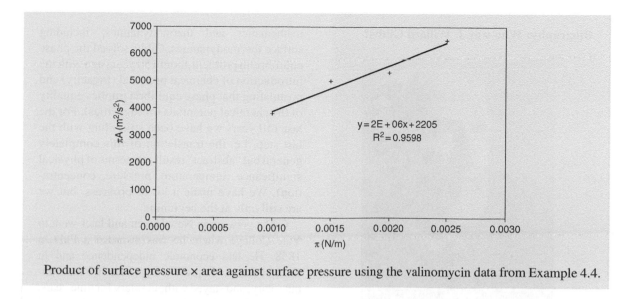

Product of surface pressure × area against surface pressure using the valinomycin data from Example 4.4.

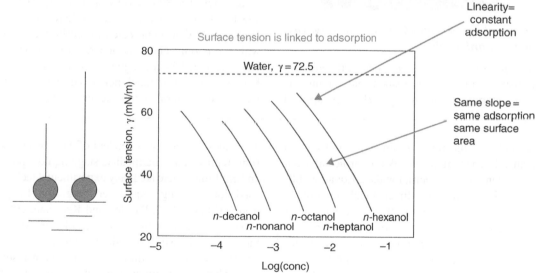

Figure 4.3   *Surface tension–log (concentration) plots for aqueous solutions of five medium-chain alcohols in water. The linearity in the figure indicates constant adsorption, while the fact that all alcohols have the same slope indicates that we have the same adsorption and thus the same surface area per molecule, as could be also physically expected (the polar head oriented towards water and the non-polar hydrocarbon chain is directed towards the air). Adapted from Jonsson et al. (2001), with permission from John Wiley & Sons, Ltd*

There are many more applications of the Gibbs equation:

1. Calculation of the mole area^{-1} (adsorption) of a surfactant from the rather easily measured surface tension–concentration data (Figure 4.3). Direct measurement of adsorption, e.g. via measuring concentration change or surface analysis can be difficult.

Biography: Who was J. Willard Gibbs?

Reproduced from http://en.wikipedia.org/wiki/ Josiah_Willard_Gibbs

J. Willard Gibbs (1839–1903), one of the greatest scientists of all times, is responsible for numerous developments in many fields of applied mathematics and thermodynamics, including surface thermodynamics. Gibbs solved the phase equilibrium problem about 150 years ago with the introduction of chemical potential (fugacity) and postulating that phase equilibria implies equality of the chemical potentials (or fugacities). For the last 150 years we have been struggling with the last step, i.e. the translation of this completely general but "abstract" result into terms of physical significance (temperature, pressure, concentration). We have made a lot of progress, but we are still only at the beginning.

Gibbs grew up in New Haven and later went to Yale College where he was awarded a PhD in 1858. He had economic independence and in 1871 he accepted a professor position in Yale without salary, and stayed without salary for nine years. He had light teaching duties and he was not tempted by an offer of $1800 by Bowdan College. Only in 1880 did he finally decided to leave to John Hopkins. At that point Yale offered him a salary, which was only about ⅓ of what John Hopkins was offering. His colleagues at Yale said: "John Hopkins can get on much better without you than we can. We cannot". Gibbs never left Yale.

2. Calculation of the molecular area (area molecule^{-1}) of an adsorbed molecule like a surfactant from surface tension–concentration data, Equation 4.8.
3. Calculation of the two-dimensional equation of state of films/solutions if the surface tension–concentration equation is known (Chapter 7).
4. Derivation of the dependency of surface tension of solutions with the concentration if a specific theory of adsorption is assumed, e.g. it can be shown that the Langmuir theory (discussed in Chapter 7) yields the Szyszkowski equation.

4.7 Monolayers

A special case of adsorption is that related to the behaviour of monomolecular films (gas, liquid, solid) (Figure 4.4). In solid films, the molecules are oriented densely close to the surface, while in the gas films the molecules are more independent of each other. Liquid films represent an intermediate stage liquid (spreading and gas films. There are many experimental techniques used for studying monolayers, e.g. the Langmuir trough (Shaw, 1992; Pashley and Karaman, 2004).

The behaviour of these films can be expressed via the so-called two dimensional equations of state (π–A), surface (spreading) pressure against surface area. There are many such equations, which will be discussed in the context of adsorption in Chapter 7 (Table 7.1).

Such films can be created by heavy alcohols, heavy acids, etc. which, like surfactants, are amphiphilic molecules and reduce the surface tension of water. "Negative" adsorption exists for salts that avoid surfaces. This decrease is greater the larger the hydrophobic part, i.e. the more carbon atoms in the molecular chain. The heavier alcohols decrease the surface tension more than the lighter ones.

Four states of monolayers (films)

(a)

Gas films

(b)   (c)

Dilute
liquid
films

Solid films

(d)   (e)

Condensed
liquid
films

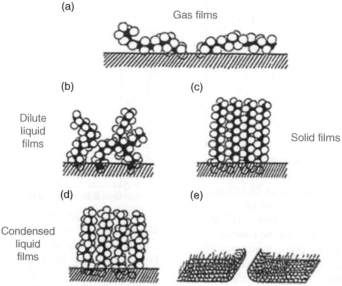

Figure 4.4 *Monolayers, or otherwise monomolecular films, are very thin films (of about one molecule on the interface, as the name implies) of polar compounds, e.g. heavy alcohols or acids. These compounds tend to pack themselves on the surface and thus reduce the surface tension of water. The polar part faces water and the non-polar sticks out. This is as in the case of surfactants, with the only difference that we have here an extreme case of adsorption at liquid interfaces, where all molecules are placed at the interface. Moreover, unlike surfactants, we do not have micelles. Cadenhead (1969). Reproduced with permission from American Chemical Society*

These monolayer films find many applications, e.g. reduction of water evaporation from lakes, estimation of the molecular weight of proteins (Equation 4.10), study of biological (cell) membranes, enzymes and the impressive Langmuir–Blodgett monolayer films, often studied with a technique called the Langmuir trough (Pashley and Karaman, 2004) (see case study below).

As mentioned, one exciting application of such insoluble monolayers is that they can considerably reduce the evaporation of water from lakes and rivers. There are several requirements that such compounds should fulfil. The film-forming compounds should create easily a film, they should be cheap (as large quantities may be needed), non-toxic for humans and for the aquatic environment and the film should be easily reformed when destroyed by wind, birds or other disturbances. Examples of such compounds are hexadecanol and octadecanol. This application is of great importance in areas where water resources are being reduced. This method is extensively used with success in many countries, e.g. USA (south-west states), Australia and Israel, where the loss of water due to evaporation has been halved.

The surface pressure/area equations that describe such films are the two-dimensional analogues of the well-known three-dimensional equations of state. As previously discussed, see also in Chapter 7 (Table 7.1), the ideal gas film equation of state (Equation 4.9) can be derived from the Gibbs adsorption equation, assuming that surface tension decreases linearly with concentration (Example 4.3). Such a dependency is a realistic picture for several (rather simple) cases, e.g. dilute surfactant or alcohol solutions and in general when the surface tension is not far away from the water surface tension. Liquid (and solid) films require more complex two-dimensional equations of state, e.g. van der Waals and Langmuir. Figures 4.5 and 4.6 present some surface pressure–area plots for liquid and solid films and discuss some of their important characteristics.

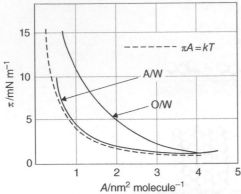

Figure 4.5 *Behaviour of a CTAMB (cetyl(trimethyl) ammonium bromide) film at air–water (A/W) and oil–water (O/W) interfaces. The behaviour of this film on the air–water interface approaches very much that of the ideal film behaviour. This is not the case at an oil–water interface, since the oily phase interacts with the hydrocarbon chains of CTAMB, reducing the intermolecular attractions and, thus, the repulsive forces dominate. Shaw (1992). Reproduced with permission from Elsevier*

The similarities between equations of state in two dimensions (surface pressure–area) and three dimensions (pressure–volume) are evident also if we plot the compressibility factor expressed either in the 3D ($PV/R_{ig}T$) or 2D forms. The trends of the former against reduced pressure and of the latter against surface pressure are evident, as shown by Shaw (1992).

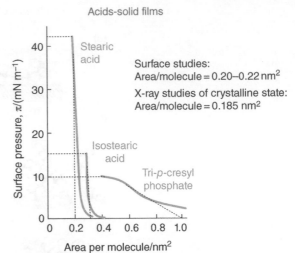

Figure 4.6 *Palmitic, stearic and other aliphatic acids create solid films at room temperature. At high surface area values, the molecules of acids are not separated but co-exist in the form of two-dimensional aggregates (clusters) at the surface. Owing to this cohesion, the surface pressure is low but increases dramatically when the molecules are tightly packed together. The surface area for these organic acids is, irrespective of the size of the molecule, about 0.2–0.22 nm² per molecule. This value is very close to the area of a solid crystalline acid structure obtained from X-ray diffraction measurements (equal to 0.185 nm² for stearic (octadecanoic) acid). Adapted from Atkins (1998), with permission from Oxford University Press*

Case study. Langmuir–Blodgett films

Near the end of the nineteenth century Agnes Pockels (1891), working in the kitchen of her parents' home, discovered how to manipulate oil films on water and thus developed the first surface film balance. Her results led Lord Rayleigh (1899) to suggest that the films were only one molecule thick, thus providing the foundations for the study of insoluble monolayers. Devaux, Hardy and Langmuir in the following years all contributed a great deal in the study of such monolayers.

If a solid surface is passed vertically through a floating monolayer, the monolayer may deposit on the solid substrate (Figure 4.7). The process is

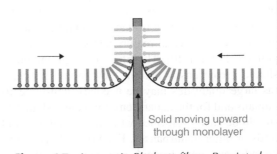

Figure 4.7 *Langmuir–Blodgett films. Reprinted from Barnes and Gentle (2005), with permission from Oxford University Press*

known as Langmuir–Blodgett or LB deposition after the scientists who discovered and developed the technique. Repeated passages of the substrate through the monolayer may lead to the deposition of additional layers on the surface, thus forming multilayer films. However, not all substrates and not all monolayers are suitable. For satisfactory deposition, the surface pressure should be sufficiently high that the monolayer is in a condensed state. A surface pressure between 20 and 40 mN m^{-1} is normally used.

There are three possible types of deposition. Once the first monolayer has been deposited on the substrate, subsequent movements in and out of the surface often lead to the deposition of an additional layer on each passage. This is known as Y-type deposition. Other types of depositions are possible (X and Z). The definitions of these deposition types refer to the behaviour during deposition, not to the structure of the film that is formed. Much work has been done using monolayers of carboxylic acids and their salts. For good quality films careful preparation of the substrate is essential. Selection of the appropriate pH and the type and concentration of salt in the subphase affect the type of film that is formed. The highly ordered layer structure of LB films is a major consideration in potential applications.

LB film deposition is unique in the control offered for producing highly organized molecular arrays of controlled, nanometre-scale thickness, and therefore plays a part in the quest towards molecular electronics (diodes and transistors using the smallest possible elements with single or very few molecules).

Great interest has led to the development of Langmuir troughs with various mechanisms for the LB deposition of, for example, alternate layers of two different amphiphiles. In addition, techniques have been developed for overcoming problems related to the deposition of a second metal layer on top of a LB film already deposited on a metal substrate.

Other applications involve photo-semiconductors, extending the frequency range of laser devices and chemical and biological sensors.

4.8 Conclusions

Surface and colloid chemistry have only a few fundamental equations that are always true (no exceptions) and control many practical phenomena. All these general equations were developed more than a century ago. These equations are the Young equation for contact angle, the Young–Laplace, Kelvin, Gibbs adsorption and the Harkins spreading coefficient as well as the work of adhesion. The latter two equations were seen in Chapter 3 and they can be considered "simple" surface energy balances. These six equations are summarized in Table 4.1 together with some comments.

The contact angle defines the angle between a liquid and a solid. The mathematical relationship between contact angle and the interfacial tensions (liquid, liquid–solid, solid–vapour) is given by the Young equation. Contact angles and surface tensions of liquids can be used for estimating the theoretical work of adhesion via the Young–Dupre equation. This equation has both advantages and shortcomings and, if an appropriate theory for the interfacial tension is available, we should use the general Dupre equation to estimate the theoretical work of adhesion. More applications will be presented in Chapter 6.

The Young–Laplace equation gives the pressure difference across a curved surface and its most important application is in the derivation of the Kelvin equation. The Kelvin equation gives the vapour pressure of a curved surface (droplet, bubble) compared to that of a flat surface. The Kelvin equation has numerous applications, e.g. in the stability of colloids (Ostwald ripening), condensation in capillaries and in explaining nucleation phenomena.

Adsorption is one of the most central topics in the science of colloids and interfaces with applications in the stability of colloidal dispersions, cleaning, catalysis, surface modification and biotechnology. The Gibbs equation for the adsorption is the basic tool in the field and it can be, in principle, applied to all interfaces. We can estimate the adsorption from surface tension–concentration data, thus linking these important concepts. At low concentrations, e.g. dilute aqueous alcohol or surfactant solutions, surface tension

Table 4.1　Six fundamental equations in colloid and surface science

Theory	Equation	Comments
Young–Laplace equation (pressure difference across a curved surface)	$\Delta P = \dfrac{2\gamma}{R}$	Applied in the derivation of the Kelvin equation
Kelvin equation for vapour pressure of a droplet (curved surface)	$\ln\left(\dfrac{P}{P^{sat}}\right) = \dfrac{V^{l}}{R_{ig}T}\dfrac{2\gamma}{R}$	Explains the Ostwald ripening destabilization mechanisms in colloids
Young equation for the contact angle	$\cos\vartheta = \dfrac{\gamma_{sv}-\gamma_{sl}}{\gamma_{l}}$ or $\cos\vartheta = \dfrac{\gamma_{s}-\gamma_{sl}}{\gamma_{l}}$ provided that the spreading pressure is zero (almost always true for low surface energy surfaces like polymers)	Two properties are readily measurable, two are not. Mostly used together with theories for the solid–liquid interfacial tension. Very important in wetting and adhesion studies (Chapter 6)
Harkins spreading coefficient (for an oil (O) in water (W))	$S = \gamma_{WA}-(\gamma_{OA}+\gamma_{OW}) = W_{adh,OW}-W_{coh,O}$	Can be applied both to liquid–liquid and liquid–solid interfaces (see also Chapter 6)
Work of adhesion	$W_{adh} = \gamma_{A}+\gamma_{B}-\gamma_{AB}$ Between solid/liquid (spreading pressure ignored): $W_{adh} = W_{sl} = \gamma_{s}+\gamma_{l}-\gamma_{sl}$ $= \gamma_{l}(1+\cos\vartheta)$	The first form is called Dupre and the second one, valid for solid–liquid interfaces, is the Young–Dupre equation
Gibbs adsorption equation – relationship between adsorption and change of surface tension with concentration	$\Gamma_{i} = \dfrac{n_{i}^{\sigma}}{A} = -\dfrac{C_{i}}{R_{ig}T}\dfrac{d\gamma}{dC_{i}} = -\dfrac{1}{R_{ig}T}\dfrac{d\gamma}{d\ln C_{i}}$	Possibly the most important of these general equations. Numerous applications related to adsorption in general, e.g. surfactants and polymers. Often used in the simple form shown here based on the concentrations. For ionic surfactants, we should use "$2R_{ig}T$" instead in the denominator

decreases linearly with concentration and the two-dimensional ideal gas law is derived. It can be used, for example, for estimating the molecular weight of proteins.

Appendix 4.1 Derivation of the Young–Laplace equation

We will derive the Young–Laplace equation, which in general terms gives the pressure difference across a curved surface, for the specific case of a liquid spherical drop, having a radius R and a surface tension γ. The pressure inside the droplet is designated as P_1 and the pressure outside as P_2.

According to the surface tension definition:

$$W = \gamma A = \gamma\left(4\pi R^2\right) \qquad (4.11)$$

For a very small change in the radius, the work change is:

$$dW = \gamma dA = \gamma d\left(4\pi R^2\right) = 8\gamma\pi R dr \qquad (4.12a)$$

The work change can be also calculated from mechanics as:

$$dW = F dx = \left(P_1 - P_2\right)\left(4\pi R^2\right) dr \qquad (4.12b)$$

Of course, both ways of calculating the work change should be identical: thus by equating (4.12a) and (4.12b) we get:

$$\Delta P = P_1 - P_2 = \frac{2\gamma}{R}$$

which is the Young–Laplace equation for spherical droplets.

Appendix 4.2 Derivation of the Kelvin equation

We provide here a derivation of the Kelvin equation based on the Young–Laplace equation and phase equilibrium principles.

We consider, as in Appendix 4.1, a liquid spherical drop, having a radius R and a surface tension γ. The pressure inside the droplet is designated as P_1 and the pressure outside as P_2. We want to develop an expression of the vapour pressure P_2 as a function of the radius and physical properties of the liquid.

Our starting point is the Young–Laplace equation (Equation 4.4):

$$\Delta P = P_1 - P_2 = \frac{2\gamma}{R} \tag{4.13}$$

Using the differential form, this equation can be written as:

$$dP_1 - dP_2 = d\left(\frac{2\gamma}{R}\right) \tag{4.14}$$

The condition for liquid–gas equilibria in terms of chemical potential is:

$$\mu^l = \mu^g \Rightarrow d\mu^l = d\mu^g \tag{4.15}$$

The chemical potential change is given for a one component system as:

$$d\mu = VdP - SdT = VdP \tag{4.16}$$

where the latter equation is valid in the case of constant temperature.

Combining Equations 4.15 and 4.16 we have:

$$V^g dP_2 = V^l dP_1 \tag{4.17}$$

Equations 4.14 (Young–Laplace) and 4.17 (gas–liquid phase equilibria) can be now combined:

$$V^g dP_2 = V^l \left[dP_2 + d\left(\frac{2\gamma}{R}\right)\right] \Rightarrow$$
$$d\left(\frac{2\gamma}{R}\right) = \left(\frac{V^g - V^l}{V^l}\right)dP_2 \tag{4.18}$$

Before integrating Equation 4.18 to determine the vapour pressure P_2 we have to make some simplifying assumptions:

i. at low pressures, the liquid volume is much smaller than the gas volume [$V_l <<< V_g$];
ii. the ideal gas equation is valid [$P_2 V_g = R_{ig}T$].

Thus, Equation 4.18 can be simplified as:

$$d\left(\frac{2\gamma}{R}\right) = \left(\frac{V^g}{V^l}\right)dP_2 = \frac{R_{ig}T}{V^l}\frac{dP_2}{P_2} \tag{4.19}$$

We will now integrate Equation 4.19 between the limits of flat surface (R equals infinity) where the vapour pressure is equal to the normal saturated pressure of the liquid (P^{sat}) and a specific radius R where the vapour pressure is simply denoted as P.

Thus, we have:

$$\int_{P^{sat}}^{P} \frac{dP_2}{P_2} = \int_{r=\infty}^{r} \frac{V^l}{R_{ig}T}d\left(\frac{2\gamma}{R}\right) \Rightarrow$$
$$\ln\left(\frac{P}{P^{sat}}\right) = \frac{V^l}{R_{ig}T}\frac{2\gamma}{R} = \frac{M}{\rho_l R_{ig}T}\frac{2\gamma}{R}$$

Appendix 4.3 Derivation of the Gibbs adsorption equation

In deriving the Gibbs adsorption equation we need to define an interface or interfacial area. As the interface in real systems is an area of small but non-zero thickness (of about 10 molecular diameters), the exact position of the interfacial area is not precisely defined and depends on the position of two planes, called AA' and BB'. We often define these planes so that the properties of phase a are uniform up to AA' and the properties of phase b are uniform after BB'. Between AA' and BB' the properties vary between the values of the phases a and b. We assume for convenience the existence of a plane SS' parallel to those of AA' and BB'. We call the SS' a surface (or interfacial) phase, a 2D phase without volume. The total properties, number of mole, entropy, internal energy or Gibbs free energy of the system are the sum of the values of all three phases, a, b and the interfacial phase.

The starting point is the thermodynamic definition of the surface tension based on the Gibbs energy:

$$\gamma = \left(\frac{\partial G}{\partial A}\right)_{T,P} \quad \text{i.e. } dG = \gamma dA \qquad (4.20)$$

We recall the definitions of the Gibbs energy in relation to enthalpy and internal energy:

$$G = H - TS \text{ (Gibbs energy)} = (U + PV) - TS \qquad (4.21)$$

where U = internal energy, H = enthalpy and S = entropy.

For an open system $G(T, P, n_1, n_2,$ etc.), the change of Gibbs energy is expressed as the changes with respect to pressure, temperature and number of mole of the components:

$$dG = \left(\frac{\partial G}{\partial P}\right)_{T,n_i} dP + \left(\frac{\partial G}{\partial T}\right)_{P,n_i} dT + \sum_i \left(\frac{\partial G}{\partial n_i}\right)_{T,P,n_j \neq i} dn_i$$

$$= VdP - SdT + \sum_i \mu_i dn_i \qquad (4.22)$$

The change of Gibbs energy at the surface (σ) should also include the effect of surface tension and so the above equation is written as:

$$dG^\sigma = V^\sigma dP - S^\sigma dT + \gamma dA + \sum_i \mu_i^\sigma dn_i^\sigma \qquad (4.23)$$

At equilibria the chemical potentials of the compounds in all phases including the surface phase are equal.

As the surface phase has no volume at constant temperature Equation 4.23 can be then expressed as:

$$dG^\sigma = \gamma dA + \sum_i \mu_i^\sigma dn_i^\sigma \qquad (4.24)$$

Thus, the Euler theorem gives for the function of Gibbs energy at a surface that:

$$G^\sigma = \gamma A + \sum_i \mu_i^\sigma n_i^\sigma \qquad (4.25)$$

or:

$$dG^\sigma = \gamma dA + A d\gamma + \sum_i \mu_i dn_i^\sigma + \sum_i n_i^\sigma d\mu_i \qquad (4.26)$$

Thus, combining (or equating) Equations 4.24 and 4.26 we obtain:

$$\sum_i n_i^\sigma d\mu_i + A d\gamma = 0 \qquad (4.27)$$

Equation 4.27 can be written in the following equivalent form, also considering the definition of adsorption (as number of mole per area):

$$-d\gamma = \sum_i \Gamma_i d\mu_i \Rightarrow \Gamma_{i,1} = -\left(\frac{\partial \gamma}{\partial \mu_i}\right)_{T,P,\mu_j \neq 1} \qquad (4.28)$$

This is the Gibbs equation for (the relative) adsorption of compound i (with respect to compound 1).

For a binary system, if the adsorption of compound 1 (e.g. the solvent) is zero then the relative adsorption of compound 2 (the solute) is given as:

$$\Gamma_2 = -\frac{d\gamma}{d\mu_2} \qquad (4.29)$$

When the change in the chemical potential is expressed as a function of the activities (α_2) or even simpler as a function of concentration (C_2) (for ideal liquid solutions), then we obtain the well-known forms of the Gibbs equation:

$$\Gamma_2 = -\frac{1}{R_{ig}T}\frac{d\gamma}{d\ln\alpha_2} = -\frac{\alpha_2}{R_{ig}T}\frac{d\gamma}{d\alpha_2} \qquad (4.30)$$

$$\Gamma_2 = -\frac{1}{R_{ig}T}\frac{\mathrm{d}\gamma}{\mathrm{d}\ln C_2} = -\frac{C_2}{R_{ig}T}\frac{\mathrm{d}\gamma}{\mathrm{d}C_2} \qquad (4.31)$$

Problems

Problem 4.1: Vapour pressures for water and hexane droplets with the Kelvin equation

What is the increase in vapour pressure (over that of a flat surface) for droplets of water and *n*-hexane at 20 °C? What do you observe? For which liquid is the effect more pronounced and why?

Make an application for the following radii values:

10^{-7}, 10^{-5}, 10^{-9} and 0.5×10^{-9} m.

The physical data for water and *n*-hexane, which are required, are given in the following table:

Liquid	Surface tension (mN m^{-1})	Liquid volume (cm^3 mol^{-1})
Water	72.8	18.0
n-Hexane	18.4	130

Problem 4.2: The capillary rise as a method for measuring the surface tension of liquids

The rise of liquids in thin capillaries is a well-known phenomenon (partially) due to surface phenomena and at the same time it represents a very efficient and simple method for obtaining experimentally the surface tension of liquids (upon measuring their rise and knowing the dimensions of the capillary tube). First of all, orient yourself (see, for example, Pashley and Karaman, 2004; Shaw, 1992, etc.) on how the capillary rise method works.

Then calculate the height (h) of capillary rise in the air–water system (surface tension = 72 mN m^{-1}) for tube radii of 0.1, 0.01 and 0.001 mm. Repeat for an oil–water system with a surface tension of 30 mN m^{-1} and density difference equal to $0.15\,\mathrm{g\,cm}^{-3}$. Assume that water completely wets the solid surface in both cases.

Problem 4.3: Amphiphilic compounds

The figure below (Shaw, 1992) shows the surface tension of various aqueous alcohol solutions against concentration at 20 °C.

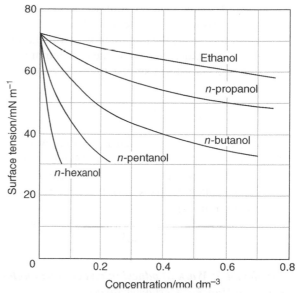

Shaw (1992). Reproduced with permission from Elsevier

1. a. What is an ideal gas film and how is mathematically represented? Mention one application of ideal gas films.
    b. For which alcohol can we expect ideal gas film over extended concentrations?
2. Estimate for *n*-butanol and at concentration equal to 0.1 mol L^{-1}:
    a. the surface pressure;
    b. the adsorption and the area (in nm^2) occupied by one butanol molecule.

Problem 4.4: Adsorption and Surface tension for aqueous solutions

The Gibbs adsorption equation is a fundamental tool in surface science, which enables us to calculate adsorption phenomena on both solid and liquid surfaces.

1. What is the Gibbs adsorption equation? Mention some of its practical applications.
2. Explain briefly how the area of a surfactant molecule can be calculated using the Gibbs adsorption equation.

In several cases, the adsorption from aqueous (and other type) solutions can be described by the Langmuir isotherm:

$$\Gamma = \frac{\Gamma_{max}kC}{1+kC}$$

where C is the concentration and k a parameter characteristic of the system involved.

3. Show that, using the Langmuir isotherm, the surface tension depends on solute concentration as follows, via a form similar to the Szyszkowski equation:

$$\gamma = \gamma_0 - R_{ig}T\Gamma_{max}\ln(1+kC)$$

where γ_0 is the surface tension of pure water.

4. How does the surface tension depend on concentration, if the adsorption isotherm follows the so-called Henry's law (valid at very low concentrations)?

$$\Gamma = \Gamma_{max}kC$$

Problem 4.5: Work of adhesion and contact angles from the Hansen/Beerbower theory

The liquid (l)–solid (s) interfacial tension can be given by the Hansen–Beerbower expression:

$$\gamma_{sl} = \left(\sqrt{\gamma_s^d}-\sqrt{\gamma_l^d}\right)^2 + \left(\sqrt{\gamma_s^p}-\sqrt{\gamma_l^p}\right)^2 + \left(\sqrt{\gamma_s^h}-\sqrt{\gamma_l^h}\right)^2$$

Show that, when the spreading pressure is zero, the contact angle for a solid–liquid interface based on the above equation is given by the equation:

$$\cos\vartheta = -1 + 2\left(\sqrt{\gamma_l^d\gamma_s^d}+\sqrt{\gamma_l^p\gamma_s^p}+\sqrt{\gamma_l^h\gamma_s^h}\right)/\gamma_l$$

Show also that, under the same assumption, the theoretical work of adhesion for a solid–liquid interface is given by the equation:

$$W_{sl}^{adh} = 2\left(\sqrt{\gamma_l^d\gamma_s^d}+\sqrt{\gamma_l^p\gamma_s^p}+\sqrt{\gamma_l^h\gamma_s^h}\right)$$

Problem 4.6: Spreading and contact angles

The interfacial tension between water and solid hexadecane is 53.8 mN m^{-1}, and the surface tensions of water and the oil are 72.8 and 27.5 mN m^{-1}, respectively. The contact angle of water on paraffin wax is 105° at 20 °C.

1. Calculate the contact angle of water in hexadecane and compare it to the experimental value for paraffin wax. Discuss the result.

2. Calculate the work of adhesion and the spreading coefficient of water on paraffin wax. Comment on the results.

Problem 4.7: Molecular weight of biomolecules from surface pressure data

The techniques used to study spread monolayers of insoluble substances can also be applied to the study of protein films. Proteins adsorb and denaturate at high-energy air–water and oil–water interfaces because unfolding allows the polypeptide chains to be oriented with most of their hydrophilic groups in the water phase and most of the hydrophobic groups pointing away from the water phase. Protein films tend to be gaseous at low concentrations, thus permitting relative molecular weight determination. Based on this assumption, estimate the molecular weight of the protein haemoglobin from the following measurements of the surface pressure for various areas at 25 °C. Haemoglobin is regarded as amphiphilic:

Surface pressure (mN m^{-1})	0.28	0.16	0.105	0.06	0.035
Area (m^2 mg^{-1})	4.0	5.0	6.0	7.5	10.0

Compare your answer with a value of 68 000 g mol^{-1} determined from sedimentation measurements.

Problem 4.8: Adhesion and spreading on solid–liquid interfaces

The Harkins equation for the spreading of one liquid (e.g. oil) on another liquid (e.g. water) is also valid for the spreading of a liquid on a solid surface.

1. Show that the spreading coefficient is given by the equation:

$$S = \gamma_l(\cos\vartheta - 1)$$

2. Show that the spreading coefficient is given as the difference between the work of adhesion of the solid–liquid interface and the work of cohesion of the liquid:

$$S = W_{adh,sl} - W_{coh,l}$$

What is the physical meaning of this equation with respect to spreading? Compare this with the result for liquid–liquid interface (Chapter 3).

3. When a solid (which originally is not in contact with a liquid) is completely immersed in a liquid, then the liquid–gas interface remains unchanged. This happens, for example, for powders immersed in liquids. For this process, which is often called *immersional wetting*, we define the work of immersion (W_{im}) as the difference between the initial and the final energy of the system. Show that the work of immersion is related to the spreading coefficient via the equation:

$$W_{im} = S + \gamma_l$$

For which values of contact angle is the immersion process spontaneous?

References

P.W. Atkins, 1998. Physical Chemistry, 6th ed. Oxford University Press.

G.T. Barnes, I.R. Gentle, 2005. Interfacial Science – An Introduction. Oxford University Press.

J. Bentley, G.P.A. Turner, 1998. Introduction to Paint Chemistry and Principles of Paint Technology. Chapman & Hall.

D.A. Cadenhead, 1969. Monomolecular Films at the Air-Water Interface: Some Practical Applications, Industrial & Engineering Chemistry, 61(4), 22–28.

J.M. Holmgren and J.M. Persson, 2008. M.Sc. Thesis, Lund University.

J. Israelachvili, 1985. Intermolecular and Surface Forces, Academic Press.

B. Jonsson, B. Lindman, K. Holmberg, B. Kronberg, 2001. Surfactants and Polymers in Aqueous Solution, John Wiley & Sons, Ltd, Chichester.

X.M. Li, D. Reinhoudt, M. Crego-Calama, 2007. What do we need for a superhydrophobic surface? A review on the recent progress in the preparation of superhydrophobic surfaces, Chem. Soc. Rev., 36, 1350–1368.

R.M. Pashley, M.E. Karaman, 2004. Applied Colloid and Surface Chemistry. John Wiley & Sons, Ltd, Chichester.

D. Shaw, 1992. Introduction to Colloid & Surface Chemistry. 4th edn, Butterworth-Heinemann, Oxford.

R.N. Wenzel, Ind. Eng. Chem., 1936, 28, 988.

5

Surfactants and Self-assembly. Detergents and Cleaning

5.1 Introduction to surfactants – basic properties, self-assembly and critical packing parameter (CPP)

Surfactants are amphiphilic compounds (Figures 5.1 and 5.2), consisting of a "polar" head (ionic –anionic- or cationic- or polar group) and a hydrophobic part. The hydrophobic part is typically a hydrocarbon chain of varying length; it can be also a fluorocarbon or a dimethylsiloxane chain.

Surfactants adsorb on liquid–air and oil–water interfaces and decrease surface and interfacial tensions and the corresponding works of adhesion. Thus, they facilitate the cleaning of various surfaces. Important characteristics in surfactant science, and for understanding their action in cleaning and other applications, include both parameters of the surfactants (CMC, Krafft point, hydrophilic/lipophilic balance) and of the (often aqueous) solution they are in (temperature, salts, co-surfactants, time). We will discuss all of these parameters in this chapter.

The surfactants can be classified according to the nature of the "head", i.e. their hydrophilic group (anionic, non-ionic, cationic), see Table 5.1, or according to their applications. Figure 5.3 shows some commercial products containing surfactants. Sodium dodecyl sulfate (SDS) is one of the most widely used ionic surfactants, while the most famous non-ionics are the poly(ethylene oxide)s or polyoxyethylenes. Notice (Table 5.1) the way these latter surfactants are abbreviated according to the number of hydrophobic and hydrophilic groups, an abbreviation useful in calculations of their properties with, for example, group contribution methods (as shown, for example, in Appendix 5.2).

Surfactants have many applications: as important parts of detergents for cleaning, as emulsifiers, foaming agents or stabilizers for colloidal dispersions, in various applications in biotechnology and catalysis and as components in many complex products, e.g. paints and coatings as well as in lotions, shampoos, etc. In many cases mixed surfactants are used in commercial products.

Especially for cleaning, the synthetic surfactants/ detergents have with time replaced natural soaps, which do not work well in acid solutions and hard

Introduction to Applied Colloid and Surface Chemistry, First Edition. Georgios M. Kontogeorgis and Søren Kiil.
© 2016 John Wiley & Sons, Ltd. Published 2016 by John Wiley & Sons, Ltd.
Companion website: www.wiley.com/go/kontogeorgis/colloid

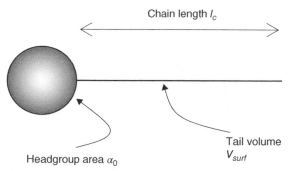

Chain length l_c

Headgroup area α_0

Tail volume V_{surf}

Figure 5.1 *Amphiphilic nature of surfactants. We typically represent the hydrophilic groups of surfactants as spheres and the hydrophobic hydrocarbon chains as stalks (lines); the stalks are mobile*

water. Soaps have great sensitivity to pH and they cannot tolerate commonly encountered ions. Commercial detergents and other formulations contain surfactants such as the ones shown in Table 5.1.

A decrease of water surface tension is observed up to a certain concentration, called the critical micelle concentration (CMC). Above the CMC, surfactants create new liquid-like often near spherical structures, called micelles (Figure 5.2). Micelles are small semi-spherical agglomerates, all about the same size, where (in ordinary micelles) the non-polar tails are inside and the polar heads are outside. The radius of the micelles is of the order of the length of the surfactant molecules: a few nanometres.

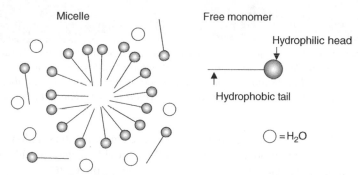

Micelle

Free monomer

Hydrophilic head

Hydrophobic tail

○ = H_2O

Figure 5.2 *Close to CMC (critical micelle concentration), micelles are spherical or semi-spherical "liquid-like" aggregates in a very dynamic situation; the life-time of micelles is 1–100 ms and the occupancy time of one surfactant in a micelle is 10–1000 μs!*

Table 5.1 *Examples of some common surfactants. Non-ionic surfactants are the second biggest (after the anionic ones) family of surfactants worldwide and the biggest (over 50%) in Europe. They can be easily mixed with other types of surfactants. Ionic surfactants are widely used and cheap. The ionic groups can pull, due to electrostatic interactions, large hydrocarbon chains with them in solution. For example, the palmitic acid is insoluble in water but the sodium palmitate (its ester with sodium) fully ionizes and dissolves in water (above the Krafft temperature)*

Category	Name	Structure
Anionic	Sodium stearate	$CH_3(CH_2)_{16}COO^-Na^+$
	Sodium oleate	$CH_3(CH_2)_7CH=CH(CH_2)_7COO^-Na^+$
	Sodium dodecyl sulfate (SDS)	$CH_3(CH_2)_{11}SO_4^-Na^+$
	Sodium dodecyl benzene sulfonate	$CH_3(CH_2)_{11}C_6H_4SO_3^-Na^+$
Cationic	Dodecylamine hydrochloride	$CH_3(CH_2)_{11}NH_3^+Cl^-$
	Hexadecyl(trimethyl)ammonium bromide (CTAB)	$CH_3(CH_2)_{15}N(CH_3)_3^+Br^-$
Non-ionic	Polyethylene oxides (C_xE_y or C_xEO_y)	$CH_3(CH_2)_{x-1}(OCH_2CH_2)_yOH$
	Nonylphenyl ethoxylates (X_{EO} is the number of ethylene oxide groups)	$CH_3(CH_2)_8-C_6H_4-(OCH_2CH_2)_{X(EO)}OH$

Figure 5.3   *Detergents remain the most important use of surfactants. Some commercial examples are shown here. Surfactants play a major role in cleaning the huge variety of different stains (types of dirt), for both hard and soft surfaces. The components in soil and stains are not well-defined. Both contain any number of species, in varying quantities. Even so, we can put them into several categories, which require different methods of treatment.*

Why is this happening? Let us follow the process of adding surfactant molecules in an aqueous solution. Originally, some molecules enter the solution but most of them stick on the (water–air) surface. During this process the surface tension decreases (exactly as happens for alcohols or acids). When, however, all the surfaces are "saturated" (complete), the surfactants will find other ways to minimize the energy; this happens by the creation of (semi-spherical) liquid-like aggregates, called micelles. These aggregates are of enormous importance in surface science. Micellization is an alternative to the adsorption mechanism for minimizing the system's energy. Micelles are created when we have reached a certain concentration of surfactants that is called the critical micelle concentration (CMC). Then, the surface tension is constant, and thus the CMC resembles a thermodynamic transition state. However, CMC is *not* rigorously a thermodynamic transition state – it roughly depends on the method used for its estimation (e.g. surface tension, conductivity and solubilization) and the approach employed to determine the discontinuity point. As discussed next, a plot of surface tension versus concentration and the fact that the surface tension of the solution becomes constant after CMC provide a very useful method for measuring (estimating) the CMC.

A very important characteristic of surfactant molecules, which quantifies their hydrophilic and hydrophobic character, is the so-called CPP (critical packing parameter). CPP is a geometric parameter of the surfactant defined as (Figure 5.1):

$$CPP = \frac{V_{surf}}{\alpha_0 l_c} \qquad (5.1)$$

where V_{surf} denotes the tail chain (or chains) volume, l_c is the critical tail chain length and α_0 is the head-group area at the head-tail interface.

The CPP depends significantly on salt, pH, temperature, double bonds and double chains. Especially, salts have a profound effect as they lower the effective head area, thus increasing the CPP.

Approximate expressions for V_{surf} and l_c of surfactants as a function of the carbon number chain have been presented (Pashley and Karaman, 2004; Israelachvili, 1985; Hamley, 2000):

$$V_{surf} = (0.0274 + 0.0269n)m$$
$$l_c = 0.154 + 0.1265n$$
(5.2)

where n is the number of carbon atoms in the hydrocarbon (hydrophobic) chain and m is the number of hydrocarbon (hydrophobic) chains. In Equation 5.2 V_{surf} is expressed in nm^3 and l_c is in nm.

The CPP value is directly connected to the shape/structure of the micelles. It can be proved for example that surfactants having CPP values below $1/3$ will create spherical micelles (see Example 5.2). Higher CPP values result to other structures, as shown in Tables 5.2 and 5.3.

Some of the non-spherical structures, which are observed at concentrations higher than CMC, are liquid crystalline structures like hexagonal or lamellar, some of which resemble the structure of biological membranes.

There are other approaches for characterizing the hydrophobic–hydrophilic tendencies of surfactants (and other molecules). One useful approach, often employed in the study of emulsions, is the so-called HLB (hydrophilic–lipophilic balance) which is described in Appendix 5.2. All of these ways of assessing the hydrophilicity/hydrophobicity of molecules are different expressions of more or less the same concept and because of this they are correlated to each other (see, for example, Cheng *et al.*, 2003, 2005). Some correlations are shown in Appendix 5.2.

5.2 Micelles and critical micelle concentration (CMC)

Let us further consider micelles. Although not always correct, we often picture micelles as spherical or semi-spherical aggregates (Figure 5.2). They are almost spherical at low concentrations, close to CMC, but can have different forms at higher concentrations (hexagonal or lamellar phases). We believe that they are more or less liquid-like in a very dynamic state. There are many surfactant molecules in each micelle, often about 50–100. This number is called the aggregate (or aggregation) number (see below).

The CMC can be measured by observing the change of several properties, e.g. surface tension with concentration (Figure 5.4). Other properties like osmotic pressure and electrical conductivity also change at CMC. The conductivity increases after CMC and this is a useful CMC estimation method for ionic surfactants. Turbidity and detergency also increase up to the CMC. Solubilization also increases above CMC, as many compounds are readily soluble in the micelles. All of these observations further verify the existence of micelles after CMC and indicate the importance of micelles in cleaning (detergency).

As mentioned, CMC is not, strictly speaking, a thermodynamic transition state; its value roughly depends on the estimation method. We observe in most cases that there is a small range of concentrations (range

Table 5.2 *Relationship between CPP values and micelle structure. See also Appendix 5.1 for some useful relationships on surface and volume of well-known geometrical structures, of relevance to micellar structures*

CPP value	Micelle structure	Example
<1/3	Spherical micelles	Single-chained surfactants, e.g. SDS in low salt
1/3–1/2	Cylindrical micelles	Single-chained surfactants with small head group areas, e.g. SDS and CTAB in high salt concentrations, non-ionic surfactants
1/2–1	Flexible bilayers	Double-chain surfactants with large head group areas, e.g. phosphatidyl choline (lecithin), dihexadecyl phosphate
~1	Planar bilayers	Double-chain surfactants with small head group areas
		Anionic surfactants in high salt concentrations, e.g. phosphatidyl ethanolamine
>1	Reversed or inverted micelles	Double-chain surfactants with small head group areas
		Non-ionic surfactants
		Anionic surfactants in high salt concentrations, e.g. cardiolipin + Ca^{2+}

Adapted from Israelachvili (1985).
SDS: sodium dodecyl sulfate, CTAB: hexadecyl(trimethyl)ammonium bromide.

Table 5.3   *Use of the critical packing parameter (CPP) to predict aggregate structure*

Lipid	Critical packing parameter $v/a_0 l_c$	Critical packing shape	Structures formed
Single-chained lipids (surfactants) with large head-group areas: *SDS in low salt*	< 1/3	Cone	Spherical micelles
Single-chained lipids with small head-group areas: *SDS and CTAB in high salt, nonionics*	1/3–1/2	Truncated cone	Cylindrical micelles
Double-chained lipids with large head-group areas, fluid chains: *Phosphatidyl choline (lecithin), Phosphatidyl serine, Phosphatidyl glycerol, Phosphatidyl inositol, Phosphatidic acid, sphingomyelin, DGDG[a] dihexadecyl phosphate, dialkyl dimethyl ammonium salts*	1/2–1	Truncated cone	Flexible bilayers, vesicles
Double-chained lipids with small head-group areas, anionic lipids in high salt, saturated frozen chains: *phosphatidyl ethanolamine, phosphatidyl serine + Ca^{2+}*	~1	Cylinder	Planar bilayers
Double-chained lipids with small head-group areas, nonionic lipids, poly *(cis)* unsaturated chains, high T: *unsat. phosphatidyl ethanolamine, cardiolipin + Ca^{2+} phosphatidic acid + Ca^{2+} cholesterol, MGDG[b]*	>1	Inverted truncated cone or wedge	Inverted micelles

[a] DGDG, digalactosyl diglyceride, diglucosyl diglyceride.
[b] MGDG, monogalactosyl diglyceride, monoglucosyl diglyceride.
Reprinted from Israelachvili (1985), with permission from Elsevier.

of CMC) where these changes take place rather than a single CMC transition point.

Sometimes, we also observe a minimum in the plot of surface tension against concentration, which is due to small amounts of impurities (e.g. alcohols) that can be present in surfactant formulation, which adsorb at the surface and lower further the surface tension.

Plots like the one shown in Figure 5.4 combined with the Gibbs adsorption equation provide an intuitive explanation for micellization. Let us apply the Gibbs adsorption equation at concentrations after the CMC. Then we see that we have no further adsorption and as more surfactants may still enter the solution the system finds other ways to minimize the energy via the creation of this special type of liquid-like compounds, which we have called micelles.

The number of surfactants in a micelle is called the "aggregation number" (N_{agg}) and is about 50–100 for spherical micelles (Figure 5.5) but can be higher for other structures, depending also on the surfactant type (Figure 5.6). The aggregation number can be estimated from the ratio of the volume of the micelle to the volume of a surfactant molecule or, equivalently, as the surface area of a micelle divided by the surface area of the surfactant head:

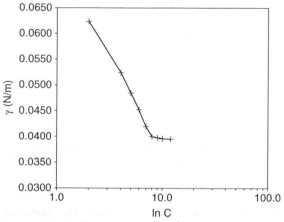

Figure 5.4 *Surface tension (N m^{-1}) versus logarithm of concentration (C in mol m^{-3}) for an aqueous solution of sodium dodecyl sulfate (SDS) at 20 °C. This plot can be used to estimate the CMC (critical micelle concentration). The CMC is the point where the decrease of surface tension with the logarithm of concentration stops and the surface tension becomes constant (independent of concentration). At CMC and beyond the surface tension is much lower than that of pure water, often with a value around half of that of water. From Kontogeorgis and Folas (2010). Reprinted from Kontogeorgis and Folas (2010), with permission from John Wiley & Sons, Ltd*

$$N_{agg} = \frac{V_{micelle}}{V_{surfactant}} = \frac{A_{micelle}}{A_{surfactant}} \qquad (5.3)$$

The aggregation number increases with decreasing CMC and with increasing salt concentration, as shown for two surfactant families in Table 5.1.

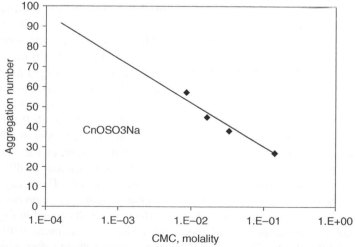

Figure 5.5 *Correlation results for $C_nH_{2n+1}SO_4Na$ (C_nOSO_3Na). Points are experimental data at 25 °C and the line is a regression result. From Cheng (2003, Phd Thesis), Technical University of Denmark*

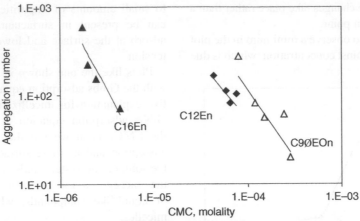

Figure 5.6   *Aggregation number of non-ionic polyethylene oxide surfactants. Points are experimental data at 25 °C and the lines are regression results. The abbreviations are as shown in Table 5.1. From Cheng (2003, Phd Thesis), Technical University of Denmark*

Table 5.4   *CMC and aggregation numbers for some families of ionic surfactants*

Surfactant	Solvent	CMC (mM)	N_{agg}
SDS	Water	8.1	58–80
	0.02 M NaCl	3.82	94
	0.03 M NaCl	3.09	100
	0.10 M NaCl	1.39	112–91
	0.20 M NaCl	0.83	118–105
	0.40 M NaCl	0.52	126–129
Dodecyl	Water	5.60	87
pyridinium	0.0025 M KI	4.53	90
iodide	0.0050 M KI	3.87	94
	0.0100 M KI	2.94	124
C_8E_6	Water	9.9	41
$C_{10}E_6$	Water	0.95	260
$C_{12}E_6$	Water	0.068	1400 at 35 °C
			400 at 25 °C

Data are from various sources presented by Hunter (2001) and Jonsson *et al.* (2001).

In some cases, the aggregation numbers can be quite high, as can be seen for some non-ionic surfactant families in Figure 5.6.

The properties (e.g. cleaning and stabilizing capabilities) of surfactants depend on both solution properties (temperature, time, presence of salts and co-surfactants) and the characteristics of the surfactants, especially CMC, the Krafft point (see

Section 5.3) and their chemistry. The surfactant chemistry and especially the balance between hydrophobic and hydrophilic parts are quantified using tools like the CPP or HLB (critical packing parameter, hydrophilic–lipophilic balance) as explained in the previous section and in Appendix 5.2.

CMC decreases with increasing chain length of the hydrophobic part (longer tails cause the surfactant to be less soluble and it starts working at lower concentrations), slightly with decreasing hydrophilic chain length, adding salt (for ionics), see Table 5.4, or co-surfactants.

Salts do not influence equally the CMC for non-ionic and ionic surfactants. For non-ionics, salts do not have an important influence. However, for ionic surfactants, salts result in a drastic decrease of CMC, as illustrated for some typical systems in Table 5.4. This is due to the reduction of the repulsions between the charged head groups of the surfactants. The CMC of SDS without salt is (at 25 °C) 8.1 mM; it is decreased to 5.6 mM with just 0.01 mol L^{-1} NaCl, while 0.3 mol L^{-1} of the same salt results in a CMC as low as 0.7 mM.

The temperature, on the other hand, can have various effects, which may be different for non-ionic and anionic surfactants. This is explained later. The effect of chain length is more pronounced for the non-ionic surfactants, as illustrated in Figure 5.7. The decrease of CMC is pronounced until there are 18 carbon atoms in the chain, after which CMC tends to be

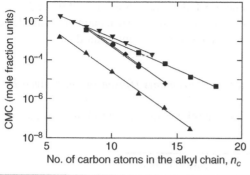

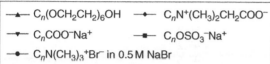

Figure 5.7 CMC decreases (almost linearly) with increasing chain length of the hydrophobic tail for all types of surfactants. This decrease of CMC with chain length is much more pronounced for the non-ionic surfactants, e.g. $C_n(OCH_2CH_2)_6OH$ than for ionic ones. As the CMC is related directly to detergency (cleaning capability) and we would like to use as little surfactant as possible, this makes non-ionic surfactants very attractive, but still, although they are used often in mixtures with anionics, the ionic (anionic) surfactants are much more popular. Economics is not the main reason (see Section 5.4)! Adapted from Jonsson et al. (2001), with permission from John Wiley & Sons, Ltd

rather constant, possibly due to coiling of the hydrophobic chains in the water phase (Shaw, 1992).

Although CMC decreases monotonically with chain length, an optimum length is often found to be about 12–16, as heavier surfactants diffuse very slowly in the solution. CMC and the amount of surfactant (detergent) are important but so is time! Formation of the adsorbed layer of the micelles is not an instantaneous process but depends on the diffusion rate of the surfactant from the solution to the interface. It may take several seconds for the solution to attain its equilibrium surface tension.

The linear decrease of CMC with the number of carbon atoms in the hydrophobic chain is a general phenomenon and is observed for very many families of surfactants. Jonsson *et al.* (2001), Myers (1991, 1999) and Hunter (2001) present linear correlations for CMC with carbon number, including parameters for the corresponding equations for a long list of surfactant families (non-ionic and ionic ones). Most of the CMC measurements and equations are at 20–30 °C but a few data are available also at higher temperatures (40–60 °C).

Mattei, Kontogeorgis and Gani (2013) have recently shown that the CMC for a wide range of non-ionic surfactants can be predicted by a group contribution method using first, second and third-order groups. The agreement is very satisfactory, as

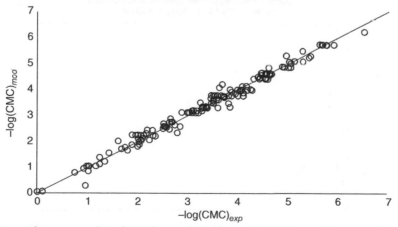

Figure 5.8 Experimental versus predicted (via the model) values of CMC for a wide range of non-ionic surfactants (150 data points). The CMC values are predicted by a group contribution method. Reprinted from Mattei et al. (2013), with permission from American Chemical Society

shown by the parity plot in Figure 5.8. Mattei, Kontogeorgis and Gani. (2014) have extended this approach to other surfactant properties as well.

Example 5.1 illustrates how the CMC and other properties of a non-ionic surfactant are calculated from surface tension–concentration data.

Example 5.1.   CMC and surfactant area from surface tension measurements (Based on data given by Shaw, 1992).

The following surface tensions were measured for aqueous solutions of the non-ionic surfactant $CH_3(CH_2)_9(OCH_2CH_2)_5OH$ at 25 °C:

Concentration, 10^4C (mol L^{-1})	Surface tension (mN m^{-1})
0.1	63.9
0.3	56.2
1.0	47.2
2.0	41.6
5.0	34.0
8.0	30.3
10.0	29.8
20.0	29.6
30.0	29.5

1. Determine the critical micelle concentration and calculate the area occupied by each adsorbed surfactant molecule at the critical micelle concentration.
2. Calculate the surface pressure at the CMC as well as the area per molecule at the CMC that corresponds to an ideal gas film. Do we have an ideal gas film at CMC for this system?

Solution

We plot the surface tension against the concentration ($\times 10^{-3}$) see figure below. The point where the surface tension becomes approximately constant with respect to concentration is the CMC, which in this case is about 9.0×10^{-4} mol L^{-1}.

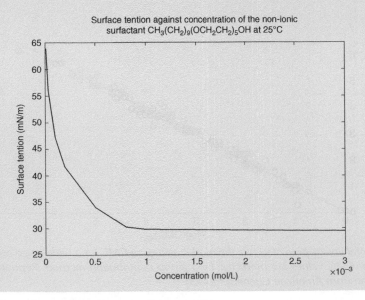

To calculate the area occupied by a surfactant molecule at the CMC we will use the Gibbs adsorption equation:

$$\Gamma = \frac{n_i^\sigma}{A} = -\frac{C_i}{R_{ig}T}\frac{d\gamma_i}{dC_i} = -\frac{9\times 10^{-4}}{8.314\times 298}\left(-\frac{(39.3-29.8)\times 10^{-3}}{(10-0)\times 10^{-4}}\right) = 3.4509\times 10^{-6}\, \text{mol}\,\text{m}^{-2}$$

where R_{ig} is in J mol^{-1} K^{-1}, the surface tension is in mN m^{-1}, and the concentration is expressed in mol L^{-1}. The surfactant molecular area at CMC is (see Equation 4.8) is thus:

$$A = \frac{10^{18}}{N_A\Gamma} = 0.48119\, \text{nm}^2 \text{ or } 48.119\times 10^{-20}\, \text{m}^2 \text{ molecule}^{-1}$$

The surface pressure at CMC is:

$$\pi = \gamma_0 - \gamma = 72.8 - 29.8 = 43.0\, \text{mN}\,\text{m}^{-1}$$

Hence, the area for an ideal gas film is given by the ideal gas law shown previously (Example 4.3), which per molecule can be written using the Boltzmann constant (k_B):

$$A = \frac{k_BT}{\pi} = 9.61189\times 10^{-20}\, \text{m}^2 \text{ molecule}^{-1}$$

This value is of the same order of magnitude but significantly lower than the one calculated from the Gibbs equation, thus, as expected, we do not have an ideal gas film. The plot of the surface tension with respect to concentration is not (close to) linear over the whole range from pure water up to CMC.

Alternatively, as shown below, the surface tension can be plotted against the logarithm of concentration. This is the recommended approach since, in this way, the CMC and especially the slope can be easily determined.

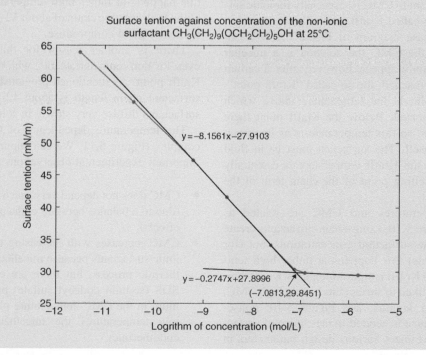

Surface tention against concentration of the non-ionic surfactant $CH_3(CH_2)_9(OCH_2CH_2)_5OH$ at 25°C

$y = -8.1561x - 27.9103$

$y = -0.2747x + 27.8996$

$(-7.0813, 29.8451)$

Surface tention (mN/m)

Logrithm of concentration (mol/L)

To calculate the area occupied by a surfactant molecule at the CMC, we will use the Gibbs adsorption equation:

$$\Gamma = \frac{n_i^\sigma}{A} = -\frac{1}{R_{ig}T}\frac{d\gamma_i}{d\ln C_i} = -\frac{1}{8.314 \times 298}(-8.1561) = 3.2903 \times 10^{-6} \text{ mol/m}^2$$

(Notice that the slope of −8.1561 should be multiplied by 10^{-3} to transform the units from mN into N and thus have all the values in SI units. The units of the concentration do not affect the results – see also the Gibbs equation in its previous form. Recall that 1 J = 1 N m.)

Thus, the surfactant molecular area at CMC is:

$$A = \frac{10^{18}}{N_A \Gamma} = \frac{10^{18}}{6.02 \times 10^{23} \times 3.2903 \times 10^{-6}} = 0.5049 \text{ nm}^2 \text{ or } 50.49 \cdot 10^{-20} \text{ m}^2 \text{ molecule}^{-1}$$

The surface pressure at CMC is:

$$\pi = \gamma_0 - \gamma = 72.8 - 29.8451 = 42.9549 \text{ mN m}^{-1}$$

The area for an ideal gas film is, thus, given by the ideal gas law derived in Example 4.3, which per molecule can be written using the Boltzmann constant (k_B):

$$A = \frac{k_B T}{\pi} = 9.5860 \times 10^{-20} \text{ m}^2 \text{ molecule}^{-1}$$

5.3   Micellization – theories and key parameters

Equally important to CMC is, especially for ionic surfactants, the so-called Krafft point. This is shown in the general phase diagram of Figure 5.9. We can see that after the CMC the solution is a micellar one. Micelles only appear, however, after a certain temperature is reached, the so-called "Krafft point". The Krafft point is the temperature above which micelles are created. Below the Krafft point there are no micelles, so this temperature is as important as the CMC itself. The surfactant must be in fluid form and thus the Krafft temperature is essentially close to the melting point of the chain term of the ionic surfactant.

Krafft temperatures and CMC are related as shown in Figure 5.10. Long-chain surfactants create micelles at low surfactant concentrations, but (for ionic surfactants) this happens at rather high temperatures (high Krafft temperature). Low Krafft temperatures are linked to surfactants having high CMC, i.e. low chain lengths, see Figure 5.10; consequently, a balance is needed in the choice of surfactants so that we meet various design parameters in

addition to good cleaning (economy, effectiveness and time). This is because, long-chain ionic surfactants create micelles at low concentrations, but this happens at rather high temperatures. A length of the hydrophobic chain of about 12–14 or 16 atoms is often a good compromise.

There is another reason for this compromise, even for non-ionic surfactants, which do not have a Krafft point. As mentioned previously, an optimum surfactant chain length is about 12–16, as heavier surfactants diffuse very slowly in solution.

The temperature dependency of CMC is rather complex (Figure 5.11. We can summarize the most important experimental observations as follows:

- CMC does not depend very much on temperature (due to a balance between enthalpic and entropic effects);
- CMC increases with increasing temperature for ionic surfactants because micellization is an exothermic process, but there are exceptions, e.g. SDS (sodium dodecyl sulfate) presents a minimum in the CMC–temperature plot and, thus, at low temperatures the micellization must be endothermic;

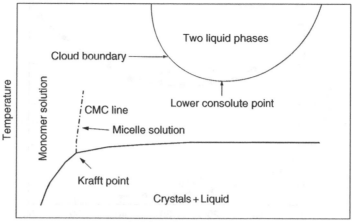

Figure 5.9 *General phase diagram of a surfactant solution, showing the CMC line, the Krafft point (temperature) and the lower consolute point (or lower critical temperature). As can be seen, the phase behaviour of aqueous surfactant solutions is rather complex and various phases are distinguished. At high concentrations, we can see various special surfactant phases (hexagonal, lamellar, cubic). These are called liquid crystalline phases and although there are 18 different types, the three mentioned are the most important. Many of these complex structures have found exciting applications (e.g. liquid crystal displays and study of biological membranes)*

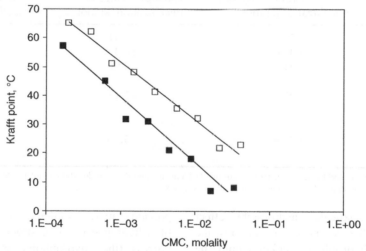

Figure 5.10 *Krafft–CMC relationship for two families of surfactants (black: sodium alkyl sulfates, white: sodium alkyl sulfonates). A balance is needed to satisfy as much as possible all parameters involved, including low CMC (less consumption of surfactant) – not very high temperatures (thus saving of energy) – rather fast cleaning (thus avoiding very heavy surfactant molecules that diffuse slowly from solutions). A chain-length of about 12–14 (maybe up to 16) carbon atoms fulfils in many cases most of these conditions. Reprinted from Kontogeorgis and Folas (2010), with permission from John Wiley & Sons, Ltd*

- CMC decreases with increasing temperature for non-ionic surfactants, e.g. the polyoxyethylenes.

In conclusion, as shown in Table 5.5, micellization is either an endothermic (positive enthalpy change), e.g. for SDS and non-ionic polyoxyethylenes, or an exothermic process, but most importantly it is an entropic phenomenon with high positive values of the entropy change, which are typically attributed to the changes of the structure of water in the presence of surfactants

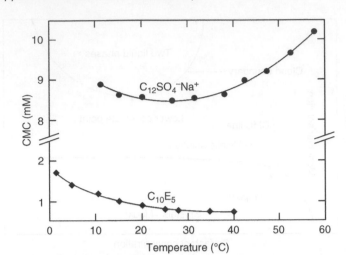

Figure 5.11   *CMC–temperature dependency of SDS ($C_{12}SO_4^-Na^+$) and a typical non-ionic surfactant ($C_{10}E_5$). Adapted from Jonsson et al. (2001), with permission from John Wiley & Sons, Ltd*

Table 5.5   *Enthalpy, entropy and Gibbs free energy change during the micellization process for several ionic and non-ionic surfactants at room temperature*

Surfactant	Gibbs energy change of micellization (kJ mol^{-1})	Enthalpy change of micellization (kJ mol^{-1})	Entropy change of micellization (J K^{-1} mol^{-1})
SDS	−21.9	+2.51	+81.9
$C_{12}E_6$	−33.0	+16.3	+49.3
Dodecyl pyridinium bromide	−21.0	−4.06	+56.9
N,N-Dimethyl-dodecylamine oxide	−25.4	+7.11	+109.0
N-Dodecyl-N,N-dimethyl glycine	−25.6	−5.86	+64.9

From various sources (Hiemenz and Rajagopalan, 1997; Prausnitz, Lichtenthaler and de Azevedo, 1999; Hunter, 2001) as summarized by Kontogeorgis and Folas (2010). Reprinted with permission from John Wiley & Sons, Ltd.

and micelles, i.e. the so-called hydrophobic effect (Chapter 2). As applied to surfactant solutions, the influence of the hydrophobic effect could be described as follows (Figure 5.12): Water molecules in liquid water are connected with hydrogen bonds and are "forced" into smaller even more structured cavities if some surfactant molecules are added in the solution. Thus, water molecules due to their very strong tendency to be with each other ("and far away from the hydrophobic enemies") create (in the presence of hydrophobic compounds) a higher degree of local order than in pure liquid water (entropy decrease). When these surfactants leave the solution to create micelles, the liquid water assumes its

previous "more disordered" (or less ordered, if you like) hydrogen bonding structure. Thus, the whole process (the "hydrophobic effect") leaves water molecules the freedom to "freely interact" with each other and results in a high positive entropy change of water, which possibly counter-balances the (possibly) negative entropy of change of the surfactant molecules.

Another explanation (of less importance than the hydrophobic effect) is that the hydrocarbon chains of surfactants have a much higher freedom of movement inside the micelles than in the solution; consequently, there is a positive entropy change of the surfactants themselves.

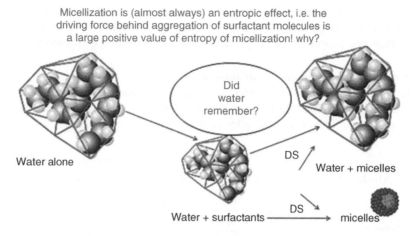

Micellization is (almost always) an entropic effect, i.e. the driving force behind aggregation of surfactant molecules is a large positive value of entropy of micellization! why?

Did water remember?

Water alone

DS

Water + micelles

Water + surfactants ⟶ DS ⟶ micelles

The protagonist is WATER!!

Figure 5.12 *The micellization process in surfactant solutions is an entropic effect. This is explained by the hydrophobic phenomenon and the increase of entropy of the solution during the micelle creation (see also Table 5.5).*

The energy (Gibbs, enthalpy) and entropy changes during the micellization process can be predicted if the CMC and its dependency on temperature are known, as shown in Equation 5.4:

$$\Delta G = R_{ig} T \ln(\text{CMC})$$

$$\Delta S = -\frac{dG}{dT} = -R_{ig} T \frac{d\ln(\text{CMC})}{dT} - R_{ig} \ln(\text{CMC})$$

$$(5.4)$$

$$\Delta H = \Delta G + T \Delta S = -R_{ig} T^2 \frac{d\ln(\text{CMC})}{dT}$$

where $R_{ig} = 8.314$ J mol^{-1} K^{-1} and CMC is transformed from mol L^{-1} into mole fractions by dividing the values which are provided in mol L^{-1} by 55.5 mol L^{-1} (water's molarity at 25 °C).

Assuming that the enthalpy of micellization is almost constant within a certain temperature range, then the quantity ln(CMC) can be seen to be linear with respect to $1/T$. The enthalpy of micellization could be obtained from the slope of such a plot.

As an alternative to experimental measurements, CMC can be predicted using thermodynamic models like the group contribution activity coefficient model UNIFAC (Flores *et al.*, 2001; Chen, 1996; Cheng,

2003; Cheng, Kontogeorgis and Stenby, 2002; Kontogeorgis and Folas, 2010). In this way micellization is treated as a simple phase separation of the surfactants into a micelle phase (only containing associated surfactants, no water present) and an aqueous phase containing both water and free-surfactants. The unassociated (free) surfactant concentration remains practically constant after CMC, as all new surfactants go into the micelles. Satisfactory results have been reported, especially for non-ionic surfactants, e.g. poly(ethylene oxide)s, including the CMC dependency with the chain length of the hydrophobic and the hydrophilic parts of the surfactant. Importantly, existing UNIFAC models with published parameter tables do not perform very well for aqueous non-ionic surfactants of the polyoxyethylene type. When, however, the oxyethylene group (OCH$_2$CH$_2$) is defined as a separate group and the parameters are fitted to phase equilibrium data for low molecular weight (non-surfactant) solutions, very good results are obtained. The results are as good as when the parameters values are taken directly from the CMC data. More details are found in pertinent literature (e.g. Cheng, Kontogeorgis and Stenby, 2002; Flores *et al.*, 2001).

As mentioned, CPP has a special role in the studies of surfactant solutions, among other reasons due to its connection with the micellar structure. One of the most well-known surfactants, SDS, forms often (semi)

spherical micelles. We investigate some properties of SDS and spherical micelles in general in Example 5.2. Example 5.3 presents properties of a typical non-ionic surfactant which, as mentioned previously, can accommodate in the micellar structure more surfactant molecules, compared to the ionic surfactants.

Example 5.2.   *Critical packing parameter of surfactants.*
1. Show that the CPP (critical packing parameter) of surfactants in spherical micelles is equal to or less than $1/3$.
2. It is experimentally found that SDS (sodium dodecyl sulfate) forms (almost) spherical micelles. Show that the aggregation number of SDS micelles is approximately equal to 56.

Solution
1. Using the definition of the aggregation number (Equation 5.3), we can derive the following relationship between the surfactant volume and area with the radius of spherical micelles:

$$N_{agg} = \frac{V_{micelle}}{V_{surfact}} = \frac{A_{micelle}}{A_{surfact}} \Rightarrow$$

$$\frac{4\pi R^3}{3V_s} = \frac{4\pi R^2}{\alpha_0} \Rightarrow$$

$$\frac{V_s}{\alpha_0} = \frac{R}{3}$$

Using the CPP definition, we get:

$$CPP = \frac{V_s}{\alpha_0 l_c} = \frac{R}{3l_c}$$

As in the general case the radius of the micelle (R) is equal to or less than the critical length of the surfactant, then the condition for a spherical micelle is that CPP has to be equal or less than $1/3$. The value of CPP = $1/3$ is obtained exactly when $R = l_c$.

Alternatively, the CPP can be estimated as (assuming a "cone" type surfactant in a spherical micelle; the surfactant having a height = micelle radius, R):

$$CPP = \frac{V_s}{\alpha_0 l_c} = \frac{\frac{1}{3}\alpha_0 R}{\alpha_0 l_c} = \frac{R}{3l_c}$$

(See Appendix 5.1 for some useful geometrical relationships.)

2. For SDS, $n = 12$ (and $m = 1$), thus the volume of the surfactant molecule is:

$$V_s = 0.0274 + 0.0269n = 0.3502 \, nm^3$$

The maximum length of the surfactant molecule is:

$$l_{max} = 0.15 + 0.1265n = 1.668 \, nm$$

Since SDS forms almost spherical micelles, in the limiting case where $R = l_{max}$, we know (see previous question) that CPP = $1/3$:

$$CPP = \frac{V_s}{\alpha_0 l_c} \Rightarrow$$

$$\frac{1}{3} = \frac{0.3502}{\alpha_0 1.668} \Rightarrow \alpha_0 = 0.6298 \, nm^2$$

Hence, the aggregation number is:

$$N_{agg} = \frac{A_{micelle}}{A_{surfact}} \Rightarrow$$

$$N_{agg} = \frac{4\pi R^2}{\alpha_0} = \frac{4 \times 3.14 \times 1.668^2}{0.6298} = 56$$

Alternatively, the aggregation number can be estimated using the definition based on the volume of micelle and surfactant leading to the same result:

$$N_{agg} = \frac{V_{micelle}}{V_{surfact}} = \frac{4\pi R^3}{3V_s}$$

Example 5.3.   Critical packing parameter and aggregation number of non-ionic surfactants.
For the non-ionic surfactant of Example 5.1, estimate the CPP parameter and based on this provide an estimation of the micellar shape and aggregation number.

Solution
From Example 5.1, we have estimated the surface area to be equal to 0.5049 nm².
   We know that we have ten carbon atoms in the hydrophobic chain and, using Equation 5.2, we can calculate the volume and length of a single surfactant molecule to be about 0.2964 nm³ and 1.415 nm, respectively. Thus, we have for CPP:

$$CPP = \frac{V_s}{\alpha_0 l_c} = \frac{0.2964}{0.5049 \times 1.415} = 0.415$$

A CPP value between 0.33 and 0.5 indicates cylindrical micelles, as expected for non-ionic surfactants. Then, using Equation 5.3 for the aggregation number and the geometric relations of Appendix 5.1, we can estimate first the height of the cylindrical micellar structure:

$$N_{agg} = \frac{V_{micelle}}{V_{surfactant}} = \frac{A_{micelle}}{A_{surfactant}} \Rightarrow$$

$$\frac{\pi r^2 h}{V_{surf}} = \frac{2\pi r(r+h)}{\alpha_0} \Rightarrow \frac{\pi(1.415)^2 h}{0.2964} = \frac{2\pi 1.415(1.415+h)}{0.5049} \Rightarrow$$

$$h = 6.8961 \, nm$$

Substituting this value to either form of the aggregation number we get a value equal to 146, which is higher than the one we found for SDS (Example 5.2) and for other ionic surfactants (but is consistent with the cylindrical shape of the micellar structure, which can accommodate more surfactants).

5.4   Surfactants and cleaning (detergency)

How do we clean the various surfaces? How do surfactants work? We recall that surfactants are crucial elements but *not* the *only* component of commercial detergent formulations. Detergents contain also builders, possibly enzymes and other components.

A detergent is a very complex chemical product. Its cleaning ability depends on several factors: type of surfactant and other compounds (chemistry), the concentration of contaminants, the mechanical energy used, the time, the temperature, the type of surface, etc.

Nevertheless, surfactants play a major role in cleaning of the huge variety of different stains (types of dirt) and for both hard and soft surfaces as they solubilize and remove the dirt (fats, stains, etc.) of various surfaces. We can roughly summarize the major mechanisms of surfactant cleaning in the following three steps, as illustrated in Figure 5.13:

1. improving the wetting of surfaces by decreasing the surface tension of water/aqueous solution (water alone cannot in most cases provide good wetting);
2. good adhesion on substrate and dirt and reduction of the interfacial tensions between dirt(d)-water(w) and solid substrate(s)-water and thus reducing the corresponding work of adhesion, since the interfacial tensions between dirt–water and solid–water are reduced ($W_{adh,ds} = \gamma_{sw} + \gamma_{dw} - \gamma_{sd}$);
3. solubilization of dirt inside micelles, which ensures that the dirt is removed with little mechanical help, thus avoiding deposition of the dirt back on the surface.

The properties of surfactants can be manipulated as we saw earlier by changing the proportion of the hydrophobic and hydrophilic parts, the temperature, the pH of the solution (salts), using mixtures of surfactants, etc.

Foam formation is related to the decrease of water–air interfacial tension and this is *not* directly related to the detergency. However, psychologically, it is often

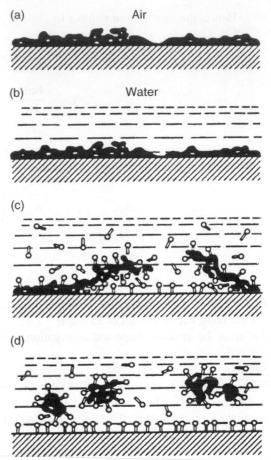

Figure 5.13   *The three major mechanisms in the cleaning of surfaces via surfactants: (a) The solid substrate with the dirt prior to adding water and surfactants, (b) wetting with water alone, (c) decrease of adhesion with the presence of surfactants and (d) solubilization inside micelles. Shaw (1992). Reproduced with permission from Elsevier*

important. For this reason, non-ionic surfactants (which have otherwise excellent cleaning potential) are often not used alone because they do not produce much foam. Thus, typically, most detergents are mixtures of non-ionic and anionic surfactants. Mixed surfactants are almost always used in commercial cleaning formulations.

Detergency is often connected to CMC and CPP and many experiments indicate that detergency increases with concentration especially up to CMC and is often best at CPP values around 1. Other important parameters affecting the performance of detergents are pH, presence of additives, nature of contaminants, type of surface, time/temperature and the mechanical energy used.

Mixed surfactants are often employed in commercial formulations both for cleaning and stabilization. Simple additive mixing rules can be used to estimate the CMC of the mixed surfactant system from knowledge of the CMC's of the individual surfactants and the concentration (in mole fraction) of the surfactant in the system.

One of the most well-known uses of surfactants is in detergent formulations used in washing machines. A very exciting presentation on the significance of washing machines in improving the living standards in the Western world is provided in the following link by professor Hans Rosling (http://www.ted.com/talks/hans_rosling_and_the_magic_washing_machine.html).

5.5   Other applications of surfactants

Besides their use in detergents and in dispersions as stabilizers, surfactants find many more applications. One interesting application is their solubilization capability inside micelles. Aqueous solutions of surfactants above CMC can solubilize organic compounds that are otherwise insoluble in water by incorporating them inside the micelles. For example, the dye xylenol orange (a colourant) dissolves sparingly in pure water but gives a deep red solution with SDS at concentrations above CMC. This method can be used experimentally for determining (measuring) the CMC. The area of solubilization varies, depending on the various interactions (electrostatic, hydrophobic). Hydrocarbons and non-polar compounds solubilize inside the micelle, while fatty acids and alcohols solubilize in the area close to the heads of the surfactants of the micelle. The phenomenon of solubilization inside micelles finds widespread applications in pharmaceuticals, detergents and catalysis.

Another exciting application is in using surfactants in biotechnology for separating proteins (Prausnitz, 1996; Woll and Hatton, 1989) (Figure 5.14).

Reverse micelles can be used as mini-reactors for enzyme-catalysed reactions. These reactions require an aqueous environment but often the reactants are only sparingly soluble in water. Consequently, the formation of reverse micelles in an organic solvent has been suggested. The reverse micelle contains the enzyme in a tiny pool of water. Reactants dissolved in the organic solvent diffuse into the reverse micelle and products diffuse out. For the design of

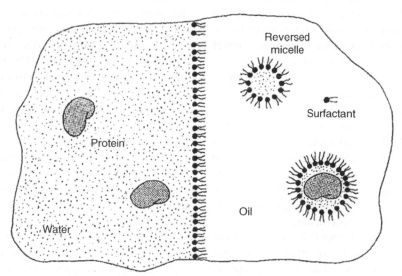

Figure 5.14   *Representation of the protein partitioning in a reversed micellar extraction processes. Reprinted from Woll and Hatton (1989), with permission from Springer Science and Business Media*

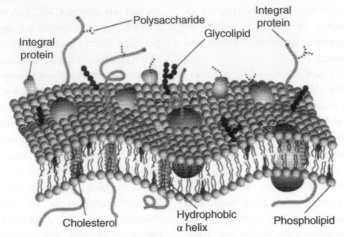

Figure 5.15 *Cell membranes and surfactants. Reprinted from Hamley (2000), with permission from John Wiley & Sons, Ltd*

such reactors, it is necessary to know the distribution coefficient for the enzyme between the reverse micelle and a continuous aqueous phase. This coefficient depends on numerous variables especially the electric charge of the enzyme, which depends on pH and the surfactant concentration. Woll and Hatton (1989) have developed simple yet successful correlations of the distribution coefficient against pH which can be used for calculations/interpretations of literature data and they are thus useful in designing novel separations based on reverse micelles.

Finally, surfactants play a crucial role in the study of many biological systems, e.g. biological membranes (Figure 5.15). Biological membranes contain lipid contents that vary from as little as 25% up to 80% by weight. The lipids contained therein are phospholipids, which in water are situated in such a way so that the hydrophilic heads are close to water and the hydrophobic tails are converging so that they can avoid water. Such double phospholipid-layers are present in biological (cell) membranes.

Cell membranes consist essentially of a bimolecular layer of lipids with the hydrocarbon chain orientated towards the interior and the hydrophilic groups on the outside. As with micelles, this organization is primarily the result of hydrophobic bonding. However, this simple lipid biomolecular model needs

to be modified so that proteins can both adsorb and penetrate this bimolecular layer.

5.6   Concluding remarks

Surfactants are amphiphilic compounds (with colloidal dimensions) that adsorb on liquid–air and oil–water interfaces and decrease surface and interfacial tensions and the corresponding works of adhesion (substrate/water, dirt/water). The decrease of water surface tension is observed up to a certain concentration, called the CMC (critical micelle concentration). After the CMC, the surfactants create new liquid-like near spherical structures, called micelles. At even higher concentration we observe more complex (liquid crystalline) structures, e.g. hexagonal or lamellar, some of which resemble the structure in biological membranes. The CMC can be measured by observing the change of several properties, e.g. surface tension with concentration.

The micellization is either endothermic or exothermic but above all it is an entropic phenomenon with very high positive values of the entropy change, which are typically attributed to changes of the structure of water and the so-called hydrophobic effect.

CMC decreases with increasing chain length (of the hydrophobic part), a little with decreasing hydrophilic part length, with addition of salt or co-surfactants but the temperature can have various effects, sometimes different for non-ionic and anionic surfactants. The CMC decreases with lowering the temperature for ionics and with increasing temperature for non-ionics. The Krafft point is the temperature above which micelles are created and is important for ionic surfactants. Low Krafft temperatures are linked with surfactants having high CMC, i.e. low chain lengths, and thus a balance is needed in the choice of surfactants to ensure all necessary design parameters in addition to good cleaning (economy, effectiveness, and time).

Surfactants have many applications; they are present in detergents for cleaning of both soft and hard surfaces, as emulsifiers, foaming agents or stabilizers for colloidal dispersions, in various applications in biotechnology, e.g. separation of proteins in reversed micelles, and catalysis and as components in many complex products, e.g. paints and coatings.

The properties (e.g. cleaning and stabilizing capabilities) of surfactants depend on both solution properties (temperature, time, presence of salts and co-surfactants) and their own characteristics, especially CMC, the Krafft point and their chemistry. The surfactant chemistry and especially the balance between hydrophobic and hydrophilic parts is quantified using tools like the CPP or HLB (critical packing parameter, hydrophilic–lipophilic balance, respectively). For example, it is often observed that detergency increases with concentration especially up to CMC and is often best at CPP values around 1. We will meet the important concept of CPP again in Chapter 7 where we will see that surfactant adsorption on solid surfaces is connected to CPP.

Surfactants clean surfaces first of all by ensuring good wetting and adhesion (decreasing the work of adhesion between dirt and substrate) and by solubilization of the dirt into the micelles.

Appendix 5.1 Useful relationships from geometry

Name	Shape	Volume	Area
Sphere		$\dfrac{4\pi R^3}{3}$	$4\pi R^2$
Cone		$\dfrac{1}{3}a_0R - \dfrac{1}{3}\pi r^2 h$	$\pi r^2 + \pi rs - \pi r(r+s)$ $s = \sqrt{h^2 + r^2}$
Cylinder		$\pi r^2 h$	$2\pi r(r+h)$
Truncated cone		$\dfrac{1}{3}\pi h\left[r^2 + R^2 + Rr\right]$	$\pi(R+r)S + \pi(R^2 + r^2)$ $S = \sqrt{(R+r)^2 + h^2}$

Appendix 5.2 The Hydrophilic–Lipophilic Balance (HLB)

As will be seen in the chapter on emulsions (Chapter 12), the hydrophilic–lipophilic balance (HLB) is a very useful parameter which, in conjunction with several other semi-empirical tools (like the Bancroft rule), can be used in the design of emulsions.

The HLB is a rather empirical parameter. Values below 8 indicate hydrophobic surfactants (emulsifiers), while values above 8 (or 10) indicate hydrophilic surfactants.

The most well-known method for estimating the HLB is the group contribution method of Davies and Rideal:

$$HLB = 7 + \sum_i n_i HLB_i \qquad (5.5)$$

where n_i is the number of groups of a particular type in the molecule being considered.

The group values for HLB_i given in Table 5.6 are used.

According to the values shown in Table 5.6, and as expected, ionized emulsifiers will have high HLB values. This can be attributed to the strong interaction with water molecules exerted by the charged head groups. The values were obtained by studying the behaviour (e.g. the type of emulsion formed) of selected surfactants in mixtures and referring the results to the HLB scale. The latter was established with certain commercial surfactants called SPANs and TWEENs, which are surfactants based on cyclic ethers of mannitol, polyethoxylated to various degrees.

There are a few special estimation methods as well, e.g. for nonyl phenyl ethoxylates:

$$HLB = 20(17 + 44X_{EO})/(220 + 44X_{EO}) \qquad (5.6)$$

where X_{EO} is the number of ethylene oxide groups.

There is also a method based on a, rather loosely defined, oil–water partition coefficient, $K_{oil/w}$:

$$HLB = 7 + 0.36 \ln K_{oil/w} \qquad (5.7)$$

For mixed surfactants, HLB is estimated from an additive mixing rule:

$$HLB = \sum_{i=1}^{n} w_i HLB_i \qquad (5.8)$$

where n = number of emulsifiers and w_i = weight percentage fraction of an emulsifier.

CPP and HLB are not the only methods used to assess the hydrophilicity/hydrophobicity of a substance. In fact, in thermodynamics, a widely used method is the so-called octanol–water partition coefficient:

$$K_{OW_i} = \lim_{x_i \to 0} \left(\frac{C_i^O}{C_i^W} \right) = 0.151 \frac{\gamma_i^{W,\infty}}{\gamma_i^{O,\infty}} \qquad (5.9)$$

where C_i is the concentration of compound i in octanol (O) or water (W) phase and γ_i^∞ is the infinite dilution activity coefficient of the compound in the octanol (O) or water (W) phase. This method is used especially in environmental studies, where $\log(K_{ow})$ values above 4 indicate dangerous pollutants.

As expected, all these methods (HLB, CPP and K_{ow}) are connected to each other (Cheng, 2003; Cheng et al., 2005). These cited references also showed that K_{ow} can be used to correlate toxicity and the bioconcentration factor of various small alcohol ethoxylate surfactants. HLB is also linked to CMC, as shown for one surfactant family in Figure 5.16.

Table 5.6 *Group-contribution values for estimating the HLB according to Davies-Rideal method (Equation 5.5). When hydroxyl or ester groups are parts of a sorbitan molecule, special values are used as indicated.*

Group	HLB	Group	HLB
-SO_4Na	38.7	-OH (free)	1.9
-COOK	21.1	-OH (sorbitan)	0.5
-COONa	19.1	Sulfonate	11
-N (tertiary amine)	9.4	CH, CH_2, CH_3	–0.475
Ester (sorbitan)	6.8	$-CH_2CH_2O$	0.33
Ester (free)	2.4	$-CH_2CH_2CH_2O$	–0.15
-COOH	2.1	$-CF_2, -CF_3$	–0.87
-O-	1.3		

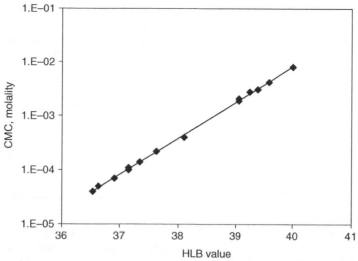

Figure 5.16   *CMC of polyoxyethylene sodium alkyl sulfates ($C_nH_{2n+1}(OC_2H_4)_mSO_4^-Na^+$) as a function of the corresponding HLB values. The line represents the equation $log(CMC) = -28.726 + 0.666HLB$. From Cheng (2003, Phd Thesis),* Technical University of Denmark

---

Example 5.4.   Estimation of HLB from the Davies–Rideal method.
Estimate the HLB for an emulsifier mixture of the following two non-ionic (poly(ethylene oxide)s) surfactants $C_{12}E_{30}$ and $C_{16}E_4$, and a mixture containing 60% of the most hydrophilic and 40% of the most hydrophobic surfactant. How would you characterize the surfactant mixture?

Solution
The HLB of the two given non-ionic surfactants (poly(ethylene oxide)s) will be estimated with the group contribution method by Davies and Rideal (Table 5.6, Equation 5.5) using the information on the structure of the surfactants (see Table 5.1):

$$HLB\ (C_{16}E_4) = 7 + 16(-0.475) + 4 \cdot 0.33 + 1.9 = 2.62$$
$$HLB\ (C_{12}E_{30}) = 7 + 12(-0.475) + 30 \cdot 0.33 + 1.9 = 13.1$$

The first is the most hydrophobic surfactant, thus the HLB of the mixture is:

$$HLB = \sum_{i=1}^{n} w_i HLB_i = 0.6 \times 13.1 + 0.4 \times 2.62 = 8.91$$

which is a (rather) hydrophilic surfactant (above the limiting value of 8).

Problems

Problem 5.1: Critical micelle concentration (CMC) and surfactant area from surface tension measurements (data from Hamley (2000) and from Shaw (1992))
The following surface tensions were measured for an aqueous solution of sodium dodecyl sulfate (SDS) at 20 °C:

Concentration, 10^3C (mol L^{-1})	Surface tension (mN m^{-1})
0	72.0
2	62.3
4	52.4
5	48.5
6	45.2

7	42.0
8	40.0
9	39.8
10	39.6
12	39.5

1. Determine the critical micelle concentration (CMC).
2. The Krafft temperature of SDS is reported to be between 10 and 15 °C. What is understood by the Krafft temperature?
3. Explain the difference between the Gibbs adsorption equation for non-ionic and ionic surfactants.
4. Calculate the surface pressure at the CMC as well as the area per molecule at CMC which corresponds to an ideal gas film.
5. Calculate (in nm^2) the area occupied by a surfactant molecule (actually each adsorbed dodecyl sulfate ion) at the critical micelle concentration. Do we have an ideal gas film at CMC for this system? Comment on the answer!
6. Using the value you have estimated in the previous question for the SDS area, provide an approximate estimation of the shape of the micellar structure created by SDS under these conditions.
7. Discuss briefly the effect of salt concentration on the CMC and on the critical packing parameter of SDS.

Problem 5.2: Energetics of the micellization process (adapted from Hamley (2000))

1. Calculate the values of the thermodynamic properties, Gibbs free energy, enthalpy and entropy of micellization, for the following surfactants from the information below for $T = 298$ K.

Surfactant	C at CMC (mol L^{-1})	d ln C (at CMC)/dT (K^{-1})
$C_6H_{13}S(CH_3)O$	0.437	−0.01436
$C_8H_{17}S(CH_3)O$	0.0281	−0.01056
$C_{10}H_{21}S(CH_3)O$	0.00188	−0.007314

2. Show that ln(CMC) is a linear function of the number of carbon atoms and evaluate the constants of the equations. Comment on the results.

Problem 5.3: Aggregation number of ionic surfactants

The CMC of some sodium alkyl sulfates are given in the table below together with the experimental value of the aggregation number (N_{agg}) for one of them.

Surfactant	T (°C)	CMC (mol L^{-1})	N_{agg}
C_6NaSO_4	25	0.42	17
C_8NaSO_4	25	0.22	
$C_{12}NaSO_4$	25	8.2×10^{-3}	
$C_{14}NaSO_4$	40	2.05×10^{-3}	

Estimate the aggregation number of $C_{14}NaSO_4$ (sodium tetradecyl sulfate) and compare your estimation to the experimental value (80).

Hint: make a plausible assumption for the shape of the micelles.

Problem 5.4: Aggregation number and shape of non-ionic surfactants and micelles

The following data is available for the (single-chain) non-ionic surfactant C_{13}Ethoxylate:

Surfactant	CMC (mol L^{-1})	Surface tension at CMC (mN m^{-1})	N_{agg}
C_{13}Ethoxylate	0.027–0.112	35	171

1. What structure do you expect for the micelles formed by this non-ionic surfactant? Why?
2. Estimate the CPP value of the surfactant molecules. What is the possible shape of the surfactant molecules?

Problem 5.5: Properties of a cationic surfactant

The figure below shows the surface tension–concentration plot of a well-known cationic surfactant, CTAB (hexadecyl(trimethyl)ammonium bromide) from an aqueous solution at 35 °C. Owing to surface hydrolysis, only the ammonium part of the head adsorbs at the interface.

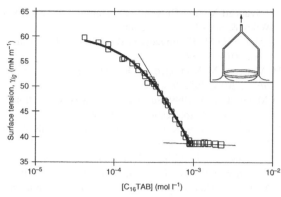

Goodwin (2004). Reproduced with permission from John Wiley & Sons, Ltd.

1. Estimate the critical micelle concentration.
2. Estimate the adsorption of CTAB at CMC as well as the area occupied by a CTAB molecule.
3. What is an approximate value of the critical packing parameter (CPP) of a CTAB molecule and, based on this, suggest the expected shape of the micellar structure which will be formed by CTAB molecules.
4. How many CTAB surfactant molecules do you expect to be present in one micelle?

Problem 5.6: Properties of surfactants

1. Which of the following statements is *wrong* about the CMC (critical micelle concentration)?
   a. It decreases with increasing hydrophobic chain length.
   b. It decreases with adding salt (for ionics).
   c. It increases with increasing hydrophilic length.
   d. It decreases with temperature for all known surfactant families.
2. For the nonionic surfactant C_9EO_5 you have previously calculated, from surface tension–concentration data, a surfactant (head group) area equal to 0.505 nm^2. Which micellar structure do you expect for this surfactant?
   a. spherical
   b. planar bilayer
   c. cylindrical
   d. inverted micelles

3. The table below shows CMC and Krafft temperatures for four sulfate esters.

Surfactant	CMC (ppm)	Krafft temperature (°C)
C12	2300	10
C14	700	28
C16	190	44
C18	50	62

   Which sulfate ester will you choose if you are interested in cleaning with the minimum amount of surfactant?
   a. C12
   b. C14
   c. C18
   d. C16
4. Explain briefly (max five lines) which are the major mechanisms that can describe the act of surfactants in cleaning.
5. Explain briefly (max five lines) the role of the hydrophobic effect in relation to the micellization process.
6. Explain briefly (max five lines) how the surface area of a surfactant on a solid surface can be estimated from adsorption data.
7. Explain briefly (max five lines) how the micellar structure of surfactants can be determined based on surface tension–concentration measurements.

References

C-C. Chen, AIChE J., 1996, 42, 3231.

H. Cheng, G.M. Kontogeorgis, E.H. Stenby, Ind. Eng. Chem. Res., 2002, 41(5), 892.

H. Cheng, G.M. Kontogeorgis, E.H. Stenby, Ind. Eng. Chem. Res., 2005, 44, 7255.

H. Cheng, 2003. Thermodynamic Modeling of Surfactant Solutions. Phd Thesis. Technical University of Denmark.

M.V. Flores, E.C. Voutsas, N. Spiliotis, G.M. Eccleston, G. Bell, D.P. Tassios, J. Colloid and Interface Science, 2001, 240, 277.

J. Goodwin, 2004. Colloids and Interfaces with Surfactants and Polymers. An Introduction. John Wiley & Sons Ltd, Chichester.

I. Hamley, 2000. Introduction to Soft Matter. Polymers, Colloids, Amphiphiles and Liquid Crystals, John Wiley & Sons Ltd, Chichester.

P.C. Hiemenz, and R. Rajagopalan, 1997. Principles of Colloid and Surface Science, 3rd edn, Marcel Dekker.

R.J. Hunter, 2001. Foundations of Colloid Science. 2nd ed., Oxford University Press, Oxford.

J. Israelachvili, 1985. Intermolecular and Surface Forces, Academic Press.

B. Jonsson, B. Lindman, K. Holmberg, B. Kronberg, 2001. Surfactants and Polymers in Aqueous Solution, John Wiley & Sons Ltd, Chichester.

G.M. Kontogeorgis, G.K. Folas, 2010. Thermodynamic Models for Industrial Applications: From Classical and Advanced Mixing Rules to Association Theories, John Wiley & Sons Ltd, Chichester.

M. Mattei, G.M. Kontogeorgis, R. Gani, *Ind.Eng. Chem. Res.*, 2013, 52(34), 12236.

M. Mattei, G.M. Kontogeorgis, R. Gani, *Fluid Phase Equilibria*, 2014, 362, 288.

D. Myers, 1991 & 1999. Surfaces, Interfaces, and Colloids. Principles and Applications, VCH, Weinheim.

R. M. Pashley, M.E. Karaman, 2004, Applied Colloid and Surface Chemistry, John Wiley & Sons Ltd, Chichester.

J.M. Prausnitz, Fluid Phase Equilibria, 1996, 116, 12.

J.M. Prausnitz, R.N. Lichtenthaler, E.G. de Azevedo, 1999. Molecular Thermodynamics of Fluid-Phase Equilibria, 3rd edn, Prentice Hall International.

D. Shaw, 1992. Introduction to Colloid & Surface Chemistry, 4th edn, Butterworth-Heinemann, Oxford.

J.M. Woll, T.A. Hatton, Bioprocess Engineering, 1989, 4, 193.

6

Wetting and Adhesion

6.1 Introduction

Adhesion on solid surfaces can be desirable or unde-sirable depending on the context. Adhesion of dirt on the surfaces is of course not desirable and this is exactly what we combat when we clean surfaces. As we discussed previously, two things are crucial for cleaning: the decrease of work of adhesion between dirt and substrate, which is accomplished due to the surfactants (which decrease the interfacial tensions of substrate–water and dirt–water), and the creation of micelles that "entrap" the dirt after it is removed. Adhesion of marine organisms on the coat-ing of a ship is also a very negative phenomenon, called "biofouling". This is why ship-coatings need to be "antifouling coatings" to avoid biofouling. Bac-terial fouling is, in general, a very serious problem in numerous cases: corrosion of oil platforms, fouling of heat exchangers, tooth decay, contact lens infections, food production plants, etc. On the other hand, for paints, glues, coatings, adhesives, fibre–matrix composites, etc. good adhesion between the fluid and the solid substrate is required.

Thus, we understand that wetting and adhesion phenomena are of immense importance, irrespec-tive of whether we want them, for example, for paints and glues or not, e.g. in cleaning! It is important to understand when they happen, how they happen and how we can change them, if needed, in one or the other direction. We have laid the foundations in Chapters 3 and 4 by presenting the theories for interfacial tension, and the con-cepts of the work of adhesion and contact angle. We have also presented the associated Young equation for the contact angle. We will now com-bine all of these concepts and see how they can be used to provide useful insight in the understanding of solid surfaces and the associated wetting and adhesion phenomena. We will also see that theoret-ical analyses are rarely adequate for obtaining a full picture. Especially, adhesion is a complex phenomenon and we will discuss the role of the

Introduction to Applied Colloid and Surface Chemistry, First Edition. Georgios M. Kontogeorgis and Søren Kiil.
© 2016 John Wiley & Sons, Ltd. Published 2016 by John Wiley & Sons, Ltd.
Companion website: www.wiley.com/go/kontogeorgis/colloid

mechanical properties of the materials and other phenomena in an attempt to arrive at a more complete picture of adhesion.

6.2 Wetting and adhesion via the Zisman plot and theories for interfacial tensions

There are two important approaches for studying wetting phenomena: the Zisman's plot and the associated concept of the critical surface tension and the use of interfacial theories. These two approaches are presented in this section. The approach based on interfacial theories is somewhat more complex but provides more information. When combined with the Young equation for the contact angle (Chapter 4), interfacial theories can be used to calculate not only the solid surface tension but also the work of adhesion and the characterization–profile of the solid surface. This is very useful in several applications, e.g. identifying possible contaminants which may be present, for surface modification and in adhesion studies.

6.2.1 Zisman plot

The Zisman plot (Fox and Zisman, 1950 and 1952) is an empirical technique, widely used by industry, which is based on the observation of Fox and Zisman that a plot of the cosine of contact angle against liquid–gas surface tension is often linear (see Figure 6.1 for an example).

Mathematically, it is represented by the equation:

$$\cos\vartheta = 1 \quad \text{when} \quad \gamma_l = \gamma^{crit} \qquad (6.1)$$

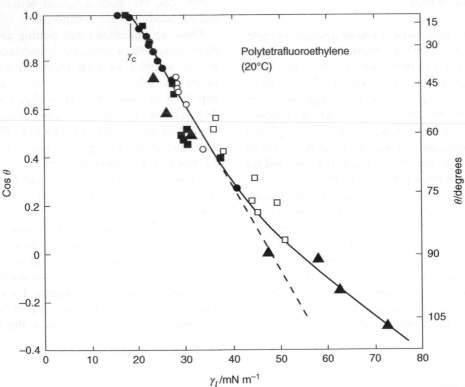

Figure 6.1 *Zisman plot for Teflon (PTFE). Reprinted from Zisman (1964), with permission from American Chemical Society*

Table 6.1 Values of the critical surface tensions of some solids, γ^{crit} (mN m^{-1}) at 20 °C

Solid	γ^{crit}	Solid	γ^{crit}
Polyhexafluoropropylene	16.2	Poly(vinylidene chloride)	40
Polytetrafluoroethylene (PTFE, Teflon)	18.5	Poly(ethylene terephthalate) (PET)	43
Polytrifluoroethylene	22.0	Nylon 6,6	43
Poly(dimethylsiloxane)	24.0	Poly(acrylonitrile)	44
Poly(vinylidene fluoride)	25.0	Cellulose – from wood	36–42
Poly(vinyl fluoride)	28.0	Cellulose – from cotton	42
Butyl rubber	27.0	Wool	45
Polyethylene (PE)	31.0	Urea-formaldehyde resin	61
cis-Polyisoprene	31.0	Polyamide–epichlorohydrin resin	52
cis-Polybutadiene	32.0	Casein	43
Polystyrene (PS)	33.0	Starch	39
Polyvinyl alcohol	37.0	Resorcinol adhesives	51
Poly(methyl methacrylate) (PMMA)	39.0	Aluminium	~500
Poly(vinyl chloride) (PVC)	39.0	Copper	~1000

According to Zisman's plot, complete wetting is achieved if $\gamma_l \leq \gamma^{crit}$.

In connection with the Zisman plot we define the critical surface tension, γ^{crit}. The critical surface tension is the surface tension of the liquid that yields exactly zero contact angle (or $\cos \vartheta = 1$) with the solid, e.g. $\gamma^{crit} = 18$ mN m^{-1} for Teflon in Figure 6.1.

Liquids having surface tensions lower than the critical surface tension will wet the surface. For example, for Teflon shown in Figure 6.1, only liquids that have surface tension lower than 18 mN m^{-1} will completely wet the polymer.

We should emphasize that the critical surface tension is a (semi-)empirical parameter and is *not* the surface tension of the solid, although it is close to this value, as will be seen in Section 6.2.2. Extensive critical surface tension data are available for many polymers and other solids, e.g. as shown in Table 6.1. As expected, the fluorocarbon surfaces have the lowest critical surface tensions, ranging from only 6 mN m^{-1} [-CF$_3$] to 28 mN m^{-1} [-CFH-CH$_2$-]. The hydrocarbon polymeric surfaces have critical surface tensions around 30 mN m^{-1}, while chloro- and nitrated hydrocarbon surfaces have higher critical surface tensions, around 40–45 mN m^{-1}. The exact value depends on the precise polymeric surface considered.

The Zisman plot is very useful in wetting studies. Polymers with low critical surface tensions, e.g. PE and especially Teflon (PTFE), are wetted by very few liquids. These surfaces need, thus, to be modified for improving wetting or surfactants should be used. In general, it is more difficult to wet plastics compared to clean metals and ceramics (materials which, when clean, have surface tensions of the order of 500 mN m^{-1} or more and are called high-energy solids).

There are, however, some complications and peculiarities with the Zisman plot. First, it is recommended to construct the Zisman plot with pure liquids. Mixtures or surfactant solutions may alter the solid surface due to adsorption. Moreover, the choice of test liquids used to construct the Zisman plot is observed to affect the results, and different values of the critical surface tensions may be obtained depending upon whether polar, non-polar or hydrogen-bonding fluids are used. The use of mostly non-polar/polar (but not hydrogen bonding) liquids is suggested. Finally, most Zisman plots have been developed for low-energy surfaces like polymers; it is much more difficult to construct a Zisman plot for high energy surfaces like metals due to adsorption problems (e.g. water or oil).

William Albert Zisman (1905–1986)

Courtesy of ramé-hart instrument co.

William Albert Zisman was an American chemist and geophysicist, arguably one of the greatest figures in surface science, and a worthy successor of "surface science legends" like Langmuir, Rideal and Harkins. He had worked for many years in the US Naval Research Laboratory, and at some point he was the head of the entire Chemistry Division. A Chair of Science was created for him in 1969 to honour his scientific achievements.

He is the author of over 150 publications in surface chemistry and about 40 patents on lubricants and protective treatment of surfaces. He has always tried to apply his research to solving practical problems, e.g. lubricants and protection of weapons and vessels from the very wide range of temperatures, weather and corrosive conditions encountered by the Navy.

In a volume published by the American Chemical Society in 1964, Frederick Fowkes writes about Zisman, who was the recipient of the Kendall Award:

"One interesting example occurred recently when an aircraft carrier nearing completion in the Brooklyn Naval Yard caught fire and in the process of putting out the fire all of the expensive and delicate electronic equipment was damaged by smoke, oil and sea water. Dr. Zisman's group was called in, and thanks to their background of practical applications in surface chemistry, they devised a cleaning solution which displaced the contamination from the electrical apparatus, leaving it in as good condition as new. This one application saved the Navy millions of dollars".

Zisman's research was often guided by the practical problems and needs of the Navy and US Department of Defense but he genuinely believed in the importance to pursue long-term research that can find profound applications in the future. Zisman had indeed a talent for combining fundamental with applied research. He has once replied about inclusion in a manuscript of some speculative material: "There's not much reason to climb a mountain if you aren't going to look around when you reach the top" (Shafrin, 1987).

William Zisman was the recipient of numerous awards for his great scientific achievements. The Zisman plot is still widely used in practical applications related to wetting and adhesion phenomena. An account of his research achievements is summarized by Shafrin (1987).

6.2.2 Combining theories of interfacial tensions with Young equation and work of adhesion for studying wetting and adhesion

The second approach is particularly useful and builds on the theories of interfacial tensions (Chapter 3) and concepts of contact angle/Young equation (Chapter 4) presented previously. When these theories are applied to solid–liquid interfaces and combined with Young's equation for the contact angle, the following results are obtained (ignoring the effect of the spreading pressure):

Young and Fowkes:

$$\cos\vartheta = \frac{2\sqrt{\gamma_s^d \gamma_l^d}}{\gamma_l} - 1 \qquad (6.2)$$

Young and Owens–Wendt:

$$\cos\vartheta = \frac{2\sqrt{\gamma_s^d\gamma_l^d}}{\gamma_l} + \frac{2\sqrt{\gamma_s^{spec}\gamma_l^{spec}}}{\gamma_l} - 1 \qquad (6.3)$$

Young and Girifalco–Good:

$$\cos\vartheta = 2\phi\left(\frac{\gamma_s}{\gamma_l}\right)^{1/2} - 1 \qquad (6.4)$$

Young and Hansen/Beerbower:

$$\cos\vartheta = -1 + 2\left(\sqrt{\gamma_l^d\gamma_s^d} + \sqrt{\gamma_l^p\gamma_s^p} + \sqrt{\gamma_l^h\gamma_s^h}\right)/\gamma_l \qquad (6.5)$$

Young and van Oss–Good:

$$\cos\vartheta = -1 + \left(2\sqrt{\gamma_s^{LW}\gamma_l^{LW}} + 2\sqrt{\gamma_s^+\gamma_l^-} + 2\sqrt{\gamma_s^-\gamma_l^+}\right)/\gamma_l \qquad (6.6)$$

All the above equations (6.2–6.6) can be written in the following general form using the concept of the work of adhesion:

$$\cos\vartheta = -1 + \frac{W_{adh,sl}}{\gamma_l} = -1 + 2\frac{W_{adh,sl}}{W_{coh,l}}$$

The expressions for the work of adhesion from the various theories are given in Table 3.6.

An immediate first application is that we can obtain, using these theories, plots similar (or alternative to) Zisman plot, as shown in Figure 6.2, for the

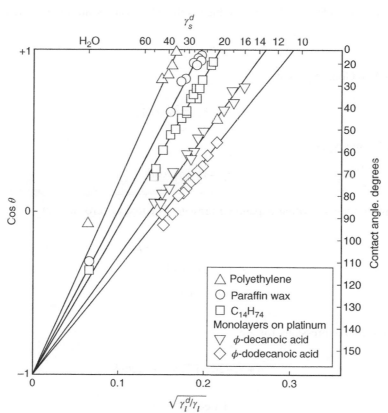

Figure 6.2 Combination of Fowkes with the Young equation applied on several low energy surfaces (see Equation 6.2). Reprinted from Zisman (1964), with permission from American Chemical Society

Fowkes equation. From the slopes and intercepts, we can calculate the surface tensions of the solids. This is also illustrated in Example 6.1 where we see that the connection between critical surface tension and solid surface tension depends on the theory chosen.

Example 6.1.   *Critical and Solid Surface tensions – Are they the same?*
It is often stated that Zisman's critical surface tension γ^{crit} is numerically close to the surface tension of the solid γ_s. Under some simplifying assumptions and more specifically:

1.   When the Girifalco–Good equation of the interfacial tension is used together with the Young equation, show that:

$$\gamma_s = \frac{\gamma^{crit}}{\phi^2}$$

2.   When the Fowkes equation of the interfacial tension is used together with the Young equation, show that:

$$\gamma^{crit} = \gamma_s^d$$

Which assumption is made in both cases? Which is the additional assumption in the case of the Fowkes equation?

Solution
1.   For the Girifalco–Good equation:

$$\gamma_{ls} = \gamma_l + \gamma_s - 2\phi\sqrt{\gamma_l\gamma_s}$$

Using the Young equation:

$$\gamma_{sv} = \gamma_{ls} + \gamma_l \cos\vartheta$$

and assuming that the spreading expanding pressure is equal to zero, we get:

$$\pi_{sv} = \gamma_s - \gamma_{sv} \rightarrow \gamma_{sv} = \gamma_s$$
$$\gamma_s = \gamma_{ls} + \gamma_l \cos\vartheta$$
$$\gamma_{ls} = \gamma_s - \gamma_l \cos\vartheta$$

Equating the two expressions for γ_{ls}:

$$\gamma_l + \gamma_s - 2\phi\sqrt{\gamma_l\gamma_s} = \gamma_s - \gamma_l \cos\vartheta$$

$$\gamma_s = \frac{\gamma_l(1 + \cos\vartheta)^2}{4\phi^2}$$

We have complete wetting:

$$\vartheta = 0 \rightarrow \cos\vartheta = 1$$

$$\gamma_l = \gamma^{crit}$$

$$\gamma_s = \frac{\gamma^{crit}}{\phi^2}$$

2. Using the Fowkes Equation:

$$\gamma_{ls} = \gamma_l + \gamma_s - 2\sqrt{\gamma_l^d \gamma_s^d}$$

Assuming once again that the spreading pressure is zero and by using a similar method:

$$\gamma_s^d = \frac{\gamma_l^2 (1 + \cos\vartheta)^2}{4\gamma_l^d}$$

We have again wetting and assuming that the solvent is a non-polar, non-hydrogen bonding solvent:

$$\gamma_l^p = \gamma_l^h = 0 \rightarrow \gamma_l = \gamma_l^d$$

$$\gamma_s^d = \gamma^{crit}$$

The most important application, however, is to use Equations 6.2–6.6 for surface characterization, as illustrated already in Figure 3.16. This means using experimental data for contact angles and surface tensions from liquid–liquid interfaces to calculate the surface tensions of the solids and their "components", e.g. non-polar and polar parts.

To do so, we must perform contact angle measurements for different liquids on the same solid. The minimum number of required data depends on the theory. In the case of Owens–Wendt we need data for two liquids on the same solid, while the most advanced theories (Hansen/Beerbower and van Oss–Good) require data for at least three liquids on the same solid. Graphical methods can also be used, e.g. the Owens–Wendt equation can be written in the form:

$$\frac{(1 + \cos\theta) \cdot \gamma_l}{2\sqrt{\gamma_l^d}} = \sqrt{\gamma_s^{spec}} \cdot \sqrt{\frac{\gamma_l^{spec}}{\gamma_l^d}} + \sqrt{\gamma_s^d}$$

The slope and the intercept will yield the specific and the dispersion part of the surface tensions of the solid.

The Owens–Wendt theory often performs better than Fowkes equation for polar systems, but problems have been often reported, e.g. for systems like ethanol–water and acetone–water, which are erroneously predicted to be as immiscible as benzene–water. Nonetheless, Owens–Wendt is an old theory, well-established in certain fields, with extensive parameter tables and many successful applications, particularly for polymers. Table 6.2 shows some parameters of the Owens–Wendt equation for liquids and some polymers.

We now illustrate how the surface characterization approach can be used based on the van Oss, Chaudhury and Good (1987) approach (hereafter referred to as van Oss–Good).

Equation 6.6 can be equivalently written for the work of adhesion as:

Table 6.2 *Owens–Wendt theory parameters for certain liquids and some polymers. In mN m^{-1}*

Liquid	γ	γ^d	γ^{spec}
n-Hexane	18.4	18.4	0
Dimethyl siloxane	19.0	16.9	2.1
Aniline	42.9	24.2	18.7
Trichlorophenyl	45.3	44	1.3
Glycol	48	33.8	14.2
Formamide	58.2	39.5	19
Glycerol	63.4	37	26
Water	72.8	21.8	51

Polymer	γ	γ^d	γ^{spec}
Polyethylene	35.7	35.7	0
Polystyrene	40.7	34.5	6.1
Polytetrafluoroethylene	20.0	18.4	1.6
Poly(methyl acrylate)	41.0	29.7	10.3
Poly(hexyl methacrylate)	30.0	27.0	3.0
Poly(ethylene terephthalate)	44.6	35.6	9.0
Polyamide 66	46.5	32.5	14
Poly(dimethyl siloxane)	19.8	19.0	0.8

$$W_{adh} = \gamma_l(1 + \cos\vartheta)$$

$$= 2\sqrt{\gamma_s^{LW}\gamma_l^{LW}} + 2\sqrt{\gamma_s^+\gamma_l^-} + 2\sqrt{\gamma_s^-\gamma_l^+} \quad (6.7)$$

The purpose is to estimate the solid surface tension and its components from experimental contact angle data and data from liquids.

First, we can estimate the LW part of the solid from Equation 6.7 using a non-polar/non-hydrogen bonding liquid (i.e. $\gamma_l = \gamma_l^{LW}$):

$$\gamma_s^{LW} = \frac{\gamma_l(1 + \cos\vartheta)^2}{4} \quad (6.8)$$

In a typical case, the contact angle data and liquid surface tensions of three liquids on the same solid are used to estimate the solid surface tensions and its components (London–van der Waals or LW and acid–base or AB parameters). Two of the liquids must be polar and one non-polar (or actually apolar, i.e. with zero Lewis AB component). The LW part of the solid will be estimated directly from the properties of the non-polar (apolar) liquid (surface tension and contact angle), Equation 6.8.

Water is typically used as one of the three fluids used and is considered to have equal LA-LB (acid–base) components (=25.5 mN m^{-1}) and a LW (van der Waals) value of 21.8 mN m^{-1} (recall the dispersion value obtained from Fowkes equation, Chapter 3, Example 3.3). Water is thus often considered as a "reference fluid" in this type of calculation and all other AB values are "relative to those of water". Other "reference" values for the van Oss–Good theory are discussed in Chapters 3 and 15.

There are many ways of estimating the solid surface tension components and sometimes different triplets of liquids can give somewhat different values. There is much research on the subject and graphical methods have been also proposed for obtaining the van Oss–Good surface parameters for solids. As can be seen from the values in Table 3.4, the base (−) values obtained for most solids with the van Oss–Good theory are typically rather high – much higher than the acid (+) values – and this observation should be considered when evaluating the results. Of course these very high LB values are also a limitation of the method. More "balanced" values are obtained using the Della Volpe and Siboni parameters (see Table 3.5).

6.2.2.1 Discussion

An alternative approach to estimate the van Oss–Good parameters and at the same time an exciting application of the van Oss–Good theory has been presented by McCafferty from the Naval Research Laboratory in the USA (2002). He estimated van Oss–Good parameters for two polymer surfaces (PMMA and PVC) using both direct and graphical methods as well as both using the original "test" values and the revised values by Della Volpe and Siboni (1997, 2000). The results are summarized in Table 6.3.

The graphical method is based on a mathematical manipulation of Equation 6.7, which gives:

$$\frac{\left[\dfrac{(1+\cos\vartheta)\gamma_l}{2} - \sqrt{\gamma_s^{LW}\gamma_l^{LW}}\right]}{\sqrt{\gamma_l^+}} = \sqrt{\gamma_s^+}\sqrt{\frac{\gamma_l^-}{\gamma_l^+}} + \sqrt{\gamma_s^-}$$

Table 6.3 *Summary of the results for the van Oss–Good theory's acidic (+) and basic (−) parameters of the two polymeric surfaces, as calculated by various methods*

			Poly(vinyl chloride), PVC		Poly(methyl methacrylate), PMMA	
			γ_s^+	γ_s^-	γ_s^+	γ_s^-
van Oss–Good						
MI	W	F	0.33	4.8	0.0040	16.1
MI	W	EG	0.15	4.1	0.13	19.3
MI	W	G	0.78	6.3	0.11	19.0
Average			0.42	5.1	0.080	18.1
Linear plot			0.24	3.5	0.071	19.2
Della Volpe (1997)						
MI	W	F	3.6	1.0	0.36	5.2
MI	W	EG	0.18	1.4	0.024	6.7
MI	F	G	0.095	0.56	0.78	0.43
MI	EG	G	0.016	1.13	0.0066	1.8
Average			0.97	0.77	0.29	3.5
Linear plot			0.0025	0.101	0.16	2.7
Della Volpe (2000)						
MI	W	FG	0.26	1.0	0.074	6.8
MI	F	EG	0.25	0.95	0.083	7.5
MI	EG	G	0.11	0.10	0.0027	1.0
Average			0.21	0.68	0.053	5.1
Linear plot			0.19	0.29	0.038	4.4

MI = methylene iodide, W = water, F = formamide, EG = ethylene glycol, G = glycerol.
Adapted from McCafferty (2002).

The following conclusions are extracted:

- Different results, though similar trends and qualitative agreement among the various approaches are observed.
- There is relatively good agreement between the graphical approach and the "analytical" classical method for estimating the van Oss–Good parameters of the solids.
- Neither of the solid surfaces display high acidic (+) values, but the acidic polymer (PVC) has higher acidic values than the basic polymer surface (PMMA).
- PMMA is a very basic surface (large basic parameter), while for PVC the two contributions are more or less balanced. However, considering that the methods tend to give rather large basic (−) components, PVC can be considered much more acidic surface than PMMA. Basic surfaces

clearly have higher basic (−) values than acidic surfaces.

As evident from the above, the van Oss–Good theory parameters for solids depend greatly on the liquids chosen and the exact calculation method as well as the values of the experimental contact angle data. Nevertheless, trends and average values are presented for several solids in the literature. Table 6.4 presents values for some solids together with the critical surface tension, when available. There is good correspondence between the solid surface tension and the critical surface tension in some cases, but there are differences for some solids, as might have been anticipated.

All interfacial theories such as Fowkes, Owens–Wendt, Hansen and van Oss–Good can in principle be applied to all types of various interfaces (liquid–liquid, solid–liquid and even solid–solid ones).

Table 6.4 *Parameters of the van Oss–Good theory for some polymers (in mN m^{-1}). Notice the high values of the base contribution to the surface tension in all cases, exactly as was the case for the liquids based on which these surface tensions are estimated (Table 3.4). Where available, the Zisman critical surface tensions are also shown (also expressed in mN m^{-1})*

Solid	γ	γ^{LW}	γ^{AB}	γ^+	γ^-	γ^{crit}
Polypropylene	25.7	25.7	0	0	0	
Polypropylene – corona treated	33	33	0	0	11.1	
Polyethylene	33	33	0	0	0	31
Polystyrene	42	42	0	0	1.1	33
Nylon	42.8	38.6	4.2	0.4	11.3	43
Cellulose	49.2	44	5.2	0.28	24.3	42
Amylose	52.6	42.3	10.3	0.64	41.4	

However, all theories have advantages and shortcomings (see, for example, Balkenende *et al.*, 1998), but based on numerous investigations the van Oss–Good is possibly the best tool to-date, even though many companies have working experience with other theories as well, especially the Owens–Wendt (particularly for applications related to polymers).

The van Oss–Good method has been criticized because:

- the convention of equal LA-LB parameters for water seems arbitrary;
- almost always too large LB values are obtained;
- different "triplets" of liquids yield often different LW/AB parameters for solids;
- it cannot always describe phenomena associated with strong chemical interactions.

Nevertheless, the theory has found widespread use in describing interfacial phenomena (interactions) involving polymers, paints, proteins and other complex systems (polymer surface characterization, CMC determination of surfactants, protein adsorption, cell adhesion, enzyme–substrate interactions).

Much more about the theories for estimating the interfacial tension is presented in Chapter 15.

Finally, an alternative approach to assess wetting is via the so-called wettability envelopes discussed, for example, by Reynolds (2010) and Hansen (1970, 2000). These envelopes can be created and expressed in different ways. Moreover, they can be used in the context of several of the aforementioned theories of interfacial tension. For example, Reynolds (2010) used the Owens–Wendt model and showed that a wettability envelope can be constructed by plotting the dispersive against the polar contribution to the surface tension for various liquids applied on the same substrate. Liquids that will (partially) wet the surface (contact angle less than 90°) are inside the envelope and those that do not are outside, whereas the boundary indicates exactly contact angle equal to 90°, i.e. the border of wetting and dewetting. A similar approach has been given by Hansen (1970, 2000) but he presented wetting envelopes where the polar solubility parameter is plotted against hydrogen bonding solubility parameter. A partial circle created indicates again the area of wetting. Alternatively, the polar and hydrogen bonding contributions to the surface tension, obtained from Hansen–Beerbower equation, could be used, as they are directly related to the solubility parameters. Hansen illustrated the success of this method for a wide range of paint surfaces and other substrates.

6.2.3 Applications of wetting and solid characterization

There are many applications related to contact angles and solid-surfaces:

1. If wetting is not optimum, there are basically two different directions that can be followed. One way is to "improve" the surface (this can be further studied via atomic force microscopy (AFM)

Table 6.5 *Polymer surface tensions estimated with various methods (in mN m^{-1})*

Polymer	From melt data	From contact angle data	From critical surface tension
Polyethylene	35.7	33.2	31.0
Polystyrene	40.7	42.0	33–36
PET	44.6	43.0	40–43
Nylon 6,6	46.5	39.3	42–46
PDMS	19.8	—	24.0
PVAC	36.5	—	37.0
Teflon	22.6	19.0	18.5

for controlling the roughness) in order to increase its (critical) surface tension and an alternative approach is to reduce the surface tension of the liquid via addition of a surfactant.

Of course, in many cases, e.g. umbrellas, raincoats or novel non-wetting or self-cleaning materials, we want to have hydrophobic or even super-hydrophobic surfaces. The contact angles should have the maximum possible values in these cases. Alternatively we wish to have as low critical surface tensions as possible. There are many application examples of such novel surfaces, by TCNANO and others. Nature can provide inspiration for making such superhydrophobic materials, as illustrated by Scientific American (March and September 2003) about waterproof coats.

Sometimes we want the exact opposite of hydrophobic surfaces, we would like to have hydrophilic surfaces, and so it is of interest to know which functionalities afford such surfaces. While there is no phenomenon rigorously known as the "hydrophilic effect" (Israelachvilli, 1985), the hydrophilic (water loving) groups are those that prefer to be in contact with water rather than with each other and they are often hygroscopic (i.e. take up water from vapour). Strongly hydrated ions (carboxylate, sulfonate, phosphate, trimethyl ammonium) are hydrophilic, as are certain molecules and groups, e.g. small alcohols, glycerol, sugar, poly(ethylene oxide), amine, amine oxide, sulfoxide and phosphine oxide groups.

These hydrophilic surfaces are wetted by water and repel each other in aqueous solutions.

Certain polar groups (alcohol, ether, amine, amide, ketone, aldehyde, mercaptan) are not hydrophilic when they are attached to a long hydrophobic chain. If the hydrophobic group is very long, it can completely neutralize a small hydrophilic group. Thus, it seems that hydrophilic and hydrophobic interactions are not, unlike the dispersion interactions, additive.

2. The various approaches mentioned previously in this chapter can be used to estimate the surface tensions, e.g. of various polymers. Thus, polymer surface tensions have been reported using interfacial theories and contact angle data, from the critical surface tension (Zisman plot) and from extrapolating melt data to room temperature. There is relatively good agreement among the various methods for many polymers, within 4–5 mN m^{-1}. Some typical results are shown in Table 6.5.

3. The Young–Dupre equation is an expression for the "ideal thermodynamic" work of adhesion. This work can be related to but is typically numerically very different from the "true" or "practical" work of adhesion (Sections 6.3 and 6.4).

All interfacial theories can be used to estimate the theoretical work of adhesion and, in combination with the Young equation for contact angle, also the solid components and solid surface tensions, which cannot be directly measured. Thus, the theories can be used to understand wetting and adhesion phenomena and for characterizing solid surfaces.

Characterization of solid surfaces is very important and we may get useful information about the surface, compare "similar surfaces", e.g. various types of steel or epoxies, get ideas about the modification of the surface required for tailor-made applications, and can even help determining the nature of possible surface contaminants (upon comparing the surface tension we obtain with that of an "ideal, clean" surface of the same type as the one studied). The characterization of solid surfaces is, moreover, often the first step in understanding and improving adhesion, which is discussed in Sections 6.3 and 6.4.

Example 6.2 shows how wetting can be analysed and how solid surface tensions and components can be estimated for Teflon using the Hansen theory.

Example 6.2.　*Characterization and wetting of Teflon surfaces.*

Measurements of contact angles for several liquids on a poly(tetrafluoroethylene) surface (Teflon) at 20 °C gave the following results:

Liquid	Contact angle (°)	Surface tension (mN m^{-1})
Water	107	72.8
Propylene carbonate	76	40.7
Hexachloro butadiene	59	35.3
Di(2-ethyl hexyl) adipate	63	30.6
Dibutyl ether	35	22.8
n-Octane	33	21.8

1.　Estimate the critical surface tension of Teflon using the Zisman plot.
2.　Based on the answer to the previous question, is Teflon an easy material to wet? How can we improve the wetting of Teflon surfaces?
3.　Using the Hansen equation for the surface tension (Equation 3.14), estimate the dispersion, the polar and the hydrogen bonding contributions of the surface tension of propylene carbonate ($\gamma_d, \gamma_p, \gamma_h$).
4.　Estimate the dispersion, polar and hydrogen bonding surface tension contributions ($\gamma_d, \gamma_p, \gamma_h$) of Teflon. You can assume that the polar and hydrogen bonding (surface tension) contributions of Teflon (γ_p, γ_h) are equal.

　　Compare the overall value of the surface tension of Teflon to the critical surface tension, as obtained from Zisman plot (answer to question 1). What do you observe?

　　Note: If one of the surface tension contributions (dispersion, polar, hydrogen bonding) is zero, then you can ignore, in Equation 3.14, the square root term where this contribution is present.

Solution

1.　We construct, based on the given data, a plot of the cosine of the contact angle as a function of the surface tension of the liquid (Zisman plot, e.g. Figure 6.1). We notice that the available information covers almost the complete range of contact angles–surface tensions, so only slight extrapolation is required. The critical surface tension is obtained by simply reading the value of the surface tension (on the *x*-axis) for which the cosine of the contact angle is 1 (i.e. when the contact angle is 0). Then, we read:

$$\gamma^{crit} = 18 \text{ mN m}^{-1} \text{(approximately)}$$

2. The critical surface tension of Teflon is extremely low. In fact Teflon is one of the materials having among the lowest critical surface tensions (Table 6.1). Thus, Teflon is very difficult to wet by ordinary liquids, most of which have surface tensions above 20 mN m^{-1}. The wetting can be improved either by using surfactants that can lower substantially the surface tension of liquids or by special surface treatment which can increase the surface tension of Teflon, e.g. by adding polar groups, plasma treatment, etc.

3. The (total) surface tension of propylene carbonate is (see table in the problem) 40.7 mN m^{-1}. According to the Hansen equation for the surface tension, we have:

$$\gamma = \gamma_d + \gamma_p + \gamma_h = 0.0715 V^{1/3}\delta_d^2 + 0.0715 l V^{1/3}\delta_p^2 + 0.0715 l V^{1/3}\delta_h^2 = 40.7 \qquad (6.2.1)$$

or:

$$\gamma = \gamma_d + \gamma_p + \gamma_h = 0.0715 \times 85^{1/3}9^2 + 0.0715 l \times 85^{1/3}9.6^2 + 0.0715 l \times 85^{1/3}2^2 = 40.7$$

The values of V and (Hansen) solubility parameters are obtained from the tables in Appendix 3.1. From the above equation, we calculate the l-value, which best fits the data. This is $l = 0.504$. Using this value and Equation 6.2.1, we can estimate the three surface tension contributions of propylene carbonate:

$$\gamma_d = 25.46$$
$$\gamma_p = 14.58$$
$$\gamma_h = 0.63$$

All values are in mN m^{-1}.

We note that the major contribution comes from the dispersion forces. The polar forces also contribute substantially, while the hydrogen bonding contribution value has the smallest value.

4. Using the Hansen equation for the interfacial tension, we have the following equation for the contact angle as a function of the surface tension contributions:

$$\gamma_l(1 + \cos\vartheta) = 2\left(\sqrt{\gamma_l^d\gamma_s^d} + \sqrt{\gamma_l^p\gamma_s^p} + \sqrt{\gamma_l^h\gamma_s^h}\right) \Rightarrow$$
$$\cos\vartheta = -1 + 2\left(\sqrt{\gamma_l^d\gamma_s^d} + \sqrt{\gamma_l^p\gamma_s^p} + \sqrt{\gamma_l^h\gamma_s^h}\right)/\gamma_l \qquad (6.2.2)$$

The surface tension contributions of Teflon will be estimated based on Equation 6.2.2, applied twice, for the octane/Teflon and propylene carbonate/Teflon systems. From the octane/Teflon system, we can estimate the dispersion surface tension contribution of Teflon, since the other two terms apparently vanish (no polar and hydrogen bonding contributions of octane). Thus, for octane/Teflon we have:

$$\cos(33) = 0.8386 = -1 + 2\left(\sqrt{21.8\gamma_s^d} + 0\right)/21.8 \Rightarrow$$
$$\gamma_s^d = 18.42\,\text{mN\,m}^{-1}$$

Using this value and again Equation 6.2.2, this time for propylene carbonate/Teflon, we can estimate the polar and hydrogen bonding surface tension contributions of Teflon, assuming that they are equal to each other:

$$\cos(76) = 0.242 = -1 + 2\left(\sqrt{25.46 \times 18.42} + \sqrt{14.58\gamma_s^p} + \sqrt{0.63\gamma_s^h}\right)/40.7 \Rightarrow$$

$$\gamma_s^p = \gamma_s^h = 0.61\,\mathrm{mN\,m^{-1}}$$

We observe that the total surface tension of the (solid) Teflon as calculated by adding the three surface tension contribution is 19.64 mN m^{-1}, which is rather close to the value of the critical surface tension (as could be expected – the critical surface tension is a measure of the solid surface tension). Moreover, we observe that the polar and hydrogen bonding surface tension contributions are both extremely small compared to the dispersion contribution, which is by far the dominant one for Teflon.

While Teflon is a polymer that is particularly difficult to wet with a low critical surface tension, as we see in this problem, other fluoro-materials such as perfluoro-acid surfaces are equally or even harder to wet as they have very low values of the critical surface tension. This can be seen in Figure 6.3. Thus these materials can form the basis of super-hydrophobic and potential self-cleaning surfaces, graphically illustrated in Figure 6.4.

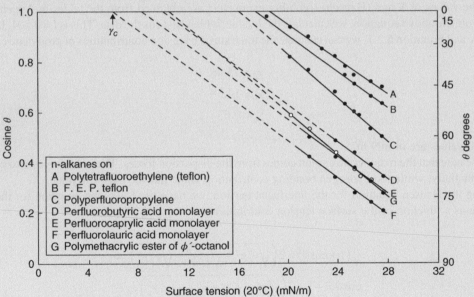

Figure 6.3 *Zisman plots for different fluoro-materials. Some of them have critical surface tensions much lower than that of Teflon. Reprinted from Zisman, (1964), with permission from American Chemical Society*

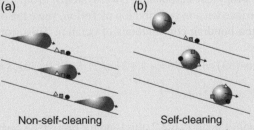

Figure 6.4 *Schematic image of non-self-cleaning (a) and self-cleaning (b) surfaces. Holmgren and Persson (2008) with data from Li et al. (2007). Reproduced with permission from J. M. Holmgren and J. M. Persson*

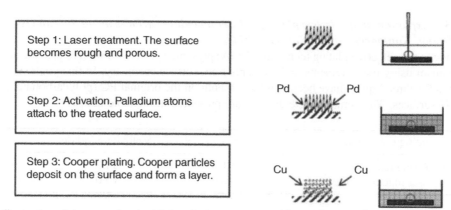

Figure 6.5   *Illustration of the steps for the LISA technique. Reprinted from Zhang* et al. *(2011), with permission from Taylor & Francis*

In the next example, we will use the van Oss–Good theory to explain the selective plating of laser machines. This example is based on the experiments and modelling approach presented by Zhang, Kontogeorgis and Hansen (2011). First, the background of the problem and the procedure are briefly presented.

LISA (laser induced selective activation) is a new technique for the selective metallization on polymers and is used in 3D moulded interconnect devices (3D-MIDs). This finds applications in, for example, manufacturing of antenna and circuit boards. Figure 6.5 presents the three steps of the technique. Step 1 is the laser treatment on the polymer surface in a medium of water, after which the surface becomes as porous as a sponge. Step 2 is the activation of the laser machined area using an activation solution composed mainly of $PdCl_2$ and $SnCl_2$. The material is rinsed by alcohol to remove dirt and improve the wetting of the laser machined area by water. Finally, in step 3 the metallization of the laser track is accomplished in a commercial autocatalytic electroless copper (or nickel or palladium) bath. Metal layers such as nickel and gold can be deposited on top of copper, if needed.

In LISA, the laser only modifies the substrate surface to make the surface porous, but no extra metal seeds need to be embedded. The activation step provides active sites, and only copper deposits on those sites. The mechanism of the Pd/Sn activation has been studied by many research groups, and several explanations have been reported.

The problem that needs to be explained is why, after rinsing, the activation colloids still exist on the laser machined area, but not on the non-machined area. There could be several explanations (chemical bonds between the Pd and the laser machined surface, etc.) but no convincing answers are available. One hypothesis is that the laser treated surface area has different wetting characteristics than the non-machined area. Or, in other words, from a surface energy point of view, the work of adhesion for the activation solution with the laser machined area is larger than the corresponding work of the non-machined area. It was shown that most laser machined surfaces produced much higher contact angle than the non-machined surface, for both water and activation solution. The conclusion from the contact angle measurement seems to be that the laser treated surface "likes" the activation bath less than the non-treated surface. This provides new insight into the possible mechanism and it has been decided to further study the wetting characteristics via assessment of the interaction between the liquid and solid surface. Some of the measurements and analysis with the van Oss–Good theory as well as the conclusions are shown in Example 6.3.

Example 6.3. Explanation of selective plating of laser machines surfaces using the van Oss–Good theory (based on Zhang, Kontogeorgis and Hansen, 2011).

With reference to the selective plating technique (LISA) presented above, we wish to provide an explanation of the mechanism using measurements and concepts from surface chemistry. In this context, contact angle measurements for three liquids have been conducted, both on the original PC (polycarbonate) and the laser machined PC surfaces. The contact angle data (°) are presented in the table below.

Advancing contact angle (°)	Ethylene glycol	Methylene iodide	Water
On PC original	41.8	34.7	86.0
On Laser machined PC surface	6.0	19.0	145.8

Using the van Oss–Good approach estimate the surface energies and the surface energy components of both the original PC and the laser machined PC surface. Then, discuss the results, especially whether they can provide an explanation for the selective metallization technique on the PC surfaces.

To obtain a picture that is as complete as possible use both the original (Table 3.4) and the more recent (Table 3.5) values for the parameters of the van Oss–Good approach for the three test liquids.

Solution

According to the van Oss–Good approach, the surface tension can be expressed as the sum of contributions from the London/van der Waals (LW) and Lewis acid (+)–Lewis base (–) (AB) interactions:

$$\gamma = \gamma^{LW} + \gamma^{AB} = \gamma^{LW} + 2\sqrt{\gamma^{+}\gamma^{-}} \tag{6.3.1}$$

Thus, the first term includes dispersion, polar and induction effects, while the latter includes "asymmetric" interactions, e.g. hydrogen bonding. Equation (6.3.1) is valid for both solid and liquid interfaces. Three liquids and contact angle data must be used for determining the three parameters of a solid surface, and thus characterising the surface.

The van Oss–Good method parameters for the test liquids are available in Tables 3.4 (original values from van Oss, Chaudhury and Good, 1987) and Table 3.5 (from Della Volpe and Siboni, 1997, 2000) of Chapter 3.

According to the original parameters, the acid and base values of water are the same, while the Della Volpe and Siboni values assign different values to the acid and base parts of water and, thus, they treat the "reference water fluid" in different ways.

The work of adhesion can be, using the van Oss–Good theory, estimated as:

$$W_A = \gamma_l(1 + \cos\vartheta) = 2\sqrt{\gamma_s^{LW}\gamma_l^{LW}} + 2\sqrt{\gamma_s^{+}\gamma_l^{-}} + 2\sqrt{\gamma_s^{-}\gamma_l^{+}} \tag{6.3.2}$$

Thus, using Equation 6.3.2 and experimental contact angle data for three liquids on the same solids, the three surface components of the solid as well as its total surface tension can be estimated.

As mentioned, water is typically one of the three liquids used. The calculations can be simplified if one of the three liquids can be assumed to have only LW contributions since, as based on its contact angle with the solid, the LW part of the solid can be estimated as:

$$\gamma_s^{LW} = \frac{\gamma_l(1 + \cos\vartheta)^2}{4} \tag{6.3.3}$$

In this work, methylene iodide is used for this purpose, as the reported total and LW contributions are almost identical. The other two liquids chosen are water and ethylene glycol, for which we have experimental data. This is a good choice of test liquids as we cover a large possible variation in interactions and surface tension values.

The estimated values of the LW, acid–base contributions and the total value are tabulated below for both surfaces (PC and PC laser) and using all three set of values of the test liquids.

mJ m^{-2}	Della Volpe (1997)	PC laser
	PC	
γ_S^{LW}	42.2	48.1
γ_S^+	0.5	4.7
γ_S^-	0.6	16.6
γ_S^{AB}	1.1	17.7
γ_S^{Total}	43.3	65.8

mJ m^{-2}	Della Volpe (2000)	PC laser
	PC	
γ_S^{LW}	42.2	48.1
γ_S^+	0.2	3.0
γ_S^-	0.3	25.1
γ_S^{AB}	0.6	17.3
γ_S^{Total}	42.8	65.4

mJ m^{-2}	Van Oss–Good	PC laser
	PC	
γ_S^{LW}	42.2	48.1
γ_S^+	0.7	10.4
γ_S^-	0.8	70.1
γ_S^{AB}	1.5	54.0
γ_S^{Total}	43.7	102.1

One of the serious criticisms of the van Oss–Good method is related to the convention of equal acid/base parameters for water (used by the authors), which seems arbitrary. In addition, large base values are obtained and different "triplets" of liquids may yield different surface parameters for solids. For all these reasons, we have carried out our investigation using three different sets of parameters for test liquids, as mentioned previously. We see that, in all cases, the laser machined PC surface has larger polar components than the original surface, while the apolar (LW) parts are rather similar. This means that the laser machined surface has a large surface tension in total, and thus larger adhesion work can be obtained between the laser-machined surface and the activation solution.

It can be calculated that the work of adhesion of the activation solution with laser machined surface is much larger than with the non-treated surface. Thus, the activation solution may tend to stay more on the laser machined surface than water does. Moreover, the laser machined surface has a higher basic surface tension component, while the original surface has similar basic and acid parts. As the activation solution is acidic (pH = 1), this observation provides additional confidence for the favourable adsorption (adhesion) of the activation solution on laser machined areas.

In conclusion, the surface analysis showed that the laser machined surface shows more attraction to the activation solution. This (partially) explains why, after rinsing with water, there are still activation drops on the laser machined area but not on the original polymer surface. This explanation is useful, especially when the chemical bonding mechanism is still not available. Nevertheless, our analysis does have limitations as the Young equation is based on ideal smooth surfaces. When porosity or roughness are included, factors such as interlock forces and maybe even chemical bonding may contribute to the adhesion work between the activation solution and the surface.

One of the most impressive and large-scale "surface chemistry experiments" is the preservation of the VASA ship in Sweden. This is described in the next case study.

The next example illustrates the use of the Zisman plot and of the Hansen equation to characterize a PS surface using surface tension data. Finally, the contact angle of a liquid with the PS surface is estimated.

Case study. Preservation of the VASA Ship

Courtesy of Statens Maritima Museer/Swedish National Maritime Museums.

An exciting application of surface chemistry is related to the preservation of the VASA ship in Sweden. The ship was saved by spraying the wood with a water–PEG (poly(ethylene glycol)) mixture. This is how the Swedish removed the water from the ship, after they took it from the sea in 1961, 333 years after it sunk. When the VASA came to the surface, each kilo of wood contained 1.5 kg of water. A total of 580 tones of water had to be removed from the whole ship, otherwise the VASA would crack apart. PEG has the capability to spread on and displace water from wood (due to its high solubility in water), thus preventing shrinkage and cracking. Preservation started in 1962 and continued for 17 years! The experiment was a success but still the ship is extremely fragile. The VASA ship will, however, be preserved forever!

PEG is typically used in preservation of ships and other similar objects (*Dansk Kemi*, 2003).

Example 6.4.   *Wetting and adhesion in PS surfaces.*

The Zisman plot below (from Zisman, 1964) presents the cosine of contact angle against the surface tension of several liquids on a polystyrene (PS) solid surface (two different series of measurements are shown).

All measurements have been carried out at 20 °C. Some data about PS: repeating unit: $-[CH_2CH(C_6H_5)]-$; density: $1.04 \, g \, cm^{-3}$; solubility parameters (dispersion, polar, hydrogen bonding): (8.9, 2.7, 1.8) $(cal \, cm^{-3})^{1/2}$.

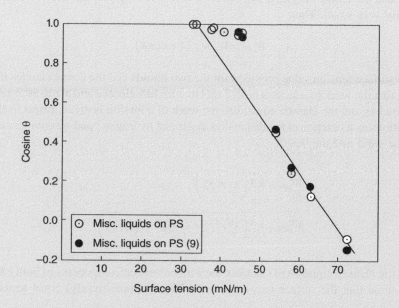

Zisman (1964). Reproduced with permission from American Chemical Society

1. What is, based on this Zisman plot, the critical surface tension of PS?

   Is it easier to wet a PS, a poly(vinyl chloride) (PVC) (40 mN m^{-1}) or a Teflon (PTFE) (18 mN m^{-1}) surface? Which of the three polymers is easiest to wet and which is the most difficult to wet? Explain briefly your answer. The numbers in parenthesis are the critical surface tension values at 20 °C.

2. Which of following liquids will completely wet PS: *n*-hexadecane (27.6), formamide (58.2), acetone (23.3), water (72.8) and 1,1-dibromoethane (39.55)? Explain briefly your answer. The values in parentheses are the surface tensions of the liquids, all in mN m^{-1}.

   Provide an estimation of the theoretical work of adhesion of *n*-hexadecane and water with PS.

3. Show, using the Fowkes and the Hansen methods for the interfacial tension, how the work of adhesion between a liquid and a solid can be estimated.

4. Estimate using the Hansen method for the interfacial tension the contact angle of formamide with PS. Compare your result with the experimental value which is 76°. Use, when needed, experimental data from the previous questions of this problem.

Solution

1. The critical surface tension can be read from the plot. It is approximately equal to 33 mN m^{-1}. PVC is the easiest surface to wet as it has the highest critical surface tension among the three polymers. Teflon, with the lowest value, is the most difficult to wet, while PS is intermediate between the two others.

2. Hexadecane and acetone both have surface tensions lower than the critical surface tension of PS and, thus, it is expected that they are able to fully wet PS. The other liquids will have a finite contact angle with a PS surface.

    The work of adhesion can be calculated from the Young–Dupre equation, as both the surface tension of the liquid and the contact angles are available. For low energy polymer surfaces, the spreading pressure is usually very small. Thus:

    $$W_{adh} = W_{sl} = \gamma_l(1 + \cos\vartheta)$$

    Using the surface tension value provided for the two liquids and the contact angles from the plot, we can estimate that the work of adhesion is 55.2 mN m^{-1} for hexadecane and about 59.7 mN m^{-1} for water.

3. Using the Fowkes and the Hansen equations, the work of adhesion is given, respectively, from the following equations as a function of the dispersion, polar and hydrogen bonding components of the surface tension of the solid and the liquid.

    $$W_{adh,sl} = 2\left(\sqrt{\gamma_l^d \gamma_s^d}\right)$$

    $$W_{adh,sl} = 2\left(\sqrt{\gamma_l^d \gamma_s^d} + \sqrt{\gamma_l^p \gamma_s^p} + \sqrt{\gamma_l^h \gamma_s^h}\right)$$

4. We will use the Hansen equation to calculate the surface tension components of both PS and formamide first. We assume that the surface tension of solid PS is (approximately) equal to its critical surface tension.

    Thus, we have:

    $$\gamma = 0.0715 V^{1/3}\delta_d^2 + 0.0715 l V^{1/3}\delta_p^2 + 0.0715 l V^{1/3}\delta_h^2 = 33$$

or:

$$\gamma = 0.0715 \times (104/1.04)^{1/3} 8.9^2 + 0.0715 l \times (104/1.04)^{1/3} 2.7^2$$
$$+ 0.0715 l \times (104/1.04)^{1.3} 1.8^2 = 33$$

The molar volume of the repeating unit is taken as the molecular weight divided by the density. The solubility parameters are given.

From the above equation, we calculate the l-value, which best fits the data. This is $l = 1.9228$. Thus, the dispersion, polar and hydrogen bonding surface tensions of PS are calculated to be 26.28, 4.65 and 2.067 mN m^{-1}, respectively. The highest contribution comes from the dispersion forces, as could be expected (PS is a hydrocarbon polymer).

Then we find the dispersion, polar and hydrogen bonding surface tension of formamide using the Hansen method:

Hence, we have (using the solubility parameters and volume in Chapter 3 Appendix 3.1):

$$\gamma = 0.0715 V^{1/3} \delta_d^2 + 0.0715 l V^{1/3} \delta_p^2 + 0.0715 l V^{1/3} \delta_h^2 = 58.2$$

or:

$$\gamma = 0.0715 \times 39.8^{1/3} 8.4^2 + 0.0715 l \times 39.8^{1/3} 12.8^2$$
$$+ 0.0715 l \times 39.8^{1/3} 9.3^2 = 58.2$$

From the above equation, we calculate the *l*-value that best fits the data. This is $l = 0.67168$. Hence, the dispersion, polar and hydrogen bonding surface tensions of formamide are calculated to be 17.2038, 26.8317 and 14.643 mN m^{-1}, respectively.

Thus:

$$\cos \vartheta = -1 + 2\left(\sqrt{\gamma_l^d \gamma_s^d} + \sqrt{\gamma_l^p \gamma_s^p} + \sqrt{\gamma_l^h \gamma_s^h} \right) \Big/ \gamma_l$$

Using the surface tension components of PS and formamide we have calculated, we can now estimate the contact angle by simple substitution to this equation:

$$\cos(\theta) = -1 + 2\left(\sqrt{26.28 \times 17.2038} + \sqrt{4.65 \times 26.83} + \sqrt{2.067 \times 14.643} \right) / 58.2 \Rightarrow$$
$$\vartheta = 72.32°$$

which is quite close to the experimental value (5% deviation).

6.3 Adhesion theories

6.3.1 Introduction – adhesion theories

Adhesion can be either wanted (paints, glues, etc.) or not (biofouling, cleaning, etc.). In either case, studies should ensure at first good wetting and also estimation of the theoretical work of adhesion via use of an interfacial theory, e.g. Owens–Wendt or van Oss–Good.

There is little doubt that adhesion is a mystery and understanding adhesion is closely related to the understanding of not just wetting phenomena but numerous more aspects. Today, adhesion control is often done largely empirically and is considered to be a combination of art and science, often learned via training and practice. However, theoretical considerations may be of help in explaining especially the problems that may appear and suggest solutions.

In somewhat more detail, the following steps can be included in adhesion studies:

1. wetting
2. thermodynamic (ideal) work of adhesion
3. theories (mechanisms) of adhesion
4. practical adhesion – measurement
5. adhesion problems
6. surface modification
7. solutions to adhesion problems.

Adhesion forces are, in most cases, short-range ones like van der Waals type and hydrogen bonding. For such forces to apply, the two surfaces, e.g. a liquid and the solid, must be in *very* close (intimate) contact (adhesion forces apply at distances less than 10 nm). Thus, very good wetting is almost always a prerequisite for good adhesion. A lack of good wetting means in most cases a lack of good adhesion. Wetting does not imply that good adhesion will follow, but good wetting comes first! Often a compromise is needed, as can be understood from Equation 4.2, as large liquid surface energies both result in poor wetting and high work of adhesion ($W_{adh} = \gamma_l(1 + \cos\vartheta)$).

Understanding adhesion implies identifying the dominant mechanism (thermodynamic adsorption, specific adsorption, diffusion, mechanical interlocking), but combinations of these theories or mechanisms are also possible. The four major mechanisms of adhesion are illustrated in Figure 6.6.

In brief:

- While specific adhesion, diffusion and mechanical interlocking mechanisms are based on specific requirements for the system, the so-called

thermodynamic adsorption theory covers all adhesion phenomena explained by the secondary van der Waals-type (vdW) and hydrogen bonding (HB) forces (which have bond energies are up to 80 kJ mol^{-1}).

- Thermodynamic adsorption is the most general theory and the only explanation in the absence of, for example, chemical promoters, high mobility of polymeric chain and porosity. It should be emphasized that the van der Waals (dispersion) forces between particles/surfaces, which are involved, are not *as* short range as those between molecules; they decrease with the third power of the distance (not the seventh power). Adhesive forces are discussed next (Section 6.3.2).
- Mechanical interlocking may be present in wood and other porous materials (Figure 6.7).

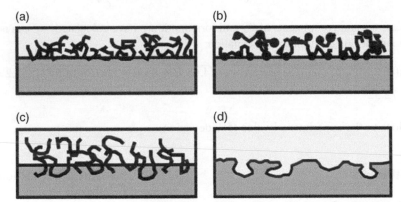

Figure 6.6 *Adhesion theories: (a) physical adhesion at the interface; (b) specific interactions (chemisorption); (c) diffusion mechanism; (d) physical interlocking. Reprinted from Myers (1999), with permission from John Wiley & Sons, Ltd*

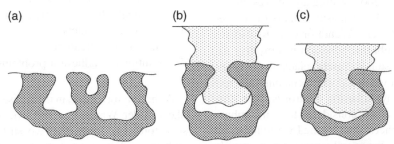

Figure 6.7 *Mechanical adhesion mechanism. A theory under debate. (a) cross-section of porous surface, (b) wet paint on surface, (c) contraction of paint film after drying. Bentley and Turner (1998). Reproduced with permission from Taylor & Francis*

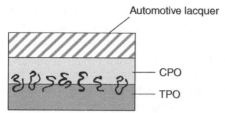

Automotive lacquer

CPO

TPO

Figure 6.8 *Diffusion adhesion mechanism*

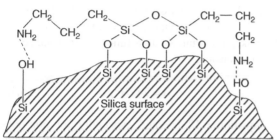

Silica surface

Figure 6.9 *Specific adhesion mechanism. In this example we see the structure and the interfacial bonding of γ-amino-propyltriethoxysilane on a silica surface. Chemical bonds (ionic, covalent, metallic) are formed across the interface via the adhesion promoters that bind to surfaces with these primary bonds. Reprinted from Chiang* et al. *(1980), with permission from Elsevier*

- Diffusion mechanism may be present in polymer–polymer systems, provided high interchain mobility and solubility are present (Figure 6.8).
- Specific adhesion or chemisorption implies that new covalent bonds at interface are created with the help of adhesion promoters (Figure 6.9). These are strong bonds with energies up to 700 kJ mol^{-1}.

Some further information on some of the major adhesion mechanisms are provided next in connection with Figures 6.7–6.9 below.

Mechanical interlocking is the dominant theory for coatings and adhesives on wood and porous materials. In these cases, it is worth trying to increase the surface roughness (or porosity) by chemical or mechanical surface treatment. However, it is under debate whether the roughness helps the adhesion due to the existence of this interlocking adhesion mechanism (Bentley and

Turner, 1998). The contribution of the mechanical interlocking to the total strength of the interface is unknown. It has been said that increase of surface roughness results also in an increase of the contact area and may also result in the improvement of the wetting of the surface by removal of contaminants.

Moreover, occasionally, as shown in Figure 6.7, problems may occur when adhesion in such porous materials is based on this mechanism if not all air of the pores is displaced by the liquid paint.

The diffusion theory explains in some cases the adhesion between polymers. This theory postulates that the adhesion is due to the mutual diffusion of polymer molecules across the interface. This requires that the polymers or their chain segments are sufficiently mobile and that the two substrates are mutually soluble, e.g. they have similar solubility parameters. If the solubility parameters are very different (incompatible polymers), then there is little chain entanglement and, thus, a very poor joint strength.

One example where this mechanism of adhesion is of importance is shown in Figure 6.8 in the case of chlorinated polyolefins (CPO) which are used as primer coatings between thermoplastic polyolefins (TPO) and automotive lacquers.

It is very important to have high molecular weight polymers in order to have high molecular (chain) entanglement for good adhesion. Low molecular weight polymers will often lead to poor adhesion (cohesive failure). The effect of molecular weight on the polymer properties, including the strength, is not linear. The property values increase up to molecular weight of about 10000–100 000 (depending on the polymer and property considered), but above that value a plateau is reached where the properties are constant (independent of molecular weight). Polymers with high molecular weights are (mechanical) stronger, and thus good adhesion is achieved when we are in the "plateau" area where the mechanical strength and the other properties have their highest value, independent of molecular weight.

Specific adhesion or chemisorption occurs when various types of primary bonds (ionic, metallic, covalent) are formed between the two surfaces, e.g. coating and substrate. This is typically achieved with the use of adhesion promoters. Adhesion promoters indeed often bind to surfaces with primary bonds, e.g. the Si–O–Si primary bonds have been detected between silane primers and glass surfaces.

Adhesion promoters are important for coating "difficult to bond" surfaces, for improving performance of the adhesion in varying environments (water, high temperatures, etc.) as well as to protect surfaces that easily absorb water and contaminants from the atmosphere. Organosilanes (Figure 6.9) are among the most well-known adhesion promoters for metals, glass and other inorganic surfaces.

In some cases, strong covalent bonds can be created between coatings and solid substrates leading to strong adhesion and thus use of adhesion promoters is not needed. One example is the adhesion of two-component polyurethane coatings to wood or substrates containing hydroxyl functionalities. There is a strong interaction between the isocyanates and the hydroxyl groups of the substrate.

6.3.2   Adhesive forces

The thermodynamic adsorption theory (dominant in many adhesion practices) is based on the secondary forces (van der Waals and acid–base ones/hydrogen bonding). The importance of these forces is discussed in this section.

Adhesive forces between macroscopic surfaces can be strong even when just van der Waals (dispersion) forces are present (as seen also in Chapter 2). This is because such forces between macroscopic particles or surfaces are much longer-range than between molecules, as illustrated below for some characteristic cases of practical significance:

Force between two-spheres:

$$F = \frac{A \cdot R}{12H^2} = 2\pi R \gamma \qquad (6.9)$$

Adhesive pressure between two plates:

$$P = \frac{A}{6\pi H^3} = \frac{4\gamma}{H} \qquad (6.10)$$

Force between sphere and plate:

$$F = \frac{A \cdot R}{6H^2} = 4\pi R \gamma \qquad (6.11)$$

In Equations 6.9–6.11, H is the distance, R is the sphere radius, γ is the surface tension and A is the so-called Hamaker constant (see Chapter 2 and also later in Chapter 11).

The value for the surface tension in Equations 6.9–6.11 depends on the situation; it is the solid surface tension for a solid in vacuum or the solid–vapour surface tension if the solid is in a vapour environment or the solid–liquid interfacial tension, if the solid is immersed in a liquid.

The last parts of Equations 6.9–6.11 are based on one of the most widely used expression for the Hamaker constant, A, as a function of distance and surface tension:

$$A = 24\pi H^2 \gamma \qquad (6.12)$$

Equation 6.12 is valid especially for non-polar materials, but for polar materials the following expression is sometimes used:

$$A = \left(\frac{4\pi}{1.2}\right) H^2 \gamma^d \qquad (6.13)$$

Equations 6.12 and 6.13 can be used in many ways, e.g. for estimating Hamaker constants from surface tension data and vice versa. Once reliable values of surface tensions and Hamaker constants are available from other techniques or measurements, these equations can be used to estimate the interparticle distance. Such calculations yield for many compounds like hydrocarbons, water and polystyrene values around 0.2–0.3 nm, which are physically reasonable.

These van der Waals forces between colloid particles or surfaces are very strong, as shown in Figure 6.10, where some numerical examples are provided. We can see that the van der Waals forces between macroscopic particles are large and not only when the bodies are in contact!

Calculations using the "ideal adhesion forces or pressures" of Equations 6.9–6.11 suggest that if good wetting can be achieved, then dispersion forces alone should be sufficient to ensure a strong adhesive bond (as these equations are derived from the van der Waals forces). Unfortunately, reality does not seem to agree with this conclusion. Ideal calculations imply thermodynamically reversible separation processes,

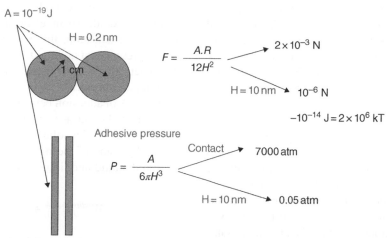

Figure 6.10 *Strength of the van der Waals forces between colloid particles/surfaces. Even for particles as small as 20 nm in radius their potential energy exceeds k_BT even at a distance of 10 nm. For two planar surfaces, the energy at contact is about 66 mJ m^{-2}, which is of the order of magnitude expected for the surface energies of solids. Calculations based on Israelachvilli (1985)*

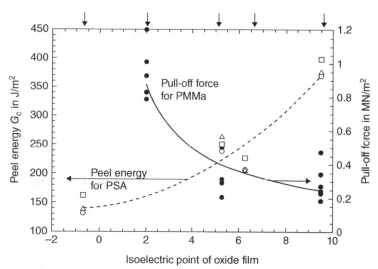

Figure 6.11 *Measured peel energy for a pressure-sensitive adhesive (PSA) and pull-off force for poly(methyl methacrylate) (PMMA) versus the isoelectric point of the oxide film. Reprinted from McCafferty (2002), with permission from Taylor & Francis*

while in reality such conditions are not attained. Fracture processes are invariably accompanied by irreversible viscoelastic processes that dissipate energy and complicate the analysis of the situation. Of course the flaws of real adhesive joints are also a problem. These strong adhesive forces are though only ideal values assuming no defects.

Polar and acid/base interactions are often very helpful in increasing the adhesion as shown in Figure 6.11.

Figure 6.11 shows an impressive result of "practical" adhesion; its interpretation via the van Oss–Good parameters is shown here (from the work of McCafferty, 2002, mentioned in Chapter 3 and Section 6.2.2). The peel energy and thus the adhesion for the acidic pressure-sensitive (commercially available) adhesive is clearly highest for the basic oxide film of aluminium. This means that there is greater affinity for this basic oxide film to donate electrons to the acidic polymer. On the other hand, the very basic PMMA is attracted by the acid oxide Si film. These increased acid–base interactions lead to increased practical adhesion. Although the measured pull-off force for PMMA includes both interfacial and viscoelastic contributions, since the polymer is the same for each metal surface, the viscoelastic contributions may be taken to be the same in all cases. Thus, differences in the measured pull-off forces may be ascribed to differences in interfacial adhesion.

We can conclude that results like the ones shown here provide direct experimental evidence of the importance of Lewis acid–Lewis base effects in practical adhesion of polymers to oxide-covered metals, and possibly several more practical cases.

While all theories for interfacial tension are approximate, there can in some cases provide us with practical information about adhesion and the strength of adhesive points, as illustrated in example 6.5. In this example we see an application where we test whether an adhesive (solid–solid) point is stable towards a liquid. To investigate that, we calculate the work of adhesion; this is the work of adhesion in the presence of a liquid, where the two solids "hypothetically" have been separated into two solid–liquid interfaces.

The work of adhesion between two solids, which is the sum of the surface tensions of the two solids minus the interfacial solid–solid tension, can be calculated by some of the methods we saw previously for liquid–liquid and liquid–solid interfaces:

$$W_{s1s2l}^{adh} = \gamma_{s1l} + \gamma_{s2l} - \gamma_{s1s2}$$

Negative values imply that the two separate solid–liquid interfaces have together lower energy than the original solid–solid interface, and thus separation of the two solid surfaces is favoured. As explained later, polymer adhesion is a very complex phenomenon, depending on both surface phenomena and mechanical properties. Nevertheless it can be, at least qualitatively, understood via the thermodynamic work of adhesion. Example 6.5 presents such an application for the resistance of an adhesive joint in the presence of liquids.

Example 6.5.   *Resistance of an adhesive joint in presence of liquids* (From Owens, 1970).
We have a solid substrate – a flame-treated polypropylene (PP) film coated with a polymer – that is a vinylidene chloride/methyl acrylate co-polymer. The surface of two solids, coating–PP, is immersed in a solution of a surfactant, SDS (this is the well-known sodium dodecyl sulfate). Is there a danger of separation of the two solids due to the surfactant? Use the data available in the table below for the three materials and assume the validity of the Owens–Wendt theory (Chapter 3).

Compound	Surface tension (mN m^{-1})	Dispersion part of the surface tension (mN m^{-1})
Polypropylene (S1)	37.6	33.5
Coating (S2)	53.6	38.9
SDS (L)	37.2	29.0

Solution

Using the Owens–Wendt theory, we have (the specific surface tensions of the two solids and the liquid are simply calculated as the difference between the total and the dispersion parts of the surface tension). All values are in mN m^{-1}:

$$\gamma_{s1s2} = \gamma_{s1} + \gamma_{s2} - 2\sqrt{\gamma_{s1}^d \gamma_{s2}^d} - 2\sqrt{\gamma_{s1}^{spec}\gamma_{s2}^{spec}}$$

$$= 37.6 + 53.6 - 2\sqrt{33.5 \times 38.9} - 2\sqrt{4.1 \times 14.7} = 3.5$$

$$\gamma_{s1l} = \gamma_{s1} + \gamma_l - 2\sqrt{\gamma_{s1}^d \gamma_l^d} - 2\sqrt{\gamma_{s1}^{spec}\gamma_l^{spec}}$$

$$= 37.6 + 37.2 - 2\sqrt{33.5 \times 29} - 2\sqrt{4.1 \times 8.2} = 0.9$$

$$\gamma_{s2l} = \gamma_{s2} + \gamma_l - 2\sqrt{\gamma_{s2}^d \gamma_l^d} - 2\sqrt{\gamma_{s2}^{spec}\gamma_l^{spec}}$$

$$= 53.6 + 37.2 - 2\sqrt{38.9 \times 29} - 2\sqrt{14.7 \times 8.2} = 1.7$$

Thus, the work of adhesion is:

$$W_{adh} = \gamma_{s1l} + \gamma_{s2l} - \gamma_{s1s2} = 0.9 + 1.7 - 3.5 = -0.9 < 0$$

As the work of adhesion in the process of adding the liquid surfactant solution is negative, there is a danger of separation between the copolymer coating and the PP substrate, in the presence of SDS.

6.4   Practical adhesion: forces, work of adhesion, problems and protection

6.4.1   Effect of surface phenomena and mechanical properties

The term "practical adhesion" implies assessment of the adhesion in practice which involves both surface phenomena and bulk mechanical properties.

The "real" work of adhesion, which is usually the product of two terms, the thermodynamic or "ideal" work of adhesion that we have already seen and a (temperature and rate-dependent) term related to the energy lost during the breaking of the adhesive bond. The latter depends on the mechanical properties of the materials.

In destructive adhesion tests, this latter term may be higher than the thermodynamic work of adhesion. However, small changes in the thermodynamic work of adhesion may result in large changes in the total (real) work of adhesion, thus both terms are of great importance. Of special importance is the method developed by Johnson, Kendall and Roberts (JKR) in the 1960s (e.g. Kendall, 1971) that relates the work of adhesion with the load, the compression modulus and the contact radius, a. The compression modulus is a function of the Young (elasticity) modulus and the Poisson's ratio:

$$P = \pi a^2 \sqrt{\frac{2\gamma E}{3t(1-2\nu)}} \qquad (6.14)$$

Clearly, both surface properties and the strength of materials are important for obtaining good adhesion. As the strength of polymers increases with molecular weight, reaching a plateau at high molecular weights, polymers with high molecular weight are preferred as adhesives – glues.

One example illustrating that it is often best to include *both* surface energies and mechanical properties in adhesion studies is shown in Figure 6.12.

(a)

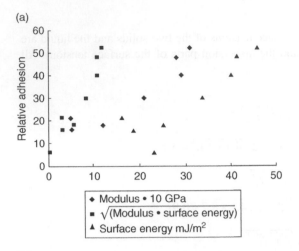

(b)

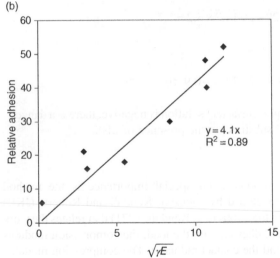

Figure 6.12 *(a) and (b). Relative adhesion of several fouling paints against surface tension, the modulus of elasticity and the square root of the product of these two parameters. Adhesion is best described by a combination of surface and mechanical properties. The polymers with the highest adhesion are Nylon, PS and PMMA and the four lowest are PDMS (poly(dimethylsiloxane)), PHFP, PTFE (Teflon) and PVF. Adapted from Brady and Singer (2000), with permission from Taylor & Francis*

In his study of various coatings for fouling release systems for ships, Brady and Singer (from Naval Research Laboratory, USA), showed that there is a strong correlation between surface energy and

resistance to bioadhesion (both in ship coatings and other environments like dental implants, artificial joints, etc.) (Brady and Singer, 2000). Polymers with low surface energies include Teflon and other heavily fluorinated materials (poly(vinylidene fluoride) (PVF) and polychlorotrifluoroethylene). Adhesion was found to decrease with decreasing surface energy until a critical value of surface energy around 22 mN m^{-1} is reached. Then, adhesion starts to increase. Clearly, surface energy is an important factor but is not the only one. The same figure shows that the coating with the lowest elastic modulus shows the lowest bioadhesion even though this does not correspond to the coating with the lowest surface energy. The conclusion from Figure 6.12 is that adhesion is better correlated with the square root of surface tension and elasticity modulus rather than any of the surface tension and modulus factors alone. Similar results have been obtained from other studies, e.g. Svendsen *et al.* (2007).

6.4.2 Practical adhesion – locus of failure

Clearly, from the above, practical adhesion depends on much more than just surface tensions:

i. bulk mechanical properties (strength) of the materials;
ii. non-ideal adhesive joints due to cracks, trapped gas/liquid.

Determination of the locus of failure is crucial for designing improvements of the adhesion; different possibilities are shown in Figure 6.13. Cohesive failure occurs more often than adhesive failure. Wet adhesion, poor wetting, migration of low molecular weight compounds and shrinkage can also cause adhesive or cohesive failure.

Let us first examine the "proper" or "ideal" adhesive joints, an ideal situation with no flaws (cracks, etc.) and no contaminants (moisture, dirt, oil). Even for ideal adhesive joints, the real work of adhesion is often at least one order of magnitude lower than the theoretical value, or even less for poorly wetted systems.

Not all adhesion problems are the same. The location of the locus of failure is quite important, as can

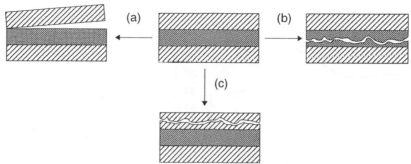

Figure 6.13 *Different possibilities for the locus of failure in adhesion studies. Cases (b) and (c) (cohesive failure in adhesive/coating and in substrate) are more common than case (a) (adhesive failure). Reprinted from Myers (1991), with permission from John Wiley & Sons, Ltd*

be seen in Figure 6.13. We may have adhesive failure, i.e. between the substrate and liquid, e.g. paint, cohesive failure inside the liquid (adhesive/paint-coating) or cohesive failure in the (solid) substrate. For a proper joint, we almost never have failure at the interface (adhesive). Cohesive failure of the weaker of the two solids is much more common than adhesive failure. However, both yield loss of adhesion.

Determination of the locus of failure is very important for obtaining a proper solution to the adhesion problem. It is not always easy to determine the locus of failure; if the failure is at a place more than 100 nm from the interface it is rather easy to find the locus, but it may be more difficult at distances of 10–100 nm.

If the failure is at the interface, then we may need to increase adhesive forces, e.g. via specific interactions, chemical bonds, enhanced entanglement, etc. If the failure is cohesive we may need to increase the strength of the materials. Finally, it is worth mentioning that the locus of failure may also depend on the rate of stress that is applied with fast stresses yielding often cohesive failure and slow ones adhesive failure.

In addition to the factors discussed in relation to Figure 6.13, we should keep in mind that most joints are not "ideal" as they may include cracks, diverse contaminants or other flaws in the adhesive layers, etc. For example, cracks close to the adhesive

layer may yield cohesive failure along the crack. Moreover, cracks propagate easily and a great deal in adsorbing environments, e.g. liquid or vapour (Myers, 1991).

6.4.3 Adhesion problems and some solutions

We can now, irrespective of the locus of failure, summarize some frequent problems associated to adhesion. These are:

- contamination of the substrate;
- low energy surface of substrates (e.g. polyolefins);
- migration of low molecular weight substances (plasticizers, surfactants, etc.) to the interface;
- coating (liquid) layer becomes smaller ("shrinkage") in some cases, e.g. curing of paints;
- adhesion loss due to water or other environmental effects ("wet adhesion").

Depending on the locus of adhesion failure and the problems presented previously, we can try to improve adhesion by suitable modification of the surface. This is a whole science by itself and we only very briefly present here the most important physical and chemical modifications that may be adopted, including several physical and chemical methods for surface modification.

The physical or mechanical methods involve cleaning of surface via organics, increase of the surface

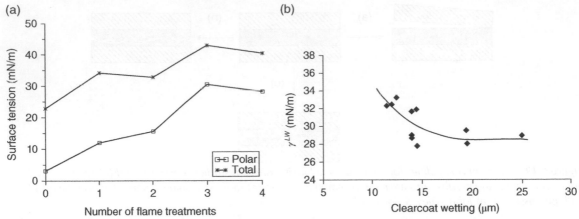

Figure 6.14 *Adhesion and surface modification. (a) Total surface tension and polar component of a PP-EPDM sample at different numbers of flame treatments. (b) LW surface tension values of various dried basecoats as a function of the clearcoat wetting. Reprinted from Osterhold and Armbruster (1998), with permission from Elsevier*

roughness or porosity (also via chemical methods) and adding of surfactants (wetting enhancement, cleaning). The chemical methods include a large variety of methods (corona, plasma, flame treatments, e.g. radiation), electrolytic treatment as well as addition of adhesion promoters. One example of the importance of surface treatment is shown in Figure 6.14.

Figure 6.14 illustrates the importance of surface tension as a parameter for the description of wetting and adhesion phenomena in the paint industry. In their work, Osterhold and Armbruster (1998) concluded that the separation of surface tension of a solid into polar and dispersion components permits understanding the type of forces active on the surface. As can be seen in Figure 6.14, many pre-treatment methods, e.g. flame treatment or plasma, increase the polar component of the surface tension by "integrating" polar groups in the surface, e.g. hydroxyl, carbonyl and carboxyl groups. This increases the total surface tension and improves the wettability and adhesion of paints to the surfaces. Figure 6.14 shows that after the third flame treatment procedure, surface tension and polarity no longer change. Over a period of time, the surface tension values remain rather constant. They decrease after three weeks. Figure 6.14b indicates that best wetting is obtained at rather high LW (van der Waals) values of the various basecoats.

Alternatively to changing the surface, we can change the liquid that is applied, e.g. via the use of surfactants, change of formulation, etc. Cleaning of the surface is of enormous importance as even the smallest "oily" dirt can lower dramatically the surface tensions of even "high energy surfaces" like steel, other metals and metal oxides and ceramics and thus create enormous wetting and most possibly also adhesion problems.

Avoiding contamination and water on the surface is crucial. When surfactants are used for cleaning (instead of organic solvents), we should pay attention to the fact that surfactants may adsorb on the metal.

The effect of water on the adhesion between coatings and metal can be very severe. The water molecules can interact with the metal surface. One way to solve this problem is by utilizing a phenomenon called "cooperative adhesion". By having multiple groups in the coating, e.g. both amine and phosphate groups, we can hope that some of them will react with water, while the others will interact with the metal (steel) surface, thus ensuring good adhesion.

There are many solution strategies depending on the problem. In a form of "how-to-do recipes" some guidelines are presented in Table 6.6 for solving specific adhesion problems. The table also illustrates

Table 6.6 *Guidelines for solving specific adhesion problems*

Problem	Suggested solutions
Against substrate contamination	Better cleaning procedure (using organic solvents or surfactants – care with adsorption!)
	Use liquids that dissolve contaminants
Low surface energy of substrate (polyolefins)	Surface oxidation (plasma, corona, flame treatments) to increase surface energy
	Increase temperature – help diffusion
	Use solvents that swell in the polymer substrate
	Add adhesion promoters
Migration of low molecular weight substances to interface	Increase compatibility of substrate–liquid
	Remove the substance by cleaning or via a liquid that dissolves it
	Crosslink surface, diffusion barrier
Loss of adhesion in water	Investigate penetration of water in the liquid
	Introduce adhesion promoters for stronger adhesion forces (organosilanes, organometallics)
	Cooperative adhesion (multiple polar groups in the coating)

Adapted from YKI institute, Stockholm, Sweden.

how/when to apply the various modification techniques mentioned above.

Paints and coatings are big business, as paints are needed for protection and decoration for numerous surfaces (in our homes, in industry, ships, etc.). In paint and coating formulations, processing and application, colloids and interfaces play a crucial role in diverse ways, as explained in the case study below.

Case study: Colloids and Interfaces in Paint Industry

Paints and coatings are complex multicomponent products consisting of the binder (typically one or more polymers), pigments, solvents and additives. Ten or more components are present in many commercial paint formulations. In many cases, we may have more than one paint-layer over a specific substrate. Many paints are colloidal systems themselves, often called "emulsion paints". Paints are often classified based on the way they are dried/cured. During drying/curing, the solvents evaporate and some binder components may chemically react. Environmental parameters like temperature may also have a significant effect at this stage. Colloid and surface chemistry play a very important role in the formulation and application of paints. We can mention several applications:

i.   Wetting and adhesion are both very much needed for good paint application and problems should be carefully understood. The reasons for what roughly is termed "wetting and adhesion problems" are not always clear. They may be due to evaporation of solvents, changes in the miscibility characteristics. These may change both the rheological and adhesion properties of paints. Another reason could be the uptake of water/chemicals by the paint film after drying or curing or the migration of low molecular weight volatile substances, e.g. plasticizers to the surface. The relationship of wetting/adhesion to CPVC (critical pigment volume concentration) can be an important parameter in these studies.

ii. A special type of paints used on ships is called "antifouling coatings" and actually the purpose here is to develop paints where fouling does not adhere on the ships surface. While the paint itself should of course have good

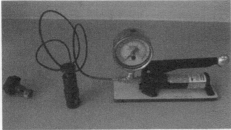

Figure 6.15 *Pull-off test for measuring the adhesion in wood coatings. Reproduced from Jensen (2008, MSc. Thesis), Technical University of Denmark*

adhesion properties. Several studies have been made (see also Section 6.4.1).

iii.   The stability of paints is very important as many are colloidal systems. This is particularly important for water-based paints which are highly unstable paints and therefore stabilized with polymers or surfactants.

In several of the above problems, colloid and surface chemistry concepts should be combined with thermodynamics, e.g. for selecting appropriate solvents and for monitoring the change of miscibility with concentration and temperature during the drying and curing process.

In a recent case study (see Svendsen *et al.*, 2007; and also Problem 6.1), in collaboration with a paint company, the adhesion of six different epoxies–silicon systems has been studied. These paints are used in marine coating systems. Some epoxies showed adhesion problems in practice while others did not. The purpose of the study was to understand the origin of these problems and whether adhesion could be described/correlated to surface characteristics, e.g. surface tensions. An extensive experimental study has been carried out including both surface analysis (contact angle measurements on the six epoxies, surface tension of silicon at various temperatures, atomic force microscopy (AFM) studies of the epoxies), as well as measurements of bulk properties (pull-off adhesion tests and modulus of elasticity). Theoretical analysis included both estimation of Zisman's critical surface tensions and surface characterization using the van Oss–Good theory.

The AFM measurements showed that most epoxies have rather smooth surfaces that are rather easy to wet.

The major conclusion was that surface effects but also mechanical properties play an important role in adhesion. The epoxies that showed the best wetting/adhesion were those with the higher critical surface tension, highest acid parameter and also the largest modulus of elasticity. For the three epoxies for which mechanical properties were studied, adhesion appears to follow the same trend as the one shown in Figure 6.12b.

Another application of interfacial phenomena in the coatings is presented in the studies of Malmos (2008) and Jensen (2008). The projects were in collaboration with a Danish wood paint company. Both investigations are about wood stains (wood paints) which were studied using "direct" adhesion measurements with the pull-off test method (Figure 6.15) and the methods of surface chemistry presented in this chapter (contact angle measurements and analysis of the results using the van Oss–Good method). Commercial products were tested in all cases.

The purpose of the first project (Malmos, 2008) was to compare adhesion characteristics of alkyd and acrylic water-based coatings also against solvent-based alkyd coatings. First, it was found via the pull-off tests that there is worse adhesion in wet wood and that solvent-based coatings showed better adhesion than water-based alkyd stains. The analysis of the results by the van Oss–Good theory was very useful and illustrated that the best adhesion was obtained for the coatings with high surface energy, high Lewis-base contribution and either a large ratio of LW/AB contributions (for alkyd stains) or a small ratio of LW/AB (acrylic stains). Moreover, the van Oss–Good surface tensions were rather close to those predicted by the Neumann theory for all coatings studied. All were in the range 32–34 mN m^{-1}. The Zisman plots gave much lower values (around 20–25 mN m^{-1} in most cases).

In the second study (Jensen, 2008), the focus was on water borne wood stains where, among others matters, the effect of alkyd content was examined. First, it was found that adhesion was, for all coatings, significantly lower in wet wood compared to dry wood. Then, it was found that the pull-off strength decreases with increased alkyd content in the coating and so did the base component of the surface tension. Thus, there was again good correlation between the "experimental" adhesion measurements with the observations from the van Oss–Good theory. The best adhesion was observed (again) with the highest overall surface energy and high base component of the coating (and thus also highest acid–base contribution to the interfacial energy).

In another project, also in collaboration with a paint company, Pardos (2008) studied two types of epoxy coatings (called "mastic" and "phenolic")

which could have potential applications in nuclear power plants. For this reason, many properties (mechanical, thermal, water resistance, corrosion, adhesion, etc.) have been measured at various temperatures. Temperature has a significant effect on many properties and the pull-off pressure in particular decreases almost linearly with temperature in the 20–60 °C region studied. Contact angle measurements have been performed in the 25–45 °C region and the data were analysed with the Owens–Wendt and van Oss–Good theories. The surface tensions of the epoxies are found to be in the region 34–40 mN m^{-1} and they are not very much affected by temperature in this narrow range. Moreover, the combined effect of mechanical and interfacial properties on adhesion has been studied. Figures 6.16 and 6.17 show some typical results and we can conclude that the Kendall theory parameter $\sqrt{(\gamma E/t)}$ (Equation 6.14) follows the same linear trend with the temperature as the experimental pull-off values for both epoxy systems, i.e. it decreases with temperature – a tendency that is independent of the film thickness. As the trend of $\sqrt{(\gamma E/t)}$ with temperature is the same as the Young's modulus, we can conclude that the influence of mechanical properties is as important – if not more – as the surface effects.

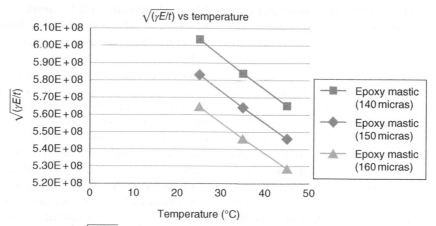

Figure 6.16 *Variation of $\sqrt{(\gamma E/t)}$ with temperature and film thickness for Epoxy Mastic. Reproduced from Pardos (2008, MSc. Thesis), Technical University of Denmark*

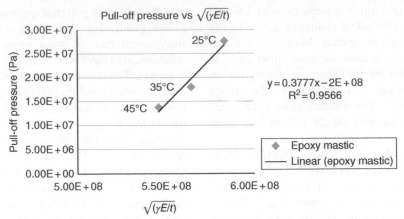

Figure 6.17 *Variation of $\sqrt{(\gamma E/t)}$ with the experimental pull-off values for Epoxy Mastic. Reproduced from Pardos (2008, MSc. Thesis), Technical University of Denmark*

6.5   Concluding remarks

Understanding wetting, lubrication, adhesion and other phenomena requires a full study and understanding of solid–liquid and solid–solid interfaces. Solid surfaces are complex (heterogeneous, contaminated, change with time, rough). Clean metals, metal oxides, glass and ceramics are high energy surfaces, while polymers and fabrics are low energy surfaces.

The key property we can measure of relevance to solid interfaces is the contact angle which, for smooth surfaces, can be related to the surface tensions of the liquid and solid and the solid–liquid interfacial tension via the Young equation. For rough surfaces we need to account for the roughness factor which can be obtained, for example, by atomic force microscopy experiments. For high energy surfaces, the solid–vapour surface tension should be corrected for adsorption phenomena (spreading pressure).

Two approaches–tools for studying wetting phenomena presented are the Zisman's plot and associated concept of the critical surface tension and the use of interfacial theories. The latter approach may be more complex but provides more information especially when combined with the Young equation for the contact angle. The work of adhesion and the characterization/profile of the solid surface can be obtained as well as ideas about what contaminants

may be present, for surface modification and adhesion. Data from liquid–liquid interfaces and contact angles are required here.

Notably, *all* theories for estimating the interfacial tension have been subjected to severe criticism. All theories have limitations and they should be used with care. Despite the problems and uncertainties associated with the "component" theories, the solid surface energies calculated from contact angles with these theories are often of practical importance in studies of wetting phenomena, surface modification and adhesion, e.g. for developing correlations between the "practical" work of adhesion with the "thermodynamic/reversible" work of adhesion, i.e. the work of adhesion calculated from the Young–Dupre equation (or based on interfacial tensions).

What is generally termed as (solid) surface characterization includes (among other things) the determination of the solid surface tension and its components (dispersion and specific/dispersion–polar–hydrogen bonding/van der Waals, acid–base, according to the theory) from experimental information (contact angle and liquid surface tensions) as well as a specific theory for solid–liquid interfacial tensions. Depending on the theory, we need information (contact angle) for one, two or three liquids on the same solid. Most advanced theories employ three liquids, e.g. the van Oss–Good and the Hansen/Beerbower.

Adhesion can be either wanted (paints, glues, etc.) or not (biofouling, cleaning etc.). In either case, studies can include, at first, estimation of the theoretical work of adhesion via use of an interfacial theory, e.g. Owens–Wendt or van Oss–Good. Understanding adhesion implies then identifying the dominant mechanism (thermodynamic adsorption, specific adsorption, diffusion, mechanical interlocking); combinations, though, are also possible.

Adhesive forces between macroscopic surfaces can be strong even when just van der Waals (dispersion) forces are present. This is because such forces between macroscopic particles or surfaces are much longer-range than between molecules. These strong adhesive forces are though ideal values assuming no defects at the surfaces. Thus, polar and acid/base interactions are often very helpful in increasing the adhesion.

Practical adhesion depends on many more things than just surface tensions, especially the bulk mechanical properties (strength) of the materials and the presence of non-ideal adhesive joints due to cracks, trapped gas/liquid.

Determination of the locus of failure is crucial for designing improvement of the adhesion. Cohesive failure is more common than adhesive failure. Wet adhesion, poor wetting, migration and shrinkage can also cause adhesive failure. There are various solutions to adhesion problems (depending on all the above) including several physical and chemical methods for surface modification.

Problems

Problem 6.1: Adhesion between paint layers based on epoxy and silicone (inspired from the work of Svendsen et al., 2007)

A new fouling-release paint by a paint company includes several layers, of which two are based on epoxy and silicone. Various epoxies have been tried because adhesion problems have been observed in certain cases. For a better understanding of the surfaces, contact angles have been measured for three liquids on the various epoxies and the results for three of them are shown in the table below.

Epoxy type	Cos (contact angle) of water	Cos(contact angle) of monoethylene glycol (MEG)	Cos(contact angle) of benzaldehyde
A	0.511	0.846	0.972
B	0.536	0.703	0.943
C	0.442	0.742	0.943

The surface tension of the silicone layer (on top of the epoxy) is 29.5 mN m^{-1}.

It is, moreover, expected that the epoxy with the highest surface tension may yield better adhesion with the silicon layer.

Using the van Oss–Good approach:

1. Estimate the LW, acid/base and the total surface tension of all three epoxies. Surface tension components for the van Oss–Good method for water and MEG are given in Table 3.4, while for benzaldehyde it can be assumed that only LW contribution exists and the surface tension is 38.5 mN m^{-1}. Comment on the values obtained for the individual components of the surface tensions for the three epoxies.

2. Which epoxy surfaces are expected to be better wetted by silicone and for which epoxy–silicon system may be the expected highest adhesion?

Problem 6.2: Characterization of a PVC surface with the Owens–Wendt theory

The contact angles of water and methylene iodide have been measured on a PVC (poly(vinyl chloride)) surface to be equal to 87° and 36°, respectively. The surface tension of water is at 25 °C equal to 72.8 with a dispersion part equal to 21.8. For methylene iodide, the surface tension is 50.8, and the dispersion part is 49.5. All surface tension values are in mN m^{-1}. Assuming the validity of the Owens–Wendt theory, calculate the surface tensions (total, dispersion and specific) of the solid PVC surface and comment briefly on the results.

Problem 6.3: Work of adhesion from the Zisman plot

The Zisman plot is sometimes expressed by the equation:

$$\cos\vartheta = 1 - \beta\left(\gamma_l - \gamma^{crit}\right)$$

where $\cos\vartheta$ is the cosine of the contact angle, γ_l is the liquid–gas surface tension, γ^{crit} is the critical surface tension and β is a positive parameter that depends on the experimental data used. We can further assume that the spreading pressure is very low and can thus be ignored.

1. Show that, under these assumptions, the work of adhesion goes through a maximum when the liquid–gas surface tension is:

$$\gamma_l = \frac{2 + \beta\gamma^{crit}}{2\beta}$$

2. Show that, in this case, the maximum value of the work of adhesion is given by the equation:

$$W_{adh} = \frac{1}{\beta}\left(1 + \frac{\beta\gamma^{crit}}{2}\right)^2$$

Problem 6.4: Characterization and wetting of Nylon surfaces

The following data is available for a specific nylon type [abbreviated as PA-6]:

Repeating unit: $-[(CH_2)_5-CO-NH]-$
Critical surface tension: 43 mN m^{-1}
Density: 1.13 g cm^{-3}
Solubility parameters (dispersion, polar, hydrogen bonding) = (9.5, 6.9, 7.1) (cal cm^{-3})$^{1/2}$

For formamide the surface tension components are available in the same order (dispersion, polar, hydrogen bonding): 16.5, 27.3 and 14.4 mN m^{-1}.

1. Which of the following liquids can wet completely PA-6: water (72.8), acetone (23.3), hexachloro butadiene (35.5), diethylene glycol (44.7)? Explain briefly your answer. The values in parentheses are the surface tensions of the liquids, all in mN m^{-1}.
2. Which of the following three surfaces is easiest to wet: PE (polyethylene), PA-6, Teflon? Write

these three surfaces in an order starting from the one that is easiest to wet. Explain briefly your answer.
3. Using the Hansen equation for the surface tension (Chapter 3), provide an estimation of the dispersion, polar and hydrogen bonding contributions of the surface tension of PA-6 ($\gamma_d, \gamma_p, \gamma_h$).
4. Estimate, based on the above, the contact angle of formamide with PA-6.

Problem 6.5: Characterization and wetting of PET surfaces

Measurements of contact angles for several liquids on a poly(ethylene terephthalate) surface (PET) at the same temperature gave the following results:

Liquid	Contact angle (°)	Surface tension (mN m^{-1})
Water	83.8	71.99
Pyridazine	34.8	49.51
Formamide	51.4	57.03
Benzonitrile	0.0	35.79
Adiponitrile	12.3	45.45
Hydrazine	75.2	66.39
Ethylene glycol	30.1	47.99
Diethylene glycol	9.7	44.77
1,2-Dichloroethane	0.0	39.55

All surface tension values provided below are at the same temperature as the values of the table (close to room temperature).

1. The critical temperature and the critical pressure are physical properties of a fluid, which means that they are characteristic for each fluid and can be measured experimentally. Is the critical surface tension also a characteristic of a solid surface? Justify your answer.
2. Estimate the critical surface tension of PET using the Zisman plot. Is it easier to wet a PET or a polystyrene (33.0 mN m^{-1}) or a Teflon (18.0 mN m^{-1}) surface? Which of the three polymers is easiest to wet and which is the most difficult to wet? Explain briefly your answer. The values in parenthesis are the critical surface tension at 20 °C.

3. Which of the following liquids will completely wet PET: acetone (23.3 mN m^{-1}), glycerol (63.3 mN m^{-1}), hexachloro-butadiene (35.5 mN m^{-1}) and methylene iodide (50.8 mN m^{-1}). Explain briefly your answer. The values in parenthesis are the surface tensions at 20 °C.

4. Can we use a paint with surface tension equal to 55 mN m^{-1} in order to paint a PET surface? Justify your answer.

5. Calculate the work of adhesion of water and formamide with PET.

6. Using the Hansen equation for the surface tension, can you conclude whether n-butyl acetate will wet the PET surface? It is given that the surface tension of ethyl acetate is equal to 23.9 mN m^{-1} and that the same parameter l could be used for all acetates.

Problem 6.6: Contact angle in solid polymeric surfaces

Estimate the contact angle of liquid methylene iodine (with surface tension equal to 50.8 mN m^{-1}) on a solid PMMA (poly(methyl methacrylate)) surface. The molar volumes (in cm^3 mol^{-1}) are 86.5 for PMMA (repeating unit) and 80.6 for the liquid. Compare the result to the experimental contact angle value, which is 41°.

Problem 6.7: Ideal and real cohesive strengths of common materials (adapted from Myers (1991, 1999))

Myers has presented the following real and ideal cohesive strengths of a few materials (at contact, distance $H = 0.4$ nm):

Material	Ideal (MPa)	Real (MPa)
PE (moulded)	180	38
PS (moulded)	210	69
Aramid yarn	7900	2760
Drawn steel	9800	1960
Graphite whisker	100 000	24 000

Myer's ideal values are based on the equation:

$$P = \frac{2.06\gamma}{H}$$

1. Starting from the equation for the potential energy based on van der Waals forces:

$$V = -\frac{A}{12\pi H^2}$$

Show that the equation for the ideal cohesive or adhesive pressure is:

$$P = \frac{A}{6\pi H^3}$$

2. Using the two different relationships between Hamaker constant and surface tension:

$$A = 24\pi H^2 \gamma \quad \text{and} \quad A = \frac{4\pi}{1.2}H^2\gamma^d$$

develop two expressions for the cohesive/adhesive pressure as a function of only the surface tension and the distance.

3. Estimate the "new ideal" values of the cohesive pressure (two different expressions) and compare them to those of Myers and the real values. Comment briefly on the results.

Problem 6.8: A multiple choice on wetting–adhesion

1. Wetting is most difficult for
   a. clean metals
   b. plastics
   c. ceramics
   d. clean glasses

2. An epoxy paint resin has a critical surface tension equal to 47 mN m^{-1} and PE (polyethylene) only 32 mN m^{-1}. Which of the following is correct?
   a. The epoxy will wet the PE surface.
   b. Adhesion is expected to be high if uncured epoxy is poured onto a PE surface and then allowed to cure.
   c. If PE is melted then strong adhesion is obtained if applied to a surface of cured epoxy.
   d. We can never obtain good adhesion between epoxy and PE.

3. The critical surface tension is:
   a. the surface tension at the critical temperature of a compound;
   b. always equal to the dispersion part of the solid surface tension;
   c. equal to the surface tension of the liquid which corresponds to zero contact angle with the solid;
   d. equal to the surface tension of the solid;

4. You want to design a waterproof (hydrophobic) surface. You are making experiments of contact angle (of water) in four surfaces. Which one do you choose?
   a. Surface A where the contact angle of water is 95° and you can decrease roughness.
   b. Surface B where the contact angle of water is 140° and you can decrease roughness.
   c. Surface C where the contact angle of water is 100° and you can increase roughness.
   d. Surface D where the contact angle of water is 140° and you can increase roughness.

5. The Fowkes, Girifalco–Good and Hansen/Beerbower are all theories for estimating the interfacial tension of solid–liquid (and other) interfaces. Which statement is always correct?
   a. All of them have the same accuracy.
   b. None of them is really good.
   c. Girifalco–Good is the best theory because it does not include separation of the surface tension into dispersion, specific, etc. parts.
   d. All are often useful when combined with Young's equation for the contact angle for characterizing a solid surface.

6. The equation:

$$\cos\vartheta = -1 + 2\left(\sqrt{\gamma_l^d \gamma_s^d} + \sqrt{\gamma_l^p \gamma_s^p} + \sqrt{\gamma_l^h \gamma_s^h}\right)/\gamma_l$$

   a. is true if the Young equation and the Hansen/Beerbower equation for the interfacial tensions are true;
   b. is always true;
   c. only true if the Young equation is correct;
   d. is never true.

7. The equation:

$$\cos\vartheta = -1 + 2\left(\sqrt{\gamma_l^d \gamma_s^d} + \sqrt{\gamma_l^{spec} \gamma_s^{spec}}\right)/\gamma_l$$

   a. is typically useful for estimating contact angles;
   b. is never useful;
   c. is useful for estimating the surface tension of the liquid;
   d. is typically used for characterizing a solid from contact angle/liquid surface tension data.

8. The theoretical work of adhesion is *always* expressed, for a solid–liquid interface, as:
   a. $\gamma_l(1 + \cos\vartheta)$
   b. $2\left(\sqrt{\gamma_l^d \gamma_s^d}\right)$
   c. $\gamma_s + \gamma_l - \gamma_{sl}$
   d. $2\left(\sqrt{\gamma_l^d \gamma_s^d} + \sqrt{\gamma_l^p \gamma_s^p} + \sqrt{\gamma_l^h \gamma_s^h}\right)$

9. Only one of the following statements is always true. Which one?
   a. The Young equation is always true using measured values of contact angles.
   b. The expanding (spreading) pressure is almost zero for most metals.
   c. Contact angle hysteresis is usually negligible for real surfaces.
   d. The advancing contact angle is higher than the receding contact angle.

10. The "real" work of adhesion between a solid and a liquid is:
    a. $\gamma_s + \gamma_l - \gamma_{sl}$
    b. $\gamma_l(1 + \cos\vartheta)$
    c. $\gamma_l(1 + \cos\vartheta) + \pi_{sv}$
    d. possibly none of the above.

11. The Owens–Wendt equation for the interfacial tension is:

$$\gamma_{ij} = \gamma_i + \gamma_j - 2\sqrt{\gamma_i^d \gamma_j^d} - 2\sqrt{\gamma_i^{spec} \gamma_j^{spec}}$$

    Then, the work of adhesion for a solid–liquid interface is:
    a. $2\left(\sqrt{\gamma_l^d \gamma_s^d}\right)$
    b. $2\left(\sqrt{\gamma_l^d \gamma_s^d} + \sqrt{\gamma_l^p \gamma_s^p} + \sqrt{\gamma_l^h \gamma_s^h}\right)$

c. $2\left(\sqrt{\gamma_l^{spec}\gamma_s^{spec}}\right)$

d. $2\left(\sqrt{\gamma_l^d\gamma_s^d}+\sqrt{\gamma_l^{spec}\gamma_s^{spec}}\right)$

12. When using the van Oss–Good theory for interfacial tensions, the parameters of the solid surface are typically estimated from:

   a. contact angle measurements for a single liquid on the solid;

   b. contact angle measurements for two liquids on the solid;

   c. contact angle measurements for three liquids on the solid;

   d. contact angle measurements for many liquids on the solid.

Problem 6.9: Work of adhesion

Assuming that you have a solid surface of high energy where adsorption phenomena are important and the spreading pressure should be explicitly considered.

1. Show that the work of adhesion is:

$$W_{adh} = W_{sl} = \gamma_l(1+\cos\vartheta) + \pi_{sv}$$

2. How should Equations 6.2 and 6.4 be modified to account for the spreading pressure?

Problem 6.10: Wetting and adhesion of PS surfaces

The plot shown in Example 6.4 presents the cosine of contact angle against the surface tension of several liquids on a polystyrene (PS) surface.

1. What is the name of this plot and what is the value of critical surface tension of PS?

2. Which of the following liquids will *completely* wet PS: mercury (485), water (72.8), *n*-octanol (27.5), *n*-hexane (18.4), perfluoro-octane (12), acetone (23.3), glycerol (63.3), hexachloro-butadiene (35.5) and methylene iodide (50.8). Justify briefly your answer. The values in parenthesis are the surface tensions (in mN m^{-1}) at 20 °C.

3. The critical surface tension of poly(methyl methacrylate) (PMMA) is 39 mN m^{-1}, of polyethylene (PE) is 31 mN m^{-1}, of poly(vinyl chloride) (PVC) is 40 mN m^{-1} and that of Teflon (PTFE) is 18. Which of the five polymers, PS, PMMA, PE, PVC and PTFE is easiest to wet and which one is the most difficult to wet? Explain briefly your answer.

   Using the Hansen method for the interfacial tension, the dispersion, polar and hydrogen bonding surface tensions of PS are estimated to equal 26.28, 4.65 and 2.067, respectively. All values in mN m^{-1}. The surface tension of formamide is equal to 58.2 mN m^{-1}.

4. Estimate using the Hansen method for the interfacial tension the contact angle of formamide with PS. Compare your result with the experimental value which is 76°.

5. Provide an estimation of the theoretical work of adhesion of formamide, of acetone, and of methylene iodide with PS.

Problem 6.11: Surface analysis of polymeric binders (adapted from Rove (2006))

An industry is interested in developing hydrophilic polymeric binders (high surface tension) to replace certain existing binders which are rather expensive.

   To do so PEG derivatives of carboxylic acids have been synthesized and allowed to react with a PUF binder (=urea modified phenol-formaldehyde). Various materials have been synthesized (1–5) by adding PEG chains in PUF. The various samples have different ratios of phthalic-acetate/PEG and various PEGs (i.e. with different molecular weights). Then, contact angles of three liquids on these materials have been measured and the results are shown in the table below (for the five new products and the existing industrial binders).

   Analyse the results using both the Owens–Wendt and the van Oss–Good theories for interfacial tensions. Compare and discuss the results. Can we conclude that the new binders have similar/higher surface tensions than the existing ones and could thus be considered useful alternatives to the existing products?

Polymer binder	Contact angle (°)		
	Water	Formamide	Diiodomethane
1	63.4 ± 0.7	38.1 ± 0.3	19.0 ± 1.7
2	51.4 ± 1.1	27.1 ± 1.2	18.4 ± 1.1
3	52.3 ± 0.8	26.2 ± 1.4	19.7 ± 2.3
4	57.7 ± 0.8	32.7 ± 2.4	20.2 ± 2.3
5	56.7 ± 0.7	29.4 ± 1.9	18.7 ± 0.4
Industry 1	30.2 ± 3.1	13.3 ± 1.6	21.2 ± 0.6
PUF	67.4 ± 1.9	45.1 ± 0.6	25.1 ± 2.3
PUF + Industry 2	27.0 ± 8.0	14.4 ± 6.3	24.8 ± 1.5

References

A.R. Balkenende, H.J.A.P. van de Boogaard, M. Scholten, N.P. Willard, Langmuir, 1998, 14, 5907.

J. Bentley, G.P.A. Turner, 1998. Introduction to Paint Chemistry and Principles of Paint Technology. Chapman & Hall.

R.F. Brady, I.L. Singer, *Biofouling*, 2000, 15(1-3), 73.

C.H. Chiang, H. Ishida, J.L. Koenig, 1980, J. Colloid Interf. Sci., 74(2), 396.

C. Della Volpe, S. Siboni, J. Colloid Interface Sci., 1997, 195, 121.

C. Della Volpe, S. Siboni, J. Adhesion Sci. Technol., 2000, 14, 235.

H.W. Fox, W.A. Zisman, J. Colloid Sci., 1950, 5, 514.

H.W. Fox, W.A. Zisman, J. Colloid Sci., 1952, 7, 109 & 428.

I. Hamley, 2000. Introduction to Soft Matter. Polymers, Colloids, Amphiphiles and Liquid Crystals, John Wiley & Sons Ltd, Chichester.

C.M. Hansen, 2000. Hansen solubility parameters. A User's Handbook, CRC Press, Boca Raton, FL.

C.M. Hansen, 1970. J. Paint Technol., 42, 550, 660.

J. M. Holmgren and J. M. Persson, 2008. M.Sc. Thesis, University of Lund.

J. Israelachvilli, 1985. Intermolecular and Surface Forces, Academic Press.

Th. Jensen, 2008. Investigation of Adhesion of Water Borne Wood Stains. M.Sc. Thesis. Technical University of Denmark.

K. Kendall, J. Phys. D: Appl. Phys., 1971, 4, 1186.

E. McCafferty, J. Adhesion Sci. Technol., 2002, 16(3), 239.

M.C. Malmos, 2008. Water Borne Wood Stains for Exterior Wood. M.Sc. Thesis. Technical University of Denmark.

D. Myers, 1991 and 1999. Surfaces, Interfaces, and Colloids. Principles and Applications. VCH, Weinheim.

M. Osterhold and K. Armbruster, Progress in Organic Coatings, 1998, 33, 197.

D. K. Owens, Appl. J. Polymer Sci., 1970, 14, 1725.

A.B. Pardos, 2008. Epoxy Coatings: Behavior with Temperature and their use in Nuclear Power Plants. M.Sc. Thesis, DTU.

P. Reynolds, 2010. Wetting of surfaces, in *Colloid Science: Principles, Methods and Applications*. 2nd edn, ed, T. Cosgrove. John Wiley & Sons, Ltd, Chichester, Chapter 10.

Th. Rove, 2006. Synthesis and Characterization of PEG Derivatives of Carboxylic Acids and Reaction in to a PUF Binder System. B.Sc. Thesis. Technical University of Denmark.

E. G. Shafrin, *Langmuir*, 1987, 3, 445–446.

D. Shaw, 1992. Introduction to Colloid & Surface Chemistry. 4th edn, Butterworth-Heinemann, Oxford.

J.R. Svendsen, G.M. Kontogeorgis, S. Kiil, C. Weinell, C., M. Grønlund, M., J. Colloid Interface Sci., 2007, 316, 678.

C.J. Van Oss, M.K. Chaudhury, R.J. Good, Adv. Colloid Interface Sci., 1987, 28, 35.

Y. Zhang, G.M. Kontogeorgis, H.N. Hansen, 2011. J. Adhesion Science & Technology, 25, 2101.

W.A. Zisman, 1964, in *Contact Angle, Wettability and Adhesion*, Advances in Chemistry Series vol. 43, American Chemical Society, Washington D.C., 1.

7

Adsorption in Colloid and Surface Science – A Universal Concept

7.1 Introduction – universality of adsorption – overview

Solids are complex materials, with inherently high energy surfaces. A spontaneous process to decrease this high surface energy (in order words to "saturate" these large surface forces) is via the adsorption of, for example, gases. In particular, high surface areas exist in porous materials, like activated carbon, alumina or, the well-known from chemistry laboratories, silica gel. These materials have "huge" surface areas (per g). This quantity is called "specific surface area" of solids, A_{spec}, and is a characteristic of solid materials. Solids are complex materials and the form, specific surface area, porosity, polarity, surface energy and chemical homogeneity, including the number of adsorption sites, are important parameters of the solid surface.

Adsorption is a universal phenomenon in colloid and surface science: We are talking about the adsorption of high molecular weight amphiphilic compounds for monolayer creation, the adsorption of gases on solids, adsorption of surfactants, polymers or proteins (biomolecules). We have adsorption on solid surfaces and/or from solutions. There are numerous applications of adsorption: stabilization, cleaning, heterogeneous catalysis including catalytic converters in cars, separations like drying, surface modification and change of properties like foaming or rheological ones. In particular for the gas adsorption we can mention additional applications such as measurement of the surface area of powders and other solid surfaces or in heterogeneous catalysis and novel high surface area materials, e.g. zeolites, which can be used for selective separations (like removing pollutants). Many porous materials, e.g. activated carbon, alumina or silica gel and zeolites, have very high specific surface areas, which can sometimes be higher than $100 \text{ m}^2 \text{ g}^{-1}$ and, on occasion, even higher than $1000 \text{ m}^2 \text{ g}^{-1}$.

7.2 Adsorption theories, two-dimensional equations of state and surface tension–concentration trends: a clear relationship

We will discuss adsorption theories in detail in this chapter. Here we offer a short overview. Many

Introduction to Applied Colloid and Surface Chemistry, First Edition. Georgios M. Kontogeorgis and Søren Kiil.
© 2016 John Wiley & Sons, Ltd. Published 2016 by John Wiley & Sons, Ltd.
Companion website: www.wiley.com/go/kontogeorgis/colloid

adsorption phenomena, especially of surfactants, polymers, proteins and the chemical adsorption of gases on solids can be well represented by the Langmuir adsorption isotherm. This equation can be expressed in a suitable linear form and we can obtain from it the two parameters of the Langmuir model (V_m and B in Equation 7.1 and Γ_{max} and k in Equation 7.4) of which one is the concentration or volume at maximum (full) coverage or the so-called monolayer coverage (Γ_{max} in Equation 7.4 and V_m in Equation 7.1). Knowledge of this monolayer coverage and of the specific surface area of the solid (A_{spec}) can help us estimate the surface area occupied by a molecule at the interface and thus the amount needed for stabilization (Equation 7.4). The specific solid surface area can be obtained from gas adsorption measurements on the same solid. The success of the Langmuir equation is unprecedented despite the drastic assumptions in its derivation (equivalent active sites on surface, no lateral interactions, only one molecule per site). In Equations 7.1 and 7.4, notice the similarities of the two forms of Langmuir's equation, for adsorption from solution and the adsorption of gases on solids.

Adsorption isotherms, surface tension–concentration equations and two-dimensional equations of the surface pressure against area are all concepts that are unified or linked via the Gibbs adsorption equation, Equation 4.7. The interrelations are presented in Table 7.1 (k_B is the Boltzmann constant).

The quality of the fit to the data is important but cannot serve alone as a criterion for selecting the best theory. This is because the ability of an equation to fit the shape of an experimental isotherm does not provide a sensitive test for the model on which the equation was based. Comparison between direct (calorimetric) experimental data of the heat of adsorption and the values obtained by the adsorption theory can provide a much more rigorous test for checking the theories.

7.3   Adsorption of gases on solids

We start our discussion with the adsorption of gases on solids, which can be either of physical or chemical nature, and both are hereafter briefly discussed. Table 7.2 summarizes the differences between physical and chemical adsorption. The physical multilayer adsorption is often faster than the chemical adsorption but with a low heat of adsorption, <10 kcal mol^{-1} versus 5–100 kcal mol^{-1} for the chemical one.

Table 7.2   *Differences between physical and chemical adsorption*

Physical	Chemical
Weak van der Waals- type forces	Strong forces, like chemical bonds
Fast reached to equilibrium, reversible	Slow, often irreversible
Multilayer (BET)	Monolayer (Langmuir)
Low heat of adsorption (<10 kcal mol^{-1})	High heat of adsorption (5–100 kcal mol^{-1})

Table 7.1   *Interrelation between surface tension–concentration equations, the adsorption isotherm theories and the two-dimensional equations of state. The surface pressure is the difference between water and solution surface tension* ($\pi = \gamma_0 - \gamma$)

Surface tension–concentration ($\gamma(C)$)	Adsorption isotherm equation (Γ)	Two-dimensional equation of state, $\pi - A$
$\gamma = \gamma_0 - \left(R_{ig} Tk\,\Gamma_{max}\right)C$ $\gamma = \gamma_0 - R_{ig}T\Gamma_{max}\ln(1+kC)$	$\Gamma = \Gamma_{max}kC$ (Henry) $\Gamma = \dfrac{\Gamma_{max}kC}{1+kC}$ (Langmuir) $kC = \dfrac{\Theta}{1-\Theta}\exp\left(\dfrac{\Theta}{1-\Theta} - \dfrac{2a\Theta}{bk_BT}\right)$ $\Theta = \Gamma/\Gamma_{max}$	$\pi A = n_i^\sigma R_{ig}T = k_BT$ (ideal gas film) $\pi = \dfrac{k_BT}{A-b}$ $\pi = \dfrac{k_BT}{A-b} - \dfrac{a}{A^2}$ (van der Waals – 2D)

7.3.1   Adsorption using the Langmuir equation

Langmuir derived his famous adsorption theory based on the assumption that the adsorption rate is proportional to the concentration in the solution and the fraction of free non-covered area of the adsorbent. The desorption rate, on the other hand, is considered to be proportional to the fraction of surface covered by adsorbate molecules. At equilibrium, the adsorption and desorption rates are equal. The Langmuir theory is derived based on many simplifying assumptions:

- "ideal" (clean, smooth, homogeneous, non-porous) surface;
- existence of active centres on the surface; each centre (site) can be free or occupied by an adsorbed molecule;
- only one molecule per active centre;
- no interactions between adsorbed molecules, i.e. homogeneous surface;
- all sites are equivalent (same heat of adsorption);
- no solute–solvent interactions;
- solute and solvent have same size;
- dilute solutions;
- monolayer coverage.

These assumptions could, in theory, very much restrict the applicability of Langmuir's equation. However, surprisingly, the Langmuir theory is of wide applicability in numerous adsorption phenomena, especially chemical adsorption and physical adsorption at low pressures and high temperatures. This is possibly due to cancellation of errors, and the fact that the Langmuir equation can fit a plateau shaped curve, which many monolayers and in general many adsorption phenomena follow. In brief, Langmuir is an excellent correlative model.

In the case of gas–solid adsorption, the mathematical representation of the Langmuir equation is often given as the fraction Θ of occupied active centres (or volume per mass of solid) over the total active centres (or the volume of monolayer) against pressure. The equation and its linear form are shown in Equation 7.1:

$$\Theta = \frac{V}{V_m} = \frac{B.P}{1 + B.P} \Rightarrow \frac{P}{V} = \frac{P}{V_m} + \frac{1}{B.V_m} \qquad (7.1)$$

The volume V is expressed in, for example, $cm^3\ g^{-1}$ (solid). P is the pressure, while V_m and B are adjustable parameters, respectively, the volume of the monolayer and the affinity parameter. The higher the B-parameter is, the higher is the affinity of a gas for the solid.

Note that a plot of P/V versus P gives a straight line. The slope is $1/V_m$ and the intercept is $1/BV_m$. At low pressures, Langmuir approaches the Henry equation ($V = BV_mP$).

Both parameters of the Langmuir's equation, B (affinity) and V_m (monolayer coverage), must be estimated from experimental data, based on the linear plot shown in Equation 7.1.

The Langmuir equation can be derived both thermodynamically and kinetically; the latter derivation is simple and assumes that the rate of adsorption is equal to the rate of desorption. The parameter B has then a meaning based on the number of collisions and the desorption rate being constant, but in practice it should be treated as an adjustable parameter.

Figure 7.1. illustrates a typical Langmuir adsorption at various temperatures (for ammonia on coal). It can be seen that adsorption typically increases with increasing pressure and with decreasing temperature. Although the quality of the fit is very good, it should be emphasized that the ability of an equation to fit the shape of an experimental isotherm does not provide a sensitive test for the model on which the equation was based. Adsorption is an exothermic process (with a negative entropy of adsorption) as the molecules are confined in two dimensions. As mentioned, direct (calorimetric) experimental data of the heat of adsorption are very useful for checking adsorption theories.

Adsorptions not described by Langmuir's theory especially the physical multilayer adsorption of gases on solids can be best described by the BET (Brunauer–Emmett–Teller) theory discussed next. BET can also be written in a linear form, for reduced pressures (P/P_o) up to about 0.4, and yield values of the monolayer coverage and of the specific solid surface area. The BET theory is a very useful tool in describing physical adsorption, although it does

(a)

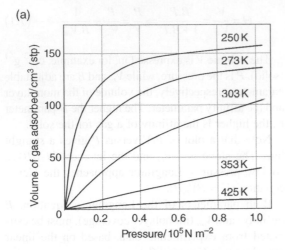

(b)

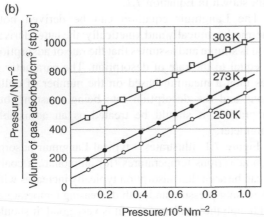

Figure 7.1 *Adsorption isotherms for ammonia on charcoal: (a) as plots of the volume of gas (per mass of solid), V, as a function of pressure (P) at constant temperature (T) and (b) in linearized Langmuir plots. Shaw (1992). Reproduced with permission from Elsevier*

(a)

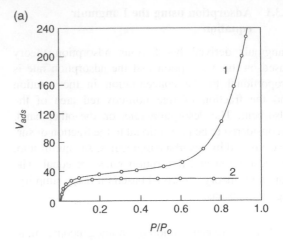

(b)

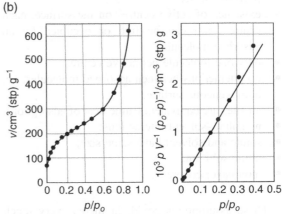

Figure 7.2 *(a) Comparison of two adsorptions: that of nitrogen on silica is a physical BET adsorption at 77 K (curve 1), while oxygen's adsorption on charcoal at 150 K (curve 2) is a chemical Langmuir adsorption. Notice that the y-axis is the pressure divided by the saturation pressure. (b) Adsorption isotherm for nitrogen on silica gel in the linearized form of the BET equation, Equation 7.2. From Shaw (1992)*

have limitations, e.g. the results are not always in excellent agreement with calorimetric measurements. A typical plot is shown in Figure 7.2. The equation and principles in BET are presented in the next section.

7.3.2 Adsorption of gases on solids using the BET equation

Stephen Brunauer (1903–1986), Paul Emmett and Edward Teller (1908–2003) developed a theoretically based adsorption theory known by the initials of their names, BET. Their purpose was to derive an equation resembling Langmuir's adsorption, which, however, can be used for multilayer physical adsorption as well, and leads to Langmuir and Freudlich (another adsorption equation) at specific limiting values. The final expression is given by Equation 7.2. Like the Langmuir equation, it is also a two-parameter adsorption isotherm. The two parameters can be obtained from a specific linear plot (shown in Equation 7.2); V_m is again the volume of monolayer coverage.

$$\frac{P}{V(P_o-P)} = \frac{1}{CV_m} + \frac{C-1}{CV_m}\frac{P}{P_o} \qquad (7.2)$$

V_m is simply calculated as $1/(\text{slope} + \text{intercept})$ of Equation 7.2.

The parameter C of BET is a measure of the difference in the heat of adsorption of the 2nd, 3rd and so on layer compared to the heat of the creation of the first layer. With the exception of the first layer, all other layers are assumed to have the same heat of adsorption. The first layer is characterized by a much higher heat of adsorption compared to the other layers.

The BET equation has found widespread use in adsorption studies. For example, the model has been used for the characterization of porous materials, i.e. getting information about the surface area and the nature of the surface. Using Equation 7.3, the specific surface area of solids can be estimated from the V_m parameter of BET:

$$A_{spec} = \frac{V_m N_A}{V_g} A_o \qquad (7.3)$$

Here V_g in this equation is the gas volume at standard temperature and pressure (s.t.p.), 1 atm and 273 K ($= 22414 \text{ cm}^3 \text{ mol}^{-1}$ or $22.414 \text{ L mol}^{-1}$). A_o is the area occupied by one gas molecule (0.162 nm^2 for the often used N_2 at 77 K, 0.138 nm^2 for Ar and 0.195 nm^2 for Kr). Nitrogen is the standard gas used in BET measurements of the specific surface area of solids. Further support of the BET theory arises when the adsorption of different gases on the same solid yields (using Equation 7.3) the same specific surface area for the solid. Based on an IUPAC Project (1970), BET is recommended as the standard for A_{spec} measurements.

Moreover, as we will see next, when the specific surface area of a solid is known (using the technique

presented here) then we can estimate, from solution adsorption data, the area occupied by a surfactant or a polymer on the same solid, A_o.

The BET theory has also its limitations:

- absence of lateral interactions between adsorbed molecules;
- valid up to relative pressures (P/P_o) of 0.4;
- not always good agreement with calorimetric measurements;
- extension to n-layer for porous materials yield complex equations that cannot be expressed in linear form.

Attempts to improve the BET theory have generally meant the introduction of additional parameters which, in most cases, cannot be evaluated independently. Thus, any improvement in fitting experimental data may be attributed to the additional flexibility conferred by another parameter. Furthermore, most improvements destroy the simplicity of the BET equation and the ease with which the experimental data can be analysed in a linear form to give values for the model's two parameters. For these reasons, BET is typically used in its well-known form, Equation 7.2.

BET-calculated surface areas of solids (using Equation 7.3) show excellent agreement with values measured by other techniques. Moreover, excellent agreement is obtained in most cases also when the specific area is measured based on the adsorption of different gases on the same solid, or measurements from different laboratories. These observations justify the use, and to some extent verify the assumptions, of the BET theory.

Example 7.1 illustrates how the BET equation is used. Example 7.2 shows how calorimetric data can be used in adsorption studies.

Example 7.1. Adsorption based on the BET equation.

The following data have been obtained for the adsorption of nitrogen on TiO_2 at 77 K:

P/P_o	0.02	0.05	0.1	0.2	0.3	0.4	0.6	0.8	0.9
volume of gas ($\text{cm}^3 \text{ g}^{-1}$)	1.2	1.5	1.9	2.4	2.7	3.0	3.8	4.6	6.7

1. Show that the adsorption isotherm is likely to be represented by the BET equation.
2. Estimate using the BET equation:

a. the volume for monolayer coverage V_m (in $cm^3 \ g^{-1}$);

b. the specific surface of solid TiO_2 in $m^2 \ g^{-1}$ if the area occupied by one nitrogen molecule is $16.2 \times 10^{-20} \ m^2$.

Solution

The V versus P/P_o plot clearly shows that the adsorption of nitrogen on titanium oxide has the BET shape.

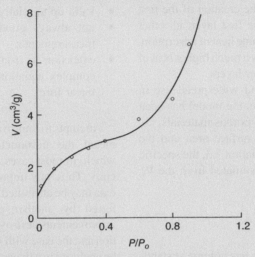

1. According to the BET theory, the plot of $P/V(P_o - P)$ versus P/P_o should be linear. If we consider all the points provided in the above table, i.e. up to very high pressures, the resulting plot is *not* linear. However, the BET equation should be constructed based only on the points up to reduced pressures of about 0.4, i.e. up to sixth or seventh point. Indeed the plot of $P/V(P_o - P)$ versus P/P_o is linear up to reduced pressure 0.4 (approximately even up to the seventh point of reduced pressure 0.6).

    The straight line is shown in the plot below.

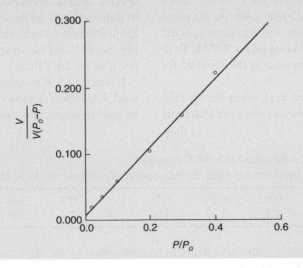

2. The straight line is represented by the equation:

$$P/[V(P_o - P)] = 0.0052999 + 0.526907(P/P_o)$$

(Notice that as only the ratio P/P_o is given, all sides of the fraction $P/[V(P_o - P)]$ should be divided by P_o.)

Then, as known, the volume for monolayer coverage $V_m = 1/(\text{slope} + \text{intercept}) = 1/(0.526907 + 0.0052999) = 1.8789$ cm^3 g^{-1}.

The specific surface area of solid TiO_2 is then calculated based on this value of V_m as:

$$A_{sp} = \frac{V_m N_A}{V} A_o = \frac{1.8789 \times (6.0225 \times 10^{23}) \times 16.2.10^{-20}}{22414} = 8.178 \text{m}^2\text{g}^{-1}$$

Example 7.2. Estimation of the heat (enthalpy) of adsorption from the Clausius–Clapeyron equation (based on data from Shaw, 1992).

The following results refer to the adsorption of nitrogen on a graphitized sample of carbon black, and give the ratio of the nitrogen pressures for temperatures 90 and 77 K which are required to achieve a given amount of adsorption:

Amount of nitrogen adsorbed (V/V_m)	0.4	0.8	1.2
P (90 K)/P (77 K)	14.3	17.4	7.8

Calculate, using the Clausius–Clapeyron equation, the constant enthalpy (heat) of adsorption for each value of V/V_m and comment on the values obtained.

Solution

Using the Clausius–Clapeyron equation:

$$\frac{d\ln P}{dT} = \frac{-\Delta H_{ads}}{R_{ig}T^2}$$

we obtain:

$$\frac{d\ln P}{dT} = \frac{-\Delta H_{ads}}{R_{ig}T^2} \Rightarrow \int_{P_1}^{P_2} d\ln P = \frac{-\Delta H_{ads}}{R_{ig}} \int_{T_1}^{T_2} \frac{dT}{T^2} \Rightarrow$$

$$\Rightarrow \ln\frac{P_2}{P_1} = \frac{-\Delta H_{ads}}{R_{ig}}\left(\frac{1}{T_1} - \frac{1}{T_2}\right) \Rightarrow$$

$$\Rightarrow \Delta H_{ads} = \frac{\ln\left(\frac{P_2}{P_1}\right) R_{ig}}{\left(\frac{1}{T_2} - \frac{1}{T_1}\right)}$$

Using as $T_1 = 77$ K and $T_2 = 90$ K and the corresponding pressure ratios available in the problem, we can calculate the heat of adsorption (in kJ mol^{-1}) at the three values of surface coverage fraction V/V_m: -11.79, -12.66, -9.104. These values (essentially two values for the heat of adsorption, around 11 and 9 kJ mol^{-1}) seem to indicate a multilayer physical adsorption.

There is a more fundamental classification of adsorption isotherms according to Brunauer, which divides the adsorption isotherms into five types. Type I is the monolayer chemical Langmuir-type adsorption and type II is the multilayer physical BET adsorption isotherm. Type III is for – rather rarely occurring – weak adsorptions and types IV and V are for capillary condensation. There are, however, some adsorption isotherms that do not fit into Brunauer's classification.

7.4   Adsorption from solution

7.4.1   Adsorption using the Langmuir equation

When we have adsorption from solutions, the adsorption typically increases with increasing concentration and with decreasing temperature.

The Langmuir theory is a very useful tool also for describing the adsorption of various compounds from solutions, as it has been used successfully for describing the chemical adsorption of gases on solids. The Langmuir equation for solutions is given below (in its original and linear forms):

Original Langmuir form:

$$\Gamma = \frac{\Gamma_{max}kC}{1 + kC} \qquad (7.4a)$$

In linear form:

$$\frac{C}{\Gamma} = \frac{C}{\Gamma_{max}} + \frac{1}{k\Gamma_{max}} \qquad (7.4b)$$

where $A_{spec} = \Gamma_{max}N_A A_0$

Typical C and Γ units are respectively mol L^{-1} and mol g^{-1}, while A_{spec} is given in m^2 g^{-1}.

If A_{spec} and Γ_{max} are known then the area occupied by a molecule (A_0) can be calculated.

When the constant plateau value has been reached, it is implied that a monolayer of the adsorbent (a surfactant or a gas) has been created on the adsorbate (solid or liquid). The parameter Γ_{max} is the adsorption at full coverage (obtained above the CMC (critical micelle concentration) for surfactant solutions). The other parameter of Langmuir, k, is a measure of the affinity of the solute for the surface, i.e. how much the solute likes the surface. The higher the parameter k is, the higher the slope and thus the higher the affinity. The two parameters of the Langmuir equation can be obtained in practice from experimental data using its linear form, Equation 7.4, and finally the specific surface area of the solid, A_{spec}, can be estimated when the area of the adsorbed molecule is known (A_0).

Alternatively, when the specific surface area of the solid is known (from other techniques, e.g. BET adsorption of gases like nitrogen on the same solid, see Equation 7.3), we can estimate the area occupied by each adsorbed molecule, e.g. of a surfactant.

Finally, from the value of the solute–surface affinity parameter k in the Langmuir adsorption equation, it is possible to determine the Gibbs energy of adsorption:

$$\Delta G_{ads} = -R_{ig}T\ln k \qquad (7.5)$$

The inverse of k ($=1/k$) is the concentration at which one half of the surface is occupied by the solute.

The Langmuir theory has been derived based on the assumption that the adsorption rate is proportional to the concentration in the solution and the fraction of free non-covered area of the adsorbent. The desorption rate, on the other hand, is assumed to be proportional to the fraction of surface covered by adsorbate molecules. At equilibrium, the adsorption and desorption rates are equal.

As we saw, several adsorption isotherm equations have actually been developed from models of the adsorption process or adsorption equilibrium together with the Gibbs adsorption equation. However, we should stress again that the observation that an equation developed from a particular model gives a good fit to the experimental isotherm is not generally sufficient to validate the model: good agreement with other properties of the adsorption, especially thermal effects, is desirable.

Example 7.3 illustrates how the Langmuir equation is used for adsorption of an alcohol from its solution in a hydrocarbon.

Irving Langmuir is one of the greatest figures in surface science. One of the most important scientific journals today is called *Langmuir*, in his honour. A short biography of this Nobel Prize winning scientist is presented next.

Biography: Irving Langmuir (1881–1957)

Reproduced from http://commons.wikimedia.org/ wiki/File:Langmuir-sitting.jpg

Irving Langmuir was born in Brooklyn, New York, on 31 January 1881. His early education was obtained in various schools and institutes in the USA, and in Paris (1892–1895). He graduated as a metallurgical engineer from the School of Mines at Columbia University in 1903. Postgraduate work in physical chemistry under Nernst in Göttingen earned him the degrees of M.A. and Ph.D. in 1906.

Returning to America, Dr Langmuir became Instructor in Chemistry at Stevens Institute of Technology, Hoboken, New Jersey, where he taught until July 1909. He then entered the Research Laboratory of the General Electric Company at Schenectady where he eventually became Associate Director.

Langmuir's studies embraced chemistry, physics and engineering and were largely the outgrowth of studies of vacuum phenomena. In seeking the atomic and molecular mechanisms of these he investigated the properties of adsorbed films and the nature of electric discharges in high vacuum and in certain gases at low pressures.

His work on filaments in gases led directly to the invention of the gas-filled incandescent lamp and to the discovery of atomic hydrogen. He later used the latter in the development of the atomic hydrogen welding process.

He was the first to observe the very stable adsorbed monatomic films on tungsten and platinum filaments, and was able, after experiments with oil films on water, to formulate a general theory of adsorbed films. He also studied the catalytic properties of such films.

Langmuir's work on space charge effects and related phenomena led to many important technical developments which have had a profound effect on later technology.

In chemistry, his interest in reaction mechanism caused him to study structure and valence, and he contributed to the development of the Lewis theory of shared electrons. He received numerous awards, of which the Nobel Prize in Chemistry (1932) has been the greatest. He received the Nobel Prize "for his discoveries and investigations in surface chemistry".

He served as President of the American Chemical Society and as President of the American Association for the Advancement of Science. After a short illness, he died on 16 August 1957.

This information was taken from the official Nobel web-page: www.nobel.se.

Example 7.3.   Adsorption from solution using the Langmuir equation.
The following data refer to the adsorption of dodecanol (from solution in toluene) onto a sample of carbon black. The specific adsorption area of the carbon black is 105 m^2 g^{-1} and is measured using the technique of the adsorption of nitrogen on the solid.

Equilibrium concentration (mol L^{-1})	Amount adsorbed (μmol g^{-1})
0.012	24.1
0.035	50.4
0.062	69.8
0.105	81.6
0.148	90.7

Show that the data fit the Langmuir adsorption isotherm equation and calculate (in Å^2) the area occupied by each adsorbed dodecanol molecule at limiting adsorption.

Solution
We will use the Langmuir equation in the form of Equation 7.4.

We plot the concentration/adsorption versus the concentration (the latter in mol L^{-1}). We use the data provided in the table and express the ratio of concentration/adsorption in units of g L^{-1} (values: 498, 694, 888, 1287 and 1632, respectively for the five points provided).

The plot is indeed linear indicating that the adsorption of dodecanol follows the Langmuir equation.

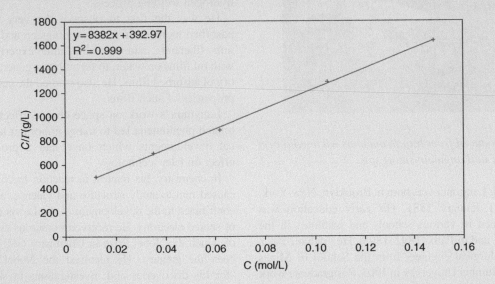

The "maximum adsorption" = 1/slope (of linear plot) = 1/8300 = 0.00012 mol g^{-1} = 120.5 μmol g^{-1}.

From the value corresponding to the monolayer, we can calculate the area occupied by each adsorbed dodecanol molecule using the equation:

$$A_{spec} = \Gamma_{max} N_A A_0$$

Thus:

$$105 = 120.5 \times 10^{-6} \times 6.0223 \times 10^{23} A_0$$

which gives the value of the dodecanol molecule area as 145×10^{-20} m^2 or 145 Å^2.

7.4.2 Adsorption from solution – the effect of solvent and concentration on adsorption

7.4.2.1 The role of solvent

Adsorption from solution on solids is, in some respects, more complex than adsorption of gases. There are several complications related to the adsorption of solutes on solid surfaces. The most important one is the role of the solvent. For example, the adsorption of stearic acid on carbon black reaches different limiting values in different solvents; ranging from 10 mmol kg^{-1} when benzene is the solvent up to more than 40 mmol kg^{-1} when cyclohexane is the solvent. The stearic acid is probably adsorbed with the acyl chain parallel to the surface and the measured adsorption values are consistent with that arrangement. The role of the solvent is indeed very important.

The effect of solvent and the competition of solute and solvent for sites on the solid surface are very important. Sometimes the effect can be predicted, as shown in the example of Figure 7.3. Here we observe the adsorption of different polar compounds on a polar surface (silica gel) but from solutions in non-polar solvents. The interaction between the solid and the acid group is nearly the same for all acids and thus the adsorption sequence is determined by the interaction between the solvent and the alkyl chains. The longer the chain of acids, the stronger the interaction (i.e. higher affinity) with the non-polar solvent, the more the acid "wants to stay" in the solution and thus the lower the adsorption.

The opposite is the case for the adsorption of carboxylic acids on a non-polar solid surface (coal) from aqueous solutions. The adsorption of acids from water on the non-polar solid coal increases with increasing chain length. Thus, the adsorption of valeric acid is much higher than that of butyric acid which

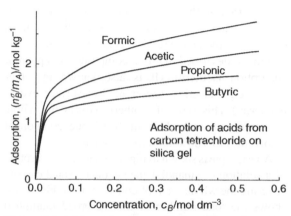

Figure 7.3 *Adsorption from solution. The adsorption of four fatty acids on a polar surface (silica gel) from a non-polar solution (carbon tetrachloride) is shown. Reprinted from Barnes and Gentle (2005), with permission from Oxford University Press*

in turn is higher than the adsorption of acetic acid. This is because of the decreasing solubility with water and increasing interaction with the solid surface.

Traube's rule describes the effect of chain length on surface tension and states that for each additional methylene group the concentration required to give a certain surface tension is reduced by a factor of 3. It turns out that the same rule can be used to predict the effect of chain length on adsorption when a homologous series of long-chain substances is examined.

Moreover, although sometimes we see a levelling out of adsorption isotherms resembling a Langmuir-type behaviour and a complete adsorbed layer on the surface of the solid, this is usually not the case (and the limiting adsorption does not usually correspond to complete coverage), due exactly to the contribution of the solvent. The solvent competes for sites on the solid surface and through its solvent power it will also compete with the surface for the solute molecules.

7.4.2.2   *Concentrated solutions are described by the so-called composite adsorption isotherms*

The previous discussion is true for adsorption from rather dilute solutions (where the concentration of the solvent in the solution is assumed to be constant). In the case, however, of concentrated solutions, this assumption does not hold and a different analysis is required. The situation is actually rather complex. This is because the competition of the other component for sites on the surface cannot be taken as constant as the concentration changes. Thus, the situation is described by the composite or apparent adsorption isotherm. (It is also known by other names, e.g. selective, preferential or surface excess isotherm.) This type of isotherm covers the whole concentration interval, from 100% one component ("solvent") to 100% the other ("solute").

An example is shown in Figure 7.4 for the adsorption of a solution of ethanol–benzene on activated carbon, covering the whole scale of concentrations. Figure 7.4 shows the total number of mol adsorbed multiplied by the change in the benzene concentration per g of solid. In dilute solutions, we have indeed adsorption of benzene, but in concentrated solutions ethanol is

adsorbed instead, while benzene is desorbed (negative adsorption). This is the reason the curve shows one maximum and one minimum. This type of curvature is very common for the adsorption of solutions on solids.

As understood, the composite isotherm does not describe the adsorbed amount of solute at the surface but rather the relative concentration of the solute (adsorbate) at the surface, i.e. the surface concentration minus the solution concentration. It can have negative values if there is an excess of solvent at the surface.

Experimentally, the investigation of adsorption from solution is much simpler than that of gas adsorption. A known mass of solid adsorbent is shaken with a known volume of solution at a given temperature until there is no further change in the concentration of the solution. This concentration can be determined by various methods involving chemical or radiochemical analysis, colorimetry, refractive index, etc. The experimental data are usually expressed in terms of the apparent (or composite) adsorption isotherm in which the amount of solute adsorbed at a given temperature per unit mass of adsorbent, as calculated from the decrease (or increase) of solution concentration, is plotted against the equilibrium concentration. The theoretical treatment of adsorption is, however,

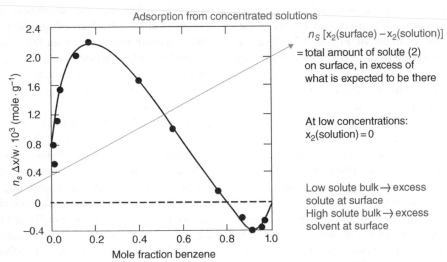

Figure 7.4   *Adsorption of a solution of ethanol (1)–benzene (2) on activated carbon, covering the whole scale of concentrations. Shaw (1992). Reproduced with permission from Elsevier*

complicated since adsorption from solution necessarily involves competition between solute(s) and solvent or between the components of a liquid mixture for the adsorption sites. Let us consider a binary liquid mixture in contact with a solid like the one shown in Figure 7.4. Zero adsorption refers to uniform mixture composition right up to the solid surface, even though (unlike zero gas adsorption) both components are, in fact, present at the solid surface. If the amount of one of the components at the surface is greater than its amount in bulk, then that component is positively adsorbed and, consequently, the other component is negatively adsorbed. Apparent, rather than true, adsorption isotherms are, therefore, calculated from changes in solution concentration. The individual adsorption isotherms can be calculated from the apparent or composite adsorption isotherm together with vapour adsorption data.

7.5 Adsorption of surfactants and polymers

7.5.1 Adsorption of surfactants and the role of CPP

The adsorption of surfactants is simpler than that of polymers and proteins and depends much on their CPP (critical packing parameter; concept presented in Chapter 5). The higher the CPP is the higher the adsorption (on hydrophobic surfaces). The nature of the surface is less important than that of surfactant. These concepts are illustrated through some examples in Figure 7.5. All factors that increase CPP (increase of hydrophobic chain or temperature, decrease of hydrophilic length, addition of salt, use of surfactants with two hydrocarbon chains, addition of surfactants with opposite charge, addition of long-chain alcohols or amines, etc.) result in increased adsorption (on a hydrophobic surface). This is because the packing of the surfactant molecules in a non-polar surface is better facilitated when the surfactant molecules have the shape of a cylinder.

Adsorption of surfactants on polar surfaces like pigments will only occur if there is an interaction between the surfactant polar head group and the surface. If this is the case and the interaction is strong then the surfactants will adsorb with the head group directed towards the surface. This will leave the surfactant hydrocarbon tail towards the aqueous solution and give the possibility for the adsorption of a second layer, similar to the first. This second adsorption will have characteristics similar to the adsorption on a hydrophobic surface. On the other hand, if the hydrophobic interaction is stronger than the interaction between head group-surface, then the surfactants will form micelles at the surface. Certain surface analysis techniques like AFM (atomic force microscopy), fluorescence, etc. can give us information on the conformation of surfactants on the surfaces. The higher the polarity of the solid surface is the higher the value of the area occupied by the surfactant molecule.

Another interesting aspect of surfactant adsorption is when two surfactants are present. In these cases we may observe what is known as competitive adsorption. Hydrophobic non-ionic surfactants are prone to displace polymers from hydrophobic (e.g. latex) surfaces but ionic surfactants and polymers, e.g. SDS and poly(vinyl alcohol) may work synergetically and may co-adsorb on polar and pigment surfaces. The adsorption of surfactants on two polymer surfaces has been studied based on CPP, e.g. the adsorption of nonyl-phenols with varying lengths on PS and PMMA latex. The adsorption is higher for the surfactants having the higher CPP values and for the most hydrophobic surface (PS).

In brief, the adsorption of surfactants on surfaces seems to depend more on the nature of the surfactant, e.g. as expressed by its CPP, rather than the nature of the surface. Generally, the higher the CPP, i.e. the more hydrophobic the surfactant, the better it adsorbs on the (hydrophobic) surface. Increasing temperature and salt concentration often also increases the adsorption (via increasing the CPP).

Many adsorption isotherms for surfactants, see Figure 7.5, can be well-represented by the Langmuir equation. The same is true for the adsorption of many liquid surfactants and polymers from solutions onto a solid surface. At this point it is of interest to recall some of the assumptions upon which the Langmuir equation has been derived. An analysis of the Langmuir equation has showed that it rests on the

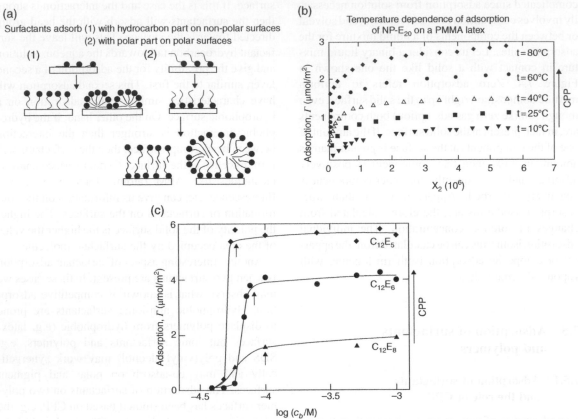

Figure 7.5   *Role of CPP in the adsorption of surfactants on solid surfaces. (a) (1) hydrophobic surfaces and (2) hydrophilic surfaces (showing cases of weak adsorption and strong adsorption – vertical surfactants). Reprinted from Jonsson et al. (2001), with permission from John Wiley & Sons, Ltd. (b) An increase of temperature increases the CPP and thus the adsorption of non-ionic surfactants (here nonyl-phenols) on a PMMA latex. The area of a surfactant molecule also decreases with increasing temperature (from 200 to 50 Å² at the two extreme temperatures). From Jonsson et al. (2001). Reprinted with permission from John Wiley & Sons, Ltd. (c) The adsorption of non-ionic surfactants onto silica is very pH sensitive. The adsorption is co-operative and is enhanced with increasing hydrocarbon chain length or decreasing the hydrophilic length. Increasing temperature or adding salts also increases the adsorption of non-ionic surfactants. The dominating factor is the structure of the surfactant not the nature of the surface. Reprinted from Tiberg et al. (1993), with permission from Elsevier*

assumptions that the surface is homogeneous (all adsorption sites are equivalent), there is only up to a monolayer adsorption of the adsorbate, there are no solute–solute or solute–solvent, e.g. surfactant–water, interactions and the molecular size of the solute and solvent is the same. The latter two assumptions are definitely not correct when considering surfactant adsorption. It has, however, been shown that these two assumptions give deviations from the Langmuir equation in opposite directions and thus

they often cancel each other, making use of the Langmuir equation fortuitous but quite useful.

7.5.2   Adsorption of polymers

7.5.2.1   General

The adsorption of polymers and other solutes from solutions on to solid surfaces is important in many practical situations:

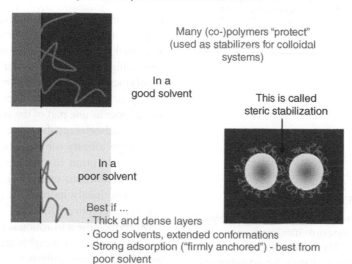

Figure 7.6 *Polymer adsorption and steric stabilization of colloid systems. Adapted from Wesselingh et al. (2007), with permission from John Wiley & Sons, Ltd*

- stability of colloidal dispersions (paints, food, cosmetics, pharmaceuticals, etc.);
- enhancing instability, e.g. water purification, mineral processing;
- surface modification (hydrophilization), e.g. food processing, bioapplications;
- surface functionalization, e.g. affinity chromatography, bio-tech applications;
- change in properties, e.g. foaming, rheology.

The adsorption of polymers is very important for understanding and controlling the steric stabilization of colloids (Figure 7.6), which is discussed in Chapter 12. As seen in Figure 7.6, the polymers adsorb on the interface (and may obtain different configurations). They enter into the liquid, and when particles come close together their polymer layers will interfere with each other, causing "steric repulsion". This type of repulsion depends not only on the properties of the polymer but also of the solvent. In a good solvent the layer will be more swollen, and more effective, than in a bad solvent. If the polymers are charged, you must also expect electrical effects. Thus, best stabilization is obtained when

the coverage of particles is complete (maximum adsorption – the plateau of Langmuir isotherms), when the adsorbed layers are thick and dense but also when the adsorption is strong (the polymer is "firmly anchored") and when extended conformations are present. The strong adsorption is favoured from a solution of a poor solvent, whereas the extended conformations exist when the solution is in a good solvent for the polymer.

Natural polymers like the protein casein in milk and many synthetic polymers like PEG and surfactants are used for stabilizing emulsions and colloidal dispersions. Many block copolymers like poly(ethylene oxide) surfactants are also used. Block and graft co-polymers are effective steric stabilizers when one type of block is soluble in the dispersion medium and the other is insoluble so that it attaches to the colloid particles.

Not only the adsorption of surfactants but also the adsorption of enzymes, other biomolecules and polymers very often follows the Langmuir isotherm (Figures 7.7 and 7.8). The adsorption of polymers and proteins is, however, more complicated than that of surfactants, as they can have various conformations.

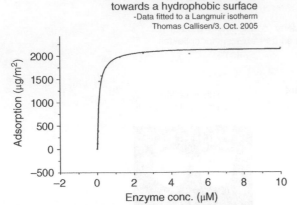

Figure 7.7 *Adsorption of a protein on a hydrophobic surface obtained by a Biacore 3000 surface plasmon resonance instrument. The data are fitted to the Langmuir isotherm. Notably, despite the apparent good fit to the Langmuir model, a more detailed understanding of the kinetics of the adsorption process relies on more rigorous approaches, e.g. phase plane analysis, as, for example, discussed by Hansen et al. (2012). Courtesy of Thomas Callisen, Novozymes*

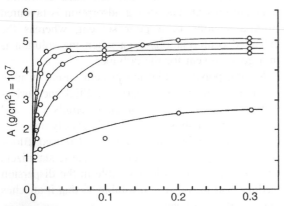

Figure 7.8 *Adsorption of polymers following the Langmuir adsorption isotherm. The adsorption typically increases with increasing polymer molecular weight. When polymers are used as steric stabilizers, high molecular weight polymers at full coverage (i.e. the plateau area) should be preferred (see also Chapter 12)*

7.5.2.2 *Molecular weight and adsorption*

The adsorption of polymers typically increases with polymer molecular weight and is higher and best (e.g. thicker, denser adsorbed layers) from a solution containing a poor solvent. For these reasons, block-copolymers are excellent steric stabilizers. Block copolymers like poly(ethylene oxide) surfactants are suitable as one part of the stabilizer has a high tendency to adsorb onto the particle surface and the other has a high affinity for the solvent.

The adsorption of polymers increases sharply at low concentrations until it reaches a plateau, exhibiting as previously indicated a typical Langmuir-type behaviour (Figure 7.8).

Moreover, the adsorption is typically higher at high polymer molecular weights and for this reason high molecular weight polymers are preferred for good adsorption. On the other hand, even if the adsorption is sharp initially, it does not necessarily mean that equilibrium and the desired constant plateau are reached very rapidly. Adsorption of macromolecules is a slow process and it may take several hours, even days, before equilibrium is achieved.

The adsorption of polymers depends on many more factors than just time and the polymer molecular weight, e.g. solution pH (an important parameter in the case of polyelectrolytes) and solid content. For a meaningful measurement of polymer adsorption, one should have control over numerous parameters: polymer molecular weight, molecular weight distribution, pH, ionic strength, presence of multivalent ions and presence of small amounts of additives.

7.5.2.3 *Role of solvent*

Of course, as mentioned, the solvent quality influences the adsorption in two ways. The first is by changes in the polymer configuration. The polymer expands in good solvents and contracts in poor solvents. Thus, a polymer sitting already at a surface will occupy a larger surface area in a good solvent and a smaller surface area in a poor one.

The second way a solvent influences the adsorption is by the stability in solution, i.e. if the polymer is not "happy" in the solution it will seek any opportunity to escape the solvent, e.g. by adsorbing on a surface. In both cases factors that decrease the solubility of the polymer in the solvent, e.g. salts and heavy alcohols, increase the adsorption of the polymer. This indicates that indeed the adsorption is a measure of the escaping tendency of the polymer from the solution.

7.5.2.4 Role of the radius of gyration

On adsorption, polymers change their conformation. Figure 7.9 illustrates various modes of polymer adsorption with only tails (a) or with loops, tails and trains (b). The tails contribute very little to the adsorbed amount but they determine the thickness of the adsorbed layer. A measure of the size (extension) of the tails is the so-called radius of gyration, R_g (see Figure 7.10). This is a very important concept, as two particles repel each other and thus we have steric stabilization only when the distance is about $2R_g$ (or lower), but the steric force is essentially zero at higher distances.

The radius of gyration (R_g) is proportional to the square root of the molecular weight of the polymer and is larger for good solvents (Figure 7.10). The

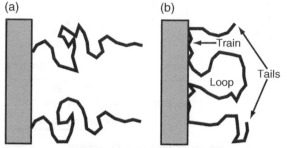

Figure 7.9 *Various modes (surface configurations) of polymer adsorption with only tails (a) or with loops, tails and trains (b). The thickness of these layers (radius of gyration) depends on various parameters (see Figure 7.10) including the polymer molecular weight but is, in many cases, of the order 5–50 nm. Reprinted from Myers (1999), with permission from John Wiley & Sons, Ltd*

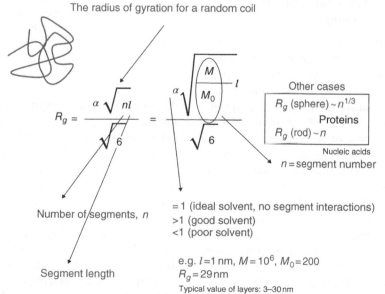

Figure 7.10 *Radius of gyration (R_g) – definition, examples and typical values. Radii of gyration can be experimentally obtained from light scattering measurements of dilute polymer solutions. Polymer molecules dissolved in good solvents have a much greater R_g than they would in the dry state*

expression shown in Figure 7.10 is for a random coil polymer, which is the most typical case. Notice that in this case R_g is proportional to $n^{\frac{1}{2}}$, where n is the segment number. Similar relationships but with other exponents are observed in the cases of sphere and (stiff) rod-like molecules. Of course, all of these are extreme cases, and there are many intermediates (partly hydrated coils, oblate/prolate shaped polymers, worm-like chains of varying flexibility and solvation). Moreover, the shape of the polymer depends on numerous factors other than the nature of polymer chain, e.g. solvent (quality), pH and electrolyte concentration.

As can be seen by these relationships, the size of the polymer molecule (as random coil) increases faster with the molecular weight than for a hard sphere molecule, e.g. certain proteins, but more slowly than for a stiff rod, e.g. nucleic acids or other highly charged polyelectrolytes at low excess electrolyte concentration. However, the conformation of a flexible polymer is not static and large fluctuations may occur.

7.5.2.5   Lewis acid–Lewis base (LA-LB) concept and polymer adsorption

Finally, we complete the discussion on polymer adsorption with the Lewis acid–base concept, which is a useful tool for evaluating the relative adsorption of polymers versus solvents on the same solid surface (Figure 7.11). The relative interactions of polymer–solid, polymer–solvent and solvent–solid must be accounted for in the case of adsorption of polymers from solutions on solids. It is important at first to know whether these compounds (solid surface, solvent, polymer) have an acidic or basic character. Notice (Figure 7.11) that very acid solvents compete (with the acidic surface) for the basic polymer and very basic solvents compete (with the basic polymer) for the acidic surface. It seems that in many cases, solvents of "balanced" acidity/basicity (i.e. not very acid, not very basic, almost neutral) are the best choices for accomplishing maximum adsorption.

Another possibility is to use the concept of the Flory–Huggins (FH) "χ" (χ)-parameter. In this way we may assign "χ-values" for the interaction with surface by considering the interaction energies for

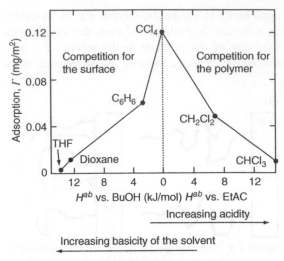

Figure 7.11 *Adsorption of the basic PMMA on the acid silica (SiO_2) surface from solutions of various solvents of varying acidic and basic characters. The figure illustrates the importance of the Lewis acid–base concept on polymer adsorption. LB solvents (e.g. benzene, esters, ethers, ketones, etc.) can adsorb on the surface instead of PMMA. In contrast, acid solvents can adsorb on PMMA, prohibiting its adsorption on the surface. Thus, adsorption is often maximum when it occurs from inert solvents. Reprinted from Jonsson et al. (2001), with permission from John Wiley & Sons, Ltd*

the polymer–solvent, polymer–surface and solvent–surface. The polymer will *not* adsorb if the FH parameter (solvent–polymer) < FH parameter (polymer–surface). This is because the polymer wants to stay in the solution with a good solvent. In the opposite case (poor solvent), the polymer will adsorb.

Finally, we can mention that various factors related to the adsorption of polymers can be measured, e.g. the adsorbed amount via ellipsometry and reflectometry, the layer thickness (R_g) via dynamic light scattering, the heat of adsorption via calorimetry, the fraction of polymer in actual contact with surface via spectroscopy (NMR, IR) and segment density distribution via SANS (small-angle neutron scattering).

A special case of adsorption is that of polyelectrolytes, for which the conformation on solids is rather flat, consisting of thin layers with often low adsorbed amount. Both the adsorbed amount and layer thickness increase with electrolyte concentration and surface charge. There is a smaller dependency of adsorption on the molecular weight and the radius of gyration increases with increasing degree of dissociation.

7.6 Concluding remarks

Adsorption is a universal phenomenon in colloid and surface science. Both adsorption of gases on solids and adsorption from solutions are of importance. Adsorption of high molecular weight amphiphilic compounds for monolayer creation, of gases on solids, of surfactants, polymers or proteins (biomolecules), find many applications.

Adsorption isotherms, surface tension–concentration equations and two-dimensional equations of the surface pressure against area are all concepts which are unified and linked via the Gibbs adsorption equation.

Many adsorption phenomena especially of surfactants, polymers, proteins and the chemical adsorption of gases on solids can be well represented by the Langmuir adsorption isotherm. This equation can be expressed in a suitable linear form and we can obtain the two parameters of the model, of which one is the concentration or volume at maximum (full) coverage or the so-called monolayer coverage. Knowledge of this monolayer coverage and of the specific surface area of the solid can help us estimate the surface area occupied by a molecule at the interface and thus the amount needed for stabilization. The specific solid surface area can be obtained from gas adsorption measurements on the same solid.

Adsorptions not described by Langmuir, especially the physical multilayer adsorption of gases on solids, can often be best described by the BET theory (which also can be written in a linear form, and yield values of the monolayer coverage and of the specific solid surface area).

Adsorption from solution on solids is more complex than adsorption of gases. The effect of the solvent and competition of the solute and solvent for sites on the solid surface are very important. Concentrated solutions are described by the so-called composite adsorption isotherms.

The adsorption of surfactants is simpler than that of polymers and proteins and depends much on their CPP. The higher the CPP value is the higher the adsorption (on hydrophobic surfaces). The CPP depends on salt, pH, temperature, presence of double bonds and double chains in the surfactant molecule. This dependency of CPP permits us to manipulate the molecular structure of the surfactant, so that we can increase its CPP and thus its adsorption on a hydrophobic surface. For non-ionic surfactants, CPP increases with increasing hydrocarbon chain length or temperature, decreasing hydrophilic chain length (head), changing the hydrocarbon from linear to branched, increasing unsaturation (double bonds) and by using surfactants with two hydrocarbon chains instead of one.

For ionic surfactants things are more difficult as, due to the electrostatic repulsion which dominates in ionic surfactants, the CPP is often (far) less than 1 for such surfactants. Thus, as it is not always enough to simply change the structure of an ionic surfactant, we should instead change its immediate environment (adding salt, surfactants with opposite charge, or alcohols/amines/non-ionic surfactants). These factors affect (increase) the CPP.

The adsorption of polymers is very important for understanding and controlling the steric stabilization of colloids, which we will discuss in detail in Chapter 12. The adsorption of polymers typically increases with polymer molecular weight and is higher and best (e.g. thicker adsorbed layers) from a solution containing a poor solvent. The opposite is the case (thin layer, essentially independent of polymer molecular weight, low adsorbed amount) when the adsorption is from a good solvent. The Lewis acid–base concept is a useful tool for evaluating the relative adsorption of polymers (for different solvents) on the same solid surface.

There are some common patterns related to the adsorption of both polymers and surfactants on solids. In both cases, the adsorption often follows the Langmuir equation and the maximum capacity can be estimated from Langmuir data and/or specific area of solid and of the adsorbed molecule at surface.

Moreover, the adsorption is maximum above CMC and increases with CPP.

Although the role of solvent has been discussed qualitatively, our discussion in this chapter has been limited to single adsorption (one solute–one surface). Multicomponent adsorption is a much more complex subject and many theories have been proposed, e.g. an extension of the Langmuir theory and different approaches like the ideal adsorbed solution theory (IAST) and the multicomponent potential adsorption theory (MPTA). They will be discussed in Chapter 14.

Problems

Problem 7.1: Relationship between the Langmuir and the BET adsorption equations

Show that the BET adsorption equation yields the Langmuir equation at the limit of very low pressures ($P \ll P_o$) and when the parameter C of the BET equation is much higher than unity.

At this limit, which is the relation of the Langmuir parameter B with the BET constant C and the saturation pressure?

Problem 7.2: Adsorption based on the Langmuir equation

The following data list volumes of ammonia (reduced to s.t.p.) adsorbed onto a sample of activated charcoal at 0 °C:

Pressure (kPa)	6.8	13.5	26.7	53.1	79.4
Volume (cm³ g⁻¹)	74	111	147	177	189

Show that the data fit a Langmuir adsorption isotherm expression and evaluate the constants.

Problem 7.3: Adsorption from solution – the role of solvent (Shaw (1992). Reproduced with permission from Elsevier)

The figure below (from Shaw, 1992) shows adsorption of three organic acids from solutions onto two different solids. Part (b) is adsorption on coal from aqueous solutions and (a) represents adsorption onto silica gel from toluene solutions.

Explain qualitatively the difference in the order of the adsorption of the three acids on the two solid surfaces. What is the role of the solvent in the two cases?

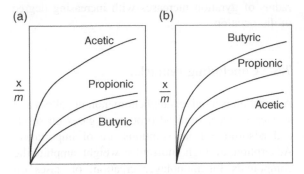

From Shaw (1992). Reproduced with permission from Elsevier

Problem 7.4: Surface area measurements via the BET equation

The BET equation for the adsorption of nitrogen on silica at 77 K can be expressed by the equation (fitted in the range of reduced pressures between 0.05 and 0.3):

$$\frac{P}{V(P_o - P)} = 2.85 \times 10^{-4} + 0.0257 \frac{P}{P_o}$$

where the units of the slope and of the intercept are g cm⁻³.

1. Estimate the volume for monolayer coverage V_m (in cm³ g⁻¹) and the parameter C of BET equation.
2. Estimate the specific surface of solid silica in m² g⁻¹ if the area occupied by one nitrogen molecule is 16.2×10^{-20} m² and compare the results to those from other laboratories (Hiemenz and Rajagopalan, 1997).

Problem 7.5: Surface analysis and two-dimensional equations of state

The following surface tensions were measured for aqueous solutions of *n*-pentanol at 20 °C:

C (mol L^{-1})	0	0.01	0.02	0.03	0.04	0.05	0.06	0.08	0.10
Surface tension (mN m^{-1})	72.6	64.6	60.0	56.8	54.3	51.9	49.8	46.0	43.0

1. Calculate the surface pressure (in mN m^{-1}), the adsorption (in mol m^{-2}) and the area occupied per molecule (in m^2) for bulk concentrations of 0.02, 0.04 and 0.08 mol L^{-1}.
2. Plot the surface pressure/area curve for the adsorbed *n*-pentanol monolayer and compare it with the corresponding curve for an ideal gaseous film.

Problem 7.6: Competitive adsorption of surfactants on solids

There are several measures of the balance between hydrophilic and lipophilic parts of amphiphilic molecules like surfactants. Two of them are the well-known "structure-based" CPP (critical packing parameter) and the empirical HLB (hydrophilic–lipophilic balance). HLB is used mostly in emulsion studies (low values below 8 indicate poor water solubility and high values, for example, above 8 indicate good water solubility). The HLB and CPP of surfactants are, as expected, related, see the plot below.

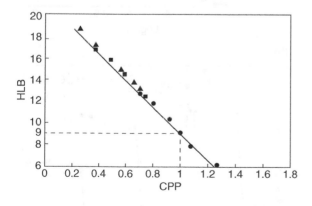

We would like to study the competitive adsorption of an anionic surfactant (SDS), which is known to form spherical micelles, and a non-ionic one (ethoxylated nonyl phenol with ten ethylene oxide units) on a hydrophobic latex surface.

The HLB of ethoxylated nonyl phenols can be estimated from the equation:

$$HLB = 20(17 + 44X_{EO})/(220 + 44X_{EO})$$

where X_{EO} is the number of ethylene oxide groups.

Two laboratories provide you with two completely different results for the relative concentrations of the two surfactants on the surface:

NPE$_{10}$/SDS: 90/10
NPE$_{10}$/SDS: 10/90

Which one of the two would you expect to be correct and why?

Problem 7.7: Adsorption of benzene on graphite

Benzene adsorbed on graphite is found to obey the Langmuir isotherm to a good approximation (chemical adsorption). At a pressure of 1.00 Torr (mmHg) the volume of benzene adsorbed on a sample of graphite was found to be 4.2 mm^3 g^{-1} and at 3.0 Torr it was 8.5 mm^3 g^{-1}. Assume that one benzene molecule occupies 30 Å^2. Estimate the specific surface area of the graphite.

Problem 7.8: Measuring the isosteric enthalpy of adsorption

The data below show the pressures of CO needed for the volume of adsorption (corrected to s.t.p) to be equal to 10.0 cm^3. Calculate the adsorption enthalpy at this surface coverage.

T (K)	200	210	220	230	240	250
P (mmHg)	30.0	37.1	45.2	54.0	63.5	73.9

Problem 7.9: Adsorption of butanol on water

The following plot represents surface tension against concentration data for the butanol–water system at 298 K against the concentration of butanol in the solution. The data can be represented by the equation shown in the plot.

Can the Langmuir equation represent the adsorption of butanol on water over an extensive concentration range?

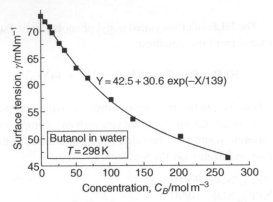

Reprinted from Barnes and Gentle (2005), with permission from Oxford University Press

Problem 7.10: Adsorption of acids and surfactants from aqueous solutions

The following figures show surface tension–concentration plots and adsorption isotherms of various compounds (non-ionic surfactants and organic acids) from aqueous solutions. The arrows indicate CMC.

1. Explain qualitatively the adsorption of organic acids on carbon.
2. Which parameter of the solutes (acids, surfactants) could be used to describe qualitatively the adsorption in all these cases? Discuss the results.

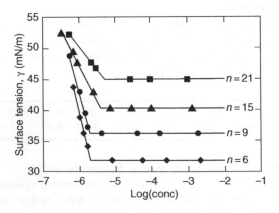

Jonsson et al. (2001). Reproduced with permission from John Wiley & Sons, Ltd

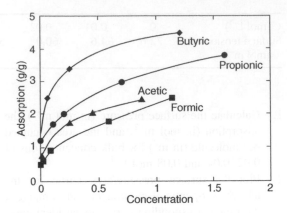

Jonsson et al. (2001). Reproduced with permission from John Wiley & Sons, Ltd

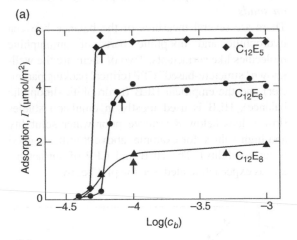

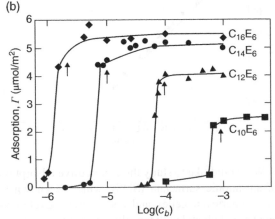

Jonsson et al. (2001). Reproduced with permission from John Wiley & Sons, Ltd

Problem 7.11: Adsorption of SDS on polystyrene latex surface

The plot below shows the adsorption of the anionic surfactant SDS (from an aqueous solution) on the hydrophobic PS latex surface at three different salt concentrations.

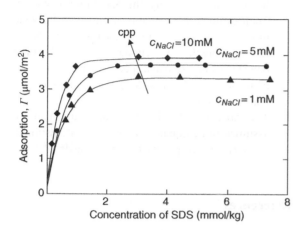

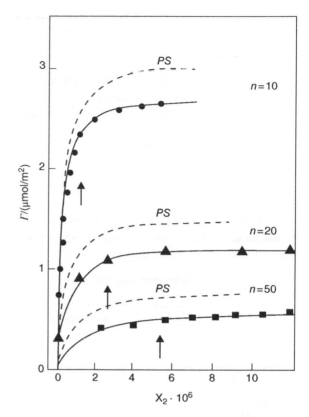

Jonsson et al. *(2001). Reproduced with permission from John Wiley & Sons, Ltd*

1. Calculate in $Å^2$ the area occupied per surfactant molecule (at monolayer level). Plot the area against the NaCl concentration.
2. How does CPP change with increasing salt concentration? Why?
3. Which CPP values can be estimated at the three salt concentrations? Do you expect differences in the micellar structure at the three concentrations? Make reasonable assumptions.
4. How would the adsorption of SDS on PS change with CPP and why?

Problem 7.12: Adsorption of nonyl-phenyl-ethoxy-lates on two different polymer surfaces

The figure below shows the adsorption of three nonyl-phenyl-ethoxylates (non-ionic surfactants) with different ethoxylate numbers at two polymer surfaces (PMMA and PS). The polarity of the two polymers is 0.3 (PMMA) and 0.2 (PS), expressed on a relative polarity scale. The arrows indicate the CMC values.

Jonsson et al. *(2001). Reproduced with permission from John Wiley & Sons, Ltd*

1. Estimate the area occupied per surfactant molecule (in $Å^2$) for the three surfactants on the two different polymer surfaces. Provide reasonable estimates for the CPP values of the three surfactants and their expected micellar structure.
2. Explain qualitatively (and using the concepts of CPP, CMC and Gibbs free energy of micellization) the observed differences in the adsorption among:
   - the three different surfactants;
   - the two different surfaces.Which is the most dominating factor governing adsorption behaviour in this case (surfactant or surface type)?

Problem 7.13: Adsorption of gases on solids

The following data have been obtained for the adsorption of nitrogen on a sample of silica SiO_2 at 77 K:

Relative pressure (P/P_o)	0.02	0.05	0.1	0.2	0.3	0.4	0.6	0.8
Volume of gas (cm^3 g^{-1})	121	165	187	215	248	270	330	484

Show that the data follow the BET isotherm (for relative pressures below 0.4) and estimate the specific (per g) surface area of the solid for this silica sample. Assume that the molecular area occupied by one nitrogen molecule is known to be 16.2 10^{-20} m^2.

Problem 7.14

The figure below shows the adsorption of a binder in a paint (solution) onto TiO$_2$ pigments (with a surface area of 14 m^2 g^{-1}). Estimate the adsorbed amount of the binder corresponding to one monolayer of a binder molecule per m^2 of pigment.

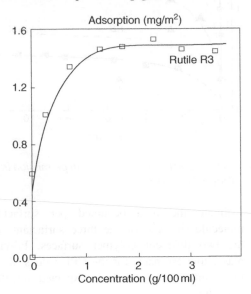

Problem 7.15: Composite adsorption isotherms

The following adsorption isotherm curves are individual (true) adsorption isotherms for the adsorption from liquid mixtures of benzene and ethanol on dispersed carbon powders.

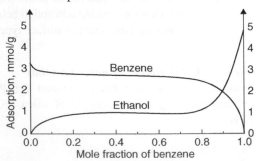

1. Explain the difference between apparent and individual adsorption isotherms.
2. Show that for benzene the adsorbed amount follows rather closely the Langmuir isotherm. Then estimate the amount of adsorbed benzene in mol per g of carbon powder for a monolayer.
3. Explain why the apparent adsorption isotherm of benzene can be zero at a specific concentration of benzene/ethanol. Provide an estimation of the concentration.
4. The area of one benzene molecule can be assumed to be equal to 40×10^{-20} m^2. Estimate the surface area per g of carbon powder.

References

G.T. Barnes, I.R. Gentle, 2005. Interfacial Science – An Introduction. Oxford University Press.

R. Hansen, H. Bruus, T.H. Callisen and O. Hassager (2012), *Langmuir* 28, 7557–7563.

P.C. Hiemenz, and R. Rajagopalan, 1997. Principles of Colloid and Surface Science, 3rd edn, Marcel Dekker, New York.

B. Jonsson, B. Lindman, K. Holmberg, B. Kronberg, 2001. Surfactants and Polymers in Aqueous Solution, John Wiley & Sons Ltd., Chichester.

D. Myers, 1991 & 1999. Surfaces, Interfaces, and Colloids. Principles and Applications. VCH, Weinheim.

D. Shaw, 1992. Introduction to Colloid & Surface Chemistry. 4th edn, Butterworth-Heinemann, Oxford.

F. Tiberg, B. Lindman, M. Landgren, *Thin Solid Films*, 1993, 234: 478.

J.A. Wesselingh, S. Kiil, M.E. Vigild, 2007. Design and Development of Biological, Chemical, Food and Pharmaceutical Products, John Wiley & Sons Ltd, Chichester.

8

Characterization Methods of Colloids – Part I: Kinetic Properties and Rheology

8.1 Introduction – importance of kinetic properties

This chapter will present the very important kinetic properties of colloids, and their motion under the influence of gravitational and centrifugal fields. Kinetic properties are important for many reasons, for example they provide methods for determining the molecular weight (molar mass) of colloidal particles, their size and shape – i.e. for the characterization of colloidal particles. Many of the methods we will see in this chapter have found widespread use in the study of biological and polymeric molecules.

Even though particles do not have a single size (they usually exhibit a distribution of sizes), we often roughly characterise a particle as having a "certain size". As discussed previously, colloids are larger than small molecules, but small enough that they cannot be seen with an ordinary (optical) microscope. Their dimensions range from a little above one nanometre, to about one micrometre although even larger particles of say up to ten micrometres are also in the colloid family.

When we are dealing with a new type of material/ colloidal dispersion, it is usually a good idea to "have a look" at the particles, often using advanced microscopic techniques (see Chapter 9) and also evaluate various characteristics, summarized in Figure 8.1. Particles are "complex"; they are not always spherical, thus we may need to know more than "just the diameter", they may have complex shapes, they seldom have a single size but a size (and molecular weight) distribution and they can be present as agglomerates or aggregates. However, it is always true that the better our system is characterized, the better our understanding and the easier it is to manipulate the formulation and optimize the final product.

8.2 Brownian motion

We now proceed with a thorough investigation of the kinetic properties of colloidal particles. Our discussion starts with the Brownian motion (Figure 8.2).

The motion of individual particles in colloids is continuously changing direction as a result of random

Introduction to Applied Colloid and Surface Chemistry, First Edition. Georgios M. Kontogeorgis and Søren Kiil.
© 2016 John Wiley & Sons, Ltd. Published 2016 by John Wiley & Sons, Ltd.
Companion website: www.wiley.com/go/kontogeorgis/colloid

Colloidal particles-what would we
like to know?

- What do they look like - shape

- Size - key dimensions

- Molecular weight

- Molecular weight distribution

- Charge

- Dimensions of adsorbed layers

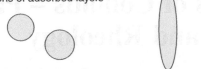

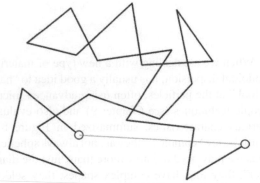

Figure 8.1   *Key characteristics of colloid particles*

Table 8.1   *Diffusion coefficients, D, of some molecules in water at 20 °C*

Compound	M (kg mol^{-1})	D (m^2 s^{-1})
Sucrose	0.342	4.59×10^{-10}
Lysozyme	14.1	1.04×10^{-10}
Haemoglobin	68	6.9×10^{-11}
Collagen	345	6.9×10^{-12}

equation for the mean Brownian displacement of a particle from its original position along a given axis as a function of time (t) and the diffusion coefficient (D):

$$\bar{x} = \sqrt{2Dt} \qquad (8.1)$$

The same equation can be derived in an independent way from Fick's first law of diffusion (see, for example, Shaw, 1992). According to Fick's first law, the rate of diffusion dn/dt of a solute across an area A, known as the diffusive flux, is proportional to the concentration gradient of the solute, dc/dx (dn is the amount of solute crossing the area A in time dt).

The diffusion coefficients are of the order 10^{-9} m^2 s^{-1} for ordinary low molecular weight liquids, but for colloids they are of the order of 10^{-11} m^2 s^{-1}. Typically, a tenfold increase in the size of a particle leads to a tenfold decrease of the diffusion coefficient. For rather large particles it is in practice difficult to measure these very low diffusion coefficients.

The Einstein equation is useful in many ways, e.g.:

- determination of the diffusion coefficient of a colloidal substance by measuring its average Brownian displacement with the help of an advanced microscope (e.g. an ultramicroscope or so-called dark-field microscopy, where observation can be made with horizontal illumination against a dark field);

- in combination with Stokes equations (Equations 8.3 and 8.4), J. Perrin (Nobel Prize in Physics 1926) used the equation for the first accurate estimation of the Avogadro number.

Table 8.1 gives some typical values of the diffusion coefficient. Very often the diffusion coefficients are expressed in cm^2 s^{-1}.

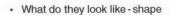

Figure 8.2   *Schematic illustration of the Brownian motion of a colloid particle. The straight line between the two small circles represents the linear displacement of the particle after a given time*

collisions with the molecules of the suspending medium, of other particles and the walls of the containing vessel. Each particle pursues a complicated and rather irregular zig-zag path, like the one shown in Figure 8.2. This random motion is called "Brownian motion", after the botanist Robert Brown who first (in 1827) observed it while studying with the microscope a suspension of grains in water. The smaller the particles are, the more evident is their Brownian motion. When the colloid particles are spherical, the Brownian motion is a simple movement leading to diffusion, but for more complex shapes the particles may also have rotational movement.

Treating Brownian motion as a three-dimensional "random walk", Einstein derived an important

Example 8.1.   Brownian motion.
Colloids in a dispersion are in constant motion due to Brownian motion. How long does it approximately take for a haemoglobin molecule to move 100 μm along a given axis at 20 °C?

Solution
This is calculated using Equation 8.1 in a rearranged form:

$$t = \frac{\bar{x}^2}{2D}$$

The diffusion coefficient of a haemoglobin molecule is given in Table 8.1 to 6.9×10^{-11} m^2 s^{-1}. Inserting this into the above equation yields:

$$t = \frac{\left(100 \times 10^{-6} \text{ m}\right)^2}{2\left(6.9 \times 10^{-11} \text{ m}^2\text{s}^{-1}\right)} = 72.5 \text{ s}$$

8.3   Sedimentation and creaming (Stokes and Einstein equations)

8.3.1   Stokes equation

Let us now study the sedimentation (or creaming) of colloidal particles in a dispersion medium under the effect of gravity. The other two forces that act, as can be seen in Figure 8.3, are both opposing gravity and are due to the buoyancy of the liquid and the viscous drag force, which is typically considered to be proportional to the velocity of the particle. We first present an important equation for the viscous drag force. We emphasize again that colloidal particles are not always spherical or even near-spherical. They can have various shapes and this issue can affect the calculations, as discussed later.

The viscous drag force is considered to be proportional to the velocity of the particle:

$$F_v = f \cdot u \qquad (8.2)$$

The proportionality factor, f, is called the friction(al) coefficient and has dimensions of "mass/time" (e.g. kg s^{-1}). Note that in some areas of the technical sciences a dimensionless friction coefficient can be used, e.g. in courses on unit operations dealing with pipe flow. Equation 8.2 does not

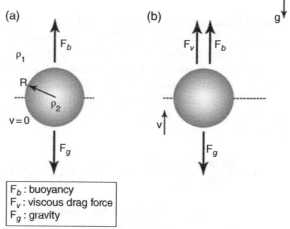

Figure 8.3   *Forces on colloidal particles: (a) only gravitational forces are present and (b) viscous drag is also included*

depend on the geometry of the particle, but the friction coefficient is often hard to calculate and in many cases we can only calculate the ratio m/f (mass/friction coefficient) and not f independently.

However, in one specific case (spherical particles), an equation for f can be obtained. This is the equation derived by G. G. Stokes in 1850 for spherical particles in very dilute dispersions. Both the viscosity of the dispersion medium and the radius of the particle enter this equation:

Example 8.2.   Stokes law.
A can of recently stirred paint is placed on a shelf during the winter season. The paint contains only one pigment type, TiO_2, with an average particle size of 0.2 μm and a density of 4230 kg m^{-3}. The density of the liquid phase (mainly binder and solvent) is 1200 kg m^{-3} and the viscosity is 1 kg m^{-1} s^{-1}. Calculate the sedimentation velocity of a pigment particle in the top paint layers in moving towards the bottom of the paint can, which has a height of 20 cm. Hindered sedimentation (particle interaction) can be neglected.

Solution
It is assumed that Stokes law is valid. The sedimentation velocity is then given by Equation 8.4:

$$u = \frac{d_p^{\;2} g \left(\rho_{TiO_2} - \rho_{medium} \right)}{18 \eta_{medium}} = \frac{(0.2 \times 10^{-6} \text{ m})^2 (9.81 \text{ ms}^{-2})(4230 \text{ kgm}^{-3} - 1200 \text{ kgm}^{-3})}{18 (1 \text{ kgm}^{-1} \text{s}^{-1})} = 6.6 \times 10^{-11} \text{ ms}^{-1}$$

This is a very low sedimentation velocity caused by the small particle size and the fairly large viscosity (compared to, for example, water). The value is quite uncertain, as a calculation of the Re_p number will show, but here it is enough to see that sedimentation due to gravity is not going to be a practical problem during the winter.

$$f_0 = 6 \pi \eta_{medium} R \qquad (8.3)$$

where R is the radius of the particles and η_{medium} is the viscosity of the dispersion medium.

The Stokes equation governs creaming and sedimentation of colloids in dilute dispersions. It is useful for rule-of-thumb estimations. The Stokes equation for the sedimentation velocity, u, can be then derived based on Equation 8.3 and the force balance of Figure 8.3:

$$u = \frac{d_p^2 g \left(\rho_p - \rho_{medium} \right)}{18 \eta_{medium}}, \quad 10^{-4} < Re_p < 1 \qquad (8.4)$$

where Re_p is the particle Reynolds number ($Re_p = \rho_{medium} d_p u / \eta_{medium}$) and d_p is the diameter of the particle or droplet.

Despite its many assumptions:

- spherical particles;
- motion of particles is extremely slow;
- extremely dilute colloidal dispersions.

for spherical colloidal particles undergoing sedimentation, diffusion or electrophoresis, deviations from Stokes' law usually amount to less than 1% and can be neglected (except for concentrated dispersions, where deviations can be significant due to

hindered sedimentation). Often, the Re number is smaller than the lower boundary in Equation 8.4, but we will still use the Stokes equation, despite the uncertainty, because we have no other ways of estimating the sedimentation velocity.

8.3.2   Effect of particle shape

What cannot be neglected is the effect of shape on the frictional coefficient and this issue is discussed here. Colloidal particles are often non-spherical. They may have different shapes, as shown in Figure 8.4, and thus the Stokes equation cannot be used in the form shown in Equation 8.4.

An additional complexity is that as the particle is moving, it is "solvated", i.e. it incorporates molecules from the dispersion medium, e.g. water (hydration). The friction coefficient is affected by both non-sphericity and solvation. Separate evaluation of the contribution of asymmetry and hydration (solvation) is often difficult, but an assessment of the "joint effect" of these two factors is shown for one specific case in Figure 8.5 for proteins in water.

With reference to Figure 8.5, we discuss the specific application of proteins in aqueous dispersions, studied by Oncley in 1941 (see Shaw, 1992). He has computed frictional ratios for ellipsoids of varying degrees of asymmetry and hydration. The

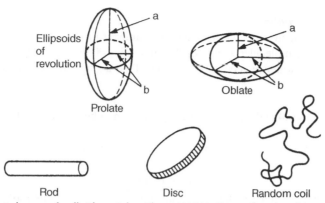

Figure 8.4   *Different shapes of colloid particles. Shaw (1992). Reproduced with permission from Elsevier*

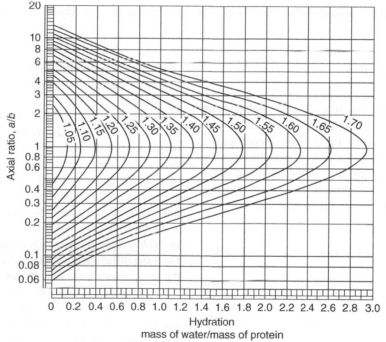

Figure 8.5   *Effect of shape and solvation on the friction coefficients of proteins in water (aqueous protein dispersion); a/b refers to the ratio of the two dimensions defining the particle as shown in Figure 8.4 (a/b = 1 is the spherical shape). The discrete variable in the figure is the ratio f/f$_0$, where f$_0$ is f (friction coefficient) for a spherical, unsolvated particle. Shaw (1992). Reproduced with permission from Elsevier*

resulting contour diagram shows the combinations of axial ratio and hydration that are compatible with given frictional ratios. The frictional ratio f/f_0 is defined as the ratio of the actual frictional coefficient, f, to the frictional coefficient of the equivalent unsolvated sphere, f_0 (from Equation 8.3). Thus, this ratio is a measure of the combination of asymmetry and solvation. The plot of Figure 8.5 can be used to determine the shapes of proteinic particles in aqueous dispersions.

Example 8.3.   Hydration number.
A particular protein in an aqueous dispersion has a shape number (a/b) equal to 0.4. The frictional ratio can be calculated to be 1.40 and the molecular weight of the protein is 100 kg mol^{-1}. What is the hydration number for the protein? How many water molecules are "attached" to a single protein molecule?

Solution
Using Figure 8.5 with $a/b = 0.4$ and $f/f_0 = 1.40$, the hydration number can be read to be approximately 0.9 g water per gram of protein.

   Converting the hydration number into a molar ratio (equivalent to a number ratio) gives:

$$\frac{n(\text{water})}{n(\text{protein})} = \frac{0.9 \text{ g water}}{\text{g protein}} \frac{\left(10^5 \dfrac{\text{g protein}}{\text{mol protein}}\right)}{\left(18 \dfrac{\text{g water}}{\text{mol water}}\right)} = 5000 \frac{\text{water molecules}}{\text{protein molecule}}$$

Hence, one large protein molecule has about 5000 water molecules "attached" due to hydration forces.

8.3.3   Einstein equation

We saw that the frictional coefficient that is needed in all the above calculations is a difficult parameter to obtain for non-spherical molecules. Shape and solvation affect the values of the frictional coefficient. Recognizing that the driving force for diffusion is thermodynamic in origin and assuming dilute ideal dispersions, Einstein combined Fick's law and the equation for the drag force and derived the very simple equation:

$$Df = k_B T \qquad (8.5a)$$

which is known as Einstein's diffusion law. This is another very important equation proposed by Einstein and can be used for calculating the mass of colloid particles:

$$f = \frac{k_B T}{D} \Rightarrow m = \frac{k_B T u}{D\left(1 - \dfrac{\rho_1}{\rho_2}\right)g} \qquad (8.5b)$$

where 1 = medium and 2 = particle or droplet. Equation 8.5b is general and no assumptions for the shape of the particles have been made. It is very useful as it can be used to determine the friction coefficient f from the much more easily obtained (and from various experimental techniques) diffusion coefficient D.

   In combination with a general force balance, Einstein's diffusion law results in Equation 8.5b, which permits the estimation of the mass of each particle. Thus, upon combining diffusion experiments (for obtaining D) and sedimentation (gravitational) experiments (for obtaining u), we can estimate the mass of colloidal particles without any assumption about their shape. Finally, due to Einstein's diffusion law ($Df = k_B T$), the ratio f/f_0 is equal to D_0/D, where D_0 is the diffusion coefficient of a system containing the equivalent unsolvated spheres.

   As expected, the larger the diffusion coefficient, the lower the drag force. Of course, Einstein's diffusion law can be combined with Stokes equation for f and the resulting equation is called Stokes–Einstein law (Problem 8.1). Together with the equation for the Brownian displacement, it was used by Perrin for early, rather accurate calculations of the Avogadro number.

   The ratio f/f_0 can be estimated, based on the Stokes and Einstein equations:

$$\frac{f}{f_0} = \frac{\left(\dfrac{k_B T}{D}\right)}{6\pi\eta_{medium}R} \qquad (8.5c)$$

Moreover, since $m = M/N_A$, Equation 8.5b can be used to estimate the molecular weight (M) of a colloid particle:

$$M = \frac{R_{ig}Tu}{D\left(1-\dfrac{\rho_1}{\rho_2}\right)g} \Rightarrow M = \frac{R_{ig}Ts}{D\left(1-\dfrac{\rho_1}{\rho_2}\right)} \qquad (8.5d)$$

where $s = u/g$, which is expressed in this way to compare with the expressions from the ultracentrifuge (see Section 8.4 (1 = medium and 2 = particle or droplet)). It is assumed that the particle is one big molecule (e.g. a protein). Some colloidal particles, e.g. latex particles in water-based coatings, can be made up of many polymer molecules and this needs to be taken into account in the analysis if the method is used for such particles.

8.4 Kinetic properties via the ultracentrifuge

The sedimentation method based on gravity has a practical lower limit of about 1 μm. Smaller colloidal particles sediment so slowly under gravity that the effect is disturbed by the mixing tendencies of diffusion and convection. Even the largest macromolecules have effective radii of not much more than 10^{-8} m (10 nm), so that their sedimentation cannot be observed in the gravitational field of earth.

By employing centrifugal forces instead of gravity, the application of sedimentation can be extended to the study of colloidal systems and has been extensively used for the characterization of substances of biological origin like proteins, nucleic acids and viruses. It is easy to realize accelerations much higher than gravity acceleration (g) and it is often customary to express the resulting acceleration as a multiplier of g, e.g. 1000g. An ultracentrifuge, shown schematically in Figure 8.6, is a high-speed centrifuge equipped with a suitable optical system, for recording sedimentation profiles, and can reach very high accelerations of the order of $10^5 g$.

Initially, a particle is immersed in a liquid at a given distance x from the axis of the centrifuge head (or rotor). During centrifugation, the particle will be exposed to three forces: centrifugal, buoyancy, and viscous drag (Figure 8.3). The centrifugal force is proportional to the angular acceleration squared, ω^2. Contrary to sedimentation under earthly gravity, the sedimentation velocity in a centrifuge is not constant, but increases as the particle moves towards the bottom of the sedimentation tube. This is why we introduce the quantity called the sedimentation coefficient (s), which is an important characteristic of the system considered:

$$s = \frac{u}{\omega^2 x} = \frac{\dfrac{dx}{dt}}{\omega^2 x} = \frac{\ln(x_2/x_1)}{\omega^2(t_2-t_1)} \qquad (8.6)$$

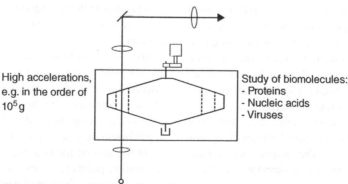

Figure 8.6 *Schematic illustration of an ultracentrifuge. The sample cell is placed between the dashed lines. On top, the motor is seen and at the bottom the resistor, which is connected by a cup to the electric circuit*

Example 8.4.   *Angular velocity and acceleration.*
A typical numerical example is given here for estimation of the angular velocity and the acceleration of the centrifugal field (remember that $g = 9.80$ m s^{-2}).

The ultracentrifuge uses 60000 rpm =1000 rps (f_{rpm}) and the angular velocity of the centrifuge (in rad s^{-1}), ω, is expressed as:

$$\omega = 2\pi f_{rpm}$$

The distance from the axis of rotation, r, is 6 cm. The angular acceleration, a, can now be calculated as:

$$a = \omega^2 r = \left(2\pi 1000 \text{ s}^{-1}\right)^2 (0.06 \text{ m}) = 2.36 \times 10^6 \text{ ms}^{-2} = 240000g$$

With such high accelerations even molecules as small as sucrose can sediment at measurable rates.

Sedimentation experiments using an ultracentrifuge are usually presented based on the obtained values of the sedimentation coefficient (s). For a particle moving in a circle of radius R with an angular velocity of the centrifuge ω, then the acceleration in the direction of this radius is $\omega^2 R$.

In the above definition of s, x is the distance of the particle from the rotation axis, and it has dimensions of time. It is often expressed in "Svedberg" units (1 S $= 10^{-13}$ s), in honour of T. Svedberg who pioneered this field and introduced the use of ultracentrifugation in the study of colloidal systems.

The final expression of the sedimentation coefficient shown in Equation 8.6 is obtained if we measure x as a function of time and integrate between two specific distances and their respective times.

The sedimentation coefficient depends on both concentration and temperature, which must be both reported. Often, s is measured at various concentrations and then extrapolated to zero concentration (infinitely dilute dispersion). Plots of ln x as a function of time at known rotation speeds can be used to determine the sedimentation coefficient, s.

Equations 8.7–8.9 show how the molecular weight (M) can be estimated based on centrifugation and related experiments. The reason for having different expressions for the molecular weight is explained below.

The ultracentrifuge can be used in two distinct ways for investigating suspended colloidal material.

In the sedimentation velocity method a high ultracentrifugal field (up to about 400 000g) is applied and the displacement of the boundary set up by sedimentation of the colloidal particles is measured from time to time. In the sedimentation equilibrium method, the colloidal solution is subjected to a much lower centrifugal field, until sedimentation and diffusion (mixing) tendencies balance one another and an equilibrium distribution of particles throughout the sample is attained.

Both methods offer a way for determining the molecular weight (molar mass) of the dispersed particles, as shown in Equations 8.7 and 8.8, each having its own advantages and shortcomings. The derivation of these two equations is left as exercise (Problem 8.8).

The first method (sedimentation velocity) requires an independent determination of the diffusion coefficient, which is avoided in the second method. Thus, in the second method (sedimentation equilibrium) we do not need to know the shape/solvation of particles. Molecules as small as sugars have been studied with the sedimentation equilibrium method. The disadvantage of the second method is that the establishment of equilibrium may take as long as several days.

In the second method, we can either use two concentration points (c) at two different distances (x) or using a sample of data make a plot of $\ln(c_2/c_1)$ against $(x_2^2 - x_1^2)$. The slope of the curve will give the molecular weight of the investigated compound.

8.4.1 Molecular weight estimated from kinetic experiments (1 = medium and 2 = particle or droplet)

Sedimentation equilibrium experiments:

$$M = \frac{2R_{ig}T \ \ln\left(\frac{c_2}{c_1}\right)}{\omega^2 \left(1-\frac{\rho_1}{\rho_2}\right)(x_2{}^2 - x_1{}^2)} \quad \text{(centrifuge)} \quad (8.7)$$

- low fields;
- slow method (days sometimes needed for equilibrium);
- no need to know D or the shape/solvation of the particles.

8.4.2 Sedimentation velocity experiments (1 = medium and 2 = particle or droplet)

$$M = \frac{R_{ig}Ts}{D\left(1-\frac{\rho_1}{\rho_2}\right)} \quad \text{(centrifuge)} \quad (8.8)$$

where $s = u/g$.

- needs very high fields up to 400 000g;
- fast method;
- requires knowledge of the diffusion coefficient, D;
- shape and solvation not required.

Gravitational experiments can also use the "sedimentation equilibrium" principle (equating sedimentation and diffusion fluxes). The molecular weight is then given by the equation:

$$M = \frac{R_{ig}T \ \ln\left(\frac{c_2}{c_1}\right)}{\left(1-\frac{\rho_1}{\rho_2}\right)g(x_2 - x_1)} \quad \text{(gravity)} \quad (8.9)$$

As stated by Shaw (1992), care must be taken to ensure that the system under investigation remains uncoagulated. This applies to any technique used to determine molecular or particle masses.

8.5 Osmosis and osmotic pressure

The next topic in this chapter, possibly familiar from other physical chemistry courses, is the phenomenon of osmosis and the associated osmotic pressure. This is very important in colloid science.

Osmotic pressure measurements offer a very convenient method for estimating the molecular weight of macromolecules (in the range 10 000–1 000 000 g mol^{-1}) as described by Shaw (1992). Osmosis takes place when a solution and a solvent (or two solutions of different concentrations) are separated from each other by a semipermeable membrane, i.e. a membrane that is permeable to the solvent and not the solute. The tendency to equalize the concentrations (chemical potentials) on either side of the membrane results in a net diffusion of solvent across the membrane. The counter-pressure necessary to balance this osmotic flow is termed osmotic pressure (Figure 8.7).

Using osmotic pressure measurements, see also Figure 8.8, the molecular weight of macromolecules can be estimated:

$$M = \frac{R_{ig}T}{\lim_{c\to0}\left(\Pi/c\right)} \quad \Pi = \rho g \Delta h \quad (8.10)$$

The osmotic pressure of a solution is generally described as a function of concentration, c (mass

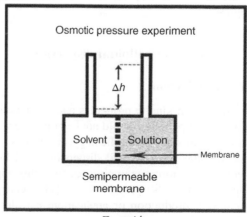

Osmotic pressure experiment

Δh

Solvent | Solution

Membrane

Semipermeable membrane

$\Pi = \rho g \Delta h$

Figure 8.7 *Concept of osmosis and osmotic pressure*

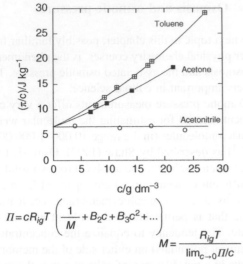

$$\Pi = cR_{ig}T\left(\frac{1}{M} + B_2c + B_3c^2 + \dots\right)$$

$$M = \frac{R_{ig}T}{\lim_{c\to 0}\Pi/c}$$

Figure 8.8 *Osmotic virial equation and the estimation of molecular weight from osmotic pressure data. The polymer is PMMA (poly(methyl methacrylate)) in various solvents at 20 °C. Toluene is a good solvent, acetone is a poor solvent, and in acetonitrile the polymer forms globular structures. From the y-axis intercept on the figure (c = 0) the molecular weight can be estimated to be about 348 000 g mol^{-1}. Adapted from Hamley (2000), with permission from John Wiley & Sons, Ltd*

of solute/volume of solution), via the so-called osmotic virial equation. *M* is the molecular weight of the solute and B_2, B_3, are the second and third virial coefficients (from thermodynamics). The molecular weight can be determined by extrapolating to zero the near-linear Π/c versus *c* plot (see example in Figure 8.8).

8.6   Rheology of colloidal dispersions

8.6.1   Introduction

A discussion on kinetic properties and characterization of colloid particles would not be complete without a few words about the rheology of colloids.

The rheology of colloidal dispersions is a huge subject and of enormous importance for many applications, e.g. of paints or food colloids. When changes occur during production or application, e.g. during the drying of paints, we may expect changes in the overall rheology of the system (due to solvent

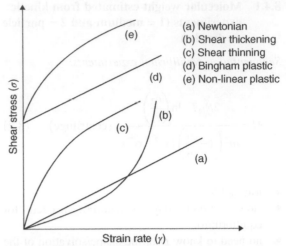

Figure 8.9 *Shear stress–strain rate plots of various fluids. For shear-shinning liquids (e.g. paint), viscosity is reduced with increasing shear rate and vice versa for shear thickening liquids (e.g. quicksand). Water and petrol are Newtonian liquids. Bingham plastics, where a yield stress is required for flow, can be, for example, toothpaste and butter*

evaporation combined with possible presence of areas of limited miscibility).

Mathematics may be less important than an appropriate qualitative understanding of the basic underlying phenomena. The basic characteristics of the rheology of dispersions can be summarized as:

- huge subject (complex mathematics, but this is often of less importance);
- huge difference between dilute and concentrated dispersions;
- it depends on liquid (dispersion medium) viscosity, concentration, size and shape of particles, interactions between particles – dispersion medium, polydispersity;
- large effect of adsorbed polymers and/or double layers;
- gives (in)direct information on structure and forces of the dispersion.

Rheology is often studied in shear stress–strain rate plots (Figure 8.9). The variables are related through:

$$\sigma = \eta\dot{\gamma} \qquad (8.11)$$

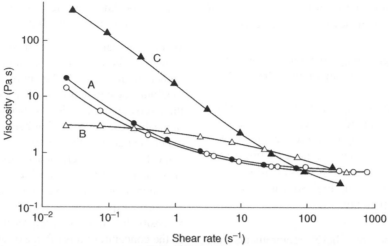

Figure 8.10 *Rheology of various paints. A = solvent-borne and B,C = water-borne (with two different thicknesses added). Hunter (1993). Reproduced with permission from Oxford University Press*

Some few "simple" liquids like water, petrol, glycerol, ethanol and dilute aqueous solutions follow Newtonian behaviour (constant viscosity value in Equation 8.11). Very few colloidal dispersions follow Newtonian behaviour, i.e. constant viscosity, and moreover the range of shear rates of interest is very broad. Deviations from Newtonian behaviour are even more pronounced in the case of concentrated dispersions and if the particles are asymmetric and/or aggregated. Formation of structure throughout the system and orientation of asymmetric particles cause non-Newtonian flow.

Some colloidal systems, e.g. butter, margarine, ketchup and toothpastes, follow Bingham pseudo-plastic behaviour, i.e. you need to put some force/stress with a knife and then flow occurs. Paints and polymers typically (should) exhibit a shear thinning behaviour, i.e. they become less viscous when stirred (but they can thicken again). Many suspensions, emulsions and foams are shear thinning (often when we have asymmetric particles), while certain structures like clay and quicksand are shear thickening. Shear thinning and plasticity are often hard to distinguish.

Most shear-thinning fluids like paints and certain polymer solutions are also thixotropic, i.e. the viscosity decreases with time (at constant shear rate). Yoghurt is also an example of a thixotropic liquid. The opposite, i.e. rheopectic liquids (viscosity increases with time),

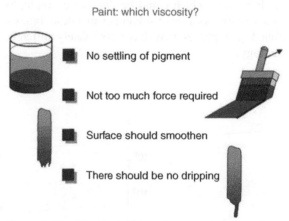

Figure 8.11 *Rheological requirements in paints. From Hans Wesselingh (personal communication)*

are rare. Paints represent a classical example of shear-thinning dispersions (Figure 8.10). Many water-based paints are oil-in-water emulsions, where the oil is a polymer that also contains pigments, while the water may also have dissolved polyalcohols to assist emulsion stability and control the drying of the paint. The rheological requirements (Figure 8.11) are quite complex and experimental observations point out that the paints should be highly shear-thinning (and often also thixotropic).

Before and during application, the viscosity of paint has to comply with several requirements:

1. It should be high enough to avoid settling of pigment particles in the can (it must be stable in the can).
2. It should not be so high as to make application difficult (easily spreading over the surface).
3. It should be low enough to allow smoothening of the surface after application, i.e. easy flow of the droplets and elimination of brush marks.
4. It should be sufficiently high so that no dripping occurs during painting.

Moreover, some of these requirements are conflicting.

Many problems often observed during the drying of the paints are often attributed to changing rheological behaviour which could be due, for example, to immiscibility because of solvent evaporation. Immiscible liquid phases may have completely different rheology.

8.6.2 Special characteristics of colloid dispersions' rheology

When colloidal particles are dispersed in a liquid, the flow of the liquid is disturbed and the viscosity (η) is higher than that of the pure liquid (dispersion medium), designated as η_0. The problem of relating the viscosities of colloidal dispersions with the nature of the dispersed particles has been the subject of many experimental and theoretical investigations. In this respect, viscosity increments, especially the relative, specific and intrinsic viscosities are of greater significance than absolute viscosities. These functions are shown in Equations 8.12a and 8.12b. The intrinsic viscosity has units of reciprocal concentration but as the concentration is often expressed as the volume fraction of the particles (=volume particles/total volume), then the intrinsic viscosity is dimensionless.

Relative viscosity:

$$\eta_r = \frac{\eta}{\eta_0} \tag{8.12a}$$

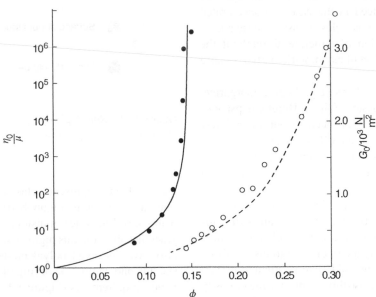

Figure 8.12 *Dimensionless low-shear limiting viscosities (η_0/μ) of a polystyrene latex (radius of 34 nm) in NaCl (5 × 10^{-4} M) as a function of particle concentration, ϕ. Filled circles represent the low shear viscosity. Open circles refer to the static shear moduli, G_0 (not of relevance in this context). Reprinted from Russel et al. (1989), with permission from Cambridge University Press*

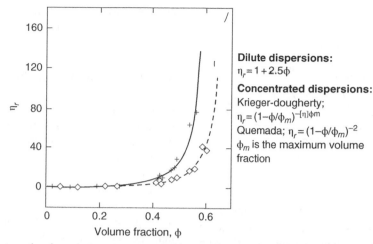

Figure 8.13 *Expressions for the relative viscosity of dilute and concentrated dispersions. Reprinted from Van der Werff and de Kruif (1989), with permission from AIP Publishing LLC*

Intrinsic or limiting viscosity:

$$[\eta] - \lim_{c \to 0} \left(\frac{\eta_{spec}}{c} \right) = \lim_{c \to 0} \left(\frac{1}{c} \frac{\eta - \eta_0}{\eta_0} \right) \quad (8.12b)$$

The viscosity increases linearly with concentration for dilute dispersions but after a certain critical volume fraction (often around 60–70%) it increases very much with concentration, as shown in Figure 8.12 and as discussed in relation to drying of latex dispersions by Kiil (2006).

Various equations have been proposed to represent these behaviours; some are shown in Figure 8.13. Einstein's equation (1906) is valid for very dilute (volume fraction below 0.05) suspensions of smooth, rigid, spherical and uncharged particles. In this simplified case, the shear rate, the sizes of the spheres and the range of sizes present do not affect the viscosity, only the phase volume is relevant. The validity of Einstein's equation has been confirmed experimentally for very dilute suspensions of glass spheres, certain fungi and polystyrene particles, in the presence of sufficient electrolyte to eliminate charge effects. The non-applicability of the Einstein equation at moderate concentrations is mainly due to an overlapping of the disturbed regions of flow around the particles. Several equations have been proposed to allow for this, some of which add a quadratic term with respect to volume in Einstein's equation.

However, for real dispersions and emulsions, even dilute ones, the viscosity is expected to be lower than that given by Einstein's equation. Droplets in emulsions may with shearing change shapes to ellipsoids and even burst at high shear rates to form smaller droplets. Moreover, since small droplets are less easily deformed than larger droplets, the viscosity of an emulsion depends on particles size.

Finally, notice that the constant "2.5" in Einstein's equation which is valid for spherical particles is the smallest possible value, i.e. values for elongated particles and other shapes are higher. Thus, having dispersions with spherical particles is a way to have low viscosities.

The Mark–Houwink–Staudinger (MHS) equation provides a relationship between the intrinsic viscosity (Equation 8.12b) and the (average) molecular weight (M) of (synthetic) polymers:

$$[\eta] = KM^a$$

$$a = 1 \quad \text{Rigid rod–like polymers} \quad (8.13)$$

$$a = 0.5 \quad \text{Flexible random coils}$$

Thus, the molecular weight of polymers can be estimated from viscosity measurements (of a series of dilute solutions). However, at low concentrations the relative viscosity is very close to unity and we

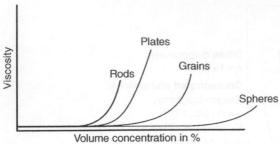

Figure 8.14 *Effect of particle shape on colloidal dispersions*

need data of high precision, which requires great deal of care, cleanliness and excellent temperature control as the viscosity changes exponentially with temperature.

We further discuss here the MHS equation. The terms K and a are semi-empirical constants that depend on the polymer and solvent. The parameter K depends on the nature of both the solvent and the solute and must be determined experimentally by measuring the molecular weight of at least one sample by some independent method.

The parameter a is predominantly dependent on the shape of polymer molecules and the quality of the solvent; it is $a = 0.5$ in a theta solvent and can rise up to 0.8 in better than theta conditions. When $a = 1$, then the average weight determined from this method is exactly the weight average.

Viscosity (rheological properties in general) are important in the application of latex and other paints (Pashley and Karaman, 2004). As Equation 8.13 also indicates, polymer solutions where the polymers are coils/spherical polymer particles are much less viscous than if the polymers have a rigid rod-like format (see also Figure 8.14). This facilitates application.

It has already been mentioned and it is further illustrated in Figure 8.14 that the viscosity depends a great deal on the shape of the particles. It also depends on the adsorbed polymers, for sterically stabilized dispersions, especially for polyelectrolytes, as the electrical double-layer (Chapter 10) also travels with the particle. Actually, the charges on dispersion or emulsion particles affect the viscosity in three different ways: the already mentioned distortion of the electrical double layer by the shear flow, double layer repulsion between particles and – for polyelectrolytes – the changes in particle dimensions with changes of pH or ionic strength.

8.7   Concluding remarks

Estimating the size, shape or molecular weight of colloidal particles is important for many types of products. There are many different analysis methods available, depending on the level of detail required, the time that can be used for a measurement and the cost of conducting the analysis. In particular, for biological and polymeric molecules there are many possibilities. The ultracentrifuge is very important and is now accessible in small versions that will fit on a laboratory bench. Osmotic pressure measurements are also a widely used technique for macromolecules.

In this chapter, the rheology of colloids has also been discussed and it was shown that very complex behaviours can be observed depending on the shape and concentration of the colloid particles. In addition, dispersions can show thixotropic (time-dependent) behaviour. The rheology of colloids is important for products such as shampoo, paint, foods and lotions.

Problems

Problem 8.1: Stokes–Einstein equation and Brownian motion

1. Starting from Einstein's equation for the diffusion coefficient and Stokes equation for the friction coefficient, show that:

$$D = \frac{R_{ig}T}{6\pi\eta_{medium}RN_A} \quad \text{(Stokes} - \text{Einstein equation)}$$

2. Then show that the Brownian displacement of a particle from its original position is given as a function of its radius by the equation:

$$\bar{x} = \sqrt{\frac{R_{ig}Tt}{3\pi\eta_{medium}RN_A}}$$

3. What is the average displacement in 1 min of a spherical particle along a given axis produced by Brownian motion? The particle has a radius equal to 0.1 μm and is suspended in water at 25 °C.
4. The diffusion coefficient of glucose in water is 6.81×10^{-10} m² s⁻¹ at 25 °C. The density of glucose is 1.55 g cm⁻³ and it can be assumed that the molecule is spherical. Estimate using the Stokes–Einstein equation the molecular weight (molar mass) of glucose and compare the result to the "true experimental" value (=180.2 g mol⁻¹).

Problem 8.2: The Avogadro number from measurements of the Brownian motion (Shaw (1992). Reproduced with permission from Elsevier)

The table below shows some diffusion coefficients and Brownian displacements calculated for uncharged spheres in water at 20 °C.

Radius	D at 20 °C (m² s⁻¹)	$\bar{x}$ after 1 h
1 nm	2.1×10^{-10}	1.23 mm
10 nm	2.1×10^{-11}	
100 nm	2.1×10^{-12}	123 μm
1 μm	2.1×10^{-13}	

When the Brownian displacements are experimentally determined, as above via an ultramicroscope, then they can be used to determine the Avogadro number. This was first done in 1908 by Perrin for fractionated mastic and gamboge suspensions of known particle size. Perrin calculated Avogadro number values varying between 5.5×10^{23} mol⁻¹ and 8×10^{23} mol⁻¹. Later, Svedberg (1911) calculated an Avogadro number value equal to 6.09×10^{23} mol⁻¹ from observations on monodispersed gold sols of known particle size. The correct determination of Avogadro's constant from observations on the Brownian motion provides striking evidence in favour of the kinetic theory.

Repeat the calculations of Perrin and Svedberg using the above data. How close are the Avogadro number values you obtained to the correct one?

Problem 8.3: Kinetics and creaming in emulsions

An emulsion formulation, in a container having a height of 10 cm, has a continuous phase, with the following characteristics: density = 1.1×10^3 kg m⁻³, viscosity = 0.0015 kg m⁻¹ s⁻¹. The emulsion contains both oil droplets (diameter = 0.5×10^{-6} m, density = 0.93×10^3 kg m⁻³) and calcium salt particles (diameter = 14×10^{-6} m, density = 2.71×10^3 kg m⁻³).

1. Where (towards the top/upwards or bottom/downwards of the container) are the droplets of oil and salt moving?
2. How much time is required for this movement of both the oil (in days) and salt (in minutes) droplets?

Problem 8.4: Molecular weight of polymers from osmotic pressure data

The following osmotic pressures were measured for solutions of a sample of polyisobutylene in benzene at 25 °C.

Concentration (g per 100 cm³)	Osmotic pressure (cm solution)
0.5	1.03
1.0	2.10
1.5	3.22
2.0	4.39

The solution density is 0.88 g cm⁻³ in each case. Calculate an average relative molecular mass.

Problem 8.5: Molecular weight and shape of proteins from sedimentation data

The sedimentation and diffusion coefficients for myoglobin in a dilute aqueous solution at 20 °C are 2.04×10^{-13} s and 1.13×10^{-10} m² s⁻¹, respectively. The partial specific volume of the protein is 0.741 cm³ g⁻¹, the density of the solution is 1.00 g cm⁻³ and the coefficient of viscosity of the solution is 1.00×10^{-3} kg m⁻¹ s⁻¹.

Calculate:

1. The relative molecular mass
2. The frictional ratio f/f_0 of this protein. What is the possible shape of a dissolved myoglobin molecule?

Problem 8.6: Molecular weight of proteins using ultracentrifuge experiments

The molecular weight (molar mass) of another protein (β-lactoglobulin) is being investigated. In a sedimentation equilibrium experiment a solution of this protein is centrifuged at 11000 rpm (rotations per minute) and the temperature was 25 °C. The following data was recorded.

Distance from axis of rotation (cm)	Protein concentration (g L^{-1})
4.90	1.30
4.95	1.46
5.00	1.64
5.05	1.84
5.10	2.06
5.15	2.31

The specific volume of the protein is 0.75 cm^3 g^{-1} and the density of the solution (assumed constant) is 1.0 g cm^{-3}.

Calculate the molar mass of the protein from the sedimentation data.

Problem 8.7: Sedimentation velocity experiment of bovine serum albumin (BSA)

The sedimentation of bovine serum albumin (BSA) in water was monitored at 25 °C. The initial location of the solute surface was at 5.50 cm from the axis of rotation, and during centrifugation at 56850 rpm it receded as follows:

Time (s)	r (cm)
0	5.50
500	5.55
1000	5.60
2000	5.70
3000	5.80
4000	5.91
5000	6.01

The specific volume of the protein is 0.735 cm^3 g^{-1} and the diffusion coefficient is 6.1 × 10^{-11} m^2 s^{-1}.

1. Calculate the sedimentation coefficient.
2. Provide an estimate of the molecular weight of this protein using this sedimentation velocity data.

Problem 8.8: Molecular weight from equilibrium measurements in centrifugal or gravity fields

1. In equilibrium sedimentation experiments, either due to centrifugal or gravity forces, the sedimentation flux ($A \cdot C \cdot dx/dt$) is equal to the diffusion (as given by Fick's law, $A \cdot D \cdot dC/dx$). Derive the follow expressions for the molecular weight, when the experiment is carried out in a centrifugal field or in a gravitational field:

$$M = \frac{2R_{ig}T \, \ln(c_2/c_1)}{\omega^2 \left(1 - \frac{\rho_1}{\rho_2}\right)\left(x_2^2 - x_1^2\right)} \quad \text{(centrifuge)}$$

$$M = \frac{R_{ig}T \, \ln(c_2/c_1)}{\left(1 - \frac{\rho_1}{\rho_2}\right)g(x_2 - x_1)} \quad \text{(gravity)}$$

2. In the sedimentation velocity experiments, the governing equation is the same as with the classical sedimentation experiment (due to gravity forces alone) with the only difference that the acceleration of gravity, g, should be substituted by the angular acceleration $\omega^2 x$. Show that the molecular weight is then calculated as:

$$M = \frac{R_{ig}Ts}{D\left(1 - \frac{\rho_1}{\rho_2}\right)}$$

where s is the sedimentation coefficient:

$$s = \frac{u}{\omega^2 x} = \frac{\frac{dx}{dt}}{\omega^2 x} = \frac{\ln(x_2/x_1)}{\omega^2(t_2 - t_1)}$$

Problem 8.9: Virus analysis

A scientist is conducting some investigations on a strain of the very dangerous Ebola virus, which has caused many deaths in Africa. It is crudely assumed that the individual virus molecule is coiled up to a spherical particle with a radius of 4 nm. The density of the virus is 1133 kg m^{-3}. The medium where the virus is dispersed has a viscosity of 10^{-3} kg m^{-1} s^{-1} and a density of 1000 kg m^{-3}. The temperature is 20 °C.

1. Calculate the diffusion coefficient of the Ebola virus in the medium.
2. Sedimentation velocity experiments are conducted in an ultracentrifuge. The sedimentation coefficient is measured to be 2 S (Svedberg units). Calculate the molecular weight of the Ebola virus. Comment on the result.
3. Which characterization methods would you suggest the scientist to use to evaluate the assumption stated in relation to the Ebola virus? You need to read Chapter 9 to answer this last question.

References

I.W. Hamley, Introduction to Soft Matter: Synthetic and Biological Self-Assembling Materials, Revised Edition, John Wiley & Sons Ltd, Chichester, 2000.

R.J. Hunter, Introduction to Modern Colloid Science, Oxford Science Publications, Oxford, 1993.

S. Kiil, Prog. Org. Coat., 57(3) (2006) 57(3) 236–250.

R.M. Pashley, M.E. Karaman, Applied Colloid and Surface Chemistry, John Wiley & Sons Ltd, Chichester, (2004).

W.B. Russel, D.A. Saville, W.R. Schowalter, Colloidal Dispersions, Cambridge University Press, Cambridge, 1989.

D.J. Shaw, Colloid & Surface Chemistry, 4th ed., Butterworth and Heinemann, Oxford, (1992).

J.C. Van der Werff, and C.G. Kruif, Hard-sphere colloidal dispersions: The scaling of rheological properties with particle size, volume fraction, and shear rate, J. Rheology, 33(3) (1989), 421–454.

9

Characterization Methods of Colloids – Part II: Optical Properties (Scattering, Spectroscopy and Microscopy)

9.1 Introduction

Many methods are available for estimating particles' dimensions and size distribution (Figure 9.1). Often a combination of various approaches is needed. Advanced methods (e.g. measurements based on certain types of microscopy or scattering) are often very useful, although other "simpler" methods (e.g. based on sedimentation or backscattering) may be preferred for continuous measurements of the particle size, e.g. in a production line.

A popular "straightforward" characterization method is microscopy. Although they cannot be used easily in all cases, microscopic techniques can be used not only for the optical observation of various surfaces and particles but also for the precise estimation of the dimensions of colloidal particles. We need to use (for colloidal particles) the advanced electron microscopic (SEM or TEM) or probe microscopic methods (STM and AFM) as the simple optical microscope cannot be used for colloids due to its low resolution, which at best can cover the upper limit of colloidal dimensions (a few micrometres).

Table 9.1 shows the resolution and some few characteristics and applications of these advanced methods.

The resolution of most of these methods covers the whole colloidal domain from a few nm up to 5 or 10 μm. In this chapter, the most important of the various optical characterization methods are presented and discussed.

9.2 Optical microscopy

Optical microscopy refers to direct optical methods employed to study structures and surfaces of particles or droplets. Microscopy utilizes the light reflected or refracted ("bent") or fluorescence from the object under investigation. Conventional optical microscopy can only access the upper end of the colloid size range with a resolution between 0.5 and 1 μm. A magnification of 1000× is the usual maximum limit available on an optical microscope and the depth of field (distance in front of and beyond the subject that appears to be in focus) is about 1 mm.

Introduction to Applied Colloid and Surface Chemistry, First Edition. Georgios M. Kontogeorgis and Søren Kiil.
© 2016 John Wiley & Sons, Ltd. Published 2016 by John Wiley & Sons, Ltd.
Companion website: www.wiley.com/go/kontogeorgis/colloid

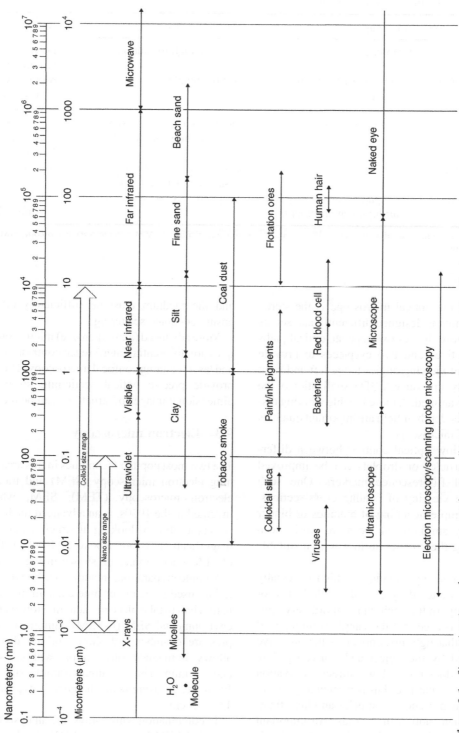

Figure 9.1 Size estimations of common particles as well as methods for characterizing colloid particles. Berg (2010). Reproduced with permission from World Scientific

Table 9.1   *Advanced microscopy techniques*

Method	Resolution	Advantages	Disadvantages	Applications
Optical/ light	2 μm	Simple, cheap	Not very high resolution	Polymer spherulites Liquid crystals
SEM	5 or 10 nm	Excellent visualization of structures	Not quantitative	Monolayers, structures in general
TEM	1 or 2 nm	Very high resolution	Instability to e-beams, sample preparation can be tedious	Heavy metals Solid crystals Block copolymers
STM	0.1 nm		Only conductive surfaces, small sample area (100 nm^2)	Monolayers
AFM	<1 nm	Topography can be mapped, air/liquid environment	Very sensitive, small sample	Micelles, polymers, surface forces

Abbreviations: SEM = scanning electron microscopy, TEM = transmission electron microscopy, AFM = atomic force microscopy, STM = scanning tunnelling microscopy.

A variant of the optical microscope is the stereo microscope, which is designed differently and serves a different purpose. It uses two separate optical paths with two objectives and two eyepieces to provide slightly different viewing angles to the left and right eyes. In this way it produces a 3D visualization of the sample being examined. A more flexible working distance and greater depth of field are important qualities for this type of microscope.

Sometimes, low optical contrast between different phases, particles or droplets can be improved by the use of fluorescence markers. One such example is the viewing of coating cross-sections, where, for example, leaching of biocides or binder degradation by sunlight can be more easily seen after contacting the coating surface with a suitable marker.

Another microscopy technique, which is actually based on light scattered by colloids, is dark-field or ultra-microscopy. In this technique, an ordinary optical microscope is used, but the sample is illuminated in such a way that light does not enter the objective unless scattered by the object under investigation. This technique does now allow a direct observation of, for example, a particle, but is particularly useful for detecting the presence of particles and investigating the Brownian motion of colloids. An important requirement is that the refractive index of the colloids

and the medium must be sufficiently different to ensure adequate scattering.

More advanced (and expensive) microscopy techniques are differential interference contrast microscopy and laser scanning confocal microscopy. The latter can provide precise optical sectioning so that three-dimensional images of structures can be recorded.

9.3   Electron microscopy

The two most popular techniques in this area are scanning electron microscopy (SEM) and transmission electron microscopy (TEM). SEM, which was invented in the 1930s, is an advanced method, which is very useful for looking at structures at very high magnifications (from 10× to 500 000×). Examples of SEM micrographs are shown in Figure 9.2.

Sample preparation, including carbon or gold sputtering, used to be rather time consuming and examination had to take place in vacuum. However, modern environmental SEM (E-SEM) equipment employs a pressure chamber and short working distance and allows immediate observation without sputtering and vacuum. The maximum sample size that will fit in the measurement chamber is typically about 15×15 cm^2.

In combination with X-rays, the analysis technique is called SEM-EDX, where EDX stands for energy

Example 9.1. Investigation of coating cross-section.

A chemically active antifouling coating is used to protect the underwater hull parts of ships against biofouling (growth of plant and animal organisms). Bulk carriers, oil tankers, container ships and pleasure crafts all need an antifouling coating to avoid a prohibitive drag resistance from biofouling.

However, for reasons of environmental protection, the coating should contain as little biocide and organic solvents as possible, yet still provide a smooth hull over many years of ship operation.

In the optimization of such products, what is a useful structure characterization method?

Solution

Upon seawater immersion of an antifouling coating, soluble pigments dissolve and leach from the coating into seawater, leaving behind a porous polymer matrix. Simultaneously, the binder also reacts with seawater and slowly dissolves (this is more or less the same process that takes place when a soap bar dissolves). A useful characterization method is SEM, but some sample preparation is required to obtain a cross-section view on the coating. Therefore, it must be cut with a fine plane, a so-called microtome. A SEM image of the coating cross-section taken from Kiil and Dam-Johansen (2007) is shown here.

Reprinted from Kiil and Dam-Johansen (2007), with permission from Elsevier.

It can be seen that soluble pigments (the white dots in the figure) have leached into seawater from the upper part of the coating forming a so-called pigment-leached layer with porosity. The thickness of the leached layer can be measured. Long fibres, for reinforcement, are also visible in the coating.

dispersive X-ray analysis. Using SEM-EDX, detection of heavy elements in selected areas can be conducted. Quantitative analysis, however, is difficult. An example is the establishment of a concentration profile of Cu or Sn in a coating cross-section as shown in Figure 9.3.

Another important technique for high magnification is transmission electron microscopy (TEM). The first TEM was built by Max Knoll and Ernst Ruska in 1931 and the first commercial edition was available in 1939. The principle is a beam of electrons that transmits through an ultrathin specimen, thereby interacting with the specimen as it passes through. From the interaction of the electrons transmitted, an image is formed. The image is subsequently magnified and recorded. The magnification is impressive, allowing for visualization of even a single row of atoms. TEM is used extensively in the fields of cancer research, catalysis and materials science, but there are also many other applications. An example of a TEM recording is shown in Figure 9.4.

(a)

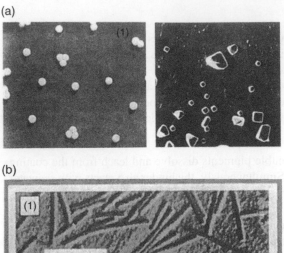

(b)

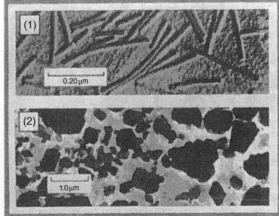

Figure 9.2 *Scanning electron micrographs (SEM) of various particles. (a) Polystyrene latex ((1) 50 000× magnification) and AgCl ((2) 15 000× magnification) particles are seen. Both size distribution and aggregation (maybe due to crosslinking) of polystyrene (PS) particles can be distinguished. Notice that the PS dispersion is monodisperse. Shaw (1992). Reproduced with permission from Elsevier. (b) Two different types of particles that represent extreme variations from spherical particles are seen ((1) shows tobacco virus particles and (2) shows clay particles). The clay (sodium kaolinite) particles have a mean diameter of 0.2 μm. Hiementz and Rajagopalan (1997). Reproduced with permission from Taylor & Francis*

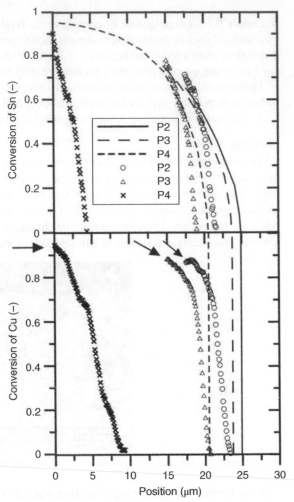

Figure 9.3 *Example of concentration (conversion) profiles of Cu and Sn in three coatings after 47 days of seawater exposure. P2, P3, and P4 are short names for the coatings investigated. Symbols are experimental data obtained using SEM with EDX analysis and lines are model simulations (not relevant for the purpose of this discussion). The arrows indicate the position of the coating surface. The latter changes position over time because the coating polishes (is reduced in thickness) in seawater. Reprinted from Kiil et al. (2001), with permission from American Chemical Society*

9.4 Atomic force microscopy

One of the most popular choices of microscopy is atomic force microscopy (AFM), which is a modification of the earlier scanning tunnelling microscope

(STM). AFM is a powerful imaging and surface–force measuring technique that is well-suited for studying complex technical systems (Figure 9.5). An extremely sharp tip is used to track the surface. A topographical image is generated from the forces acting between the

(a)

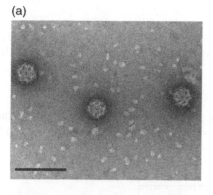

(b)

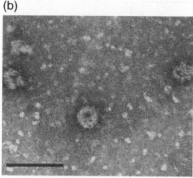

Figure 9.4 *Transmission electron microscope (TEM) image of a virus: (a) non-treated dispersion and (b) dispersion after an electrochemical deactivation treatment. Scale bars show 100 nm. Shionoiri et al. (2015). Reproduced with permission from Elsevier*

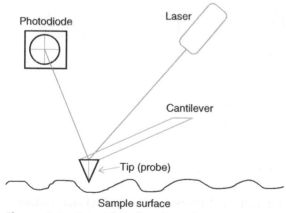

Figure 9.5 *Principle of atomic force microscopy (AFM). The movement of the tip over the surface can be followed by the use of a laser beam and a photodiode*

tip and the surface. AFM is capable of very high resolution of sample features in three dimensions, in some cases even down to atomic resolution. Unlike other surface characterization techniques (e.g. TEM), AFM measurements do not need to be carried out in vacuum but can take place in air or liquid. Moreover, the samples do not need to be treated or coated. However, it should be emphasized that the interpretation of AFM images needs to be done carefully – in reality, the image is not purely a function of the sample surface, but represents a convolution of the forces between the tip and the surface. It also represents a very small section of a macroscopic surface, which is not necessarily representative of the entire sample. Despite these facts, AFM has revolutionized the study of surfaces at high resolution, and is an incredibly useful and versatile tool. This popularity is attributed to several factors, which are illustrated in Table 9.1.

9.5 Light scattering

All materials can scatter light and this phenomenon is known as the Tyndall effect. However, colloids are capable of intense light scattering and therefore exhibit extreme turbidity. The white appearance of milk is a classic example.

Light is scattered in all directions, when it meets an object (e.g. a colloid particle) of dimensions within an order of magnitude (i.e. a factor of 10) or so of its own wavelength. The intensity, angular distribution and polarization of the light scattered depend on the size and shape of the scattering particles, the interactions between them, and the difference in refractive indices of the particles and the dispersion medium. This is exploited in light-scattering measurements to measure particle size distributions and has found wide applications. The method is fast, can use both dry and wet samples and requires very little preparation.

Scattering increases with particle size and therefore this property can be used to follow the progress of particle aggregation (coagulation) and thus study the stability of a colloidal dispersion.

Particles, and structures in general, can also scatter X-ray and neutrons and several techniques are based on those principles (Table 9.2).

Example 9.2. *Particle size distributions of pigments.*

Pigments are used in plastics, coatings, tires and cosmetics for several reasons. They can, for example, add colour, improve mechanical properties, or act as biocides. Very often it is of interest to know the particle size distribution of pigments directly in a product (e.g. a coating).

Solution

There are several methods available for this measurement, but a very easy and fast one is laser diffraction using, for example, the Malvern Mastersizer. An example of such a recording is shown below (Kiil and Dam-Johansen, 2007).

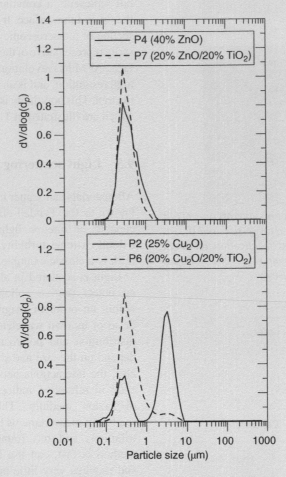

Reprinted from Kiil and Dam-Johansen (2007), with permission from Elsevier

The figure shows particle size distributions (frequency function) of pigments (Cu_2O, ZnO and mixtures Cu_2O/TiO_2 and ZnO/TiO_2) in coatings (named P2, P4, P6 and P7). It can be seen that ZnO contains mainly submicron particles; Cu_2O has a bimodal distribution of sizes with one peak in the submicron domain.

Table 9.2 *Scattering methods of importance in colloid science*

Method	Characteristics – what can we do	Applications
SALS	Molecular weight, second virial coefficient Particle shape, size distribution Colloid stability (aggregation)	Polymer, proteins, micelles, particles
DLS (PCS)	Diffusion coefficient	Polymer solutions
SAXS	Radius of gyration of polymers	Small particles (5–100 nm) Monolayers
SANS	Radius of gyration of polymer chains in melt	Concentrated dispersions, polymers, soft materials

Abbreviations: SALS = small-angle light scattering; DLS = dynamic light scattering (also called PCS = photoelectron spectroscopy or QELS = quasi-elastic light scattering); SAXS = small-angle X-ray scattering; SANS = small-angle neutron scattering.

Table 9.3 *Major spectroscopic methods of importance in colloid science*

Method	Characteristics – what can we do	Applications
ESCA/XPS	Atoms at surface How they are bonded	Paints, paper surfaces, polymers, pigments, minerals, metals
NMR	Identify compounds Copolymer composition	Polymer solutions Adsorption on colloids
(FT)IR	Chain branching of polymers, tacticity	Monolayers

Abbreviations: ESCA/XPS = electron spectroscopy for chemical analysis (X-ray photoelectron spectroscopy); NMR = nuclear magnetic resonance; FT-IR = Fourier-transformed-infrared radiation. Fourier transformation is an algorithm used to treat signals.

X-ray and neutron scattering techniques are advanced and, as the name implies, require sources of X-rays and neutrons; the latter are typically obtained from a nuclear reactor. SAXS and SANS are for very specialized purposes, typically in fundamental research, and are rarely used in product development.

9.6 Spectroscopy

Besides the microscopic methods mentioned above, there are various spectroscopic methods that can be used for analysis of colloidal systems or surfaces. They are summarized in Table 9.3.

The methods involve the analysis of the interaction of matter with radiation. In general, spectroscopic methods give information about the identity (what atoms are there) and the chemistry (how they are bonded together) of the surface. Surface diffraction methods, as we have seen, give information about the ordering and the structure in the surface layers (i.e. how the atoms are arranged).

There is little doubt that one of the most versatile and powerful spectroscopic methods is electron spectroscopy for chemical analysis, ESCA (also termed X-ray photoelectron spectroscopy, XPS), developed in the 1950s by the Swedish scientist Kai Siegbahn (winner of the Nobel Prize in Physics for this discovery in 1981). The method has been long commercialized and is currently extensively used for surface analysis. ESCA provides information on the relative amounts of different elements on the surface of materials, with a depth of analysis of only 2–10 nm. Information on the chemical environment for the elements, e.g. oxidation state, can also be obtained for many samples. Various types of materials (organic, inorganic, surfaces, powders, fibres) can be analysed with this technique. A combination of ESCA and AFM can help us identify whether differences in properties over a surface are due to heterogeneity in topography/surface structure or to chemistry.

Other spectroscopic techniques, such as those employing nuclear magnetic resonance (NMR) and infrared (IR) radiation, can also be used for chemical analysis and identification, but will not be discussed here.

9.7   Concluding remarks

Being able to characterize colloids and surfaces is an important part of this field. In a given investigation, information can be required on particle size and shape, surface morphology and chemistry, phases present, as well as degree of aggregation. In this chapter, the most important analysis techniques for these applications have been briefly presented within microscopy, light scattering and spectroscopy. There are many more specialized methods available that may become useful for a given application. Many of these are expensive to acquire and often only available at universities, in large companies or in specialized firms.

Problems

Problem 9.1
Which analysis techniques would you use to characterize the following colloidal products: paint, ice cream, shampoo, and lipstick?

Problem 9.2
Is scanning electron microscopy (SEM) a useful analysis technique for soft polymer-based wound care products?

References

J.C. Berg, An Introduction to Interfaces and Colloids – The Bridge to Nanoscience, World Scientific, New Jersey, 2010.

A. Hellgren, A. Meurk, Atomic force microscopy, *Compendium: Surface and Colloid Chemistry in Industry*, 24–28 January 2000, Ytkemiska Institut (Institute for Surface Chemistry), Sweden.

P.C. Hiemenz, R. Rajagopalan, Principles of Colloid and Surface Chemistry, 3rd edn, Marcel Dekker, 1997.

S. Kiil, K. Dam-Johansen, Prog. Org. Coat., 60, 238–247 (2007).

S. Kiil, C.E. Weinell, M.S. Pedersen, K. Dam-Johansen, *Ind. Eng. Chem. Res.*, 40, 3906–3920, (2001).

D.J. Shaw, Introduction to Colloid & Surface Chemistry, 4th edn, Butterworth–Heinemann, Oxford, (1992).

N. Shionoiri, O. Nogariya, M. Tanaka, T. Matsunaga, T. Tanaka, *J. Hazard. Mater.*, 283, 410–415, (2015).

10

Colloid Stability – Part I: The Major Players (van der Waals and Electrical Forces)

10.1 Introduction – key forces and potential energy plots – overview

Colloidal dispersions have a high interfacial area and are therefore unstable; they have to be *stabilized*. This is their most crucial property and we need to understand it so that we know how to change it, if necessary. Colloidal stability can be understood by considering the various types of forces between the particles (Figure 10.1). The most important ones are the van der Waals (vdW) forces, which are usually attractive and are the cause of instability and the, usually, repulsive forces due to electrical charges on the particles. The DLVO theory that we will present later considers only the vdW and the electrical forces, which are often the most important ones in our discussion. Steric forces due to adsorption on the particles of polymers or oligomers can also be very important for stability (see Figure 10.4 below). We will discuss steric forces in Chapter 12 together with emulsions, where the steric forces play a crucial role. All these forces act at distances of the order of nanometres. Even so they have a direct effect on macroscopic properties of dispersions and aggregates.

The *colloid stability* of a dispersion refers to small particles that do not aggregate (in the practical lifetime of the dispersion), whereas *mechanical stability* refers to dispersed particles that do not sediment or cream, as discussed in Chapter 8.

We typically use the potential energy–distance plot (*V–H* curves) to represent graphically the behaviour of a colloidal system (Figure 10.2). Minima indicate attraction and instability and a maximum indicates stability and repulsion. Often there are two minima, one at high distances indicating the reversible flocculation and one at very short distances and when/if this latter minimum is reached we have an irreversible coagulation of the colloidal system. In-between we may have at intermediate distances enough repulsion between the particles to achieve (temporarily) stability. A more extensive discussion is provided in Appendix 10.1.

The interaction (potential) energy between macroscopic bodies, V, separated by a distance, H, is defined as:

$$V(H) = -\int_{\infty}^{H} F \, dH \qquad (10.1)$$

Introduction to Applied Colloid and Surface Chemistry, First Edition. Georgios M. Kontogeorgis and Søren Kiil.
© 2016 John Wiley & Sons, Ltd. Published 2016 by John Wiley & Sons, Ltd.
Companion website: www.wiley.com/go/kontogeorgis/colloid

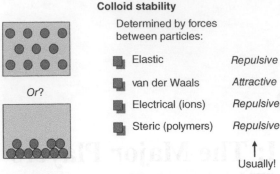

Colloid stability

Determined by forces between particles:

■	Elastic	*Repulsive*
■	van der Waals	*Attractive*
■	Electrical (ions)	*Repulsive*
■	Steric (polymers)	*Repulsive*

↑
Usually!

Figure 10.1 *Colloidal dispersions are inherently unstable systems and in the long run the attractive forces will dominate and the colloidal system will destabilize. However, colloid stability depends on the attractive van der Waals and the repulsive electrical or steric (polymeric) forces. The repulsive forces stabilize a dispersion if they are larger than the van der Waals (vdW) ones (and the total potential is larger than the "natural" kinetic energy of the particles). Surfaces are inherently unstable and the van der Waals forces "take the system" back to its stable (minimum surface area) condition and contribute to instability (aggregation)*

and represents the work done in bringing the bodies from a distance of infinity to the separation distance *H*.

The interparticle force is given by the derivative of the interaction energy:

$$F = -\frac{dV}{dH} \qquad (10.2)$$

Israelachvili (2011) describes how this definition for a force law arose historically. Equation 10.2 arises from Equation 10.1 by using Leibnitz's rule of differentiation of integrals. Negative *F* and *V* imply attraction while positive *F* and *V* mean repulsion.

The stability of colloidal dispersions is often described quantitatively via the DLVO (Derjaguin–Landau–Verwey–Overbeek) theory which accounts in a simple additive way for both the (usually) attractive van der Waals (vdW) energies, V_A, and the (typically) repulsive electric/electrostatic or double-layer energies, V_R:

$$V = V_R + V_A \qquad (10.3)$$

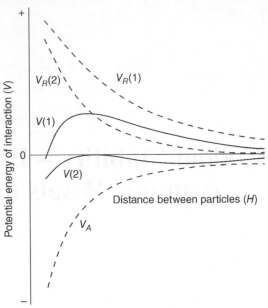

Figure 10.2 *Total potential energy (V)–interparticle distance (H) curves. Various combinations of attractive and repulsive forces are shown here. In case V(1) repulsive forces dominate and we have a stable dispersion, but in case V(2) (weaker repulsive forces) the attractive forces dominate at all separations and thus we have an unstable dispersion. In the figure, V(1) = V_R(1) + V_A and V(2) = V_R(2) + V_A. Shaw (1992). Reproduced with permission from Elsevier*

The vdW energies decrease with the first (sometimes the second or 3/2) power of the interparticle distance (see Table 2.2, Chapter 2), while the electrostatic (double layer) energies decrease exponentially with distance (see, for example, Equation 10.4). Both are much more long-range compared to the intermolecular forces.

The electrostatic forces are due to the fact that most particles are charged inside a medium, especially a polar one like water, which has a high relative permittivity. The origin of the surface charge can be complex and there are many mechanisms for this, e.g. adsorption of ions from the solution or dissociation of surface groups. Other repulsive forces, which help stability, exist especially at low interparticle distances (e.g. the so-called hydration or steric forces).

10.1.1 Critical coagulation concentration

The curve $V(2)$ in Figure 10.2 illustrates the situation where the salt (electrolyte) concentration in the medium is sufficiently high to result in $V = 0$ and $F = -dV/dH = 0$ at the same time. This maximum point is equivalent to the critical coagulation concentration, CCC (see also Section 10.8).

Both the attractive vdW and the electrostatic forces:

- are long-range (vdW forces dominate at high and small distances, electrostatic forces may dominate at intermediate distances);
- depend on the characteristics of the particles and the medium;
- depend greatly on the shape of the particles/surfaces (e.g. sphere, plates, cylinders), their size and proximity.

Many expressions exist, as we see later (see also Table 2.2 in Chapter 2), but in the simple case of equal-sized spherical particles of equal and small surface potentials (less than 25 mV for 1 : 1 electrolytes) and small electric double layer overlap the DLVO theory takes the following form (see also Figure 10.3):

$$V = V_R + V_A = 2\pi R \varepsilon \varepsilon_0 \psi_0^2 \exp(-\kappa H) - \frac{AR}{12H} \quad (10.4)$$

In DLVO the total potential energy is expressed as simply the sum of the repulsive and the attractive energy (Equation 10.3) and in Equation 10.4 DLVO is written for simple equal-sized spherical particles. According to the DLVO theory, prevention (or promotion) of flocculation can be achieved through the control of the balance of attractive and repulsive

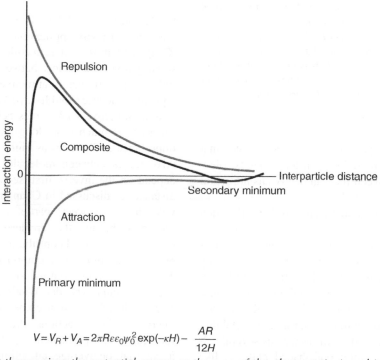

$$V = V_R + V_A = 2\pi R \varepsilon \varepsilon_0 \psi_0^2 \exp(-\kappa H) - \frac{AR}{12H}$$

Figure 10.3 *DLVO theory gives the potential energy as the sum of the electrostatic (repulsive) and the van der Waals (attractive) forces. The equation shown here is for equal-sized spherical particles ($H \ll R$). If the particles interact over a (liquid) medium (as opposed to vacuum), the Hamaker constant (A) becomes an effective Hamaker constant. The secondary shallow minimum (of a few $k_B T$) at rather high separations indicates reversible flocculation. The aggregates are rather loose (weak) and are called flocs. This secondary minimum has been confirmed experimentally and often can disappear with a small energy input, e.g. gentle stirring*

(a)

(b)

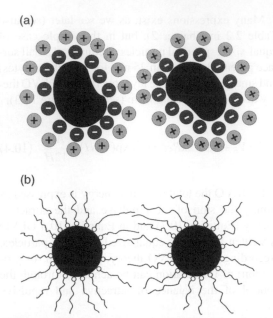

Figure 10.4 *Schematic representation of the electrical double-layer forces (a) and of the steric (or entropic) stabilization (b). The latter is obtained using adsorption of oligomers or polymers. The "hairy" surface prevents flocculation of the particles. This is the reason for very good stability of solvent-based coatings, where binder molecules can adsorb on the pigment particles*

forces between the particles. This theory states that the stability of a colloidal suspension depends solely on the relation between the van der Waals attraction and the electrostatic repulsion.

The secondary minimum that results due to the often weakly bounded flocs (loose aggregates) deserves more discussion. For very small particles (radius less than about 10 nm) the secondary minimum is not deep enough to get flocculation. If the particles are larger, flocculation in the secondary minimum may cause observable effects. Secondary minimum flocculation is considered to play an important role in the stability of certain emulsions and foams as well as several colloidal systems containing odd-shaped particles like iron oxide and tobacco mosaic virus.

Notice that the electrostatic forces are typically repulsive due to the overlap of the double layer between particles. They are always repulsive between particles of the same kind. They are proportional to the second power of the surface potential (in most cases)

and they decrease with decreasing Debye (double-layer) thickness (κ^{-1}). This is a key parameter, crucial for controlling the double-layer forces (Figure 10.4).

The Debye thickness decreases with increasing electrolyte concentration and it decreases more for the high-valency electrolytes. The surface potential is estimated, sometimes indirectly, via electrokinetic experiments (see Section 10.6). Through these experiments we can measure the electrophoretic mobility, which for very small or very large particles can be related by theory to the so-called zeta potential (Hückel and Smoluchowski equations, Equation 10.10). The zeta potential is approximately equal to the surface potential.

10.2 van der Waals forces between particles and surfaces – basics

While polar and acid–base interactions are often very important in many applications related to interfaces (Chapter 6), most studies in colloids related to colloid stability involve a balance between the van der Waals (vdW) forces (typically attractive) and the (typically repulsive) electric (double-layer) forces. The (macroscopic) attractive vdW forces between particles and surfaces are much more long-range (they can be of importance at distances as high as 100 nm) than the vdW forces between molecules and the potentials depend on the inverse first, second or 3/2 power of distance, as discussed in Chapter 2 (Table 2.2). The vdW forces between particles are important and long-range because they are derived from the summation of the individual contributions of the vdW forces between all molecules composing these particles. This can be carried out as, to a first approximation, such molecular van der Waals forces are additive. The origin of attraction comes from three types of intermolecular attractions (Shaw, 1992): (1) two molecules with permanent dipoles orient in such a way that, on average, attraction occurs; (2) dipolar molecules induce dipoles in other molecules so that attraction occurs and (3) dispersion forces due to polarization of one molecule by fluctuations in the charge distribution in a second molecule, and vice versa. Dispersion forces are more important. The reason for this is that dipole–dipole interactions, induced dipole interactions (and hydrogen bonds) are interactions between surface sited atoms (or molecules), whereas for dispersion forces every atom in one colloidal particle may undergo vdW interaction with every atom in the next particle. Note that hydrogen bonding only occurs between certain molecules with a high electronegativity (e.g. water and ammonia molecules) and must be treated separately when it occurs. Hydrogen bonds are short-range forces and complex (see, for example, Israelachvili, 2011, and the discussion in Chapter 2, Section 2.1).

The vdW forces are typically attractive but can be repulsive in special cases (see later). They depend on the size and shape (geometry) of particles/surfaces and their proximity but above all they depend on the nature (chemistry) of the materials and the medium via the so-called Hamaker constant, A.

For example the van der Waals force between equal-sized spherical colloidal particles of radius R (see also Chapter 2) is expressed in potential energy–distance form (V–H) as:

$$V_A = -\frac{AR}{12H} \qquad (10.5)$$

where H is the interparticle or interface distance and R is the radius of the particle. For the equation to hold, $H \ll R$. Table 2.2 (Chapter 2) shows the expression for the sphere–flat surface and surface–surface (infinite thickness) interactions.

It has been shown (Israelachvili, 2011) that for diameters beyond 0.5 nm a molecule must be considered a small particle and the equations of Table 2.2 should be used for the van der Waals forces. The effective Hamaker constant, which depends on the particles (chemical composition) and the medium, determines the effective strength of the vdW interactions.

The Hamaker constant for a given material in vacuum can have values that typically range between 10^{-19} and 10^{-21} J, with higher values for metals (e.g. 40×10^{-20} J for silver) than for organics (e.g. 6.6×10^{-20} J for polystyrene, 5.4×10^{-20} J for toluene and 4.1×10^{-20} J for hexane). This implies that metal particles like silver, gold and platinum are much "stickier" than particles made up of organic materials. Such metal particles are important technologically as they are widely used in the form of inks for printing electronic circuit boards and integrated circuits. The larger

Table 10.1 *Examples of non-retarded (distance-independent) Hamaker constants for two identical materials interacting in a vacuum at room temperature. It should be emphasized that values for the same material can differ significantly from one source to another. The Hamaker constant of water is very low due to water being a small molecule. Additional values can be found in, for example, Bergström (1997) and Israelachvili (2011)*

Material	Hamaker constant ($\times 10^{-20}$ J)
Water	3.7
Benzene	5.0
Ethanol	2.2
Mica	13.5
Diamond	29.6
Polystyrene	7.0
Teflon	3.8
Hexadecane	5.2

the number of atoms or molecules per volume material and the larger their polarizabilities, the larger the Hamaker constant. More values for Hamaker constants in vacuum are shown in Table 10.1.

When we have a colloid system in a medium other than vacuum or air, the Hamaker constant is decreased and an effective value must be used. We therefore use the symbol A_{eff} in the relevant equations. Thus, in the more realistic situation where spherical particles or surfaces are immersed in another liquid or solid medium, then the van der Waals forces can be greatly reduced. As most industrial processes are carried out in a liquid, there are typically two effects: (i) we may have adsorption phenomena on the particle surface and thus reduction of the surface energy and (ii) we have reduced van der Waals forces due to the smaller value of the effective Hamaker constant in a liquid compared to vacuum or air.

One possible way of lowering attractive forces between particles is by making their surfaces coarse. This will make the embossed parts on the particle surface act as spacers. However, this will also result in more dusting if the medium is a gas (e.g. air).

10.3 Estimation of effective Hamaker constants

As discussed in Chapter 2, the effective Hamaker constants can be obtained either by empirical combining rules or via the more rigorous Lifshitz theory which requires knowledge of the relative permittivities and refractive indices of all materials (Equation 2.8). We will now elaborate a little on these concepts.

The Hamaker constant is also related to the surface tension but this estimation method (not shown here) depends on the value chosen for the distance H. Once reliable values of surface tensions and Hamaker constants are available from other techniques or measurements, then the relevant equations can be used to estimate the interparticle distance. Such calculations yield for many compounds like hydrocarbons, water and polystyrene values around 0.2–0.3 nm, which are physically reasonable.

If the properties required in the Lifshitz theory are not available, effective Hamaker constants can be estimated via combining rules. The effective Hamaker constant of particles (1) in a medium (2) is given, as shown in Chapter 2, as:

$$A_{eff} = A_{121} = \left(\sqrt{A_{11}} - \sqrt{A_{22}}\right)^2 \qquad (10.6)$$

Equation 10.6 is true when we have the same type of particles in a medium, e.g. polystyrene particles in water. However, one simple equation for the effective Hamaker constant in the case of two different types of particles (1 and 3) in a medium 2 is:

$$A_{eff} = A_{123} = \left(\sqrt{A_{11}} - \sqrt{A_{22}}\right)\left(\sqrt{A_{33}} - \sqrt{A_{22}}\right) \qquad (10.7)$$

Thus, in the (often encountered) case of particles of the same type or when air/vacuum are the medium (zero Hamaker constant for vacuum and almost zero for air), Equations 10.6 and 10.7 lead always to a positive Hamaker constant and attractive van der Waals forces, which is the most usual case. However, in the case of unlike particles and when the Hamaker constant, A, of the medium has a value in between that of two different types of particles, the effective Hamaker constant can be negative, which implies repulsive van der Waals forces. In this case one material interacts more strongly with the medium than with the second body.

These repulsive vdW forces find many applications, e.g. they can be used for screening solubility in blend-solvent mixtures. Another application of negative Hamaker constants is the engulfment

process. If $A < 0$ then the particle (1) would be rejected from medium (3) and if particle (2) is liquid then particle (1) will be engulfed. Engulfment is an important process in biological systems, e.g. part of our body's defence mechanism where white cells engulf and remove foreign cells like bacteria in a process called phagocytosis. After the engulfment, particle (1) is in a medium (2) interacting with surface (3), having a positive effective Hamaker constant.

Equation 10.7 shows that the van der Waals interactions are weak when the particles and the dispersion medium are chemically similar, which implies that A_{11}, A_{22} and A_{33} are of similar magnitude and thus A_{123} will have a low value (even close to zero). This approach (suitable choice of a medium for the particles) provides us a methodology of reducing the Hamaker constant to such an extent that we can achieve a stable dispersion.

Some examples of effective Hamaker constants are given in Table 10.2. Notice that the Hamaker constant decreases in a medium compared to air and that in some cases negative effective Hamaker constants (repulsive vdW forces) exist.

The Hamaker constant can be calculated directly from London's theory (Chapter 2, Equation 2.6). The Hamaker constant values range typically between 0.4 and 4×10^{-19} J. The relatively constant values for different compounds arise because the parameter C (see Equation 2.6) is roughly proportional to the square of the polarizability which is roughly proportional to the square of the volume (or the inverse of number density), as discussed in Chapter 2. Thus the whole product (the Hamaker constant) is roughly constant, which is, of course, a gross oversimplification.

A more rigorous estimation of the effective Hamaker constant for mixtures avoids simplified combining rules and uses the Lifshitz theory (based on relative permittivities and refractive indices; see Israelachvili (2011) and Chapter 2, Equation 2.8). The Lifshitz theory is particularly useful for calculating the van der Waals force for any surface and in any medium, also because it relates the Hamaker constant with the material properties (relative permittivity and refractive index). Thus, the theory shows how the van der Waals forces can be changed via changing the Hamaker constant. The Lifshitz theory is a continuum theory, i.e. the dispersion medium, typically water, is

Table 10.2 *Some examples of positive and negative effective Hamaker constants. The magnitude of the van der Waals attraction increases with the difference in dielectric properties between the medium and the particles. Fused quartz is glass containing silica in amorphous form. The Hamaker constant for air is here taken to be 0*

Interacting media (1-2-3) 2 is the medium	A ($\times 10^{-20}$ J) Particle 1	A ($\times 10^{-20}$ J) Medium 2	A ($\times 10^{-20}$ J) Particle 3	A_{eff} ($\times 10^{-20}$ J)
Pentane–air–pentane	3.8	0	3.8	3.8
PS–water–PS	6.6	3.7	6.6	0.063
Pentane–water–pentane	3.8	3.7	3.8	0.28
Water–pentane–water	3.7	3.8	3.7	0.08
Fused quartz–air–fused quartz	6.3	0	6.3	6.3
Fused quartz–water–fused quartz	6.3	3.7	6.3	0.63
Fused quartz–air–fused quartz	6.3	0	6.3	−0.87

considered to have uniform properties and many-body effects are included. However, at short distances – of a few molecular diameters – the discrete molecular nature of the dispersion medium cannot be ignored.

This rigorous Lifshitz theory of estimating the Hamaker constant yields, moreover, values less affected by the retardation effects, which can be of some significance at larger distances. Potential energy values at distances above about 10 nm will be overestimated due to neglecting the finite time (so-called transit time) required for the propagation of electromagnetic radiation between particles. To include retardation effects, a distance-dependent Hamaker constant must be used. A description of this approach can be found in, for example, Israelachvilli (2011).

10.4 vdW forces for different geometries – some examples

There are more general expressions for the vdW forces between particles/surfaces depending on the geometry and proximity of the particles. A comprehensive list is given in Chapter 2 (Table 2.2). We see that not only size but also the shape of the particles (surfaces in general) play a very important role in vdW interactions (as is the case for the electrostatic repulsions, see later). Although (near-)spherical particles are quite common for colloidal dispersions, other shapes are also possible, e.g. calculations between flat plates are relevant for the stability of thin soap films.

We close this part of the chapter with some examples. One example of the van der Waals attractive

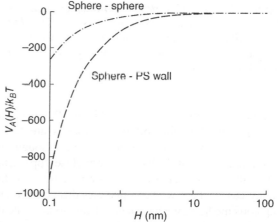

Figure 10.5 *Attractive interaction between tetradecane (C_{14}) spheres and between spheres and the polystyrene (PS) wall they are contained in. Adapted from Goodwin (2009), with permission from John Wiley & Sons, Ltd*

forces (actually potential) is shown in Figure 10.5 for the interaction between tetradecane (C_{14}) spheres with each other and with the polystyrene (PS) wall they are contained in (at close approach). The equations for the van der Waals forces (sphere–sphere, sphere–plate; see Table 2.2) were used. There are strong interactions in both cases, even at distances of tens of nanometres, stronger in the case of sphere–wall interactions. This is due to the expression used but also because the net Hamaker constant for the latter interaction is larger. Although both are simple hydrocarbons (C_{14} and PS), the polymer is a denser molecule with a large amount of aromatic character giving a different electronic polarizability.

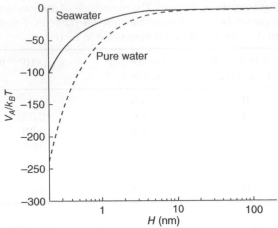

Figure 10.6 *Oil (C$_{14}$) droplets in water and seawater. In seawater, the oil droplets are less sticky. The difference is due to the Hamaker constants in the two liquids. Adapted from Goodwin (2009), with permission from John Wiley & Sons, Ltd*

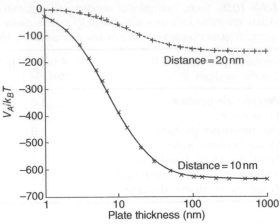

Figure 10.7 *Attractive forces between plates of finite thickness. The two inter-plate distances used are provided in the figure. The Hamaker constant used is A$_{11}$ = 10^{-19} J (vacuum) and the plate area is 0.1 μm^2. Adapted from Goodwin (2009), with permission from John Wiley & Sons, Ltd*

Another example is shown in Figure 10.6 where it can be seen that that oil (C$_{14}$) droplets are "less sticky" in seawater than in pure water, although an attraction of about 100$k_B T$, at very close approach, can still be considered a rather strong attraction. In many cases where we have particles dispersed in an aqueous medium, we have a significant ionic content (e.g. ionic surfactants, electrolytes or ions associated with the particle surface due to surface charge), which influences the value of a given Hamaker constant.

The final example is shown in Figure 10.7. The equation for the Hamaker constant for plates of finite thickness has been used for the calculations shown here. The plate area is 0.1 μm^2 and the Hamaker constant is 10^{-19} J. These values would be similar to those found for clay molecules. Clearly, from Figure 10.7, the finite thickness of the plate is only important for thin plates (less than about 50 nm) and also the variation of the interaction with distance is greater for thin plates.

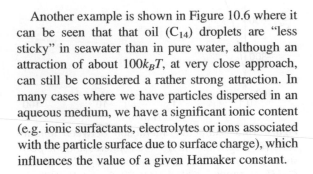

Example 10.1. Stability–Hamaker constant.

With reference to Example 2.2 and the Hamaker constants calculated in that example:

Assume that the zeta potential values for PS, alumina and zirconia particles are similar. Which of these three kinds of particles is expected to result in the most stable colloidal dispersion in water? For the most stable system, it is suggested to further improve the stability by dispersing the particles in a medium other than water. Which dispersion medium or media would you recommend using among those shown in Table 10.1?

Solution

Clearly, the lowest Hamaker value is for PS particles and, thus, they are the most stable particles in water (among the three considered here). If we should choose a solvent that provides better stability for PS particles, then we should choose a solvent that has a Hamaker constant closer to PS (=7) than water (=3.7). Possibilities from Table 10.1 are hexadecane (=5.2) or hydrocarbons in general (= about 6). In parenthesis are shown the Hamaker constants (in 10^{-20} J).

10.4.1 Complex fluids

One final remark before we continue with the electro-static forces is the issue of concentrated "soft" dispersions. As the particle concentration increases, multibody interactions become increasingly important and we have a "condensed" phase. According to Goodwin (2009), this happens when the interparticle forces produce a three-dimensional structure which is space-filling. If the Lifshitz theory (see Chapter 2) is used to calculate the effective Hamaker constant, many-body effects are taken into account, but the macroscopic behaviour of a dispersion of "soft" particles needs special attention. Concentrated dispersions can be thickened liquids and gels, which are also referred to as *complex fluids* or *soft matter*, topics of great scientific interest, which include fluid-like structures such as micelles (see Figure 10.8 and Chapter 5), microemulsions, bilayers, vesicles, biological membranes and proteins.

The most relevant technical issues for these non-rigid "materials" are a characterization of the structure and the associated macroscopic properties (e.g. rheology). However, this topic is beyond the scope of this book and the interested reader is referred to Goodwin (2009) and Israelachvili (2011) for details.

Note that when the particle concentration increases in a dispersion then the important mean interparticle distance decreases. It is, however, rather difficult to estimate the three-dimensional mean interparticle distance in a dispersion containing polydisperse

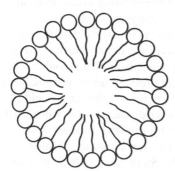

Figure 10.8 *Structure of a micelle. The diameter of a typical micelle is only a few nanometres. See Chapter 5 for details*

particles, which perhaps also have complex particle shapes. Details of such calculations can be found in, for example, Liu and Kowk (2001).

10.5 Electrostatic forces: the electric double layer and the origin of surface charge

The electrical force is due to the electrical double layer that exists around the particles in nearly all colloidal dispersions. Almost all particles are charged in aqueous (or other polar) media (Figure 10.9). The interfaces may have a positive or negative charge, but the latter (negative) is the most common. We will see that the charge depends on pH, nature of surface groups and salt concentration.

Most surfaces are negatively charged, since the smaller cations are typically more hydrated than anions (and thus the cations remain in aqueous solution, while the anions adsorb on the surface). In general, smaller ions are more hydrated due to their more intense electric field. The number of water molecules an ion "bonds" is known as the hydration number. Notably, however, these bound water molecules are not completely immobilized – they move and exchange slowly with bulk water. As suggested by Israelachvili (2011), the hydration number is more of a qualitative indicator of the degree to which ions bind water rather than an exact value. Divalent and trivalent cations are more strongly solvated (hydrated) than monovalent cations, while monovalent anions ions are weakly solvated. Hydration helps to "dissolve" the ion. Hydration of counter-ions is good because it restricts the ability to approach the particle surface and enter the Stern layer (see later). Consequently, less destabilization by hydrated ions is obtained, but ionic valency also plays an important role in itself (as shown later by the Schulze-Hardy rule).

The charge at the interface is compensated by mobile counter-ions. These are attracted electrically by the interface, but they also tend to diffuse away. The result is a diffuse *double layer* of ions (Figure 10.10).

In the diffuse double-layer model, the ionic atmosphere consists of two regions: a so-called Stern (or near-Stern) layer and a diffuse layer after that. In the Stern layer we have approximately one sharp counter-ion plane. The counter-ions dominate close to the interface due to attractions

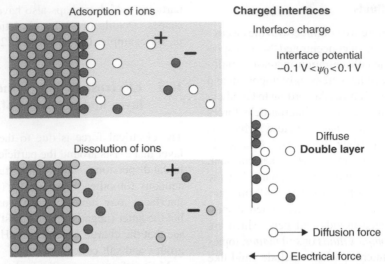

Figure 10.9   *Charged interfaces. The interface charge causes a potential difference (ψ_0) between the interface and the bulk of the fluid. This is usually a few hundredth of a volt (mV range); unfortunately, it is not directly measurable but it is often approximated by the measurable "zeta" potential. The upper part of the figure could be, for example, AgI and the lower part could be, for example, ionization of surface groups in latex or biosurfaces (proteins). Adapted from Wesselingh et al. (2007), with permission from John Wiley & Sons, Ltd*

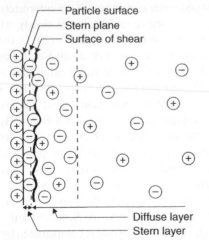

Figure 10.10   *Diffuse double-layer model. In the Stern approach of the double layer shown here, the size of the ions is considered only for the first layer of adsorbed ions, with the ions further away being treated as point charges in the Gouy–Chapman theory. Thus, we have two distinct layers: the Stern layer and the diffuse layer. The dashed line indicates the outer boundary of the diffuse layer. The particle surface shown in the figure is positively charged. Shaw (1992). Reproduced with permission from Elsevier*

with the surface. In the diffuse layer, the concentrations of counter-ions gradually decrease due to thermal motion until we reach an electroneutral solution. The exact location of the surface of shear, of great importance for the measurement of the "zeta potential" as we will see later, is rather unknown. It is considered to be quite close to the Stern plane.

The net charge of the Stern layer, diffuse layer, and the surface is zero because of electroneutrality, but if the double layer of two particles overlap the negative (or positive) charge at the Stern plane will make them repel each other. In more detail, the repelling mechanism involves the development of an osmotic pressure in the mid-plane of the double-layers overlap. The osmotic pressure arises because the concentration of ions is now higher at the mid-plane than in the bulk. This osmotic pressure will push the two surfaces apart by drawing in more solvent (water) as described by, for example, Pashley and Karaman (2004). The Stern layer, which is somewhat rigid, behaves as if it were physically part of the particle. Notice, that the thickness of the double layer is given the symbol κ^{-1} and not κ.

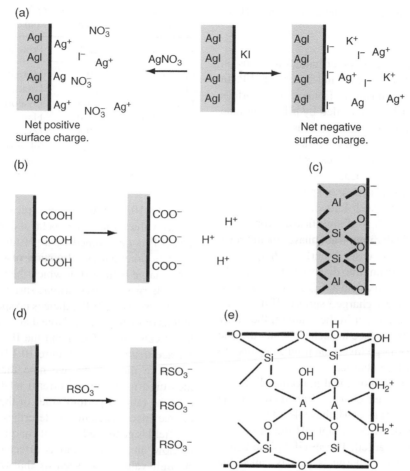

Figure 10.11 *Sources of surface charge in colloids. (a) Differential ion solubility; (b) direct ionization of surface groups; (c) isomorphous substitution; (d) specific ion adsorption; and (e) anisotropic crystals. Adapted from Myers (1999), with permission from John Wiley & Sons, Ltd*

But why are the surfaces or particles charged? Two reasons shown in Figure 10.9 are (i) the interface can adsorb ions from the surroundings or (ii) part of the interface goes into solution as ions (due to the high relative permittivity of water, as explained before). There are many more reasons and they are summarized below and graphically in Figure 10.11 (Myers, 1999):

- adsorption of surface ions, e.g. ionic crystals, AgCl, $CaCO_3$;
- dissociation (dissolution) of surface groups, e.g. latex, celluloses, biomolecules, metal oxides (TiO_2);

- chemisorption of ions, e.g. phosphates on oxide surfaces;
- reaction with OH^- and H^+, e.g. oxide surfaces;
- adsorption of ionic surfactants or polyelectrolytes (general – all surfaces);
- isomorphous substitution (clay minerals, kaolinite, mica).

The magnitude and sign of the surface charge often depend on the pH or the solution or the concentration of charge determining ions. The charge disappears at the isoelectric point (IEP) also called the point of zero charge (PZC).

The following discussion is based on Figure 10.11 (Myers, 1999):

1. The surface charge of silver halides depends on the "common ion effect" and is zero at IEP. In the case of mechanism (b) [ionization of surface groups], the IEP is controlled by pH, rather than the concentration of common ions.

    The adsorption of anionic surfactants yields negatively charged surfaces, while we obtain positively charged surfaces with the rarer cationic surfactants. One case of specific ion adsorption is via the well-known "ion-exchange" mechanism, which is of significance, for example, to biologically important bilayer membranes structures. In this case, a surface charge may result at first from the dissociation of a salt, e.g. sodium carboxylate ($-COO^- + Na+$) produces a negatively charged surface. If di- or trivalent ions are present in the solution, they may adsorb onto the surface in such a manner that the net result is a charge inversion from a negative to a positive surface charge.

2. The ionization of biological molecules like proteins (and thus their charge) depends on the pH of the solution. At low pH, the molecules are positively charged, while at high pH the molecules are negatively charged. The specific pH at which the total molecule charge is zero is called the "isoelectric point" of the protein (IEP).

3. The specific ion adsorption is another important mechanism for charging. This mechanism can result also in the change of the surface charge under certain conditions, e.g. use of surfactants.

4. Anisotropic crystals are special structures that may have either negatively or positively surface charges and sometimes may show the existence of more than one IEP.

5. In isomorphous substitution, structural ions are substituted by ions of valency one less than the original. For example, a silicon atom with valency +4 in clay may be replaced by aluminium with valency +3, producing a surface with a net negative charge. In addition, in this case we can also bring the surface to its IEP by lowering the pH.

Table 10.3 *Values of isoelectric points (IEP) (pH) of selected metal oxides and proteins*

Metal oxide	IEP	Protein	IEP
Alumina	9.1	Myoglobin	7.0
MgO	12.5	Lactoglobin	5.2
TiO_2	3.5	Ovalbumin	4.6
Silica (SiO_2)	2.0	Haemoglobin	6.9
ZnO	10.3	Myosin	5.4
ZrO_2	5.7	Fibrinogen	5.2
Fe_2O_3	6.7	Serum albumin	4.8

Table 10.3 shows some typical IEP (isoelectric point) values for some metal oxides and proteins. The IEP is determined by a pH titration: measuring the zeta potential as a function of pH. The point of zero charge is the pH at which the positive and negative charges of a zwitterionic surface are balanced. It is the same as the IEP if there is no specific adsorption of ions onto the surface. Note that some solids are surface treated (e.g. TiO_2) and the IEP depends on the surface treatment (see Figure 10.14).

Before moving on, we note that the static (i.e. measured using direct current) relative permittivity, ε, is the ratio of permittivities of the medium to that of free space (vacuum). It describes how much electric field is generated per unit charge and is a measure of how an electric field is affected by a medium. A high value of ε, such as that of water (80.2 at 20 °C) means that it is easy to charge particles in aqueous systems. The permanent dipoles of water can orient themselves and stabilize the charge on particles.

10.6 Electrical forces: key parameters (Debye length and zeta potential)

The electrical forces of particles are due to the overlap of their (diffuse) double layers (Figure 10.4). The inner Stern layer is essentially the layer of counterions with a concentration that gradually decreases in the diffuse layer. The Debye length (or screening length) is a measure of the length (thickness) of the double layer and depends on the solvent, temperature and the ionic strength of the solution. It can be changed by adsorption of polymers and charged surfactants. The Debye length decreases with increasing salt concentration, especially for multivalent ions.

The complex but rather accurate Gouy–Chapman theory (see Appendix 10.2 for a brief discussion) provides the distribution of potential and the ion concentration as a function of the distance from the surface. The potential decreases exponentially with the distance and also with decreasing Debye length.

The exact expression for the repulsive potential energy (V_R) depends on the proximity of the particles or surfaces and their shape (see e.g. Equation 10.4). There are various expressions for describing the electrostatic (typically) repulsive forces between colloidal particles but all of them indicate that the potential energy decreases exponentially with the distance and increases with increasing Debye length, relative permittivity and surface potential (see Table 10.5). Thus the key properties appearing in the various mathematical expressions are:

- the surface (or zeta, ζ), potential, ψ_o;
- the Debye length. κ^{-1}.

A potential is simply the work done in bringing a point charge from infinity to the particle surface. Potentials are always relative to ground (i.e. at an infinite distance from the surface). The surface potential is very important and is approximated by the so-called zeta potential, which can be estimated by (micro)electrophoresis experiments, at least for some special cases (small and large particles via the so-called Hückel and Smoluchowski equations). In the general case, the zeta potential is calculated from values of the electrophoretic mobility, using graphical solutions or the Henry equation, which requires a correction factor.

These two key parameters of the electrical forces (zeta potential and Debye length) are discussed next.

10.6.1 Surface or zeta potential and electrophoretic experiments

An electric double layer travels together with the particle. The surface potential is approximately equal to

the so-called zeta potential. These particles will have counter-ions and solvent molecules attached to them and this is why the two potentials are not *exactly* the same (Figure 10.12).

The zeta potential can also be defined as the potential where the centre of the first layer of solvated ions moving relative to the surface is located. Goodwin (2009) states that this is about 0.5 nm or so from the surface. A rule of thumb often cited suggests that the absolute value of the zeta potential, for stability in many practical situations, should be larger than 30 mV. However, other parameters also play a role as we will see later.

The zeta potential gives an indication of the extent to which ions from the solutions are adsorbed into the Stern layer. When the solid surface is able to ionize, e.g. oxides and proteins, the zeta potential is a measure of the extent of ionization. The Stern layer is indeed a few Å large and it reflects the finite size of charged groups as well as that of ions associated

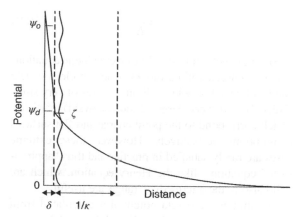

Figure 10.12 *The various "surface" potentials. Electrokinetic experiments give the "zeta (ζ) potential" at the surface of shear (as the particle moves in an electric field), which is closer to the Stern potential, ψ_d. We often consider all three potentials, especially the Stern and zeta ones to be rather close to each other. Clearly the zeta potential is, in reality, the lowest of the three. We will consider the error introduced by using the zeta potential (instead of the surface, ψ_0, or the Stern ones) to be rather small except for some special cases (in the figure the difference looks quite large, but this is only for illustrative purposes). The potential–distance dependency is linear in the Stern layer and exponential in the diffuse layer*

with surface. The zeta potential is much lower than the Stern potential in those cases where we have high potential and high salt concentrations, e.g. adsorbing ionic species. The assumption that the surface potential can be approximated by the zeta potential is nearly always made in practice:

$$\psi_o = \zeta \qquad (10.8)$$

The zeta potential is measured via the so-called electrophoresis experiment. What is measured is the velocity (in m s^{-1}) of a charged colloid particle, u, induced by an electric field with strength E in a stationary liquid (typically given in V m^{-1} or V cm^{-1}). Then, the electrophoretic mobility or electromobility is calculated and finally the zeta potential is obtained. The values of electromobility are of the order 10^{-8} m^2 V^{-1} s^{-1} for colloids.

The quantity which is measured, the so-called electrophoretic mobility (in m^2 V^{-1} s^{-1}), is defined as:

$$\mu = \frac{u}{E} \qquad (10.9)$$

Simple exact solutions of the electrokinetic equations when a charged colloid moves inside an electric field can be obtained only in the limiting cases of very small (Hückel) and very large (Smoluchowski) particles, which correspond to the point charge and flat surfaces assumptions, respectively. However, these extreme cases are rarely satisfied in practice and thus "approximate" equations, like the Henry equation, which are valid also in the intermediate range, are useful.

In summary, the zeta potential is calculated from the value of the electrophoretic mobility and the properties of the medium (viscosity, relative permittivity) according to the following equations, depending on the size of the particles:

Small particles	Large particles	General equation	
$\kappa R < 0.1$	$\kappa R > 100$		
$\psi_0 = \dfrac{3\mu\eta}{2\varepsilon\varepsilon_0}$	$\psi_0 = \dfrac{\mu\eta}{\varepsilon\varepsilon_0}$	$\psi_0 = \dfrac{1.5\mu\eta}{\varepsilon\varepsilon_0 f(\kappa R)}$	(10.10)
Hückel	Smoluchowski	Henry	

Notice that, unlike the case of the Henry equation, in the simple cases of Hückel and Smoluchowski equations the size of particles (radius) is not required. There are alternative graphical ways for "correcting" for the size of particles (Pashley and Karaman, 2004). Although the Hückel and Smoluchowski equations are very useful, it can be shown that they only cover a very small part of the colloidal domain. In most cases, corrections are needed, e.g. via the use of Henry equation or other graphical methods where the correction factor, f, can be estimated. Negative values of ψ_o can be obtained if μ is negative.

One example of the variation of the zeta potential with both pH and salt concentration is shown in Figure 10.13 for the titanium oxide particles, a well-known hydrophilic pigment, often used in paints (organic pigments are hydrophobic). The IEP where the potential is zero is around 7 in this case. The particles will aggregate close to this pH as the electrostatic repulsion will be very weak. We have positively charged surfaces at pH < IEP and negative ones at pH > IEP.

The zeta potential depends, as shown by the electrokinetic measurements, on the ionic strength. High

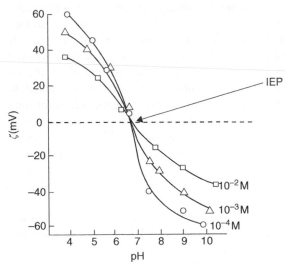

Figure 10.13 *The zeta potential decreases with increasing salt concentration and thus salts enhance instability in two ways (also by decreasing the Debye or double layer thickness). The isoelectric point, IEP, is indicated in the figure. Adapted from Hunter (1993), with permission from Oxford University Press*

salt concentrations compress the double layer and reduce the zeta potential. Stabilization can be obtained if the pH is far from the IEP and/or upon adding small charged molecules that adsorb to the surface, e.g. citric acid or upon adding polymers (block-copolymers, polyelectrolytes).

When the zeta potential is zero (at the IEP), there is no charge on the particles and flocculation will take place (viscosity will increase and coating properties be compromised). This point (and close by) should be avoided. Pure TiO_2 has an IEP of about 3.2. Figure 10.14 shows that surface treatment can substantially change the position of the IEP. At the shear plane (see Figure 10.10) the potential difference is defined to be the zeta potential.

Similar measurements to those shown in Figures 10.13 and 10.14 have been presented for many other types of particles including polymeric ones and proteins. A further example is shown in Figure 10.15 for the zeta potential of alumina particles under different conditions. We can see in Figure 10.15 that for both the negatively and positively charged alumina particles (in the presence of barium chloride and sodium sulfate), at concentrations around 10^{-3} M, the ion of sign opposite to the initial charge is able to reverse the sign of the diffuse layer and hence of the zeta potential.

Notably, there are other "electrokinetic" phenomena (sedimentation velocity, potential, electro-osmosis),

but the electrophoresis is possibly the most important one. The nature of the surface charge can be investigated by studying the dependence of the electrophoretic mobility or the zeta potential on factors such as pH, ionic strength, addition of specifically adsorbed polyvalent counter-ions, addition of surface-active agents and treatment with specific chemical reagents, particularly enzymes. Electrophoresis is a technique that is useful for many colloidal systems even for characterizing the surfaces of organisms such as bacteria, viruses and blood cells.

10.6.2 The Debye length

The other important parameter entering in electric forces, the Debye length, is defined as:

$$\kappa^{-1} = \sqrt{\frac{\varepsilon\varepsilon_o k_B T}{e^2 N_A \sum_i C_{i(B)} z_i^2}} = \sqrt{\frac{\varepsilon\varepsilon_o k_B T}{e^2 N_A 2I}} \quad (10.11)$$

where:

e = electronic or elementary (or unit) charge (1.602×10^{-19} C);
$C_{i(B)}$ = concentration of ion i in the bulk solution;
z_i = ionic valency of ion i;
I = ionic strength.

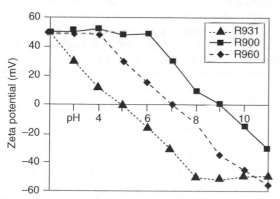

Figure 10.14 *Plots of zeta potential versus pH for three commercial TiO_2 pigments. The pigments have different pH dependencies of the zeta potential due to surface treatments. Figure adapted from internet site (no longer available)*

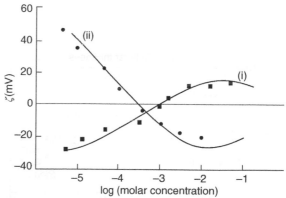

Figure 10.15 *Zeta potential (ζ) of alumina particles under different conditions: (i) negatively charged Al_2O_3 (at pH 10) in the presence of $BaCl_2$; (ii) positively charged Al_2O_3 (at pH 6.5 in the presence of Na_2SO_4). The x-axis shows the concentration of $BaCl_2$ or Na_2SO_4. Hunter (1993), with permission from Oxford University Press*

Note that the ionic valency, z_i, includes the sign of the ion charge. For example, SO_4^{2-} has $z = -2$ and Ca^{2+} has $z = +2$. The atomic valency, on the other hand, refers to the number of possible bonds an atom can form with other atoms and is always positive. The term κ (as opposed to κ^{-1}) is called the Debye–Hückel parameter. The Debye length, κ^{-1}, is often referred to as the "thickness of the double layer" even though the region of varying potential is of the order of $3/\kappa$ to $4/\kappa$ (Hunter, 1993). The Stern layer is, in most cases, much smaller than the diffuse layer and is of the order of the counter-ion diameter.

The Debye length decreases with increasing salt (electrolyte) concentration and type (as expressed by the ionic valency, z).

The thickness of the double layer is usually of the order of a few nanometres at high salt concentrations, e.g. 1 M, but it is substantially higher, up to a few hundred nanometres, at low salt concentrations, e.g. 10^{-5} M, as shown for water in Figure 10.16. Increasing salt concentration results in a decrease of the double layer thickness and consequently (see equations below) also a decrease of the double-layer (electrostatic) forces repulsion.

The summation in Equation 10.11 requires knowledge of the type of electrolyte. The equation can be simplified in specific cases, e.g. for an aqueous solution at 25 °C (κ^{-1} in nm):

$$\kappa^{-1} = \frac{\left(0.429 \text{ nm}\left(\text{mol L}^{-1}\right)^{\frac{1}{2}}\right)}{\sqrt{\sum_i C_i z_i^2}} \quad (10.12a)$$

where the concentration of the electrolyte, C, should be inserted in mol L^{-1} (M). Alternatively, as a function of the ionic strength ($I = \frac{1}{2}\sum_i C_i z_i^2$) we have:

$$\kappa^{-1} = \frac{\left(0.429 \text{ nm}\left(\text{mol L}^{-1}\right)^{\frac{1}{2}}\right)}{\sqrt{2I}} \quad (10.12b)$$

The summation requires that we know what type of electrolyte we have, for example:

$$Na_2(SO_4) \rightarrow 2Na^+ + 1SO_4^{2-}$$

$$\sum C_i z_i^2 = (2C)\cdot 1^2 + C(-2)^2 = 6C$$

Similarly, we get, for example, $6C$ also for $CaCl_2$, $8C$ for $MgSO_4$, $12C$ for $AlCl_3$ and $2C$ for NaCl.

Table 10.4 provides more expressions for the Debye length for different types of aqueous salt solutions at 25 °C. See also Problem 10.2 for the derivation.

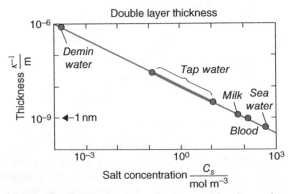

Figure 10.16 *Double layer thickness–salt relationship for water. Some typical values: sea water (about 500 mol m^{-3}), κ^{-1} is about 1 nm; tap water (about 1 mol m^{-3}), κ^{-1} is about 10 nm; demineralized water, κ^{-1} is about 1000 nm. From Hans Wesselingh (personal communication)*

Table 10.4 *Debye length for aqueous solutions at 25 °C for different types of salts (C should be inserted in mol L^{-1} and the Debye length (κ^{-1}) then comes out in nm). Note that the unit of the numerical constants in the κ^{-1}-expressions is in all cases [nm (mol L^{-1})$^{1/2}$]*

TYPE of salt	κ^{-1} expression (in nm); C in mol L^{-1}	Examples
1 : 1	$\dfrac{0.304}{\sqrt{C}}$	NaCl
1 : 2, 2 : 1	$\dfrac{0.176}{\sqrt{C}}$	$CaCl_2$ and Na_2SO_4
2 : 2	$\dfrac{0.152}{\sqrt{C}}$	$MgSO_4$
1 : 3	$\dfrac{0.124}{\sqrt{C}}$	$AlCl_3$
2 : 3, 3 : 2	$\dfrac{0.078}{\sqrt{C}}$	$Al_2(SO_4)_3$ and $Ca_3(PO_4)_2$

Example 10.2.   Zeta potential and stability of colloids.

The following data are available from a measurement of the zeta potential in an aqueous suspension of kaolin particles at 25 °C:

- diameter of the spherical particles: 0.5 micrometre (μm);
- concentration of NaCl in water: 0.003 M;
- movement of the particle: 360 μm;
- time for movement of particle: 4.6 s;
- potential of the field in a 10 cm cell: 200 V.

Moreover, the following values are available for the f correction parameter of the Henry equation for various values of κR (κ^{-1} is the Debye length and R is the radius of the particle):

κR	$f(\kappa R)$	κR	$f(\kappa R)$
0	1.000	5	1.160
1	1.027	10	1.239
2	1.066	25	1.370
3	1.101	100	1.460
4	1.133	∞	1.500

1. Calculate the electrophoretic mobility of the colloid particles.
2. Provide an estimation of the zeta potential of the particles. Mention one example of how the zeta potential can be changed.
3. Explain briefly the role of the zeta potential in the study of the stability of a colloidal dispersion.
4. The zeta potential is often of less importance in systems with organic solvents. Explain briefly why. How can a kaolin suspension be stabilized in a solution with an organic solvent?

Solution

1. The mobility will be calculated from Equation 10.9:

$$\mu = \frac{u}{E}$$

The velocity is distance over time, giving:

$$u = \frac{d}{t}$$

Combining yields:

$$\mu = \frac{d}{tE} = \frac{(360 \times 10^{-6} \ \text{m})}{(4.6 \ \text{s})\left(\dfrac{200 \ \text{V}}{0.1 \ \text{m}}\right)} = 3.913 \times 10^{-8} \ \text{m}^2\text{s}^{-1}\text{V}^{-1}$$

2. Zeta potential of the particle. For a 1 : 1 electrolyte, Table 10.4 gives:

$$\kappa^{-1} = \frac{0.304}{\sqrt{C}} = \frac{0.304}{\sqrt{0.003}} = 5.55 \ \text{nm}$$

$$\kappa^{-1} = 5.55 \text{ nm} \Rightarrow \kappa = 1.802 \times 10^8 \text{m}^{-1} \Rightarrow$$
$$\kappa R = 45.05$$

As the value of κR is neither below 0.1 (sometimes less than 1 is sufficient) or above 100, we cannot use the Hückel or Smoluchowski equations for estimating the surface or zeta potential. We have to use the Henry equation (see Equation 10.10) with a correction parameter f, which we can obtain from the available data (by interpolation):

$$\psi_0 = \frac{1.5\mu\eta}{\varepsilon\varepsilon_0 f(\kappa R)} = \frac{1.5(3.913 \times 10^{-8} \text{ m}^2\text{s}^{-1}\text{V}^{-1})(0.00089 \text{kg m}^{-1}\text{s}^{-1})}{(78.5)(8.854 \times 10^{-12} \text{ C}^2\text{J}^{-1}\text{m}^{-1})(1.39)} = 0.05407 \text{ V} = 54.1 \text{ mV}$$

The zeta-potential can be changed, e.g. by adsorption of ionic surfactants with co- or counter-ions. Non-ionic surfactants can also change the zeta-potential indirectly by pushing the "shear plane" further out.

3. The zeta potential is a measure of the potential at the surface of a particle. The higher the value of the zeta potential is, the higher the repulsive forces are, and the better is the stability of the colloidal system. The stability and repulsive forces depend of course on other factors as well especially the Debye thickness (of double layer). Values of the zeta potential above 50 mV can give good stability at electrolyte concentrations around 0.1–0.01 mol L^{-1}.

4. The kaolin suspension in an organic solvent can be stabilized sterically, i.e. using polymers (or oligomers), which have to adsorb strongly and fully cover the particles and more over be soluble (the outer part of the polymers) in the organic solvent.

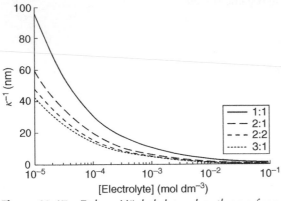

Figure 10.17 *Debye–Hückel decay length as a function of salt concentration for various types of salts. Reprinted from Goodwin (2009), with permission from John Wiley & Sons, Ltd*

The dependence of the double-layer thickness on both the electrolyte concentration and type is shown in Figure 10.17. The electrostatic force is expected to be very weak after about 2–3 Debye lengths. This underlines the importance of having a high Debye thickness to achieve good stability.

10.7 Electrical forces

Electrostatic double-layer interactions (Figure 10.18) arise when two charged surfaces are close enough for their diffuse layers to overlap. Thus, when two charged particles approach each other, their double layers begin to overlap, and they repel each other.

Electrostatic double-layer forces are always present between charged surfaces in electrolyte solutions. The counter-ions to the surface, i.e. those ions which have opposite charge to that of the surface, are attracted to the surface, whereas the co-ions are repelled. Hence, outside the surface, in the so-called diffuse layer, the concentration of ions will be different to that in the bulk solution, and the charge in the diffuse layer together with the charges on adsorbed species balance the surface charge.

However, the force falls exponentially with distance, and is effectively zero after a few times the thickness of the double layer.

Note (Figure 10.18 and Equation 10.4) that the potential energy increases proportionally to the size of the particles. Notice also the exponential functional

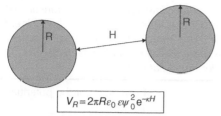

$$V_R = 2\pi R\varepsilon_0\,\varepsilon\psi_0^2\,e^{-\kappa H}$$

Figure 10.18 *Electrostatic double-layer interactions and the simplest mathematical expression for two equal-sized spheres under simplifying conditions (Debye–Hückel approximation). The repulsive forces decrease exponentially with distance and added electrolyte*

relation with distance: the electrical force is not so important at short distances, but decreases less rapidly with distance compared to the vdW forces.

The equation for the electrostatic repulsion that we see in Figure 10.18 and Equation 10.4 is valid for monodisperse spherical colloidal particles when the so-called Debye–Hückel approximation is fulfilled and when $\kappa R < 5$. The Debye–Hückel approximation can be expressed as:

$$\frac{|z\psi_o|e}{2k_BT} \ll 1 \qquad (10.13)$$

where z is the ionic valency of the co-ions present in the medium. The Debye–Hückel (linear) approximation is an evaluation of how important the electric energy is relative to the "natural" kinetic (or thermal) energy from Brownian motion. How much smaller than unity the term in Equation 10.13 should be for the approximation to be valid can be analysed from the Taylor expansion of the exponential function (see, for example, mathematical handbooks) because this is where the approximation (to a linear function) comes from and it turns out that little error is introduced if the term is smaller than 0.1. Note that the approximation, as expressed in Equation 10.13, is only strictly valid when a single symmetrical electrolyte (e.g. 1 : 1 or 2:2) is present in the medium (i.e. when $z_i = z_+ = -z_- = z$) (Shaw, 1992). Nevertheless, due to other approximations and assumptions introduced in the derivation of the Gouy–Chapman theory, and the fact that the co-ion valency only has a very weak effect on the results, Equation 10.4 is often used, even when the Debye–Hückel approximation does not quite hold [i.e. when the term in Equation 10.13 is less (as opposed to much less) than unity].

There are more "advanced" or "general" expressions, e.g. between flat surfaces, equal-sized spheres at constant surface potential or charge, dissimilar spheres, and different surface potentials. Although for colloidal dispersions the equations with spheres are more important, those with plates (flat surfaces) are of interest in the study of, for example, the stability of soap films and adhesion studies. In all cases, the repulsive electrostatic force decreases exponentially with increasing distance and with decreasing Debye length (i.e. upon adding electrolyte or for high values of the ionic valency). Some of the most well-known expressions are summarized in Table 10.5. The Reerink–Overbeek expression was originally developed by Reerink and Overbeek (1954).

There are many more equations available for different geometries and surface potentials (see, for example, Israelachvili, 2011). In all cases, the repulsive electrostatic force decreases exponentially with increasing distance and with decreasing Debye length (i.e. upon adding electrolyte or for high values of the counter-ion ionic valency). However, if we have different types of particles with different surface (zeta) potentials and/or surface charge signs, then attractive electrostatic forces can be obtained.

At very short distances of approach (Ångstrøms), H, there can be overlap of molecular orbitals. This is known as Born repulsion and these interactions are very short-range (proportional to H^{-12}).

10.7.1 Effect of particle concentration in a dispersion

When the particle concentration in a dispersion increases and the average separation distance between particles becomes similar in magnitude to the range of the diffuse layer, special attention must be paid to the background electrolyte, which in this case cannot be approximated with a particle-free electrolyte solution. The volume fraction of the particles must be taken into account and Goodwin (2009) has shown this can be done with a correction for the double layer thickness for a symmetrical electrolyte. An increasing particle concentration leads to a faster decay of the electrostatic repulsion and the effect can be expected to become significant above a few volume per cent particles, which is actually the case for most practical products.

Table 10.5   *Approximate expressions for the potential energy of electrostatic interactions, V_R, as a function of closest distance of approach, H. In the table, z is the counter-ion charge number (e.g. with NaCl as electrolyte, z is either 1 or −1, depending on whether Na^+ or Cl^- is the counter-ion to the surface). The "ionic valency" is the same as the "charge number" (both terms include the sign of the charge). More expressions are available in Israelachvili (2011), e.g. for electric double layer interaction between two cylinders or a cylinder and a flat surface*

Geometry	Expression for the potential energy $V_R(H)$
Two equal-sized spherical particles at close distances ($H \ll R$) and low potential. $\kappa R < 5$. Debye–Hückel approximation must hold: $$\frac{|z\psi_0|e}{2k_BT} \ll 1$$	$V_R = 2\pi\varepsilon\varepsilon_0 R\psi_0^2 e^{-\kappa H}$

Equal-sized spheres (general) and $\kappa R > 10$
Two cases:
    Case 1. Constant surface potential (charge controlled by the concentration of potential-determining ions in solution). Works well for all separations, H.
    Case 2. Constant surface charge (e.g. isomorphous substitution in a lattice). Should be used with caution, especially at close approach (Goodwin, 2009).
	$V_R = 2\pi\varepsilon\varepsilon_0 R\psi_0^2 \ln\left(1 + e^{-\kappa H}\right)$
	
	$V_R = -2\pi\varepsilon\varepsilon_0 R\psi_0^2 \ln\left(1 - e^{-\kappa H}\right)$

Unequal-sized spheres with different (or same) surface potential in a single electrolyte and $H \ll R_1, R_2$ (Reerink–Overbeek expression).	$V_R = \left(\frac{64\pi\varepsilon\varepsilon_0 R_1 R_2 k_B^2 T^2 \gamma_1 \gamma_2}{(R_1 + R_2)e^2 z^2}\right)e^{-\kappa H}$
Reduces to the first equation in this table for equal-sized spheres and when the Debye–Hückel approximation holds.	$\gamma_i = \dfrac{e^{\frac{ze\psi_{0i}}{2k_BT}} - 1}{e^{\frac{ze\psi_{0i}}{2k_BT}} + 1}, \quad i = 1, 2$
Two similar flat surfaces (low potential, i.e. Debye–Hückel approximation). For high potentials see Goodwin (2009).	$V_R = 2\varepsilon\varepsilon_0 \kappa\psi_0^2 e^{-\kappa H}$
Sphere-plate with same surface potential ($H \ll R$): 1. weak double-layer overlap ($\kappa R < 5$); 2. close approach and large κR.	$V_R = 4\pi\varepsilon\varepsilon_0 R\psi_0^2 e^{-\kappa H}$ $V_R = 4\pi\varepsilon\varepsilon_0 R\psi_0^2 \ln\left(1 + e^{-\kappa H}\right)$

Data from Shaw (1992), from Israelachvili (2011) and from Goodwin (2009).

10.8   Schulze–Hardy rule and the critical coagulation concentration (CCC)

The effect of salts and especially of the counter-ions on the colloid stability can be demonstrated via the so-called Schulze–Hardy rule. We will see in Chapter 11 that this rule has been verified by the DLVO theory, but it appeared many years before the DLVO. As shown in Figure 10.19, salts, even in small amounts, can reduce drastically the diffuse double layer and thus the stability of colloids. Adding salt leads to coagulation.

The added electrolyte causes a compression of the diffuse parts of the double layers around the particles and even may exert a specific effect though ion adsorption into the Stern layer. The particles coagulate when the range of double-layer repulsive interactions is sufficiently reduced to permit particles to approach close enough for the van der Waals attractive forces to predominate.

The critical coagulation concentration, often abbreviated as CCC, is the minimum concentration of an (inert) electrolyte that is just sufficient to coagulate a dispersion. By coagulate we usually mean a visible

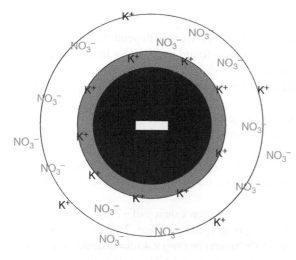

$$CCC \propto \frac{1}{z^6} \Rightarrow CCC(salt\ II) = CCC(salt\ I)\left(\frac{z_I}{z_{II}}\right)^6$$

Figure 10.19 *The role of salt in colloid stability as described via the Schulze–Hardy rule for dispersions. The critical coagulation concentration (CCC) is inversely proportional to z^6 (the exponent 6 is an average value and a reasonable number to use). This fact leads to the equation in the figure for two salts with different ionic valencies (i.e. one can calculate the CCC of one salt from CCC of another salt with a different counter-ion valency). In the drawing, the particle is negatively charged (e.g. a AgI particle), K^+ is the counter-ion and NO_3^- is the co-ion*

Table 10.6 *Critical coagulation concentrations (CCC) (in mmol L^{-1}) for various particles*

Particle	Electrolyte	CCC (mmol L^{-1})
Al_2O_3 (positively charged)	NaCl	43.5
	KCl	46
	KNO_3	60
	K_2SO_4	0.30
	$K_2Cr_2O_7$	0.63
	K_2oxalate	0.69
	$K_3[Fe(CN)_6]$	0.08
Fe_2O_3 (positively charged)	NaCl	9.25
	KCl	9.0
	$(\frac{1}{2})\ BaCl_2$	9.6
	KNO_3	12.0
	K_2SO_4	0.205
	$MgSO_4$	0.22
	$K_2Cr_2O_7$	0.195
Au (negatively charged)	NaCl	24
	KNO_3	23
	$CaCl_2$	0.41
	$BaCl_2$	0.35
	$(\frac{1}{2})\ Al_2(SO_4)_3$	0.009
	$Ce(NO_3)_3$	0.003

Data from Shaw (1992) and from Pashley and Karaman (2004).

change in the dispersion appearance. Notice the enormous dependence upon the valency (absolute value of charge) of the counter-ions. As shown in the literature (Shaw, 1992; Pashley and Karaman, 2004), see also Table 10.6, the CCC seems to depend only very weakly on other factors (other than the counter-ion valency) like the charge number of co-ions and the concentration of the particle, and even moderately on the nature of the particle itself. All these generalizations are known as the Schulze–Hardy rule.

Example 10.3.   Schulze–Hardy rule and colloid stability.

1. Consider the experimental CCC data in Table 10.6. Develop a mathematical expression for the Schulze–Hardy (SH) rule, i.e. estimate the exponent for z.
2. For a water dispersion of positively charged Al_2O_3 particles, the following values of the critical coagulation concentration (CCC) have been measured:
    CCC = 43.5 mmol L^{-1} NaCl and
    CCC = 0.63 mmol L^{-1} $K_2[Cr_2O_7]$

    Provide an estimation of the CCC for the same dispersion when the salt $K_3[Fe(CN)_6]$ is used. Which of the three salts results in the most stable dispersion?

Solution

1. The SH rule states that the critical coagulation concentration (CCC) is strongly dependent on the valency of the counter-ion and, in combination with the DLVO theory, it is found that CCC is inversely proportional to the sixth power of the valency of the counter-ion.

    For example, using the data of Table 10.6 for Al_2O_3, first for the monovalent electrolytes, we obtain: $CCC = k/1^x$

    i.e. the proportionality constant is $k = 49.8$ mmol/L (average of three values in table).

    Then for the divalent electrolyte we have $CCC = (49.8 \text{ mmol/L})/2^x = 0.54$ mmol/L which results in an x value equal to 6.5. Similarly, for the trivalent electrolyte we find $CCC = (49.8 \text{ mmol/L})/3^x = 0.08$ mmol/L, which results in an x value of about 5.9. In conclusion, an average value of the exponent is $x =$ about 6 and this is in agreement with other data for other particles like those shown in Table 10.6.

2. The SH rule would give for the 1 : 2 electrolyte a CCC value (counterion valency of +2) equal to $0.0156 \times 43.5 = 0.6786$ mmol L^{-1}, which is rather close to the value given in the problem. Thus, we can assume that the SH rule is valid. For the 1 : 3 electrolyte, then, the CCC would be $(1/3^6) \times 43.5 = 0.05967$ mmol L^{-1} i.e. approximately 0.06 mmol L^{-1}. The 1 : 1 electrolyte with the highest CCC provides the most stable dispersion.

Example 10.4. Interaction between colloid particles.

In DLVO theory the total interaction energy between two spherical colloids is given as a function of distance, under some simplifying assumptions, by Equation 10.4.

1. Explain the concepts behind DLVO theory and describe the origin of the repulsive and attractive elements of Equation 10.4.
2. Draw curves for the interactive potential between two colloids for two extreme scenarios: A system where coagulation happens spontaneously and a system where coagulation will never occur.
3. When changing electrolyte concentration the interaction curve changes shape. Draw the interaction potential curve when the colloid suspension becomes unstable.
4. Is the electrolyte concentration higher in the case of the stable solution or the unstable solution?

By electrophoresis the zeta potential of stable spherical particles of radius 60 nm was measured to be 60 mV in an aqueous monovalent symmetric electrolyte solution. The Hamaker constant is estimated to 6.3×10^{-20} J.

Estimate the concentration of the electrolyte when this colloid suspension becomes unstable. What is the name of this concentration?

Solution

1. DLVO theory approximates the total interaction energy between two identical colloids by a summation of the repulsive and the attractive contributions. The repulsive contribution (first part of the equation) originates in the interaction between the electric double layers of the two particles. The attractive contribution (second part of the equation) originates from the van der Waals interaction between the colloid bodies of the particles.
2. The curves are shown below.

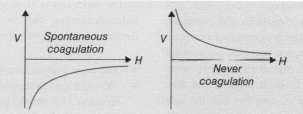

3. When changing electrolyte concentration the interaction curve changes shape. The interaction potential curve when colloid suspension becomes unstable is drawn below.

    "Minimal" conditions for instability: $V = 0$ and $dV/dH = 0$

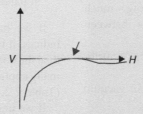

4. The electrolyte concentration is higher for the unstable solution, i.e. above CCC.
5. The conditions for instability are $V = 0$ and $dV/dH = 0$. Solving these two equations for κ^{-1} gives:

$$\frac{1}{\kappa} = \frac{A}{24\pi\varepsilon\varepsilon_0\psi_0^2 e^{-1}}$$

which equals the expression for the Debye length:

$$\frac{1}{\kappa} = \sqrt{\frac{\varepsilon\varepsilon_0 k_B T}{e^2 \sum_i C_{i(B)} z_i^2}}$$

This results in this expression for the concentration (numbers m^{-3}):

$$C = \frac{\left(24\pi / e\right)^2 \varepsilon^3 \varepsilon_0^3 \psi_0^4 k_B T}{2e^2 A^2}$$

The critical coagulation concentration for the monovalent electrolyte is then $C = 118$ mM (using Avogadro's number).

10.9   Concluding remarks on colloid stability, the vdW and electric forces

Colloid stability depends significantly on the (often) attractive van der Waals forces and the (often) repulsive electric and steric forces. DLVO theory accounts for the van der Waals and electric forces and these are the ones which are discussed in this chapter. The main characteristics of the vdW and the electric forces are summarized in this section.

10.9.1   vdW forces

The van der Waals (vdW) forces are very important in colloid stability. Specific examples where they may be crucial are in estimating adhesion of particles to

surfaces, separation of particles during dispersion, estimation of aggregation rates of particles and estimation
of the rheology of concentrated aggregated systems.

For the vdW forces between particles we summarize the following: They are due to dispersion interactions between molecules in each particle and their
calculation is based on the assumption that the total
potential is pairwise additive (presence of other mole-
cules/many-body interactions are neglected). Their
basic characteristics are:

- they are long-range and extremely strong – much
longer range than the vdW forces between
molecules;
- they are almost always attractive and thus contribute to instability;
- they are *always* attractive in air and in vacuum
(where the Hamaker constant is zero);
- they are *always* attractive between two identical
surfaces or particles;
- they depend on the size/shape, in general the
geometry of the particles (e.g. spherical, flat surfaces or cylindrical); thus there are several analytical expressions for the van der Waals forces
(Table 2.2);
- they depend on the material and the medium (via
the Hamaker constant);
- they decrease because of an intervening medium
(this can help in controlling the vdW forces!, e.g.
by changing the solvent or adding a layer containing a large amount of the solvent);
- they are weak if particles and medium are chemically similar;
- they can be repulsive (sometimes) between different surfaces (or particles) in a medium (if the
Hamaker constant of the medium has a value
intermediate to those of the particles, which
means that one material interacts more strongly
with the medium than with the second body).

Because of the extremely strong character of the
vdW forces, we could not make paints, inks, pharmaceuticals, cosmetics, many food products, emulsions,
foams, bilayer and membranes if these were the only
forces in place. The "opposite" equally strong electric
repulsive forces (and other repulsive forces, e.g. steric
or hydration) can keep colloidal systems (meta)stable.

The calculation strategy for obtaining numerical
values/assessing the strength of the vdW forces
depends on the answers to the following questions:

- For what purpose are we going to use the results?
- Do we a have a simple geometry or is the approximation to a simple geometry a limiting factor?
- Do we need to know the distance dependence of
the interaction or are we really concerned with
effects when surfaces are relatively close?
- Do we have the required material properties
at hand?

If we have relative permittivities and refractive
indices, it is better to use the Lifshitz approach rather
than the combining rules for the Hamaker constant. In
many applications, the expressions for close distances
(Table 2.2, Chapter 2) would also be sufficient.

If we understand the origin of forces between par-
ticles/surfaces, it is possible to modify them by controlling the chemical environment (chemistry) of the
dispersion. Although an accurate predictive calculation may not always be within our reach for a product,
the choice of experiments will be much better
focused.

10.9.2   Electric forces

Concerning (repulsive) electrostatic/electric forces
we can state that they:

- decrease with increasing distance exponentially
(and this is a different dependency than for the
vdW attractive forces);
- depend on the size of particles (or surfaces);
- depend on the medium (often water);
- depend on the surface (potential) – zeta potential;
- depend on the electrolyte used (concentration and
valency);
- are always repulsive between identical surfaces but
they can be attractive between dissimilar surfaces.

The understanding and control of the electrical
forces and thus the stability of a colloidal dispersion
depends a great deal on the Debye length and the
value of the surface potential as well as the charging
mechanism. It is especially crucial to control the diffuse double layer through control of the ionic strength
of the solution and the adsorption of charged

molecules, e.g. ionic surfactants and polyelectrolytes. The surface charge may be adjusted by controlling the pH or the common ions. Stability is obtained far from zero pH and far from the isoelectric point (IEP). Anionic surfactants will result in negatively charged surfaces, while the opposite is true for cationic surfactants. Polar molecules do not have charges but can still influence the diffuse double layer.

Remember, the Debye length (Debye thickness) plays an important role in the stability of colloidal dispersions. We recall that:

- It is a length (usually in nm) – a measure of the thickness of double layer.
- It depends on both electrolyte type and concentration.
- It decreases with increasing salt concentration.
- It "protects" – it helps stability! We want the Debye thickness to be as high as possible for good stability!
- The diffuse double layer fades away at a distance 2–3 times the Debye length.

In general, the methodology for understanding electrical forces involves the following steps:

- identify the source of surface charge;
- measure surface charge (zeta potential) versus pH;
- regulate surface charge (and via this the diffuse double layer), e.g. ion adsorption, adsorption of polymers especially polyelectrolytes, proteins, (ionic) surfactants, small charged molecules, surface treatment (curing, chemical modification) – pH and common ions can help in controlling the surface charge;
- stability is obtained far from pH = IEP and far from the point of zero charge (PZC);
- control salt content, temperature and medium.

Appendix 10.1 A note on the terminology of colloid stability

Figures 10.20 illustrates the various "destabilization" mechanisms in colloidal systems. The terminology can be sometimes confusing and very often we use the term "aggregation" (or "coagulation" or "agglomeration") for all the phenomena shown here, although they do represent different situations.

Coalescence, when the "bigger" particles "eat" the "smaller" ones (Ostwald ripening), is typically present only for liquid–liquid dispersions (emulsions) and will be discussed in Chapter 12. It is possibly the least understood phenomenon of the ones shown in Figure 10.20 and is completely irreversible (the droplets lose their original shape).

A coagulated aggregate separates out by creaming or sedimentation, depending on the density differences, as explained by Stokes law, which we have seen in Chapter 8.

Flocculation and creaming are reversible phenomena, while coagulation is not.

Colloidal stability is typically represented by the so-called "potential energy V–H" plots where the potential energy, V, is depicted as a function of the distance between the particles, H. Typical such plots are shown in Figures 10.3 and 10.21. The force can be obtained by simple differentiation ($F = -\mathrm{d}V/\mathrm{d}H$).

There are both similarities and differences between "interparticle" and "intermolecular" potential functions. The most important difference is, possibly, that interparticle forces are much more "long-range" than those between molecules. Let us recall the most important characteristics of this plot: the vdW attractive forces (negative V) which lead to instability (aggregation) and the repulsive electrical forces (positive V), which are due to the fact that the particles are charged, which "help the stability".

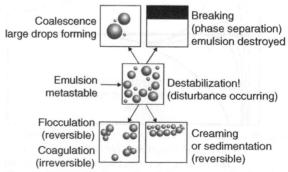

Figure 10.20 *Various destabilization methods of a colloid dispersion*

(a)

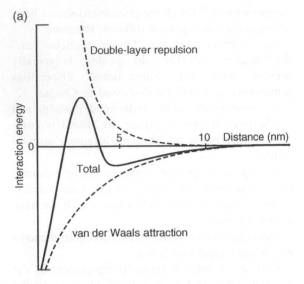

(b)

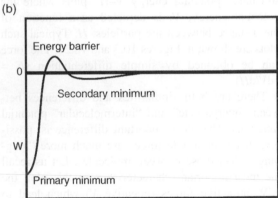

(c)

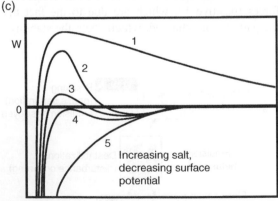

Figure 10.21 *(a)–(c) Potential energy–distance curves for different situations. For reference to curve numbers (1)–(5) in (c) see discussion of figure in main text. Adapted from Israelachivili (1985), with permission from Academic Press, Elsevier*

Manipulating colloidal stability implies knowing how we can change–influence the various forces, especially the van der Waals attractive and the electrical repulsive ones. Steric repulsion can also be important for stability, especially if the electrostatic force does not work well. If the repulsive barrier of the potential, the maximum we see in the plot, is (much) higher than the kinetic energy of particles ($=k_BT$), then the dispersion will be (kinetically) stable. How much higher? Often about $15k_BT$ or so is sufficient for good stability. Notice that at relatively large separations the van der Waals attraction is greater than the double layer repulsion giving rise to a secondary minimum (Figure 10.21). Particles aggregate with a "large" distance between them, a process called flocculation. This is, as we can see, a rather shallow minimum and, thus, flocculation is reversible and the particles can be separated by agitation.

With reference to Figure 10.21c, if the attractive potential dominates, then the colloidal suspension will *not* be stable (curve 5 and also 4). The system will minimize its energy by coagulation and will "fall" into the "well" indicated by the primary minimum, which implies an irreversible destruction.

The various profiles shown in Figure 10.21c correspond to:

1. Surfaces repel strongly; small colloidal particles remain "stable".
2. Surfaces come into stable equilibrium at the secondary minimum if it is deep enough; colloids remain "stable".
3. Surfaces come into a secondary minimum and we expect slow colloid coagulation.
4. Surfaces remain in secondary minimum or adhere and the colloids coagulate rapidly.
5. Surfaces and colloids aggregate (or coalesce if droplets) rapidly

Appendix 10.2 Gouy–Chapman theory of the diffuse electrical double-layer

The French scientist Gouy and the British one Chapman developed, about 100 years ago, a rather complex but accurate theory for the diffuse double-layer, which is a solution of the Poisson–Boltzmann

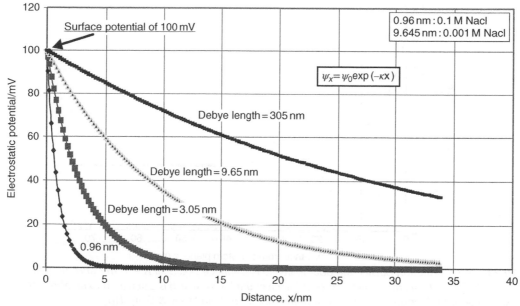

Figure 10.22   *Estimates of the decay in electrostatic potential away from a charged flat plate in a range of electrolyte solutions. Adapted from Pashley and Karaman (2004), with permission from John Wiley & Sons, Ltd*

equation for a planar layer. Their theory is based on several simplifying assumptions:

- surface flat, infinite extent, uniformly charged;
- ions in the diffuse part are point charges following the Boltzmann distribution;
- solvent influences the double-layer only via its relative permittivity, which is constant.

These assumptions are rather severe but they do not seem to affect significantly the accuracy of the theory. The second assumption implies that ions have no volumes although they do have a radii which, moreover, increases with hydration. The last assumption is definitely not correct for water. The relative permittivity decreases much nearer the surface; in addition, ions may dehydrate on a surface.

Through the Gouy–Chapman theory, we can arrive to expressions for the electrostatic potential and ion concentration as a function of distance from a charged surface and finally – what is most important for applications – the expressions for the potential energy interaction between two double layers as a function of distance between the particles or surfaces.

These are the potential electrostatic interactions due to the overlap of the diffuse double layers.

The derivations require quite some background in electrostatics and we will not repeat them here; they are presented in other textbooks (e.g. Pashley and Karaman, 2004).

In all the final equations, two important parameters appear; the "thickness" of the double layer (Debye length), κ^{-1}, and the surface (zeta) potential, ζ.

Using the Gouy–Chapman theory and solving Poisson's equation we can get, based on the assumption that the surface potential is low (i.e. multiplied by z is much lower than $k_B T$), a very simple expression for the exponential decay of the potential outside the double layer, which is shown in Figure 10.22.

For 1 : 1 electrolytes, this assumption implies that the surface potential is less than 25 mV. The κ-parameter inside the exponent is the inverse of the Debye screening length; it is a measure of the thickness of the double layer. When $x = \kappa$, then the potential has been decreased to 1/e of its original value. The Debye length is, in Figure 10.22, between 1 and 300 nm. Notice that the Debye length and the potential depend a lot (actually decrease) with increasing electrolyte concentration. Thus, the

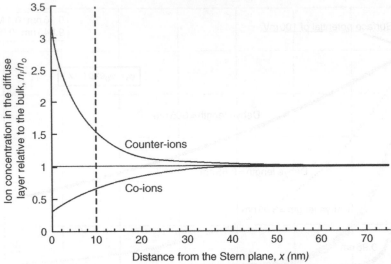

Figure 10.23 *Ion distribution outside the double layer. The surface potential is –30 mV and the concentration of NaCl is 10^{-3} M. The vertical line indicates the double layer thickness, which is calculated to be about 10 nm for this system. Reprinted from Goodwin (2009), with permission from John Wiley & Sons, Ltd*

electrostatic repulsive forces are larger (more pronounced) when the (surface) potential and the thickness of the double-layer are large. It can, thus, be concluded that electrolytes reduce the stability, as they reduce the electric double-layer forces.

The Debye–Hückel assumption, discussed earlier in this chapter, is often used when we make simplifications – approximations in the theory of the electrical double layer as many equations are largely simplified if we use this assumption. This approximation means that the surface potential should be much lower than 25 mV. The surface potential can, however, in reality for colloidal systems of practical interest be much higher (see also Figure 10.22).

When the potential is known as a function of the distance we can calculate the distribution of ions next to a charged surface immersed in an electrolyte solution. The equation describing the ion concentration next to a charge surface is given by:

$$C_i(x) = C_{i(B)} \exp\left(-\frac{z_i e \psi(x)}{k_B T}\right)$$

where $C_{i(B)}$ is the concentration of the ion in the bulk solution. Under some additional assumptions, which

will not be elaborated further here, the surface charge density (in the unit C m^{-2}) is given by:

$$\sigma_0 = \varepsilon_0 \varepsilon_r \kappa \psi_0$$

More "exact" but complex equations for the surface charge density are provided elsewhere (e.g. Pashley and Karaman 2004). The equation for ion concentration is depicted graphically in Figure 10.23. The counter-ions have an excess concentration near the interface: this decreases exponentially towards the bulk of the solution, as also the potential does. The surface in Figure 10.23 has a negative surface potential.

Problems

Problem 10.1: Stability of alumina particles in different solvents
The table below summarizes experiments of the settling behaviour of alumina particles in solvents of varying polarity which show that the dispersion stability displays a maximum (i.e. a minimum in the normalized settling rate, see Figure 10.24). The maximum stability is observed in moderately polar solvents (with relative permittivities between 20 and 45).

Solvent	Relative permittivity	A_l/k_BT (solvent)	A_{SLS}/k_BT (alumina particles in the solvent)	Observed Stability behaviour
Cyclohexane	2.02			Poor
Chloroform	4.3			Poor
2-Butanol	10.8			Good
Isopropanol	18.3			Good
Ethanol	24.3			Good
Methanol	32.4			Good
Methanol–water (40%)	66.7			Intermediate
Water	80			Intermediate

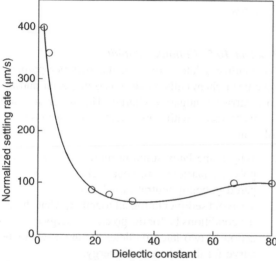

Figure 10.24 *Experimental data on the stability of alumina particles in different solvents. The normalized setting rate on the y-axis is defined as the observed settling rate times the solvent viscosity divided by the particle density minus the solvent density. The figure shows the settling behaviour of alumina particles in solvents of varying polarity. It can be observed that with an increase in solvent polarity the dispersion stability displays a maximum, which corresponds to a minimum in the normalized settling rate. Adapted from Krishnakumar and Somasundaran (1996), with permission from Elsevier*

Could this order of stability be predicted using the theory of van der Waals forces and the Hamaker constant (fill in the table first)? Comment on the results. How can you control stability for alumina dispersions?

It is given that the Hamaker constant of both alumina (which has a relative permittivity equal to 9.3) and the various liquid solvents can be given (as fraction of k_BT) by the following equation:

$$\frac{A}{k_BT} = 113.7 \frac{(\varepsilon-1)^2}{(\varepsilon+1)^{3/2}(\varepsilon+2)^{1/2}}$$

Problem 10.2: The Debye length – thickness of the double layer

1. Starting from the general equation (Equation 10.11), show that for an aqueous solution at 25 °C, the thickness of the double-layer is given in *nm* by the following equation (Equation 10.12a):

$$\kappa^{-1} = \frac{0.429}{\sqrt{\sum_i C_i z_i^2}}$$

where the concentration of the electrolyte, C, is given in mol L^{-1}.

Or alternatively, as a function of the ionization strength, I (Equation 10.12b):

$$\kappa^{-1} = \frac{0.304}{\sqrt{I}}$$

2. Show that for different electrolytes, the Debye length is given (in nm, C in mol L^{-1}) as:

$$\kappa^{-1} = \frac{0.304}{\sqrt{C}} \text{ for } 1:1 \text{ electrolytes, e.g. NaCl}$$

$\kappa^{-1} = \dfrac{0.176}{\sqrt{C}}$ for both 1 : 2 and 2 : 1 electrolytes like $CaCl_2$ and Na_2SO_4

$\kappa^{-1} = \dfrac{0.152}{\sqrt{C}}$ for 2 : 2 electrolytes like $MgSO_4$

$\kappa^{-1} = \dfrac{0.124}{\sqrt{C}}$ for 3 : 1 like $AlCl_3$

3. Calculate the "thickness" of the diffuse electric double layer for a solid surface in contact with the following aqueous solutions at 25 °C:
    a. 0.1 mol L^{-1} KCl
    b. 0.01 mol L^{-1} KNO_3
    c. 0.001 mol L^{-1} KCl
    d. 0.001 mol L^{-1} $ZnCl_2$
    e. 0.001 mol L^{-1} K_2SO_4 and 0.001 mol L^{-1} $MgCl_2$

    In your opinion, which electrolyte system may result in the most stable colloidal system and which one will possibly give the least stable colloidal system?

Problem 10.3: Debye lengths and colloidal stability

Calculate the ionic strengths and corresponding values of the Debye length for the following solutions (aqueous at 25 °C):

0.01 M Na_2SO_4

0.015 M $LaCl_3$

3×10^{-3} M $Ca(NO_3)_2$

A mixture of 10^{-4} M $La_2(SO_4)_3$ and 5×10^{-4} M $NaNO_3$

Which solution would you choose for stabilizing a colloidal system?

Problem 10.4: Zeta potential and stability of colloids

The following data are available from a measurement of the zeta potential in an aqueous polystyrene colloidal dispersion at 25 °C:

- diameter of the spherical particles: 1.00 µm;
- concentration of NaCl in water: 0.01 M (mol L^{-1});
- movement of the particle: 320 µm;
- time for the movement of particle: 2.5 s;
- potential of the field in a 10 cm cell: 140 V.

1. Calculate the electrophoretic mobility of the colloid particles.

2. Provide an estimation of the zeta potential of the particles.
3. What is the importance of the zeta potential in the study of the stability of a colloidal dispersion?
4. It is suggested to replace the NaCl solution with an aqueous $AlCl_3$ solution having a concentration of 10^{-3} M (mol L^{-1}). Assume that the electrophoretic mobility of the particles in the new solution is the same as previously, i.e. as calculated in question 1. Do you expect an improvement in the stability of the aqueous polystyrene emulsion? Why/why not? Justify your answer.

Problem 10.5: Colloidal stability

To stabilize a latex solution, the polymer particles have been chemically modified so that each particle now carries a negative charge. The water solution is exchanged with an electrolyte water-based solution.

1. Why is the latex solution more stable when the polymer particles are negatively charged, compared to being neutral (no charge)?
2. To avoid sedimentation all together, what should the conditions be for the potential energy of interaction of two latex particles? Draw a schematic curve for the interaction energy.
3. You can choose between two electrolytes: (a) 0.01 mol L^{-1} NaCl and (b) 0.01 mol L^{-1} $AlCl_3$. Which electrolyte solution would you choose for the latex, to obtain the most stable suspension? Justify your answer with an estimate of the Debye thickness of the diffuse double layer in both cases.

Problem 10.6: Electrophoresis – Determination of the zeta potential

Spherical particles of radius 0.3 µm are suspended in 0.02 mol L^{-1} of aqueous solution of KCl and they are observed to have an electrophoretic mobility of 4.0×10^{-8} m^2 s^{-1} V^{-1} at 25 °C. Calculate an approximate value for the zeta potential. Briefly mention any simplifications upon which your calculation is based and in what sense they will affect your calculations.

Problem 10.7: Properties from an electrophoresis experiment

In a microelectrophoresis experiment, a spherical particle of diameter equal to 0.5 μm is dispersed in a 0.1 mol L^{-1} aqueous solution of KCl at 25 °C. It takes this particle 8.0 s to cover a distance equal to 120 μm along one of the "stationary" levels of the cell and the potential gradient is 10.0 V cm^{-1}.

Calculate:

1. the electrophoretic mobility of the particle;
2. the probable error in this single mobility determination arising from the Brownian motion of the particle during the course of the measurement;
3. an approximate value for the zeta potential of the particle;
4. an approximate value for the charge density at the surface of shear.

Problem 10.8: Schulze–Hardy rule and colloid stability

Below are provided some measured values of the critical coagulation concentration (CCC) for a dispersion containing negatively charged gold (Au) particles:

24 mmol L^{-1} NaCl
23 mmol L^{-1} KNO$_3$
0.41 mmol L^{-1} CaCl$_2$
0.35 mmol L^{-1} BaCl$_2$

1. Comment on these experimental results. Are they in agreement with the Schulze–Hardy rule?
2. What would the CCC value be for the same dispersion of gold particles when the salt Ce(NO$_3$)$_3$ is used instead?
3. For a dispersion containing positively charged Fe$_2$O$_3$ particles, the CCC is 12 mmol L^{-1} when KNO$_3$ is used. A colloid scientist suggests that you use for the same dispersion only 2 mmol L^{-1} MgSO$_4$ and claims that you will get a stable dispersion. Do you expect the dispersion suggested by the colloid scientist to be stable or not? Justify your answer.

Problem 10.9: Repulsive – Attractive – Total potential energies – Stability of a colloid

Spherical colloidal particles of diameter 10^{-7} m are dispersed in 10^{-2} mol L^{-1} aqueous 1 : 1 electrolyte solution at 25 °C. The Hamaker constants of the particles and the dispersion medium are 1.6 × 10^{-19} and 0.4 × 10^{-19} J, respectively, and the zeta potential is −40 mV. Assume that the interparticle separation can be considered small and that all particles have the same size but that the Debye–Hückel assumption *cannot* be made (i.e. the Reerink–Overbeek expression applies, see Table 10.5).

1. Calculate the repulsive double-layer and attractive van der Waals potential energies between two of the particles when their shortest distance of approach, *H*, is 0.5, 2, 5, 10 and 20 nm.
2. Calculate the total interaction energy at the same distances as in the previous question and make a *V–H* plot (potential interaction energy versus distance).
3. Discuss the result. Is this colloidal dispersion stable?

Problem 10.10: Critical Coagulation Concentration

The DLVO interaction energy between two equal-sized spherical particles with the same surface potentials and rather small electric double layer overlap is given by Equation 10.4.

Calculate the critical value of the surface potential of the colloid that will just give the rapid coagulation (i.e. at CCC). Assume that the aqueous solutions contains 12 mM monovalent electrolyte at 25 °C. In addition, assume that the Hamaker constant for this case has a value of 4 × 10^{-20} J.

Problem 10.11: Effective Hamaker constants

You are given the following five systems:

System 1: fused quartz–air system in octane;
System 2: CaF$_2$–air in helium;
System 3: fused quartz particles in octane;
System 4: Hexadecane particles in water;
System 5: Teflon in water.

The Hamaker constants of the pure materials/fluids are all known (×10^{-20} J):

fused quartz: 6.3, octane: 4.5, CaF$_2$: 7.2, helium: 0.057, hexadecane: 5.2, water: 3.7, Teflon: 3.8, air: 3.8·10^{-5}.

1. Calculate using the combining relations the effective Hamaker constants for all five systems and compare them to the experimental (rigorously computed) Hamaker constant values, which are respectively: -0.71, -0.59, 0.13, 0.5 and 0.29 – all in 10^{-20} J. Comment on the results.

2. For which systems do you expect attractive and for which ones repulsive van der Waals forces. What is the physical meaning of the repulsive van der Waals forces?

Problem 10.12: Surface tensions and Hamaker constants

Hamaker constants are related to surface tensions.

1. Using a Hamaker constant value for hexane equal to 4.1×10^{-20} J and assuming a separation equal to 0.15 nm, provide an estimation of the surface tension of hexane. Compare it to the experimental value from direct measurements (18.4 mN m^{-1}).

2. Repeat the previous questions for water. Assume a higher intermolecular separation of say 0.33. What do you observe? Comment on the results.

Problem 10.13: Surface tensions and Hamaker constants

The following table presents Hamaker constants (A), surface tensions and liquid densities for seven normal alkanes.

Alkane	A (in 10^{-20} J)	10^3 surface tension (J m^{-2}) – exp	10^{-3} density (kg m^{-3})	H (nm)	10^3 surface tension (J m^{-2}) – predicted
n-Pentane	3.75	16.05	0.6262		
n-Hexane	4.07	18.40	0.6603		
n-Octane	4.50	21.62	0.7025		
n-Decane	4.82	23.83	0.7300		
n-Dodecane	5.03	25.35	0.7487		
n-Tetradecane	5.05	26.56	0.7628		
n-Hexadecane	5.23	27.47	0.7733		

1. Back-calculate the distance H (critical length) that would reproduce these surface tension values

using the theoretical relationship between the Hamaker constant and surface tension. Complete the table above and comment on the results.

2. Hough and White (1980) have shown that the critical length is inversely proportional to the square root of the liquid density. This relationship can be used to predict the surface tensions with only one adjustable parameter, namely, the "experimental" critical length for some reference hydrocarbon. Use the H-value for n-decane as reference and estimate the predicted critical lengths and predicted surface tensions. Complete the last column of the table above and comment on the results.

References

L. Bergström, Hamaker constants for inorganic materials, *Adv. Colloid Interface Sci.*, 70 (1997) 125–169.

J. Goodwin (2009), Colloids and Interfaces with Surfactants and Polymers, 2nd ed., John Wiley & Sons Ltd, Chichester.

R.J. Hunter, Introduction to Modern Colloid Science, Oxford Science Publishers, Oxford, 1993.

D.B. Hough and L.R. White (1980) *Adv. Colloid Interface Sci.*, 14, 3–41.

J.N. Israelachvili, Intermolecular and Surface Forces, 3rd edn, Academic Press, (2011).

S. Krishnakumar, P. Somasundaran, Colloids Surf., 117 (1996) 37–44.

Z.H. Liu, Y. Li, K.W. Kowk, Mean interparticle distances between hard particles in one to three dimensions, *Polymer*, 42 (2001) 2701–2706.

B. Müller, U. Poth, Coatings Formulation, Vincentz Network, Hannover 2006.

D. Myers, Surfaces, Interfaces, and Colloids, 2nd edn, John Wiley & Sons Ltd, Chichester, 1999.

R.M. Pashley, M.E. Karaman, Applied Colloid and Surface Chemistry, John Wiley & Sons Ltd, Chichester (2004).

H. Reerink, J. Overbeek, G. Th (1954), *Discuss. Faraday Soc.* 18, 74–84.

D.J. Shaw. Colloid & Surface Chemistry, 4th edn, Butterworth and Heinemann, Oxford, (1992).

J.A. Wesselingh, S. Kiil, M.E. Vigild, (2007) Design and Development of Biological, Chemical, Food and Pharmaceutical Products, John Wiley & Sons Ltd.

11

Colloid Stability – Part II: The DLVO Theory – Kinetics of Aggregation

11.1 DLVO theory – a rapid overview

We start this chapter with an overview of the DLVO theory and the basic characteristics of the theory we will discuss later.

The stability of colloids can be quantified with the DLVO theory which accounts (in an additive way) for the attractive van der Waals (vdW) and the repulsive electrostatic forces (or the corresponding potential energies). The DLVO theory does not account for other forces (the repulsive steric-polymeric and hydration ones, as well as hydrophobic and hydrodynamic forces).

The electrostatic double-layer force dominates at relatively medium–large separations. However, when "very" far away from the surfaces (very large separations) *and* when the surfaces are brought very close to each other, the attractive van der Waals forces (may) overcome the repulsive forces and dominate the interactions. If vdW forces dominate the surfaces will be pulled into a strong adhesive contact (attraction – instability) whereas stability is obtained in the region where the repulsion forces dominate.

If the maximum value of the potential energy is much higher than the "natural" kinetic energy of about k_BT, about $(15-25)k_BT$, then we may expect that our colloid is stable (metastable, to be precise). A secondary minimum can occur in some cases which correspond to a reversible flocculation. If the potential energy is lower than these values, and certainly when it is smaller than k_BT, then the dispersion is unstable.

Stable colloids are achieved if the Debye length (double layer thickness) is very high (i.e. low salt content, low valency ions), if the colloid particles are in a medium with high relative permittivity, if they have low (or even negative) Hamaker constants and high values of the surface or zeta potential. Control of the ionic concentration and surface charge are crucial.

The DLVO forces have been experimentally verified for many different surfaces. Some deviations may exist at rather large distances due to retardation effects as well as at close proximity of particles where solvation (hydration in case of water) forces may dominate in some cases. The hydration forces explain the stability of many lyophilic colloids, e.g. proteins and lipid bilayers.

Introduction to Applied Colloid and Surface Chemistry, First Edition. Georgios M. Kontogeorgis and Søren Kiil.
© 2016 John Wiley & Sons, Ltd. Published 2016 by John Wiley & Sons, Ltd.
Companion website: www.wiley.com/go/kontogeorgis/colloid

The critical coagulation concentration can be obtained by the DLVO theory and is found to be inversely proportional to the sixth power of the counterion valency, in agreement with the Schulze–Hardy rule.

The kinetics of coagulation are described by a second-order reaction, while the kinetics for a fast, diffusion-controlled, coagulation (in absence of a potential barrier) is independent of particle size and is very fast. The stability ratio and the Fuchs equation provide useful tools for establishing the stability and kinetics of a colloidal dispersion. The structure of aggregates containing large particles is often of the Cake-type (unwanted for pharmaceuticals), while smaller easier to redisperse particles form loose flocculants, which can sometimes be approximated using fractal geometry principles.

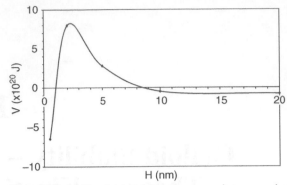

Figure 11.1 *Total interaction energy–distance plot (DLVO theory) for spherical colloidal particles of diameter 10^{-7} m dispersed in 10^{-2} mol L^{-1} aqueous 1 : 1 electrolyte at 25 °C. The Hamaker constants of the particles and of the dispersion medium are 1.6×10^{-19} J and 0.4×10^{-19} J, respectively, and the zeta potential is –40 mV. The system is stable as $V_{max}/k_BT = 19.5$*

11.2   DLVO theory – effect of various parameters

When suitable expressions are available for the attractive and repulsive potential energies, e.g. Equation 10.4, then the total potential energy (DLVO theory) can be calculated and plotted as function of the interparticle distance. High values, above $15–25k_BT$ ($k_BT = 4.12 \times 10^{-21}$ J at 25 °C), of the repulsive potential barrier indicate stable (actually metastable) colloidal systems. One example is shown in Figure 11.1.

Figure 11.2 shows the great dependency of the potential energy V on the salt concentration and the surface potential. It can be observed that V increases with both increasing double-layer thickness (decreasing salt concentration) and with increasing surface potential. However, the interpretation of surface potential is somewhat complicated. For some systems, the surface potential is adjustable by varying the concentration of potential-determining ions. The surface potential is a quantity complicated by adsorption phenomena and is in reality the potential at the "inner limit of the diffuse double layer" rather than simply the potential "at the wall". For most practical cases, we will simply assume that the surface potential is the same as the experimentally measured zeta potential (although in reality the zeta potential establishes the lower limit for the surface potential).

In both parts of Figure 11.2, the effects of added electrolyte in reducing the extent of the double layer (thickness) and lowering the potential at the surface have been artificially separated. However, in practice both effects would occur simultaneously and in a combined way reduce the double layer repulsion and consequently the stability of the dispersion. This is because the addition of electrolyte tends to lower the surface (zeta) potential and coagulation would then be expected at a lower ionic strength (see also Figure 10.13 with the zeta potential versus pH of TiO_2 at various salt concentrations).

Finally, the potential energy barrier decreases with increasing values of the Hamaker constant, A. Sometimes, the effective Hamaker constant is the variable we have least control of, i.e. if the chemical natures of the dispersed particles and the medium are (pre)determined.

According to the DLVO theory, stable colloidal dispersions ($V_{max} \gg k_BT$) are expected if one or preferably more of the following conditions are fulfilled:

a. particles with high surface (zeta) potentials;
b. low electrolyte concentration, in general low ionic strength (and especially not having

(a)

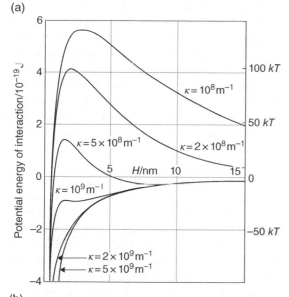

(b)

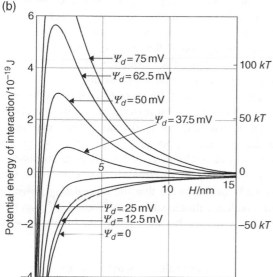

Figure 11.2 *Effect of salt concentration (decreasing debye length) (a) and Stern potential (b) on the potential curves for two interacting particles; R = 10^{-7} m, T = 298 K, z_ (counter-ion) = −1, A_{11} = 2 × 10^{-19} J, A_{22} = 0.4 × 10^{-19} J, and ε = 78.5. In (a), ψ_d = 50 mV and in (b) κ = 3 × 10^8 m. Shaw (1992). Reproduced with permission from Elsevier*

electrolytes with high valency) (i.e. high Debye length);

c. polar media like water (i.e. high relative permittivity);

d. low values of the Hamaker constant;
e. the rather unusual situation with negative Hamaker constant (leading to repulsive vdW forces) – this can happen in cases of particles of two different kinds in a third medium and if one of the particles attract more the medium than the other kind of particles;
f. low temperatures.

Notice that the control of the surface (zeta) potential and, in general, the control of the repulsive electrostatic forces require understanding of the charge mechanism. It can be controlled, depending on the dominant phenomenon behind the charge mechanism, by the control of pH, change of common or potential-determining ions, adsorption of ionic surfactants/polyelectrolytes, etc. It is especially crucial for good stability to be far from the isoelectric point (IEP). Finally, we should recall that ions can adsorb on the Stern layer, i.e. the inner part of the diffuse double layer, thus reducing the surface potential. It is especially the higher valency counter-ions that adsorb most strongly.

11.3 DLVO theory – experimental verification and applications

11.3.1 Critical coagulation concentration and the Hofmeister series

The DLVO theory is in good agreement with many empirical rules based on large volumes of experimental data such as the Schulze–Hardy rule, which was actually developed prior to the DLVO theory. This rule indicates that the CCC (critical coagulation concentration) depends largely (decreases) with the sixth power of the valency of the counter-ion. The CCC is the concentration of the electrolyte that is just sufficient to coagulate a dispersion. With some degree of approximation, it can be obtained, e.g. with reference to Figure 10.2 under the conditions where $V = 0$ and $dV/dH = 0$.

Using the conditions for CCC ($V = 0$ and $dV/dH = 0$) for aqueous dispersions at 25 °C, it can be shown that CCC is given by the following equation (Shaw, 1992):

$$CCC = \frac{3.84 \times 10^{-39} \gamma^4}{A^2 z^6} \qquad (11.1)$$

where:

$$\gamma = \frac{e^{\frac{ze\psi_0}{2k_BT}} - 1}{e^{\frac{ze\psi_0}{2k_BT}} + 1}$$

The value of γ limits to unity for rather high surface potentials. The Hamaker constant, A, must be inserted in J, CCC in mol L^{-1}, and z is the counter-ion valency (including sign). Thus, in this case, CCC is proportional to $1/z^6$. At low potentials, γ limits to $ze\psi_0/4k_BT$ (use of Taylor expansion of exponential function followed by $\psi_0 \to 0$). This leads to CCC being proportional to ψ_0^4/z^2.

Equation 11.1 can also be derived from DLVO using the Reerink–Overbeek expression for the electrostatic term (Table 10.5) and the expression for the vdW interactions between two equal-sized spheres:

1. We see a remarkable agreement with the Schulze–Hardy rule, especially at high potentials. At lower potentials, which represents the often more realistic case, the dependency of the (counter-ion) valency is less pronounced. However, the surface potential is often proportional to $1/z$, and thus we can recover the Schulze–Hardy rule. (Higher valency counter-ions adsorb strongly in the Stern layer resulting in a decrease of the surface potential.)
2. CCC for spherical particles of a given material is proportional to the third power of the relative permittivity and independent of the particle size.

Moreover, as shown by Shaw (1992), setting in these equations a typical CCC value equal to 0.1 mol L^{-1} at 25 °C and z_- (counter-ion) $= -1$ for a surface potential equal to 75 mV, we obtain a Hamaker constant equal to 8×10^{-20} J, which is within the expected values from the London/van der Waals theories (as discussed previously).

All these observations and results verify the validity of the DLVO theory, considering especially that the theoretical and experimental definitions of

CCC are somewhat different. Equation 11.1 explains why the ocean's salinity causes rivers to form deltas; a higher salt concentration can destabilize the colloid (e.g. small sand/dirt particles from river sediments).

Even within the same ionic valency the activity of the various ions will vary (a little) depending on their adsorption at interfaces, especially the Stern layer, interactions with proteins, degree of ionization of their salts, etc. It is found that for monovalent and divalent (or bivalent) ions the effectiveness of coagulating negatively (or positively) charged colloids has the order shown in Figure 11.3 (called the lyotropic or Hofmeister series). Also shown here is the order for negative ions (for coagulating a positively charged surface). Positively charged surfaces are more affected compared to negatively charged ones, in the sense that we see a stronger effect of the ion type. The explanation of this series lies in the ion hydration. For negatively charged hydrophobic colloids, the most effective cations are those which are not hydrated as the size of the hydrated ions restricts their ability to approach the particle surface and enter the Stern layer. The sequence of ions shown in Figure 11.3, however, may not always be valid. In a quite recent book by Berg (2010) on colloids he states that more recent reviews of specific ion effects suggest that different Hofmeister series' may apply under different application circumstances. This is probably also why there is not complete consistency among values in different literature sources. One should also be aware that the coagulating power of multivalent cations, such as Al^{3+} and La^{3+}, can be a function of pH because complex, highly charged

Lyotropic (hofmeister) series effectiveness of coagulation

$Cs^+ > Rb^+ > K^+ > Na^+ > Li^+$ (monovalent)

$Ba^{2+} > Sr^{2+} > Ca^{2+} > Mg^{2+}$ (divalent)

$CNS^- > I^- > Br^- > Cl^- > F^- > NO_3^- > ClO_4^-$

Figure 11.3 *Order for positive ions for coagulating a negatively charged surface and the order for negative ions (for coagulating a positively charged surface) for hydrophobic colloids. Positively charged surfaces are more affected than negatively charged ones, in the sense that we see a stronger effect of the ion type. Smaller ions are more hydrated due to a more intense electronic field*

species can be formed by the cations at basic pH values (Goodwin, 2009).

Hydrophilic colloids, e.g. polymers and proteins, are stabilized by a combination of electrostatic and solvation forces. Such colloidal systems are unaffected by small electrolyte concentrations but at very high electrolyte concentrations they may precipitate (salting-out effect). For example, gelatin has strong affinity for water even at the IEP, but casein has a weaker hydrophilic behaviour and precipitates from water close to the IEP.

But which salts should be used for protein precipitation? For hydrophilic colloids (such as proteins) the Hofmeister series (shown in Figure 11.3) is exactly inversed, as the increased hydration can result in salting-out effects. Very water soluble salts, e.g. ammonium sulfate, are preferred for protein precipitation. This is illustrated in Figure 11.4. Ions of the electrolytes dehydrate the hydrophilic colloid, i.e. remove water from the polymer. In this way, it is understood that the salting-out effect depends on the ion hydration.

The hydration of protein molecules is a very important phenomenon with many applications. One of them discussed by Dey and Prausnitz (2011) is based on the work of Vekilov and co-workers. These researchers have noticed that addition of an organic co-solvent (acetone) to an aqueous insulin solution increases the crystallization rate of insulin by almost one order of magnitude (seven times larger). This can be of importance in applications related to diabetes therapy and pharmacology. By measuring the insulin concentration at equilibrium as a function of temperature (5–30 °C) and of acetone concentration, they have calculated the enthalpy and entropy of insulin crystallization. Using these data, they have understood the thermodynamics of insulin crystallization and the role of the organic co-solvents. It has been shown that acetone prevents hydration – adsorbed water molecules – of the insulin molecule and thus good crystals are obtained (with an expected negative entropy of crystallization, which is the preferred mechanism). In general, hydration forces are very complex and active when hydration layers overlap. When overlapping, the relative permittivity of water may also change (Israelachvili, 2011).

11.3.2 DLVO, experiments and limitations

The DLVO theory is also in excellent agreement with experimentally obtained interparticle forces in many cases, e.g. between mica surfaces (Pashley and Karaman, 2004), and in other cases such as for interactions between surfactant bilayers, across soap films, between glass surfaces or quartz surfaces. The agreement between DLVO and experimental data is often very good, although additional solvation forces are often observed at separations below 1–5 nm,

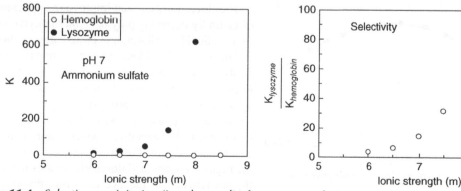

Figure 11.4 Selective precipitation (i.e. phase split) for aqueous solutions containing two proteins, lysozyme and haemoglobin at pH 7 and ammonium sulfate – the salting out effect. Ions of the electrolyte dehydrate the hydrophilic colloid, i.e. remove water from the polymer, whereby a phase split occurs where two liquids co-exist. One of these phases is very dilute in protein while the other is highly concentrated and precipitation can take place. The experimental data show that when the ionic strength of ammonium sulfate exceeds about 7.5 M excellent separation of the proteins is obtained. K is the distribution coefficient of a given protein in the two liquid phases. Reprinted from Prausnitz (1995), with permission from Elsevier

depending on the system. It is perhaps surprising that measured double-layer forces are so well described by a theory that, unlike the van der Waals force theory, contains several fairly drastic assumptions (ions considered as point charges, Poisson–Boltzmann equation remains valid at fairly high concentrations). Any deviations in the values of the forces from those expected from DLVO can usually be traced to the existence of some other forces, e.g. solvation ones, and not (so much) because of the breakdown in the DLVO interaction. These short-range solvation forces are, of course, very important especially in determining the coagulation of colloidal particles and in biological systems where short-range interactions are of paramount importance. Sometimes there are also deviations at high separations (above 5 nm) because of retardation effects.

Electrostatic stabilization works best for strongly charged colloids and at low electrolyte concentrations. An important industrial example is the latex particles in water-based paints stabilized by repulsive electrostatic forces (Pashley and Karaman, 2004 and later Example 11.4). Rapid coagulation

will be generated by even a small increase in electrolyte level by evaporation of water from the paint.

In cases, however, when electrostatic stabilization does not work well (uncharged/weakly charged particles, high salt concentration, e.g. for many biological systems, in organic solvents/low temperatures, high particle concentration) we have to make use of the "steric" stabilization which can be obtained by coating the particles with (bio)polymers, oligomers and proteins. It is important to choose polymers that have affinity for the medium (i.e. good solvent conditions), have complete coverage of the particles and full adsorption. In the opposite case, we may get bridging flocculation and actually attractive forces. The configurations and the thickness of the polymer layer are crucial factors; the latter can be considered to be approximately equal to the radius of gyration. The steric stabilization is discussed in Chapter 12 (emulsions).

An exciting application of colloid aggregation and DLVO theory is shown in the case study "Diseases caused by protein aggregation" shown next.

Case study. Diseases caused by protein aggregation (inspired from J.M. Prausnitz, 2003)

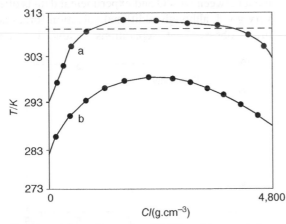

Reprinted from Prausnitz (2003), with permission from Elsevier.

The figure above shows the phase diagram for liquid–liquid system related to proteins; C is protein concentration in the eye. Diseases, such as cataract formation, are caused by protein aggregation at high protein concentrations. With age, the

protein concentration increases. The broken line is body temperature (37 °C). (a) Native γ-crystallin; (b) γ-crystallin/glutathione complex.

The figure illustrates an impressive application of naturally occurring polymers, thermodynamics (liquid–liquid equilibria) and protein aggregation.

There is much evidence to suggest that the metastable liquid–liquid region of a protein solution is responsible for cataract formation. As the eye ages, its protein concentration changes and one (or more) proteins may achieve a concentration that exceeds the saturation concentration as indicated in the figure. Following phase separation, the highly concentrated liquid phase does not settle but remains in the eye as a fine dispersion whose optical properties seriously interfere with vision. The experimental data show that when the concentration of native γ-crystallin exceeds about 100 mg cm^{-3} at body temperature (310 K, 37 °C), a second liquid phase is formed whose concentration is about 700 mg cm^{-3}. It is this second liquid phase (that may eventually become crystalline) that is responsible for cataract symptoms. Precipitation of the second phase can be avoided by adding a small amount of glutathione that forms a soluble complex with γ-crystallin to keep it in dilute solution. These experimental results provide the first step for a possible method to prevent cataracts in the human eye, a disease that strikes many millions of elderly men and women throughout the world. Similar studies have been reported for sickle-cell anaemia and for fibril formation that appears to be responsible for Alzheimer's disease.

Comparison of DLVO theory with detailed simulations

The major effect in the interaction between identical protein particles comes from the electrostatic double-layer forces (vdW forces are weak). Interestingly, these forces are indeed repulsive for monovalent cations and agree with the simulation results (points in the figures shown above). However, when the medium contains divalent salts, molecular simulation studies show that the potential of mean force is negative for some values of distance, indicating attraction between two like-charged proteins. The DLVO theory cannot show this attraction that follows from electrostatic binding of divalent ions to the oppositely charged protein particles. Improvements are needed.

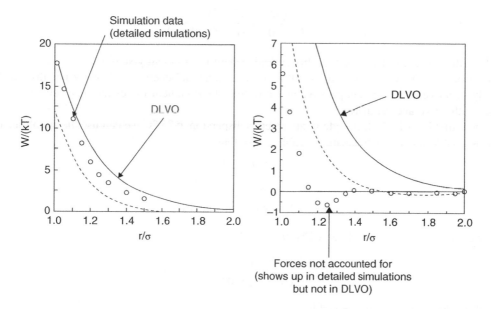

Interactions between like-charged proteins at 25 °C. The protein diameter is 0.4 nm. To the left, an ionic strength of 0.28 M and salts containing monovalent cations. To the right divalent cations at 0.31 M. The dashed line in the figure is not of relevance in this discussion. Reprinted from Prausnitz (2003), with permission from Elsevier.

Example 11.1.   Colloid stability and DLVO theory.
Spherical colloidal particles of diameter 10^{-7} m are dispersed in 10^{-2} mol L^{-1} aqueous NaCl solution at 25 °C. The Hamaker constants of the particles and the dispersion medium are 1.6×10^{-19} J and 0.4×10^{-19} J, respectively, and the zeta potential is –40 mV. Under certain assumptions, the total potential interaction energy between the particles versus the interparticle distance H is given by the plot shown in Figure 11.1.

1.   Is this colloidal dispersion stable? Justify briefly your answer.
2.   Which are the dominant forces at $H = 0.2$, 2 and 15 nm? Why?
3.   Estimate the Debye length at the conditions of the problem. How are the Debye length and the stability expected to change if the NaCl solution is replaced by an AlCl$_3$ solution of the same concentration?
4.   Draw qualitatively the V–H curve in the cases that (i) the effective Hamaker constant is changed to 0.4×10^{-20} J (all other parameters have the same values) and (ii) the zeta potential is changed to –20 mV (all other parameters have the same values).
5.   Estimate using the DLVO theory the salt concentration needed so that the colloid dispersion of the problem *just* becomes unstable, i.e. estimate the salt concentration corresponding to critical coagulation concentration (CCC) according to the DLVO theory. Explain the assumptions made in your calculations.
6.   Compare the CCC value estimated in the previous question to the CCC value obtained for NaCl in the case of negatively charged As$_2$S$_3$ particles. For these particles the CCC value is equal to 0.81 mmol L^{-1} when MgSO$_4$ is used. Explain the assumptions made in your calculations.

Solution
1.   We know that "a barrier of $(15–25)k_BT$ is sufficient to give colloid stability".
     From the V–H plot given we see that:

$$\frac{V_{max}}{k_BT} \cong \frac{\left(8.0 \times 10^{-20} \text{ J}\right)}{\left(1.381 \times 10^{-23} \text{ JK}^{-1}\right) \cdot (298 \text{ K})} = 19.5$$

Thus, there is a good chance that this colloid system is sufficiently stable.

2.   Again from the given plot, we can see that the total potential energy is strongly negative at 0.2 nm, weakly negative at 15 nm and positive at 2 nm; thus the dominant forces are, respectively, attractive, weakly attractive and repulsive.
3.   We know that for a 1 : 1 electrolyte in an aqueous dispersion at 25 °C we can use the simplified expressions available for the Debye length (Table 10.4), thus:

$$\kappa^{-1} = \frac{0.304}{\sqrt{C}} = \frac{0.304}{\sqrt{0.01}}$$

$$\kappa^{-1} = 3.04 \times 10^{-9} \text{m} = 3.04 \text{ nm}$$

If we use a 3:1 electrolyte like AlCl$_3$, then we have:

$$\kappa^{-1} = \frac{0.124}{\sqrt{C}} = \frac{0.124}{\sqrt{0.01}} = 1.24 \text{ nm}$$

and, thus, the dispersion is less stable.

4.   For the dispersion of the problem, the effective Hamaker constant is:

$$A_{121} = \left(\sqrt{A_{11}} - \sqrt{A_{22}}\right)^2 = \left(\sqrt{1.6 \cdot 10^{-19}} - \sqrt{0.4 \cdot 10^{-19}}\right)^2$$

1: particles
2: medium
thus:

$$A_{121} = 4 \times 10^{-20} \, \text{J}$$

If the Hamaker constant is decreased to 0.4×10^{-20} J (4×10^{-21} J), then the stability is improved (higher V–H curve, e.g. from form (b) to form (a)). If the zeta potential changes to -20 from -40 mV, then the stability is decreased. Then, the curve will change, see figure below, from form (b) (curve of the problem) to one of the other lower ones shown in the figure taken from Israelachvili (1985). Without additional information we cannot conclude whether the dispersion will remain stable, but in all cases the stability is decreased.

5. The CCC according to the DLVO theory is defined by the following equations (curve d in the figure).

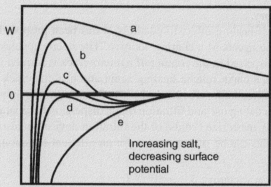

Reprinted (adapted) from Israelachvili (1985). Copyright 2015, Academic Press, Elsevier.

$V = 0$; $dV/dH = 0$.

Using the simplified forms for the repulsive and attractive forces (for equal-sized spherical particles), we have:

$$V = V_R + V_A = 2\pi\varepsilon\varepsilon_0 R\psi_0^2 e^{-\kappa H} - \frac{A_{121}R}{12H} = 0$$

$$\frac{dV}{dH} = 0 \Rightarrow H = \kappa^{-1}$$

Thus, we see that the "critical" distance where these conditions are satisfied is equal to the Debye length. Then, upon substitution to the $V = 0$ equation, we can directly calculate the Debye length corresponding to CCC (for the given values of zeta potential and Hamaker constant, see question 4):

$$\psi_0 = \sqrt{\frac{e_{ln}A_{121}}{12\kappa^{-1}2\pi\varepsilon\varepsilon_0}} \Rightarrow$$

$$\kappa^{-1} = \frac{e_{ln}A}{24\pi\varepsilon\varepsilon_0\psi_0^2} = \frac{(2.718)\left(4\times10^{-20}\,\text{J}\right)}{24\pi(78.5)\left(8.854\times10^{-12}\,\text{C}^2\text{J}^{-1}\text{m}^{-1}\right)\left(-40\times10^{-3}\,\text{V}\right)^2} = 1.29\times10^{-9}\,\text{m} = 1.29\,\text{nm}$$

Notice that the symbol e_{ln} (=2.718) here is the base of the natural logarithm.
For NaCl, this corresponds to a concentration:

$$\kappa^{-1} = \frac{0.304}{\sqrt{C}} = 1.29 \text{ nm} \Rightarrow C = 0.0555 \text{ molL}^{-1}$$

6. Using the Schulze–Hardy rule, the CCC is proportional to the sixth power of the counter-ion valency and, thus, the proportionality constant (which will be equal to the CCC for NaCl), K, is given for the data of the problem:

$$\text{CCC} = \frac{K}{Z^6} \Rightarrow 0.81 = \frac{K}{2^6} \Rightarrow K = 51.84 \text{ mmolL}^{-1} = 0.052 \text{ molL}^{-1}$$

which is slightly lower than the value estimated in question 5.

Example 11.2.   DLVO Theory.

In an attempt to provide a self-healing effect to coatings, it has been proposed to use microcapsules with chemically active core healing agents as a coating additive. The microcapsules, which must be made with a suitable wall material, are dispersed in the coating. If a micro-crack is formed in the coating during service life, the capsules near the crack burst, release healing agent and heal the crack by chemical reactions.

The spherical microcapsules (colloids) are formed by interfacial organic synthesis in water, but the capsule diameter is often too high for coating use and filtration of the dispersion to separate large from small particles is required. However, not only mesh size (voids) of the filtration device but also interactions between filtration material and microcapsules can be of importance for an efficient filtration.

Data:

Hamaker constant of water (in vacuum): 3.7×10^{-20} J;
Hamaker constant of microcapsules (in vacuum): 5.0×10^{-20} J

A capsule diameter of 10 μm only is considered.

1. Metal sieves and cellulose filters are available for the separation. Based only on a calculation of the van der Waals forces between filtration materials and microcapsules (at a colloid–surface distance of 10 nm), which of the two filtration materials would you recommend for the separation process?
   Data:

   Hamaker constant of the metal material (in vacuum): 22.0×10^{-20} J
   Hamaker constant of the cellulose material (in vacuum): 5.8×10^{-20} J

2. A third option is to use a metal sieve coated with a Teflon material. Will you expect this Teflon-coated sieve to provide a better separation than the non-coated metal sieve?
   Data: Hamaker constant of Teflon coated metal sieve (in vacuum): 3.8×10^{-20} J

3. The microcapsules have now been electrostatically stabilized and it is proposed to store the microcapsules, prior to use in coatings, at 25 °C in a 0.01 M aqueous solution of Na_2CO_3. It can be assumed that the values of the Hamaker constants do not change due to the electrostatic stabilization and the surface potential is equal to 10 mV.

   Calculate the double layer thickness and the composite (overall) potential between the microcapsules at the following interparticle distances: 0.1, 0.5, 2, 5, 10 and 20 nm.

Solution

Microcapsules for anticorrosive coatings are considered.

1. The system of microcapsules and sieve or filter is approximated by a surface–particle interaction. In this case, the van der Waals force is given by the following equation obtained from Table 2.2:

$$F = -\frac{dV_A}{dH} = -\frac{RA}{6H^2}$$

A negative force implies attraction as described by Israelachvili (2001, p. 255). We have a system with three different components so the Hamaker constant must be the effective Hamaker constant, A_{123}, leading to:

$$F = -\frac{RA_{123}}{6H^2}$$

The effective Hamaker constant is calculated as follows:

$$A_{123} = \left(\sqrt{A_{11}} - \sqrt{A_{22}}\right)\left(\sqrt{A_{33}} - \sqrt{A_{22}}\right)$$

For the microcapsules/water/metal sieve (1/2/3) we get:

$$A_{123} = \left(\sqrt{5 \times 10^{-20}\ \text{J}} - \sqrt{3.7 \times 10^{-20}\ \text{J}}\right)\left(\sqrt{22.0 \times 10^{-20}\ \text{J}} - \sqrt{3.7 \times 10^{-20}\ \text{J}}\right) = 8.65 \times 10^{-21}\ \text{J}$$

and for microcapsules/water/cellulose (1/2/3) we get:

$$A_{123} = \left(\sqrt{5 \times 10^{-20}\ \text{J}} - \sqrt{3.7 \times 10^{-20}\ \text{J}}\right)\left(\sqrt{5.8 \times 10^{-20}\ \text{J}} - \sqrt{3.7 \times 10^{-20}\ \text{J}}\right) = 1.52 \times 10^{-21}\ \text{J}$$

For comparison, the microcapsule/water/microcapsule (1/2/1) Hamaker constant is also calculated:

$$A_{121} = \left(\sqrt{5 \times 10^{-20}\ \text{J}} - \sqrt{3.7 \times 10^{-20}\ \text{J}}\right)^2 = 0.98 \times 10^{-21}\ \text{J}$$

Hence, there is a stronger attraction between capsules and filtration materials than between the capsules.

The van der Waals force for metal material:

$$F = -\frac{RA_{123}}{6H^2} = -\frac{\left(5 \times 10^{-6}\ \text{m}\right)\left(8.65 \times 10^{-21}\ \text{J}\right)}{6\left(10 \times 10^{-9}\ \text{m}\right)^2} = -7.2 \times 10^{-11}\ \text{N}$$

and for cellulose filter:

$$F = -\frac{RA_{123}}{6H^2} = -\frac{\left(5 \times 10^{-6}\ \text{m}\right)\left(1.52 \times 10^{-21}\ \text{J}\right)}{6\left(10 \times 10^{-9}\ \text{m}\right)^2} = -1.27 \times 10^{-11}\ \text{N}$$

Based on these calculations, the smallest force is obtained using the cellulose filter and it is therefore recommended to use this method of filtration.

2. Using the Hamaker constant for the Teflon coated sieve the following effective Hamaker constant (microcapsules /Teflon/water/) (1/2/3) is obtained:

$$A_{123} = \left(\sqrt{5 \times 10^{-20} \text{ J}} - \sqrt{3.7 \times 10^{-20} \text{ J}}\right)\left(\sqrt{3.8 \times 10^{-20} \text{ J}} - \sqrt{3.7 \times 10^{-20} \text{ J}}\right) = 8.07 \times 10^{-23} \text{ J}$$

This effective Hamaker constant is much lower than the one for the metal sieve and even the one for the cellulose filter and it would be a good idea to use the Teflon coated sieve instead of the non-coated metal sieve. Both possibilities were actually tried in practice by Nesterova *et al.* (2012) and the Teflon coated sieve proved to be superior. In fact, no microcapsules went through the non-coated sieve even though the microcapsules were smaller than the mesh size of the sieve.

3. The microcapsules have now been electrostatically stabilized and are stored in an electrolyte solution. The double layer thickness calculated for a 1 : 2 electrolyte at 25 °C is (see Table 10.4):

$$\kappa^{-1} = \frac{0.176}{\sqrt{C}} = \frac{0.176}{\sqrt{0.01}} = 1.76 \text{ nm}$$

To calculate the composite potential, the validity of the Debye–Hückel approximation (see Table 10.5) is first analysed using the following condition (see Table 10.4):

$$\frac{|z\psi_o|e}{k_B T} = \frac{|(-2) \times (10 \times 10^{-3} \text{ V})|(1.602 \times 10^{-19} \text{ J})}{(1.381 \times 10^{-23} \text{ J}^{-1}\text{K}^{-1})(298.15 \text{ } K)} = 0.78 < 1$$

In this equation, z is the counter-ion ionic valency and ψ_o is the surface (zeta) potential. The latter is positive and therefore the counter-ion must be CO_3^{2-}, which has an ionic valency of –2. The approximation is allowed and the composite potential can be written using the following expression for spherical particles:

$$V = V_R + V_A = 2\pi\varepsilon_o\varepsilon R\psi_o^2 e^{-\kappa H} - \frac{RA_{eff}}{12H}$$

Inserting the relevant numbers gives:

$$V = 2\pi\varepsilon_o\varepsilon R\psi_o^2 e^{-\kappa H} - \frac{RA_{eff}}{12H}$$

$$= 2\pi(8.854 \times 10^{-12} \text{ C}^2\text{J}^{-1}\text{m}^{-1})(78.5)(5 \times 10^{-6} \text{ m})(10 \times 10^{-3} \text{ V})^2 \exp\left(\frac{-H}{1.76 \text{ nm}}\right)$$

$$- \frac{(5 \times 10^{-6} \text{ m})(9.8 \times 10^{-22} \text{ J})}{12H} \frac{10^9 \text{ nm}}{1 \text{ m}}$$

$$= (2.18 \times 10^{-18})\exp\left(\frac{-H}{1.76 \text{ nm}}\right) - \frac{(4.08 \times 10^{-19} \text{ J})}{H} \quad [\text{J}]$$

where H should be inserted in nm. The value of A_{eff} was calculated in question 1 (microcapsule/water/microcapsule).

The following table can now be constructed

H (nm)	V_R (J)	V_A (J)	V (J)	V/k_BT
0.1	2.06×10^{-18}	-4.08×10^{-18}	-2.0×10^{-18}	−492
0.5	1.64×10^{-18}	-8.16×10^{-19}	8.24×10^{-19}	200
2	7.0×10^{-19}	-2.0×10^{-19}	5.0×10^{-19}	122
5	1.27×10^{-19}	-8.16×10^{-20}	4.54×10^{-20}	11
10	7.42×10^{-21}	-4.08×10^{-20}	-3.3×10^{-20}	−8
20	2.53×10^{-23}	-2.04×10^{-20}	-2.0×10^{-20}	−5

The high numerical values of V/k_BT is partly due to the rather large particle (microcapsule) size (10 μm), which is somewhat outside the colloidal domain, typically given as 1 nm to about 1 μm.

11.4 Kinetics of aggregation

11.4.1 General – the Smoluchowski model

Some general rules apply to the kinetics of coagulation of colloids, which can also be used in the specific case of emulsions discussed in Chapter 12.

When the colloid destabilization occurs, the kinetics of collapse is often fast. Let us consider that we have a colloid system containing originally n_o particles per volume. Once destabilized, the particles aggregate by colliding with each other and thus the concentration of particles drops with time. Typically this phenomenon is represented by the second-order equation, often called "Smoluchowski model":

$$-\frac{dn}{dt} = k_2 n^2 \Rightarrow \frac{1}{n} - \frac{1}{n_0} = k_2 t \qquad (11.2)$$

where:

n = number of particles per volume at some time t (m^{-3});
k_2 = reaction constant (m^3 number^{-1} s^{-1});
n_0 = number of particles at start ($t = 0$) per unit volume (m^{-3}).

According to Equation 11.2, a plot of the reciprocal of the concentration ($1/n$) should vary linearly with time t, which indeed is verified by many experiments. The coagulation rates can be measured under non-agitated conditions (perikinetic conditions) and agitated (orthokinetic) conditions. In the latter case, the rate may be up to 10000-times higher than in the former. The number of particles as a function of time can often be measured with dynamic light scattering.

11.4.2 Fast (diffusion-controlled) coagulation

If we remove the electrostatic barrier, e.g. by ion adsorption or by adding electrolyte, we can obtain a fast coagulation that is controlled by the diffusion mechanism and where we can assume that all collisions lead to adhesion of molecules. Smoluchowski has derived the following equation for the rate constant in the case of fast coagulation:

$$k_2^0 = \frac{4k_BT}{3\eta_{medium}} \qquad (11.3)$$

In this case the second-order constant does not depend on particle size, only on temperature and the viscosity of medium. This rate constant often implies rapid colloid aggregation.

Equation 11.3 is approximate but gives an idea of how fast the coagulation can be. An important measure of the speed of coagulation is the so-called half-life time (see Problem 11.5), the time when the number of particles is decreased to half:

$$t_{\frac{1}{2}} = \frac{3\eta_{medium}}{4k_BTn_o} \qquad (11.4)$$

For concentrated dispersions, the half-life time is in the milliseconds range, while for more usual colloidal systems typical values are higher, of the order of seconds to minutes.

11.4.3 Stability ratio W

An important measure of how fast a coagulation will be is the so-called stability ratio. It is defined as the ratio of encounters that stick or in other words the total collisions between particles divided by the collisions of particles which result in a coagulation. The rate of fast coagulation is the rate with no potential barrier and is limited only by the diffusion rate of the particles towards one another:

$$W = \frac{k_2^0}{k_2} \qquad (11.5)$$

It can be shown that the stability ratio is, approximately, directly related to the maximum (barrier) of the potential energy function. This is because in the slow-coagulation regime the electrolyte concentration is such that the diffuse layer is very compressed. In the more general case, when we know the potential–distance (V–H) function, the Fuchs equation can be used but this often requires numerical solutions for a given Hamaker/surface potential values:

$$W = 2R \int_{2R}^{\infty} \frac{\exp\left(\dfrac{V}{k_B T}\right)}{H^2} \, dH \qquad (11.6)$$

where R is the particle radius. The values of the stability ratio are directly linked to the stability and the following can be stated:

- $W > 10^5 \rightarrow$ easily obtained with modest potentials, about 15 $k_B T$;
- $W = 10^9 \rightarrow$ corresponds to a V of about 25 $k_B T$ (very slow coagulation, rather stable dispersion);
- Even $W > 10^{10}$ are not impossible.

If only the V_{max} value is known then the Reerink–Overbeek equation can be used:

$$W = \frac{1}{2\kappa R} e^{\frac{V_{max}}{k_B T}} \qquad (11.7)$$

A dilute dispersion that rapidly coagulated in about 1 s can become stable with coagulation times of several months if the V_{max} is about 20 $k_B T$. As a rule of thumb, we can mention that V_{max} values of about $(15–25)k_B T$ are often sufficient for stable dispersions provided that the Debye length is also rather large,

above 20 nm. These values correspond to W above 10^5.

It is always important to remember that the V_{max} poses a kinetic barrier, because the potential is lowest in the primary minimum. Eventually, even if it will take long time, the dispersion will be destroyed (complete coagulation) because some particles will have sufficient kinetic energy to overcome the barrier ($k_B T$ is an *average* of the natural kinetic energy of the particles). If there is no secondary minimum and the potential barrier is sufficiently large compared to the thermal energy ($k_B T$), then very few molecules will be able to "fall into" the primary

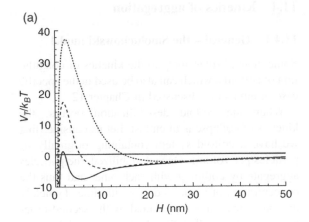

(a)

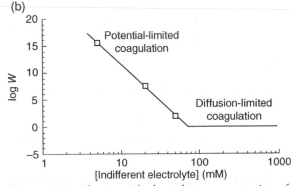

(b)

Figure 11.5 *Change with electrolyte concentration of the potential energy (a) and stability ratio (b) curves for silver bromide particles with a zeta potential equal to −50 mV and a particle radius of 50 nm. The electrolyte is symmetrical and of 1 : 1 type. In (a), the top curve corresponds to 5 mM, then follows 20 mM, and finally 50 mM. Reprinted from Goodwin (2009), with permission from John Wiley & Sons, Ltd*

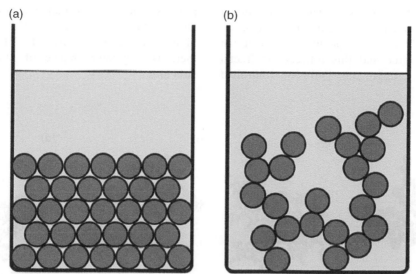

Figure 11.6 *Different structures of aggregates: (a) cake structure (hard to disperse); (b) loose flocculants (easy to disperse)*

minimum and the dispersion is kinetically stable (and there is no flocculation or aggregation).

One application is shown in Figure 11.5, which illustrates the change of potential energy and stability ratio curves for silver bromide (AgBr) particles with a zeta potential equal to –50 mV and a particle radius of 50 nm. The height of the primary maximum relative to the secondary minimum was used in the calculation, as this would be the energy required for the particles to move from a secondary minimum flocculation to a primary minimum coagulated state.

Clearly, the stability ratio changes very rapidly with electrolyte concentration. Thus, Figure 11.5 illustrates why we usually just notice a change from a stable dispersion to a rapidly aggregated one under normal laboratory conditions. Although it might appear that the electrolyte concentration of a sample is low enough for the stability to be adequate, the slow coagulation process is the type of problem that often occurs during storage (Goodwin, 2009).

11.4.4 Structure of aggregates

Long-range repulsion can yield long-range order (colloidal crystal), while we can have stable dilute suspensions even without long-range order (liquid–liquid structure). The two examples given in Figure 11.6 show flocculated/aggregated suspensions. The cake structure is hard to redisperse and is not wanted in many cases, e.g. in pharmaceutical dispersions. We often want, for example, in soils, paints, etc. to have loose flocculants which have large sedimentation volumes and are easy to redisperse. When the particles aggregate to form a continuous network structure that extends throughout the available volume and immobilizes the dispersion medium, the resulting semi-solid system is called a gel.

An important related process is peptization. In this case we may get, shortly after aggregation, dispersion with little or no agitation by changing the composition of the dispersion medium. This is achieved upon addition of multi-charge (polyvalent) co-ions (e.g. polyphosphate ions to a negatively charged coagulated dispersion), surfactants, diluting the medium or use of membranes to remove the electrolytes from the dispersion. In all cases, the repulsive potential energy is modified and a potential energy maximum (barrier) is created which acts against re-coagulation.

Very often the structure of aggregates is described using fractal geometry (Figure 11.7). This means that the mass of the aggregate is proportional to a certain exponent of its radius rather than an "ideal" value of three. The positive exponent can be lower than three and this reflects the fractal dimension of the aggregate. Diffusion-controlled processes can result in such less compact structures. On the other hand, when the stability ratio W has high values i.e. there exists a high potential barrier, there is a greater chance of obtaining a cake structure.

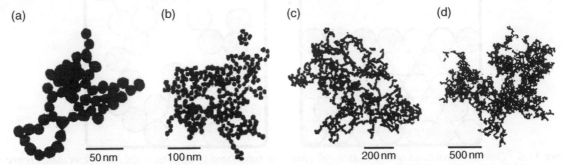

(a)   (b)   (c)   (d)

50 nm   100 nm   200 nm   500 nm

Figure 11.7 *Fractal structure of aggregates of gold particles (TEM recordings): (a)–(d) aggregates at various resolutions. Reprinted from Hiemenz and Rajagopalan (1997), with permission from Elsevier*

Example 11.3. *Flocculation of hydrosols and zeta potential.*
The measurements below are particle counts (number of particles per cm^3) made during flocculation at 25 °C of a hydrosol having spherical particles, with a diameter of 5×10^{-8} m, in a NaCl solution with a concentration 0.93×10^{-3} mol L^{-1}. The viscosity of the solution is 9.0×10^{-4} kg m^{-1} s^{-1} and the relative permittivity is 78.5.

Time (min)	Number of particles per volume (per cm^3)
0	1500×10^8
2	95×10^8
4	45×10^8
8	23×10^8
16	11.5×10^8

Moreover, the following values are available for the f correction parameter of the Henry equation for various values of κR (κ^{-1} is the Debye length and R is the radius of the particle):

κR	$f(\kappa R)$	κR	$f(\kappa R)$
0	1.000	5	1.160
1	1.027	10	1.239
2	1.066	25	1.370
3	1.101	100	1.460
4	1.133	∞	1.500

1. Calculate the "second-order" rate constant, k_2, and compare it with the value, $k_2{}^0$, calculated based on the assumption that coagulation is a diffusion-controlled process. Then calculate the stability ratio. Comment on the result. What values of the stability ratio would indicate a stable colloidal system?
2. What is the importance of the zeta potential in the study of the stability of a colloidal dispersion? Is the zeta potential of equal importance to aqueous and non-aqueous colloidal dispersions?
3. An electrophoretic experiment with this hydrosol is designed so that a non-flocculated particle moves 100 μm in 10 s. The zeta potential is determined to be 20 mV.
   What is the value of the potential gradient of the field expressed in V cm^{-1}?
   Use, when appropriate the *f*-corrections to the Henry equation.

Solution

1. A plot of $1/n$ versus t gives $k_2 = 0.912 \times 10^{-18}$ m^3(number of droplets s)$^{-1}$.
   The aggregation constant for the fast aggregation can be calculated using Equation 11.3 and it is equal to 6.095×10^{-18} m^3 (number of droplets s)$^{-1}$.
   Thus, $W = k_2{}^0/k_2 = 6.095 \times 10^{-18}/(0.912 \times 10^{-18}) = 6.7$
   This is a low value and, thus, the colloid system is possibly not very stable.
   Values of $W > 10^5$ can be obtained with modest potentials, about $15k_BT$, and the colloidal system is often (meta)stable. Really stable colloidal systems are obtained with a W of the order of 10^9, which corresponds to a V of about $25k_BT$ (very slow coagulation, rather stable dispersion).

2. The zeta potential is a measure of the potential at the surface of a particle. It is not exactly identical to the surface potential, but it is the potential obtained experimentally from (micro)electrophoresis experiments. The repulsive electrical forces are proportional to the second power of the surface or zeta potential. Thus, the higher the value of the zeta potential is, the higher the repulsive forces are, and the better the stability of the colloidal system is.
   The stability and repulsive forces depend of course on other factors as well, especially the Debye thickness (of double layer). Values of the zeta potential above 50 mV can give good stability at electrolyte concentrations of around 0.1–0.01 mol L^{-1}.
   The zeta potential is more important in aqueous (and other very polar solvent) dispersions, where electric forces (and stabilization) are important. The zeta potential is of less importance for non-aqueous (organic) dispersions, where steric stabilization is often a more effective stabilization method.
   We first calculate the Debye length. For a 1 : 1 electrolyte it is calculated as (see Table 10.4):

$$\kappa^{-1} = \frac{0.304}{\sqrt{C}} = \frac{0.304}{\sqrt{0.00093 \text{ molL}^{-1}}}$$

$$\kappa^{-1} = 9.968 \text{ nm} \Rightarrow \kappa = 1.003 \times 10^8 \text{m}^{-1} \Rightarrow$$
$$\kappa R = 2.507$$

For this low value, $f = 1.084$ (interpolation in the table).
Thus, we can calculate the electrophoretic mobility (Henry equation):

$$\psi_0 = \frac{1.5\mu\eta}{\varepsilon\varepsilon_0 f} \Rightarrow$$

$$20 \times 10^{-3} \text{ JC}^{-1} = \frac{1.5\mu(0.0009 \text{ kgm}^{-1}\text{s}^{-1})}{(78.5)(8.854 \times 10^{-12} \text{ C}^2\text{J}^{-1}\text{m}^{-1})(1.084)} \Rightarrow$$

$$\mu = 1.116 \times 10^{-8} \text{ m}^2\text{s}^{-1}\text{V}^{-1}$$

Thus, the potential gradient can be easily calculated as:

$$\mu = \frac{d}{tE} \Rightarrow E = \frac{d}{t\mu} = \frac{\left(100 \times 10^{-6}\ \text{m}\right)}{\left(10\ \text{s}\right)\left(1.116 \times 10^{-8}\ \text{m}^2\text{s}^{-1}\text{V}^{-1}\right)} = 896.05\ \text{Vm}^{-1} = 8.96\ \text{Vcm}^{-1}$$

Example 11.4. *Water-based coating.*

A coating producer is preparing a white, water-based formulation to be used for protection of exterior wood structures on buildings. The binder of the coating, a so-called alkyd, is available as an oil-in-water emulsion, containing surfactants for stabilization.

In a real situation, the coating will contain 10–15 components, but here we will use a simplified formulation consisting of 55 vol.% TiO_2 (spherical, surface-treated pigment particles), 32 vol.% emulsion, and 13 vol.% water (containing some dissolved compounds). The oil content of the emulsion is 50 vol.%. The oil droplet and TiO_2 particle diameters are 1 and 0.2 μm, respectively. The temperature is 25 °C and the viscosity of the medium (thickened aqueous phase) is 1 kg m^{-1} s^{-1}. The Hamaker constants of the oil and of TiO_2 in vacuum are 5×10^{-20} and 15×10^{-20} J, respectively. The Hamaker constant of the aqueous medium in vacuum is 3.75×10^{-20} J. The medium concentration of salt, NH_4OH (applied to control pH to basic conditions), in the coating formulation is 10^{-5} M. The relative permittivity of the medium is 80.

1: Calculate the attractive Van der Waals interaction energy between two oil droplets, between two TiO_2 particles, and between an oil droplet and a TiO_2 particle in the coating when the distance between the colloids (in all cases) is 10 nm. Comment on the result.

2: It has been decided to stabilize both pigments and oil droplets electrostatically. The surface potential for both colloids is –50 mV.

Calculate the electrostatic and composite (overall) interaction energies between an oil droplet and a TiO_2 particle in the coating when the distance between them is 10 nm.

Based on these calculations, what can you conclude more generally about the interaction of oil droplets and TiO_2 particles in the coating?

3: Describe what you think will happen to the colloids if a can of the coating is exposed to very cold winter temperatures.

4: By mistake, a much too high salt concentration is used in the coating and diffusion-controlled coagulation takes place. How long time does it take before 90% of all the colloids are coagulated?

Solution

The production of a water-based coating is considered.

1: The attractive vdW energies between the different colloids in the aqueous medium must be calculated. For equal-size spherical colloids, the following equation from Table 2.2 can be used:

$$V_A = -\frac{A_{121}R}{12H} \tag{11.4.1}$$

where the Hamaker constant, A, has been replaced by an effective Hamaker constant, A_{121}, which is valid for a liquid medium (as opposed to vacuum). The value of A_{121} can be calculated from Equation 2.7b or Equation 10.4:

$$A_{121} = \left(\sqrt{A_{11}} - \sqrt{A_{22}}\right)^2 \tag{11.4.2}$$

For oil-droplets, we get (oil–water–oil):

$$A_{121} = \left(\sqrt{5 \times 10^{-20} \text{ J}} - \sqrt{3.75 \times 10^{-20} \text{ J}} \right)^2 = 8.97 \times 10^{-22} \text{ J} \qquad (11.4.3)$$

For TiO$_2$ particles, we get (TiO$_2$–water–TiO$_2$):

$$A_{121} = \left(\sqrt{15 \times 10^{-20} \text{ J}} - \sqrt{3.75 \times 10^{-20} \text{ J}} \right)^2 = 3.74 \times 10^{-20} \text{ J} \qquad (11.4.4)$$

Using Equation 11.4.1, the energies are:

$$V_A(\text{oil-water-oil}) = -\frac{A_{121}R}{12H} = -\frac{(8.97 \times 10^{-22} \text{ J}) \left(\frac{1.0 \times 10^{-6} \text{ m}}{2} \right)}{12 (10 \times 10^{-9} \text{ m})} = -3.75 \times 10^{-21} \text{ J} \qquad (11.4.5)$$

$$V_A(\text{TiO}_2\text{-water-TiO}_2) = -\frac{A_{121}R}{12H} = -\frac{(3.75 \times 10^{-20} \text{ J}) \left(\frac{0.2 \times 10^{-6} \text{ m}}{2} \right)}{12 (10 \times 10^{-9} \text{ m})} = -3.13 \times 10^{-20} \text{ J} \qquad (11.4.6)$$

For two unequal-sized spheres (oil and TiO$_2$) and $H \gg R_1$, R_2, another equation from Table 2.2 must be used:

$$V_A = -\frac{A_{123}}{6H} \left(\frac{R_1 R_2}{R_1 + R_2} \right) \qquad (11.4.7)$$

In this case, the Hamaker constant, A, has been replaced by an effective Hamaker constant, A_{123}, which is valid for two different materials in a liquid medium. The value of A_{123} can be calculated from Equation 10.7:

$$A_{123} = \left(\sqrt{A_{11}} - \sqrt{A_{22}} \right) \left(\sqrt{A_{33}} - \sqrt{A_{22}} \right) \qquad (11.4.8)$$

Inserting numbers gives (oil–water–TiO$_2$):

$$A_{123} = \left(\sqrt{5 \times 10^{-20} \text{ J}} - \sqrt{3.75 \times 10^{-20} \text{ J}} \right) \left(\sqrt{15 \times 10^{-20} \text{ J}} - \sqrt{3.75 \times 10^{-20} \text{ J}} \right) = 5.80 \times 10^{-21} \text{ J} \quad (11.4.9)$$

Using Equation 11.4.7, the energy is:

$$V_A(\text{oil-water-TiO}_2) = -\frac{A_{123}}{6H} \left(\frac{R_1 R_2}{R_1 + R_2} \right) = -\frac{(5.80 \times 10^{-21} \text{ J})}{6 (10 \times 10^{-9} \text{ m})} \left(\frac{\left(\frac{10^{-6} \text{ m}}{2} \right) \left(\frac{0.2 \times 10^{-6} \text{ m}}{2} \right)}{\left(\frac{10^{-6} \text{ m}}{2} \right) + \left(\frac{0.2 \times 10^{-6} \text{ m}}{2} \right)} \right)$$

$$= -8.06 \times 10^{-21} \text{ J}$$

$$(11.4.10)$$

Comment: At a distance of 10 nm, it can be seen that the strongest interaction energy is between the TiO_2 particles and the lowest between the oil droplets. The oil–TiO_2 interaction energy is in between the other two. Consequently, most attention should be paid to TiO_2–TiO_2 interaction.

2: The electrostatic and composite interaction energies must be calculated. For two unequal-sized spheres with the same surface potential the Reerink–Overbeek expression (Table 10.5) is used for the interaction energy:

$$V_R = \left[\frac{64\pi\varepsilon\varepsilon_o R_1 R_2 k_B^2 T^2 \gamma^2}{(R_1 + R_2)e^2 z^2}\right]\exp(-\kappa H) \tag{11.4.11}$$

where:

$$\gamma = \frac{\exp\left(\dfrac{ze\psi_o}{2k_BT}\right) - 1}{\exp\left(\dfrac{ze\psi_o}{2k_BT}\right) + 1} \tag{11.4.12}$$

The medium is water with a salt, NH_4OH (1 : 1 salt) at 10^{-5} M and κ^{-1} is calculated from equation in Table 10.4:

$$\kappa^{-1} = \frac{0.304}{\sqrt{10^{-5}\,\text{M}}} = 96.1\ \text{nm} \tag{11.4.13}$$

Furthermore:

$$\frac{ze\psi_o}{2k_BT} = \frac{(+1)(1.602 \times 10^{-19}\ \text{C})(-50 \times 10^{-3}\ \text{V})}{2(1.381 \times 10^{-23}\ \text{JK}^{-1})(298.15\ \text{K})} = -0.973 \tag{11.4.14}$$

and:

$$\gamma = \gamma_1 = \gamma_2 = \frac{\exp(-0.973) - 1}{\exp(-0.973) + 1} = -0.45 \tag{11.4.15}$$

and:

$$\exp(-\kappa H) = \exp\left(-\frac{10\ \text{nm}}{96.1\ \text{nm}}\right) = 0.90 \tag{11.4.16}$$

Finally:

$$V_R = \left[\frac{64\pi\varepsilon\varepsilon_o R_1 R_2 k_B^2 T^2 \gamma_1 \gamma_2}{(R_1 + R_2)e^2 z^2}\right]\exp(-\kappa H)$$

$$= \left[\frac{64\pi(80)(8.854 \times 10^{-12}\ \text{C}^2\text{J}^{-1}\text{m}^{-1})\left(\dfrac{10^{-6}\ \text{m}}{2}\right)\left(\dfrac{0.2 \times 10^{-6}\ \text{m}}{2}\right)(1.381 \times 10^{-23}\ \text{JK}^{-1})^2(298.15\ K)^2(-0.45)^2}{\left(\dfrac{10^{-6}\ \text{m}}{2} + \dfrac{0.2 \times 10^{-6}\ \text{m}}{2}\right)(1.602 \times 10^{-19}\ \text{C})^2(+1)^2}\right] \times 0.90$$

$$= 1.43 \times 10^{-18}\ \text{J}$$

$$\tag{11.4.17}$$

The composite (or overall) interaction energy is given by Equation 10.1:

$$V(10 \text{ nm}) = V_R + V_A = \left(1.43 \times 10^{-18} \text{ J}\right) + \left(-8.06 \times 10^{-21} \text{ J}\right) = 1.42 \times 10^{-18} \text{ J} \tag{11.4.18}$$

In terms of $k_B T$, one obtains:

$$\frac{V(10 \text{ nm})}{k_B T} = \frac{\left(1.42 \times 10^{-18} \text{ J}\right)}{\left(1.381 \times 10^{-23} \text{ JK}^{-1}\right)(298.15 \text{ K})} = 345 \tag{11.4.19}$$

Comment: At a distance of 10 nm, the colloids considered do not flocculate or coagulate because V is positive and $V_{max}/k_B T$ is at least 345 (the exact location of $V_{max}/k_B T$ is not calculated, but it must be higher than or equal to 345), whereby the colloids are kinetically stable ($V_{max}/k_B T > 15–25$ ensures kinetic stability). Note, however, that the interaction between the pigment particles is stronger (see Q1) and thereby less stable.

3: If the water phase freezes, the particles will be pushed closer together to very low interaction distance (H) and coagulation will eventually take place (see, for example, Figure 10.3, where very low H leads to very negative interaction energies). Owing to the presence of a salt (and in real coatings also glycols), there will be some freezing point depression and a few degrees under 0 °C may be tolerated, but for very cold winters, as stated in the question, the water-based coating will become badly coagulated (and eventually one big chunk of ice). This phenomenon is similar to what happens when water evaporates upon drying after application of the coating. However, during drying of a water-based coating, coagulation is desired and inevitable, whereas coagulation from freezing is undesirable. Consequently, the coating should not be kept in a garden shed in cold countries.

4: Diffusion-controlled coagulation can be calculated using Equation 11.2:

$$\frac{1}{n} - \frac{1}{n_o} = k_2 t \tag{11.4.20}$$

Rearranging leads to:

$$t = \frac{1}{k_2} \left[\frac{1}{n} - \frac{1}{n_o}\right] \tag{11.4.21}$$

At 90% coagulation, we have that $n = (1 - 0.9)n_o$. Inserting in Equation 11.4.21 gives:

$$t = \frac{1}{k_2} \left[\frac{1}{(1-0.9)n_o} - \frac{1}{n_o}\right] = \frac{1}{k_2{}^\circ} \left[\frac{9}{n_o}\right] \tag{11.4.22}$$

It can be seen that Equation 11.4.22 is independent of particle size and all particles and droplets can be pooled in one calculation. The term $k_2{}^\circ$ is the rate constant for diffusion-controlled coagulation.

The rate constant in this case can be calculated by Equation 11.3:

$$k_2{}^\circ = \frac{4k_B T}{3\eta_{medium}} = \left(\frac{4\left(1.381 \times 10^{-23} \text{ JK}^{-1}\right)(298.15 \text{ K})}{3(1.0 \text{ kgm}^{-1}\text{s}^{-1})}\right) = 5.49 \times 10^{-21} \text{ m}^3\text{s}^{-1} \tag{11.4.23}$$

The number of particles at $t = 0$ is given by:

$$n_{o,total} = n_{o,oil} + n_{o,TiO_2} = \left(\frac{\varphi_{emulsion,coating}\varphi_{oil,emulsion}}{\left(\frac{\pi}{6}d_{p,oil}^3\right)} + \frac{\varphi_{TiO_2,coating}}{\left(\frac{\pi}{6}d_{p,TiO_2}^3\right)} \right)$$

$$= \left(\frac{0.32 \times 0.5}{\frac{\pi}{6}\left(10^{-6}\ \text{m}\right)^3} + \frac{0.55}{\frac{\pi}{6}\left(0.2 \times 10^{-6}\ \text{m}\right)^3} \right) \qquad (11.4.24)$$

$$= 3.06 \times 10^{17}\ \text{number}\left(\text{m}^3\ \text{coating}\right)^{-1} + 1.31 \times 10^{20}\ \text{number}\left(\text{m}^3\ \text{coating}\right)^{-1}$$

$$= 1.32 \times 10^{20}\ \text{number}\left(\text{m}^3\ \text{coating}\right)^{-1}$$

Note that there are 430 times more particles than droplets and practically no accuracy would be lost by disregarding the oil droplets in the above calculation.

Inserting in Equation 11.4.22 gives:

$$t = \frac{1}{k_2^o}\left[\frac{9}{n_o}\right] = \frac{1}{\left(5.49 \times 10^{-21}\ \text{m}^3\text{s}^{-1}\right)}\left[\frac{9}{\left(1.32 \times 10^{20}\ \text{number}\,\text{m}^{-3}\right)}\right] = 12.5\ \text{s} \qquad (11.4.25)$$

Consequently, it only takes about 13 s to coagulate the coating once destabilized. *Comment*: In this problem, any effect of steric stabilization (see Chapter 12) has been ignored. In fact, this is too simplified an approach. For typical commercial water-based coating systems, salt concentrations range from about 0.025 M to about 0.065 M (for a 1:1 electrolyte), which will destabilize the dispersion if only electrostatic repulsion was applied. Water-based coatings are stabilized with electrosteric stabilization, which is a combination of electrostatic repulsion and steric effects (for details related to coatings see Croll, 2002).

11.5   Concluding remarks

The stability of colloids is of crucial importance in many applications and physical phenomena; some of them are summarized here:

- stability of food, pharmaceutical and cosmetic dispersions;
- many paints, inks, liquid toners;
- understanding/curing diseases, e.g. Alzheimer's, cataract, sickle-cell anaemia;
- delta of rivers;
- waste-water treatment;
- mining industry – getting valuable minerals;

The colloid stability is determined by the balance between van der Waals (vdW) (A) and electrostatic (R) forces. The most important points are summarized here:

- vdW dominates at large and small interparticle distances (instability);

- double-layer repulsions dominate at intermediate distances (stability);
- at very small distances we can have repulsion due to overlapping electron clouds (Born repulsion);
- if $V\ (= V_A + V_R)$ is large compared to thermal energy of particles $(k_BT) \rightarrow$ stability (else aggregation);
- Both secondary and primary coagulation are observed in practice;
- a barrier of $(15–25)k_BT$ is often sufficient to give colloid stability provided the Debye length is relatively large (>20 nm).

Table 11.1 illustrates how stability can be controlled. The great value of the DLVO theory is due to the following:

- It accounts successfully for the most important (but not for all!) forces: van der Waals and electrostatic.

Table 11.1 *Parameters influencing colloid stability and how they can be controlled*

Parameter	What to control
Debye length	Salt (*T*, medium)
Dielectric constant	Polarity/hydrogen bonding of the medium
Hamaker constant	Chemical nature of particles and medium
Surface (zeta) potential	Understanding charge mechanism[a]
	Far from IEP

[a] Control pH, ionic environment, adsorption of ionic surfactants, polyelectrolytes, etc.

- It agrees with many previously collected experimental observations.
- It shows and predicts the importance of understanding the effect of the ionic environment to which colloids are exposed.
- It predicts the effects on colloid stability which are electrical in origin, e.g. added inorganic ions or surfactants, where the main result is a change in the surface potential of the particles.
- It is verified by many recent "direct" experimental measurements of forces between colloid particles.

The most important limitation of the DLVO theory is that it does not account for other important "repulsive" forces such as the steric and solvation ones at close distances. Natural and synthetic polymers, especially block and graft copolymers, can provide steric stabilization. It is best if one part of the stabilizer has a high tendency to adsorb onto the particle surface and the other has a high affinity for the solvent.

There is, thus, one important difference between stabilization by electrical and by steric forces:

- you stabilize electrically by *removing* salt;
- you stabilize sterically by *adding* polymer.

For much more on steric forces see Chapter 12.

Problems

Problem 11.1: Colloid stability

A dispersion of negatively charged AgI particles is prepared using two different aqueous electrolyte solutions at 25 °C: (a) $LiNO_3$ (165) and (b) $NaNO_3$ (140).

In parenthesis are the values of the critical coagulation concentration (CCC) in mmol L^{-1}.

1. For each electrolyte solution draw one particle and its immediate ionic environment, indicating the co-ions and counter-ions. Explain in your own words what the CCC is and why the CCC value of $LiNO_3$ is higher than that of $NaNO_3$.
2. Draw a potential energy–interparticle distance curve (*V–H*) for the interaction between two particles, in the case when the electrolyte concentration is (a) $C = 3 \times 10^{-1}$ mol L^{-1} and (b) $C = 5 \times 10^{-2}$ mol L^{-1}. Do you expect the *V–H* curves to be qualitatively the same or different for the two different salts? In which of the above cases would you describe the colloidal system as stable?
3. Estimate (in nm) and for both salts the Debye thickness of the electric double layer at the two concentrations mentioned in question 2. Explain briefly how the Debye thickness is related to the stability of colloidal systems.
4. Provide an estimation of the CCC for the same colloidal system with the AgI particles in the case when the following aqueous salt solutions are used: $Ca(NO_3)_2$ and $Al(NO_3)_3$. Explain the assumptions made in your calculations. Do you expect that a colloid dispersion of AgI particles is stable in an electrolyte solution of $Al(NO_3)_3$ having a concentration of $C = 2$ mmol L^{-1}?
5. The electrophoretic mobility of AgI particles of various dimensions was measured in a 0.01 mol L^{-1} NaCl solution at 25 °C and is found in all cases to be equal to 4×10^{-8} m^2 s^{-1} V^{-1}. Calculate an approximate value of the zeta potential for particles of radius (a) 500 nm and (b) 1 nm. For which particle size is it more likely that better stability is obtained and why?

Problem 11.2: Critical coagulation concentration and stability

1. In your own words, explain what a dispersion is.
2. A dispersion of positive Al_2O_3 particles is prepared using two different aqueous electrolyte solutions: (a) NaCl and (b) K_2SO_4.

   For each electrolyte solution draw one colloid particle and its immediate ionic environment, indicating the co-ions and counter-ions.

3. Explain in your own words what the critical coagulation concentration (CCC) is. Draw an energy potential curve for two interacting particles, in the case when the electrolyte concentration is above CCC and in the case when the electrolyte concentration is below CCC. Which of these two conditions would you describe as stable?

4. The CCC values are given in Table 10.6. Calculate the thickness of the diffuse double layer exactly at the point of CCC for the colloidal system of question 2 and for both electrolytes. Compare and discuss the results.

Problem 11.3: Charged colloids in an electrolyte

In DLVO theory the interaction energy between two colloidal particles is estimated by the summation of repulsive and attractive contributions to the interaction. In Figure 11.2a the potential curve is illustrated for different values of the double layer thickness (inverted Debye–Hückel parameter), which are related to different concentrations of a monovalent salt, e.g. NaCl.

1. Estimate the critical coagulation concentration (CCC) for this system. Is the CCC the concentration of co-ions, counter-ions or colloid particles? Is the system coagulated above or below the CCC? What is the difference between coagulation and flocculation?

2. Estimate the Debye thickness of the electric double layer, which surrounds each colloid particle, for the six concentrations of the electrolyte solution illustrated in the figure. How does the range of interaction change with the change of concentration? Can you explain the coagulation from looking at the Debye thickness?

Problem 11.4: Comparison of colloidal stability

There are four different aqueous solutions at room temperature of the same colloid dispersion of negatively charged particles, As_2S_3 (CCC values are given in parenthesis in mmol L^{-1}). The electrolytes and their concentrations are given in the table:

No of sample	Electrolyte	Concentration (mol L^{-1}) (in parentheses in mM)
Sample 1	KCl (49.5)	0.100 (100)
Sample 2	KCl (49.5)	0.001 (1)
Sample 3	K_2SO_4 (0.65)	0.001 (1)
Sample 4	$MgCl_2$ (0.72)	0.001 (1)

a. Based on a comparison of the thickness of the electric double layer, you should identify the most stable and the least stable colloid from the four samples in the above table. Draw two schematic curves for the potential energy of interaction between two colloidal particles: one for a stable system and one for an unstable system!

b. How can you argue that the colloids of sample 1, 2, and 4 are stable or not? Give a qualified estimate on the stability of sample 1, 2 and 4.

c. Which forces (attractive or repulsive) are dominant for the three samples 1, 2 and 4?

Problem 11.5: Destabilizing a latex paint and half-life times (inspired from Cussler and Moggridge (2011))

A particular latex paint contains 20% by volume polymer particles having a diameter of 0.6 μm. When this paint is spread, the water in the emulsion evaporates, the colloid becomes unstable and the particles fuse into a single smooth layer. In cases where the paint is used at temperatures below the glass transition of the polymer, the polymer particles must be plasticized so that they fuse easily. This fusion is why latex paints are hard to clean up after they dry, even though the original colloid is easily wiped up.

Paints like this are normally stabilized by surfactants, especially polyphosphates. In some cases, however, the surfactants can separate, and the colloid becomes unstable. Freezing can cause such an instability ($T = 0$ °C). Assume that the viscosity is 0.01 g cm^{-1} s^{-1}.

1. Show that if we assume that the aggregation is fast (diffusion-controlled), then the half-life time is given by the equation:

$$t_{1/2} = \frac{3\eta_{medium}}{4k_B T n_0}$$

2. If the paint becomes fast unstable, how fast does the paint agglomerate? Provide an estimate of the half-life time and comment on the result.

Problem 11.6: Kinetics of coagulation of a hydrosol

Negatively charged spherical particles with radius R flocculate at 25 °C in an aqueous Na_2SO_4 solution, with a concentration equal to 0.93×10^{-3} mol L^{-1}. The viscosity of the solution is 8.9×10^{-4} kg m^{-1} s^{-1}.

The following results were obtained by particle counting during the flocculation of this colloidal system as a function of time:

Time (min)	Particle concentration (cm^{-3})
0	100×10^8
4	40×10^8
8	25×10^8
12	18×10^8

1. Calculate the "second-order" rate constant, k_2, as well as the stability ratio for this system.

    The potential energy barrier, V, which is applied so that the particles do not approach (too much) is constant and non-zero in the interval $2R$ and $2.25R$, but for larger distances the potential is zero.
2. Estimate a value of the potential height V_{max} between $2R$ and $2.25R$. Give the value both in J and as multiplier of $k_B T$. Comment on the results.
3. Provide an estimate of the Debye length and the particle radius for this system.

Problem 11.7: Kinetics of coagulation of a hydrosol

The following results were obtained by particle counting during the coagulation of a hydrosol at 25 °C by excess of 1 : 1 electrolyte:

Time (min)	Particle concentration (cm^{-3})
0	100×10^8
2	14×10^8

4	8.2×10^8
7	4.6×10^8
12	2.8×10^8
20	1.7×10^8

1. Calculate the "second-order" rate constant, k_2, and compare it with the value, $k_2{}^0$, calculated based on the assumption that coagulation is a diffusion-controlled process.
2. Calculate the half-life time as well as the stability ratio. Comment on the results.

Problem 11.8: Coagulation of biomass after fermentation

A biotechnology company produces industrial enzymes by fermentation on a large scale. After the fermentation with enzyme-producing microorganisms, the biomass (i.e. the living cells) are killed using heat treatment and high pH. The dead biomass is removed from the enzyme-containing liquid by coagulation or flocculation followed by centrifugation. The moist biomass is finally sold as fertilizer to farmers.

The coagulation/flocculation process takes place in a so-called thickener using the coagulation agent $CaCl_2$ and adjustment of pH. The concentration of $CaCl_2$ used in the thickener is 100 mM. The thickener liquid has a viscosity of 10^{-3} kg m^{-1} s^{-1}, a density of 1000 kg m^{-3}, a relative permittivity of 80.2 and a Hamaker constant (in vacuum) of 3.7×10^{-20} J. The temperature is 20 °C. The individual cells can be assumed to be spherical, rigid and have a diameter of 1 μm. The Hamaker constant (in vacuum) of the cells considered is 6×10^{-20} J. The cell surface (phospholipid membrane) zeta potential, at the pertinent pH, is assumed constant at all times and equal to −15 mV.

Any potential influence of steric stabilization and retardation effects can be neglected.

1. Calculate the composite (overall) interaction energy between two cells in the thickener when the closest distance between the surfaces of the particles is 0.1, 1, 2, 5, 10 and 25 nm.

    Comment on the stability of the cell dispersion.
2. For economic reasons, it is suggested to replace $CaCl_2$ by $MgCl_2$ as coagulation agent and use the

new salt at the same concentration as CaCl$_2$ was used in the thickener.

Explain qualitatively (in words) how this process change is expected to affect the efficiency of the coagulation process.

3. Provide a crude estimate of the critical coagulation concentration of MgCl$_2$.

Comment on the result.

4. In an attempt to lower the salt consumption and form larger agglomerates, a small amount of an adsorbing polymer (e.g. poly(aluminium chloride)) is also added to the thickener.

Explain a potential flocculation mechanism (working principle) for the polymer–cell system (you will need to read Chapter 12 to answer this last question).

References

J.C. Berg, An Introduction to Interfaces and Colloids – The Bridge to Nanoscience, World Scientific, New Jersey, 2010.

S. Croll (2002), Prog. Org. Coat., 44, 131–146.

E.L. Cussler, G.D. Moggridge, Chemical Product Design, 2nd edn, Cambridge University Press, Cambridge, 2011.

S. Dey, J.M. Prausnitz (2011) *Ind. Eng. Chem. Res.* 50(1), 3–15.

J. Goodwin (2009), Colloids and Interfaces with Surfactants and Polymers: An Introduction, John Wiley & Sons, Chichester.

P.C. Hiemenz, R. Rajagopalan, Principles of Colloid and Surface Chemistry, 3rd edn, Marcel Dekker, New York, (1997).

J.N. Israelachvili, Intermolecular and Surface Forces, 1st edn, Academic Press, (1985).

J.N. Israelachvili, Intermolecular and Surface Forces, 3rd edn, Academic Press, (2011).

T. Nesterova, K. Dam-Johansen, L. T. Pedersen, S. Kiil (2012), Prog. Org. Coat., 75(4), 309–318.

R.M. Pashley, M.E. Karaman, Applied Colloid and Surface Chemistry, John Wiley & Sons, Chichester (2004).

J.M. Prausnitz (1995), *Fluid Phase Equilibria*, 104: 1–20.

J.M. Prausnitz (2003), *J. Chem. Thermodynamics*, 35: 22–39.

D.J. Shaw, Colloid & Surface Chemistry, 4th edn, Butterworth and Heinemann, Oxford, (1992).

12

Emulsions

12.1 Introduction

Emulsions are defined as dispersed systems for which the phases are immiscible or partially miscible liquids. The emulsions are dispersions of one liquid (oil) in another, often water (w), and are thus typically classified as oil-in-water or water-in-oil emulsions. They are relatively static systems with rather large droplets (diameter of about 1 μm). The word emulsion comes from Latin and means "to milk". Emulsions are typically highly unstable liquid–liquid systems which will eventually phase separate and require emulsifiers, most often mixed surfactants (or other substances, e.g. synthetic polymers or proteins). In milk, the emulsifier is the protein casein (Figure 12.1).

Indeed emulsions are everywhere and find extensive applications in the food, pharmaceutical, and cosmetics industries. Emulsions are highly complex multicomponent systems containing emulsifiers, foam stabilizers, surfactants, solubilizers, viscosity conditioners and numerous other compounds. About 80% of emulsion preparations in the market are of oil-in-water type. Owing to the skin feel, both cosmetic oil-in-water and water-in-oil emulsions do not contain more than approximately 30% oil. Butter is a water-in-oil emulsion, where the oil is butterfat.

In the emulsion field, we often meet "less scientific" concepts, e.g. that of "creaminess". Emulsions is a field between science and art. We will start this chapter with some basic definitions and look at some more applications. Then, we will present the most important property of emulsions, the stability, the factors that affect it, and how we can "manipulate" (influence) the stability of emulsions. We will discuss destabilization mechanisms of emulsions and use the HLB factor (as introduced in Chapter 5) as a quantitative design tool in emulsion science.

12.2 Applications and characterization of emulsions

The purpose of an emulsion is often to treat a surface of some kind, e.g. a floor, human skin or a plant.

Introduction to Applied Colloid and Surface Chemistry, First Edition. Georgios M. Kontogeorgis and Søren Kiil.
© 2016 John Wiley & Sons, Ltd. Published 2016 by John Wiley & Sons, Ltd.
Companion website: www.wiley.com/go/kontogeorgis/colloid

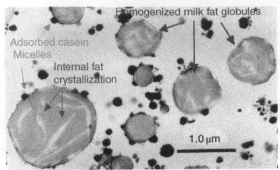

Figure 12.1 *Casein is the protein that has the function of emulsifier in milk. Without the casein (milk's "natural polymer"), the milk would be destabilized, since the fat globules, via coagulation or coalescence, would result in destruction of the colloidal dispersion. Reprinted from Walstra et al. (1984), with permission from John Wiley & Sons, Ltd*

There are many means by which to accomplish this, the most important one being delivery of droplets to some surface, for instance cosmetics and medicines to the skin, treatment of textiles, wax to floors, or herbicides to plants. An emulsion must be stable during storage, but should usually break during application. Breaking can be accomplished by a component evaporating or emulsifiers also adsorbing strongly on the target surface. There are also situations where emulsions are entirely unwanted, e.g. in liquid extraction devices or as a consequence of other multiphase mixing operations, where separation of two phases can become difficult.

Emulsions are unstable liquid–liquid systems that will eventually phase separate. They need moderate amounts of surfactants for their stabilization. On the other hand, the so-called microemulsions require also co-surfactants, e.g. small alcohols or amines. They appear to be thermodynamically stable, having small aggregates (about 10 nm) and this is why sometimes microemulsions are considered to be in-between small-droplet emulsions and swollen micelles. They are highly dynamic systems and have a high internal surface that requires a large amount of surfactants. The oil–water interfacial film can be highly curved. Since the discovery of microemulsions, these structures have attained increasing significance both in basic research and in industry. Owing to their unique properties, namely, ultralow interfacial tension, large interfacial area, thermodynamic stability and the ability to solubilize otherwise immiscible liquids, uses and applications of microemulsions have been numerous. Microemulsions are used for enhanced oil recovery, liquid fuels (low emission levels due to low temperature combustion because of water evaporation), coatings and textile finishing, lubricants and corrosion inhibitors, detergents, cosmetics, and agrochemicals (Paul and Moulik, 2001). An example of a microemulsion is water, benzene, potassium oleate and hexanol in one mixture.

Emulsions are colloidal systems and, thus, are unstable. They require surfactants (large amounts in the case of microemulsions) to reduce the interfacial tension and achieve stabilization. The work, W, required to make an emulsion is given by:

$$W = \gamma \Delta A \qquad (12.1)$$

where ΔA is the change in surface area and γ is the interfacial surface tension.

Example 12.1.   Estimation of the work required to make an emulsion.
To realise the importance of the use of surfactants (or more specifically here an emulsifier), let us consider a simple calculation for dispersing 10 mL of an oil in water to create an oil-in-water emulsion with droplets having a diameter of 1.0 μm. The oil–water interfacial tension is 52 mN m^{-1}.

Solution
The theoretical work required to make the emulsion batch must be calculated. Liquid droplets are spherical. From Equation 12.1, we obtain:

$$W = \gamma_{o/w} \Delta A = \gamma_{o/w} \frac{V_{oil}}{V_{droplet}} A_{droplet}$$

$$= \gamma_{o/w} \frac{V_{oil}}{\left(\frac{\pi}{6} d_p{}^3\right)} \left(\pi d_p{}^2\right) = \frac{6\gamma_{o/w} V_{oil}}{d_p} = 3.1 \text{ J}$$

This is the value of the work that is needed to create the huge interface, work that remains in the system. Thermodynamics will try to reduce the energy via aggregation of droplets. One way to avoid (in reality delay) the process of aggregation is by adding a surfactant and then the interfacial tension can be as low as 1 mN m^{-1}, which will result in a work value one order of magnitude lower. This is a much better situation; nonetheless, the system is thermodynamically unstable and eventually will "break" (phase separate). However, kinetic factors are also important and the production of emulsions which are stable over some period of time is a crucial issue in the design, production and use of emulsion-based products.

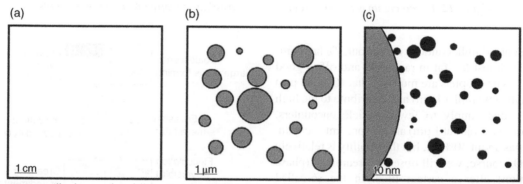

Figure 12.2 *Milk observed at different scales. Milk is an emulsion and contains fat globules. The size of fat globules can vary from 0.1 to 15 μm. There are no visible structures at the length scale used in the drawing (a). In (b), where a smaller length scale is used, fat globules can be seen. At even higher magnification (c), casein (a protein) molecules near a fat droplet can be seen*

Let us now see some more applications of emulsions, this time related to food. One typical example is milk (Figure 12.2).

Milk is *not* a homogenous white liquid, although it may look like one. However, the fact that it is not transparent is an indication that milk is an emulsion, a "colloidal system". It is actually a quite complicated mixture. If you look at milk under an electronic microscope, you will see that it contains fat globules that are about a micrometre in diameter. Even smaller are the protein (casein) molecules between the fat globules. Casein is the protein ("polymer") of the milk. Finally, milk also contains a substantial amount of dissolved substances. Among these are lactose (a saccharide or sugar) and many minerals. Milk powder consists mainly of proteins, lactose, the minerals in the milk and smaller amounts of fat: much of the fat is largely removed by centrifugation and used for making butter. Many more food products are emulsions, e.g. mayonnaise.

Without the casein (milk's "natural polymer"), the milk would be destroyed, since the fat globules, through coagulation or coalescence, would result in destruction of the colloidal dispersion. This does not happen (or actually it is being delayed) since the adsorbed casein molecules provide stabilization when the fat globules collide with each other.

Ice cream

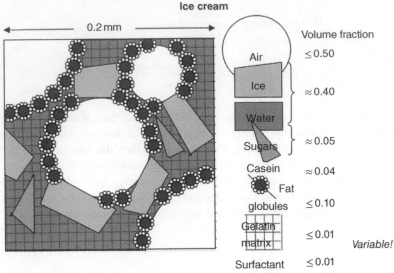

Figure 12.3   Ice cream is a complex emulsion consisting of many different components

Homogenized milk contains only about 3% fat (compared to almost 6% fat in raw milk) and offers good stability. The droplets are much smaller (0.2–0.4 μm) than in the raw milk (4 μm) and contribute to the high stability due simply to fewer particle encounters. Small droplets (about 1 μm) are also present in cream, which has about 40% fat, but the stability is relatively good. Of course, we will observe "creaming" (phase separation) after a certain time. In the so-called homogenization process, milk is pushed through small, tapered tubes or pores. As the diameter shrinks and the flow of milk remains constant, pressure builds up and fat globules break apart in the turbulence. The higher the pressure, the smaller the particles.

Mechanical shaking of cream makes the fat globules burst whereby the fat is released and butter can be formed. It is the crystals in the fat globules that cut open the globules. Bacteria help to give butter a good flavour. Ice cream is another complex emulsion (oil-in-water) (Figure 12.3).

12.3   Destabilization of emulsions

Emulsions are thermodynamically unstable. We cannot completely stop or avoid this process as the droplets collide; what is of course important is that the droplets do not adhere (stick) when they

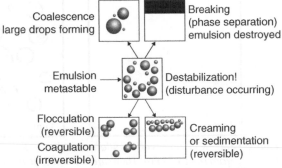

Figure 12.4   Various ways in which an emulsion can be destabilized

collide with each other. The emulsions can be destabilized in many different ways (Figure 12.4): creaming, flocculation or coagulation, and coalescence/breaking.

Creaming or sedimentation can be caused by density differences, as explained by Stokes law (see later and Chapter 8). Sometimes, many of the above phenomena are termed generally as aggregation. In the case of coalescence, the droplets must encounter each other, they then lose their original shape; coalescence is an irreversible phenomenon that eventually may lead to complete phase separation (breaking). On the other hand, flocculation and creaming are reversible. The number of individual droplets can actually be counted experimentally using, for example, laser diffraction techniques and this is a way to "control" stability even if macroscopically we cannot sometimes see any of the phenomena shown here (until it is too late!).

Ostwald ripening (see also Chapter 4) is an observed phenomenon in suspensions or emulsions that describes the change of an inhomogeneous structure over time. In other words, over time, small crystals or dispersed particles dissolve, and redeposit onto larger crystals or dispersed particles. Ostwald ripening is generally found in water-in-oil emulsions, while flocculation is found in oil-in-water emulsions. This thermodynamically-driven spontaneous process occurs because larger particles are more energetically favoured than smaller particles. This originates from the fact that molecules on the surface of a particle are energetically less stable than those in the interior. Consider, for example, a crystal of atoms where each atom has six neighbours. The atoms will all be quite stable. Atoms on the surface, however, are only bonded to five neighbours or fewer, which makes these surface atoms less stable. Large particles are more energetically favourable because more atoms are bonded to six neighbours and fewer atoms are at the unfavourable surface. According to the Kelvin equation (Chapter 4), molecules on the surface of a small particle (energetically unfavourable) will tend to detach from the particle and diffuse into the solution. When all small particles do this, the concentration of free atoms in solution is increased. When the free atoms in solution are supersaturated, the free atoms have a tendency to condense on the surface of larger particles. Therefore, all smaller particles shrink, while larger particles grow, and overall the average size will increase. After an infinite amount of time, the entire population of particles will have become one, huge, spherical particle to minimize the total surface area.

12.4   Emulsion stability

How can we understand and manipulate colloidal stability? No theory can explain all phenomena, perhaps though the most successful one is the DLVO theory, which is built on an understanding/balance of the attractive van der Waals forces and the repulsive electrostatic forces. Here, we will focus on a stabilization method, which is actually not really explained by the DLVO theory, the stabilization based on an additional compound (e.g. a polymer, oligomer or protein) known as steric (or entropic) stabilization.

As already mentioned, to obtain stable emulsions, we need to introduce a third component, known under the broad name of "emulsifier". Emulsifiers facilitate emulsification but also protect sterically via the formation of an adsorbed film on the particles that prevents coagulation or coalescence. Of course it is important to add compounds that can indeed adsorb on the surfaces (without bridging phenomena among the particles). Polymers and proteins can adsorb in many more ways and through more complex mechanisms than surfactants, thus often offering better protection than surfactants which also typically have too low molecular weights. They do not easily desorb upon collision. Although, in the case of polymers, there are possibilities for various mechanisms of adsorption, it is important for enhancing stability to cover all the area of the particle in order to avoid bridging (Figure 12.12b below). Finely divided solids can also stabilize emulsions. This phenomenon is not used commercially to prepare emulsions, but has a certain nuisance value in that it forms and stabilizes emulsions in unwanted situations. Amphiphilic particles, even if they do not make emulsification easier, migrate to the interface after the emulsion is formed and thus they improve the stability.

The factors that favour emulsion stability are summarized in Figure 12.5.

We have already seen the first two factors shown in Figure 12.5. The third is explained by the DLVO theory and is related to the thickness of the double layer; the smaller the thickness the lower the stability. The stability of O/W emulsions is enhanced if the

- Low interfacial tension
- A mechanically strong and elastic interfacial film
- Electrical double layer repulsions
- Relatively small volume of dispersed phase
- Narrow droplet size distribution

$$u_{stokes} = \frac{d_{drop}{}^2 g (\rho_{drop} - \rho_{medium})}{18 \eta_{medium}}$$

Rate of movement can be predicted via stokes equation at idealised conditions

Figure 12.5 *Factors that favour emulsion stability*

concentration of ions, particularly multivalent ions, in the aqueous phase is minimized. On the other hand, flocculation can be brought about by adding salts, particularly those with multivalent ions of opposite charge.

The last factors shown in Figure 12.5 can also be explained with help of the famous Stokes equation (also discussed in Chapter 8). The Stokes equation is valid for Newtonian fluids, dilute region (no particle interactions) and when there is no particle deformation (true if the radius is less than 1000 µm). Despite these simplifications, it can be used for qualitative calculations in emulsion science. If the amount of particles is small, then fewer encounters are expected, while narrow droplet size works against Ostwald ripening. As can be seen by the Stokes equation, the higher the viscosity of the continuous phase (medium), the lower the velocity of the particles and thus fewer encounters. In most cases we want to enhance emulsion stability. However, there are many cases, e.g. creaming processes (creation of butter from raw milk) and to prevent corrosion from emulsions formed with water-lubricating oil systems, where we actually want to destroy the emulsions. Electrostatic stabilization is often of little importance to lyophilic colloids like polymer solutions, e.g. casein and gelatin which have a high affinity for water, or for systems with high particle concentration and low relative permittivity like oil-based paints and in biological systems which have high salt concentration. In these cases a combination of other factors, e.g. steric effects and solvation, is necessary. Furthermore, for the stability of lyophobic sols can often be enhanced by the addition of soluble lyophilic material (protective agents) that adsorbs onto the particle interface. The mechanism of this steric stabilization is usually complex and many factors are involved.

As mentioned, steric stabilization can be achieved with polymers, including polyelectrolytes, e.g. proteins and biopolymers, and also with (ionic or not) oligomers. (Block-) copolymers are used much more than homopolymers, as explained previously, since one part of the copolymer adsorbs on the surface and the other is soluble with the solvent and is thus extended in the solution. Natural polymers like the protein casein in milk and many synthetic polymers like PEG (poly(ethylene glycol)s) are used to stabilize emulsions and colloidal dispersions. Many block

copolymers like poly(ethylene oxide) are also used. It is best if one part of the stabilizer has a high tendency to adsorb onto the particle surface and the other has a high affinity for the solvent.

Figure 12.6 illustrates such steric stabilization.

The polymers adsorb on the interface (and may obtain different configurations). They enter into the liquid, and when particles come close together their polymer layers will interfere with each other, causing "steric repulsion". This type of repulsion depends not only on the properties of the polymer but also of the solvent. In a good solvent the layer will be more swollen, and more effective, than in a bad solvent. If the polymers are charged, you must also expect electrical effects.

Polymer layers have an effect that is similar to that of electrostatic repulsion. There is one important difference between stabilization by electrical and by steric forces: you stabilize electrically by *removing* salt, you stabilize sterically by *adding* polymer. The stabilization mechanism in the presence of polymers or oligomers is usually complex and many factors may be involved (Figure 12.7).

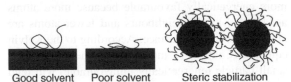

Good solvent   Poor solvent   Steric stabilization

Figure 12.6 Steric stabilization using polymers. For proper stabilization, the steric barrier needs to extent at least 10 nm from the particle surface and therefore polymers or oligomers are used (low molecular surfactants are too small). It works best with thick and dense layers, good solvents, strong adsorption and complete particle coverage

1. Enhanced double-layer repulsions for ionised stabilizers e.g. anionic surfactants
2. Reduced vdW interactions due to lowering the Hamaker constant
3. "Steric effects" (all the others...)

Figure 12.7 Stabilization mechanism in the presence of polymers or surfactants

Even if the adsorbed stabilizing agent is non-ionic it will influence the electrostatic interactions by causing a displacement of the Stern plane away from the particle surface. This will increase the range of electric double layer repulsion and thus enhance the stability.

The van der Waals (vdW) forces may also be affected. Adsorbed layers of stabilizing agent may cause a significant lowering of the effective Hamaker constant and, therefore, weakening of the interparticle vdW attraction. However, clearly, the stability of many "protected" colloidal dispersions cannot be explained solely on the basis of electric double layer repulsion and vdW attraction. This general term describes several different possible stabilizing mechanisms between adsorbed macromolecules. Any theory trying to describe the magnitude and range of the interaction between polymer layers needs to account for both the solution properties of the polymer and the conformations of the polymer at the solid–liquid interface. Repulsive steric forces for polymers in good solvents can be characterized using De Gennes scaling theory. In scaling theory, the adsorbed polymer conformation is assumed to be either a low surface coverage mushroom, in which the volume of the individual polymer is unconstrained by neighbours, or a high surface coverage brush, where the proximity of neighbour polymer chains constrains the chain volume and causes extension of the polymer into the solvent.

12.5 Quantitative representation of the steric stabilization

Figure 12.8 shows an extension of the DLVO theory to include a term for steric stabilization.

Many quantitative theories for steric stabilization have been developed in recent decades. The forces between sterically stabilized particles have been shown to be short-range with a range comparable with twice the contour length of the lyophilic chains. Electrostatic forces are often adequate to make the secondary (flocculation) minimum disappear and increase the barrier to the primary minimum, thereby decreasing the probability of aggregation. However, for many systems stabilized solely by electrostatic repulsion, a decrease in the energy barrier (e.g. via addition of counter-ions) will lead to rapid coagulation. In this case, a combination of electrostatic and steric forces is necessary for effective stabilization. Adsorbed layers of polymerss can affect stability. If they are charged, we can expect a change of the electrostatic repulsion as well. For long-chain polymers, we can expect steric repulsion when the adsorbed layers will start to penetrate. The adsorbed layer has a Hamaker constant different from that of the particle and hence the van der Waals interactions are altered. In many practical circumstances, we prepare colloidal systems with the most effective steric stabilizers we can get. We thus frequently use a uniform densely packed profile and the value of δ (the thickness of the adsorbed layer) can be measured. The interaction energy will increase very rapidly as soon as the two layers come into contact.

As shown in Figure 12.8, when the distance H between the particles is greater than (about) twice the adsorbed layer thickness (δ) then there is little or no interaction at all. However, when $H < 2\delta$, we have an interpenetration of the adsorbed layers, which results in repulsive (entropic) interactions.

Figure 12.9 shows the overlap of adsorbed layers of uniform concentrations for several geometries for layers that have a thickness of δ and a particle-to-particle separation distance of H.

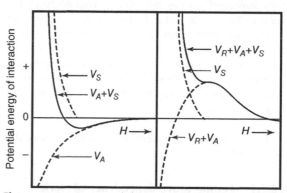

Figure 12.8 *Steric stabilization, extending the DLVO theory. For* H > 2δ, V_S = 0 *(no overlapping) and for* H < 2δ, V_S = ∞ *(overlapping); δ is the thickness of the adsorbed layer*

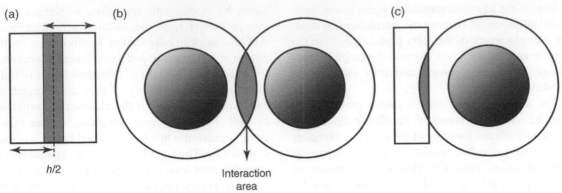

Figure 12.9 *Steric interactions – overlap of adsorbed layers for different geometries. Adapted from Goodwin (2009), with permission from John Wiley & Sons, Ltd*

$$V_s = \frac{2RT}{v_1} \left(\frac{c_2 v_2}{M_2} \right)^2 \left(\frac{1}{2} - \chi \right) v_0$$

Plate-plate: Plate-sphere:

$$v_0 = 2\delta - H \qquad v_0 = \frac{4\pi}{3} \left(\delta - \frac{H}{2} \right)^2 (3d_p + \delta + H)$$

Sphere-sphere:

$$v_0 = \frac{2\pi}{3} \left(\delta - \frac{H}{2} \right)^2 \left(3d_p + 2\delta + \frac{H}{2} \right)$$

Figure 12.10 *DLVO term for steric repulsion. The first term (the one squared in parenthesis) is simply the volume fraction of the polymer in the layer. Different expressions are obtained for the three different geometries indicated here as the overlap volume (v_0); d_p = sphere diameter (m), H = distance between two objects (m), χ = thermodynamic interaction parameter (estimation requires thermodynamic model) (dimensionless), M_2 = molecular weight of adsorbed polymer (kg mol^{-1}), v = molar volume (m^3 mol^{-1}), c_2 = surface layer concentration of adsorbed polymer (mol m^{-3}), δ = adsorbed layer thickness (m)*

The DLVO term for steric repulsion is presented in Figure 12.10.

12.5.1 Temperature-dependency of steric stabilization

The free energy of mixing of the adsorbed molecular chains, as the particles come together, can be broken

$$\Delta G_{mix} = \Delta H_{mix} - T \Delta S_{mix}$$

Figure 12.11 *Temperature-dependency of steric stabilization from thermodynamics. $\Delta G_{mix} > 0$ for repulsion*

into an enthalpic and an entropic part (as shown in Figure 12.11), and we can thus have two different types of stabilization.

It is important that the adsorbed polymer is anchored firmly to the surface so that the chains cannot desorb or move out of the way as the particles approach each other. Thus, we have repulsive steric forces if the Gibbs free energy of mixing (Figure 12.11) is positive. Clearly, from this equation, it is actually the outer parts of the polymer layers that are the most important ones in terms of stability considerations. The free energy of close approach must be positive for a sterically stabilized dispersion to be stable. There are three ways to obtain a positive free energy. When the enthalpy and entropy of close approach of the particles are both negative and $\Delta H < \Delta S$ the enthalpy change favours flocculation and the entropy term opposes it. Since the entropy term dominates the free energy then this is known as entropic stabilization and the dispersion flocculates on cooling. When the enthalpy and entropy are both positive and $\Delta H > \Delta S$ the enthalpic term dominates and this is known as enthalpic stabilization and is characterized by flocculation on heating. When ΔH is positive and ΔS is negative then both terms contribute to

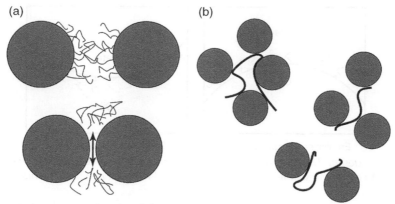

Figure 12.12 *Polymers can also destabilize: (a) depletion flocculation and (b) bridging flocculation*

stability and in principle such systems cannot be floc-culated at any accessible temperature (Croucher and Hair, 1978).

Adding polymers can result in both stabili-zation and destabilization of colloidal systems (Figure 12.12).

By adding polymers to colloids we also change the hydrodynamic features of the system. For exam-ple, small additions of adsorbing polymer can lead to partial coverage which may yield bridging floc-culation (e.g. as used in water purification). On the other hand, if no adsorption is accomplished then we have depletion, i.e. the polymer concentra-tion is lower in proximity to the surface. The reason for this is the entropic loss of the flexible polymers close to the surface. The depletion layer thickness is of the order of the radius of gyration (a measure used to describe the dimensions of a polymer chain) and depends on the concentration – it is high at high concentrations of the polymer. The depletion floc-culation is a weak process and can be reversed by dilution of the system. Depletion flocculation is used to fuse biological cells, using poly(ethylene glycol) in aqueous solution.

Bridging and depletion flocculation are two of the major mechanisms under the family called "sensitisa-tion", i.e. controlled flocculation of dispersions by addition of small quantities of materials which, if used in larger amounts, would act as stabilizing agents. Steric forces can also be attractive in some cases (at short separation) and then bridging can

occur, especially when the surface coverage of the adsorbed polymer chains is not complete, i.e. at low polymer concentrations (partial surface cover-age). Polymeric flocculants are normally used at con-centrations of the order of parts per million (ppm) (Goodwin, 2009).

Bridging leads to rather loose flocculated structures. One example is colloid solutions to which gelatin is added, or in the final stages of water purification where addition of a few parts per million of a high molar mass polyacrylamide leads to flocculation of the remaining particulates in water. It has been found that in water treatment the really effective polymeric flocculants are those bearing the opposite charge to the colloid par-ticles, where bridging flocculation can still be active. Another application is the use of polymers in soil in order to flocculate (loosely) the soil particles and thus they have an open structure that permits the free move-ment of moisture and air in the soil.

12.5.2 Conditions for good stabilization

Steric forces are extremely short range and for this reason the layers have to be thick as the forces are rather small at distances higher than twice the radius of gyration (see Chapter 7). The adsorption must be strong and complete to avoid displacement during collision, bridging or depletion phenomena. This means that the lyophobic particles should be attached strongly to the lyophobic segments of the polymer. On the other hand, steric stabilization is actually

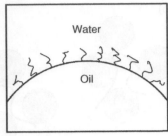 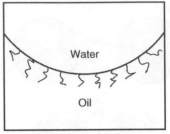

Emulsifier is wetted by water   Emulsifier is wetted by oil

Figure 12.13   *Emulsion design. The Bancroft rule: "The phase in which the emulsifier is more soluble tends to be the dispersion medium." An emulsifier soluble in water leads to an O/W emulsion, while an emulsifier soluble in oil leads to a W/O emulsion*

much helped by the lyophilic parts of the polymer that are extended far out into the medium. The role of solvent is very important, as polymers can be made to desorb and collapse at the surface simply by changing the solvent. It is the repulsive interaction of the lyophilic segments on approaching particles that prevents coagulation. Clearly, we need both "anchor" and "stabilizing" elements (segments) and this explains the wide use of block (and graft) copolymers as stabilizers.

12.6   Emulsion design

The most important types of emulsions are oil-in-water (o/w) and water-in-oil (w/o). Oil-in-oil emulsions are rare, while double emulsions (e.g. oil-in-water-in oil) are also possible. A very important issue in emulsion design is to define/design the right type of emulsion for a certain application. The emulsion type can be determined from the hydrophilic/lipophilic properties of emulsifiers, as explained by Bancroft's rule, shown schematically in Figure 12.13.

We often use mixed emulsifiers, e.g. steric ones and ionic surfactants, or we combine oil and water-soluble emulsifying agents. Other factors are, of course, also important, e.g. the relative volume of the two phases, the type of oil, electrolytes in the aqueous phase and the temperature.

The most important tool used often in conjunction with Bancroft's rule is the so-called HLB

HLB scale:

3–6   : w/o emulsions   ↑ Hydrophobic
7–9   : wetting agents
8–15   : o/w emulsions
13–15 : detergents   ↓ Hydrophilic
15–18 : solubilizers

Figure 12.14   *Emulsion design – Griffin's HLB. Rough rule: A HLB < 8 gives emulsifiers for W/O emulsions and HLB > 8 gives emulsifiers for O/W emulsions*

(hydrophilic–lipophilic balance), an empirical parameter first introduced by Griffin. HLB was discussed in Chapter 5 (Appendix 5.2). It is a parameter that tells us about the hydrophilicity–lipophilicity (hydrophobicity) of surfactants and other emulsifiers. As many common surfactants lie outside the range shown in Figure 12.14 (e.g. sodium lauryl sulfate has a value of 40 and oleic acid 1), they cannot be used directly and therefore mixed surfactants are often used.

The most popular estimation methods for HLB were shown in Appendix 5.2. For mixed emulsifiers we use the additive rule presented in Equation 5.8. This equation can also be used to calculate the required HLB of an oil mixture.

A rule of thumb states that different emulsions have similar characteristics if their emulsifiers have similar HLB numbers. For proper emulsification, the required HLB of an emulsion must match the HLB of the emulsifier mixture used. Mixed emulsifiers, e.g. non-ionic and ionic surfactants, can protect in many ways, steric (high molecular emulsifiers) and electrostatic.

Example 12.2. *Estimation of HLB value of emulsion mixture.*
An o/w emulsion is made from 20% sorbitan tristearate with an HLB value of 2.1 and 80% polyoxyethylene sorbitan monostearate with an HLB value of 14.9.

1. Calculate the HLB value for the emulsion.
2. Calculate the HLB value for the same type of emulsion, but with 30% sorbitan mono-oleate (HLB = 4.3) and 70% polyoxyethylene sorbitan monopalmitate (HLB = 15.6).

Solution
The HLB value of an emulsion based on a mixture of emulsifiers can be calculated from Equation 5.8:

$$\text{HLB}_{mix} = \sum\nolimits_{i=1}^{n} w_i \text{HLB}_i$$

Inserting numbers gives:

1. $\text{HLB}_{mix} = 0.2 \times 2.1 + 0.8 \times 14.9 = 12.3$
2. $\text{HLB}_{mix} = 0.3 \times 4.3 + 0.7 \times 15.6 = 12.21$

It can be seen that in this case the two emulsifier mixtures give practically the same HLB value.

Table 12.1 *Steric versus electrostatic stabilization*

Steric	Electrostatic
Not affected by presence of electrolytes	Electrolytes result in destabilization
Equally good in aqueous and non-aqueous dispersions	Effective especially for aqueous colloidal dispersions
Equally good in dilute and dense colloid systems	Best for dilute colloidal dispersions
Often results in reversible flocculation	Often irreversible coagulation is obtained
Good stability with temperature changes	Coagulation often upon freezing

12.7 PIT – Phase inversion temperature of emulsion based on non-ionic emulsifiers

Although HLB is an important tool in industrial studies related to emulsions, it does not account for changes of temperature, which can be very important. For this reason, Shinoda and Arai introduced about 40 years ago the so-called phase inversion temperature (PIT), which is the temperature at which an emulsion based on non-ionic emulsifiers will change from oil-in-water to water-in-oil. To carry out this experiment, equal weights of oil and water are mixed with 3–5% surfactant to make the emulsion and then the emulsion is heated until an inversion is apparent. The PIT occurs for thermodynamic reasons (for a temperature below PIT, the emulsifier is strongly polar, above PIT the emulsifier interacts much less with water). The temperature dependency

of different forces was discussed in Chapter 2. For emulsions used in detergents, it is often observed that optimum detergency happens close to PIT. As expected, HLB and PIT are related to each other. The final choice of emulsifier, however, will depend on several other factors, e.g. cost, appearance and environmental sustainability. Emulsion science uses several more empirical rules in design studies.

12.8 Concluding remarks

Summarizing, Table 12.1 shows a comparison of steric versus electrostatic stabilization. Emulsions are industrially important in many products and processes and tools to design these structures are essential. A very useful tool is the hydrophilic–lipophilic balance, HLB, which can be used to select a mixture of emulsifiers to use for a given application. Emulsions are metastable and there are several destabilization mechanisms, which may need to be considered depending on the emulsion type, application and expected lifetime. The most important mechanisms are creaming, flocculation or coagulation, and coalescence/breaking. The DLVO theory can, in many cases, be used to analyse stability of emulsions. However, steric stabilization is often used in the emulsion field and the theoretical background for this part is not as well established as the electrostatic stabilization. Practical experience or empirical experiments are more often used in industrial applications. Attention must also be paid to the temperature dependency of emulsions, which can cause inversion of emulsion phases. In the coming years, microemulsions are expected to become more important on an industrial scale.

Problems

Problem 12.1: Emulsions with hydrophilic and hydrophobic surfactants

A particular product is an o/w emulsion and is based on equal amounts (weight percent) of two surfactants, which are both hydrophilic. The first surfactant has a HLB = 13 and the other has a HLB = 11.

The company where you are currently working wishes, for various reasons, to change the formulation of the surfactants used for the stabilization of this o/w emulsion product. Of course, the product must have the same o/w emulsion characteristics.

The company suggests using this time a mixture of hydrophobic and hydrophilic surfactants, both of which are non-ionic and belong to the family of poly(oxyethylene alcohol)s, known as poly(ethylene oxide)s. The research director suggests using the surfactants $C_{12}E_{30}$ and $C_{16}E_6$ (see Table 5.1). The formulation is, thus, based on both hydrophilic and hydrophobic surfactants. How much (weight) percentage of the new hydrophilic surfactant is needed for this specific o/w emulsion?

Problem 12.2: An industrial case (HLB with different methods)

An optimal HLB for an emulsion product is reported to be 10.3.

1. What type of emulsion do we have?
2. Suddenly, one of the hydrophilic surfactants you are using in your company for this specific formulation is no longer available in the market! The hydrophobic surfactant (with HLB = 2.1) is still available and it is very cheap (irrespective of the amount employed). Researchers from the research department of your company suggest three possibilities: (i) a secret "in-house" (possibly anionic) surfactant X which has an oil–water partition coefficient equal to 2×10^9 (they do not reveal the structure!); (ii) NPE_{15} and NPE_{30} (NPE = nonyl phenyl ethoxylates where the subscript is the number of ethylene oxide groups). The management department informs you that the cost of these three surfactants is almost the same. Which one of the three surfactants (X, NPE_{15} or NPE_{30}) would you recommend to your manager for this specific product as a substitute for the surfactant that is no longer on the market?

Problem 12.3: Emulsions

1. Describe briefly the various destabilization mechanisms in emulsions. Which of these mechanisms are reversible and which are not (maximum 20 lines)?
2. Which parameters favour the stability of emulsions (maximum 1–2 lines per parameter)?
3. What is the Bancroft rule and what is its importance in the study of emulsions (maximum 10 lines)?
4. What is the HLB (hydrophilic–lipophilic balance), how can it be estimated and what is its importance in the design of emulsions (maximum 20 lines)?
5. A specific commercial emulsion (formulation) is stabilized with a mixture of 80% aromatic mineral oil (HLB = 13) and 20% paraffin oil (HLB = 10). The company you are working for wants to change this emulsifier mixture and use a mixture of non-ionic (poly(ethylene oxide)s) surfactants $C_{12}E_{30}$ and $C_{16}E_4$, of which one is hydrophilic and the other is hydrophobic. How much would you need of the hydrophilic surfactant so that the emulsion has the same characteristics as the original one? All concentrations are in weight fractions.

Problem 12.4: *Emulsions: Type and Kinetics*

An emulsion is made based on a knowledge of its phase inversion temperature (PIT) temperature.

1. What is the PIT temperature?
2. How is the emulsion made?

    The following data are available for the emulsion:
    - volume fraction of oil particles in the emulsion (volume particles/volume emulsion): 40%;
    - PIT: 60 °C;
    - droplet diameter: 0.65 µm;
    - temperature: 25 °C;
    - viscosity of water/oil-phases: $8.9 \times 10^{-4}/1.0 \times 10^{-2}$ Pa s.

3. What type of emulsion do we have? What would for this emulsion be the half-life time, if there is no potential barrier?
4. The stability of the emulsion is investigated via the following measurements of the particle concentration (at 25 °C):

Time (days)	Droplet concentration (cm^{-3})
1	2.57×10^{12}
7	1.62×10^{12}
14	1.13×10^{12}
21	0.88×10^{12}
30	0.68×10^{12}

Calculate the "second-order" rate constant, k_2, and the stability ratio.

Problem 12.5: *Emulsions and DLVO theory*

A company is asked by a customer to make a 1 m^3 batch of mayonnaise. In crude terms, the mayonnaise is an oil-in-vinegar emulsion made with a suitable emulsifier. The vinegar can be treated here as pure water. Other ingredients such as lemon juice, egg yolk, salt, pepper, mustard, citric acid and flavour enhancers are not considered unless specified below. The temperature is 25 °C. The HLB (hydrophilic–lipophilic balance) value required for the mayonnaise is 12 and the interfacial tension between the two liquid phases, in the presence of any one or more emulsifiers, can be assumed to be 1.0 mN m^{-1}. The oil content of the oil-in-water emulsion is 75 vol. % and the oil droplet diameter can be assumed to be 1 µm. The Hamaker constant of the oil phase is 5×10^{-20} J. The viscosity of water is 10^{-3} kg m^{-1} s^{-1}. Any effects of steric stabilization can be considered negligible in this problem.

1. A suitable industrial emulsifier, as an alternative to using egg yolk, is sought. It is suggested to use the emulsifier shown below, which has a long hydrocarbon tail and a hydrophilic head:

    Estimate the HLB value of this emulsifier and comment on its suitability for the mayonnaise.
2. Two other emulsifiers, which have HLB values of 6 and 15, respectively, are considered as an alternative. What should be the concentration ratio of these two emulsifiers, on an emulsifier weight fraction basis, if they are used for the mayonnaise?
3. Calculate the theoretical work required to make the emulsion batch and comment on the calculated value.
4. Calculate the attractive force between two oil droplets in the mayonnaise when the distance between them is 0.5 and 10 nm, respectively.
5. The oil droplets in the mayonnaise are electrostatically stabilized. The water phase can be taken here as an aqueous solution of NaCl. The surface potential is 10 mV. The composite (overall) potential between the oil droplets at a distance of 0.5 nm must be higher than $20k_BT$. Calculate the maximum salt concentration that can be allowed in the mayonnaise.
6. In the case of destabilization of the emulsion, how long does it take to coagulate 99 % of the oil droplets if the process can be considered to be diffusion controlled and only droplet duplets (pairs) are formed?
7. As an alternative, the mayonnaise can be produced using selected non-ionic emulsifiers. In this case, the emulsion has a phase inversion temperature of 30 °C. Explain what is meant by a phase inversion temperature and discuss whether the value provided for the mayonnaise can result in any practical problems.

References

M.D. Croucher, M.L. Hair, (1978), Macromolecules, 11(5), 874–879.

J. Goodwin, (2009), Colloids and Interfaces with Surfactants and Polymers, 2nd edn, John Wiley & Sons Ltd, Chichester.

B.K. Paul, S.P. Moulik, (2001) *Curr. Sci. (special edn: soft condensed matter)*, 80(8), 25 April 2001, 990–1001.

P. Walstra, R. Jenness, H.T. Badings, *Dairy Physics and Chemistry*, John Wiley & Sons Ltd, Chichester, 1984.

13

Foams

13.1 Introduction

In short, a foam is a dispersion of a gas in a liquid prepared using a foaming agent, which in most cases consists of one or more surfactants. These elegant gas–liquid structures, which appear in numerous products and processes, can be analysed and manipulated using many of the tools that have been introduced in earlier chapters. Foams belong to the topic of this book because of the large surfaces involved. In addition, the dimension (thickness) of the thin liquid films (so-called lamellae) present in foams fall, at least in the later part of the foam lifetime, within the colloid regime, from approximately 1 nm to 1 μm. Therefore, a foam is a system with two dimensions in the macroscopic size range and one dimension potentially in the colloidal range.

Foams are industrially important in many end-use products and also quite often as an undesired side effect in various processes. Dead plant material in seawater can also lead to excessive foaming (Figure 13.1).

13.2 Applications of foams

We will begin with some important definitions and numbers. The gas volume fraction in a foam is mostly between 0.5 and 0.9, and the gas makes up the dispersed phase, while the liquid is the continuous phase (Figure 13.2).

Two important terms are "lamella" and "Plateau border", which are the two liquid regions of a foam (Figure 13.3).

The size of the individual bubbles in a foam is typically from about one mm to several centimetres (much larger than the droplets in ordinary emulsions), but can be as small as 10 nm. In an ideal foam, with all bubbles having the same size, the foam would assume the shape of a pentagonal dodecahedron (composed of 12 regular pentagonal faces) upon crowding. The lamella thickness is about 30 μm for "young" foams and about 0.1 μm after complete gravitational liquid drainage. If a foam contains micro- or nanosize particles it is called a froth.

Introduction to Applied Colloid and Surface Chemistry, First Edition. Georgios M. Kontogeorgis and Søren Kiil.
© 2016 John Wiley & Sons, Ltd. Published 2016 by John Wiley & Sons, Ltd.
Companion website: www.wiley.com/go/kontogeorgis/colloid

Figure 13.1   *Excessive ocean foaming on the east coast of Australia due to surface active compounds (proteins) originating from degrading plant materials in combination with a severe storm. Getty Images/Chris Hyde*

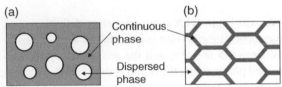

Figure 13.2   *Schematic illustration showing a "dilute" foam (a) and a "concentrated" foam (b). In the concentrated foam, the bubbles are closely packed so that their shape becomes non-spherical. Reprinted from Myers (1999), with permission from John Wiley & Sons, Ltd*

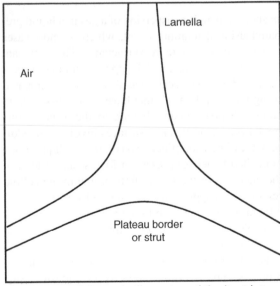

Figure 13.3   *Schematic illustration of the liquid structure in a "true" foam, where the lamella thickness is of the order of micrometres*

We now take a look at some applications of foam structures (Figure 13.4).

If we start with the food industry, foams play an important part for both appearance and taste. Ordinary bread, as an example, is a solid foam structure, whereas whipped cream is a foam according to the traditional understanding. Potato chips, e.g. Wokkles, is also a foamed food product and the design of the right beer foam is a science in itself. Froth flotation is a process for separating minerals from non-valuable rock and dirt by taking advantage of

differences in the particle hydrophobicities and requires extensive use of surface active agents. Froth flotation involves the capture of hydrophobic

(a)   (b)   (c)

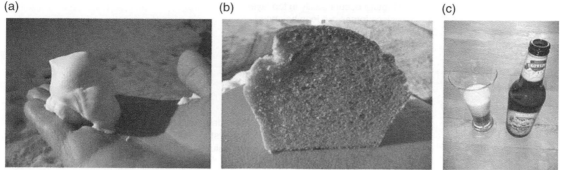

Figure 13.4 Examples of the use of foams in products: (a) shaving cream, (b) bread (solid foam) and (c) a Danish beer

particles by air bubbles and transport of the bubble–particle aggregates to the liquid surface (the air–water interface is considered to be one of the most hydrophobic surfaces known). The non-valuable materials sediment.

Foams are typically very lightweight materials. The gas is often generated upon application and therefore a large volume of foam can be formed from a small volume of liquid. This is useful for transportation of, for example, fire-fighting foams, where a foam concentrate, consisting of surfactants and various additives, is used. A "foam blanket" for fire-fighting is used to cool the fire and prevent contact with oxygen and flammable or toxic vapours. Many different types of foam blankets are available for different types of fires. To form a foam blanket, a foam concentrate and water, in combination with conventional air-aspirating or non-aspirating fire-fighting equipment, are needed. Foams with a high water content work best because the water helps in removing heat from the system. Foams are also used in cosmetic, detergent and personal care products such as shaving cream, shampoo and bubble bath. Foams are convenient to use for quick application of a lotion to the skin and are very useful for delivering pharmaceuticals to sensitive parts of the human body. Powder puffs, used by girls to apply powders to the face, can also be foams. Foam structures are also present in construction materials such as polymeric or mineral wool insulation materials and concrete.

13.3 Characterization of foams

In a foam column, various transitional structures may occur (Figure 13.5).

Near the surface, a high (>70 vol.%) gas content structure, polyhedral foam is usually formed. Below the surface is a low (<70 vol.%) gas content structure called a spherical foam (or "kugelschaum" or gas emulsion).

Overall, the foam density decreases with the height of the column. The change in structure with height can easily be seen in beer foam immediately after pouring of the beer into a glass (carbon dioxide rises from the liquid below the foam and provides the gas injection). Foam collapse usually occurs from the top to the bottom because the thin films in the polyhedral foam are sensitive to rupture by shock, temperature gradients, or vibration. There are also other degradation mechanisms, as we will see later. Spherical foams are more sensitive to liquid drainage due to thick lamellae.

The foam number, F, is often used for a crude comparison of foams and is defined as:

$$F = \frac{V_F}{V_L} \qquad (13.1)$$

where V_F and V_L are the volumes of foam and liquid, respectively. Typical foam numbers are 2–20. Polyhedral foams take on large values of F.

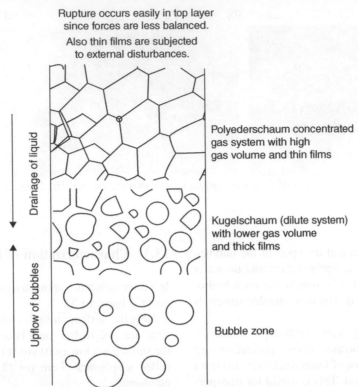

Figure 13.5 *Foam structure during formation and drainage in a vertical column. Gas is injected from below. Polyhedral foam ("Polyderschaum") has a high gas content (>70 vol.%). The liquid in polyhedral foam structure is distributed between films and Plateau borders (i.e. channels that form where films meet). Reprinted from Pugh (1996), with permission from Elsevier*

Example 13.1. Calculation of gas volume fraction from foam number.

Bubbles in freshly poured beer are typically polyhedral. Using a laboratory test, the foam number of a beer foam has been measured to a value of 4.

What is the volume fraction of gas in this beer foam?

Is the number you have calculated in agreement with a characterization of the foam as being polyhedral?

Solution

From the definition of the foam number one has:

$$F = \frac{V_L + V_G}{V_L} = 1 + \frac{V_G}{V_L}$$

Upon rearrangement this gives:

$$V_G = (F - 1)V_L$$

From the definition of volume fraction one has:

$$\varphi_G = \frac{V_G}{V_L + V_G} = \frac{1}{1 + \dfrac{V_L}{V_G}}$$

Combining the last two equations gives:

$$\varphi_G = \frac{1}{1 + \dfrac{V_L}{V_G}} = \frac{1}{1 + \dfrac{V_L}{V_L(F-1)}} = \frac{F-1}{F}$$

For $F = 4$ one gets:

$$\varphi_G(F=4) = \frac{F-1}{F} = \frac{4-1}{4} = 0.75 = 75\%$$

A polyhedral foam has a gas volume fraction above 70% and therefore the beer foam is (initially) polyhedral.

For convenience, foams have been classified into two extreme types: unstable or transient foams with lifetimes of seconds and metastable (or permanent) foams with lifetimes measured in days. Metastable foams can withstand ordinary disturbances (Brownian fluctuations), but collapse from abnormal disturbances (e.g. evaporation or temperature gradients).

Two final terms to consider are the foam ability and stability. The foam ability of a foaming agent refers to the volume of foam generated in a given process with a given concentration of a foaming agent. The foam stability refers to the life time (or half-life) of a generated foam.

13.4 Preparation of foams

Foams (or foam solutions) can be prepared from two overall processes termed condensation and dispersion. In condensation (of a gas), foam is generated from a liquid supersaturated with a gas. A beer in a can is a typical example of this. When the can is opened, the pressure is reduced and less gas (carbon dioxide) can be contained in the liquid and therefore comes out as bubbles and produces foam. Heating can

also be used as a method of gas release. When using the dispersion method, the gas is injected into the liquid in various ways (e.g. stirring of whipped cream) or bubbling through a porous plug. Conventional stirring or mixing of a foaming liquid is another way of ensuring gas entrainment and thereby foaming.

13.5 Measurements of foam stability

An often used test method for characterization of foams is the Bikerman column (Figure 13.6). Here, nitrogen or air is injected from a porous plug into the liquid to be tested and the steady foam height is measured. The Bikerman coefficient, Σ, ($= H/v$, where H is the foam height and v the linear gas velocity) expresses the average bubble lifetime in a foam before it bursts. The coefficient is independent of the gas flow rate provided that evaporation (low gas velocities) or rupture of the lamella (high gas velocities) is not significant. The Bikerman column has been used, for example, to study foam in wet flue gas desulphurization plants (see, for example, Hansen, Kiil and Johnsson, 2008). The method is not useful for high concentrations of strong surfactant

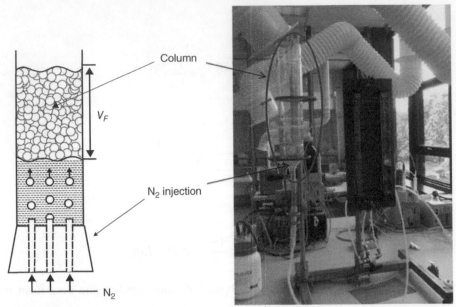

Figure 13.6   *Test of foaming agents using the Bikerman approach, where nitrogen or air is injected through a porous plug and the height of the foam layer is measured over time. The diameter of the cylinder is about 5 cm. (Left) Adapted from Siqiang et al. (2013), with permission from John Wiley & Sons, Ltd*

where the foam quickly transports itself out of the column.

Many other test methods and standards are available. Ross–Miles testing is another important test though it is often criticized for not being representative of real life foaming. In this test, a dilute solution of surfactant is dropped from a fixed height into a pool of the same dilute solution and the foam volume developed is measured.

13.6   Destabilization of foams

All foams are thermodynamically unstable in accordance with the following equation that we have seen in earlier chapters (e.g. Chapters 3 and 12):

$$W = \gamma_{GL}\Delta A \qquad (13.2)$$

where W is the work contained in the system, γ_{GL} is the gas–liquid surface tension and ΔA is the change in

surface area taking place when the foam is made. The higher the surface tension, the more work is required to produce the foam and the less stable the foam becomes.

There are several mechanisms of foam destabilization: gas diffusion, drainage and coalescence. It is important to remember that these processes are typically strongly coupled, which makes it difficult to analyse and optimize a foam in a given situation. It is the rate (kinetics) of the various processes that determines whether these are important. Each of these destabilization processes will now be discussed in more detail.

Figure 13.7 shows the various stages that foams can go through, as they mature and eventually are destroyed.

Some foams collapse due to disturbances, whereas others may go through the full number of stages. The rate at which they undergo these processes may differ greatly from one foam system to another. External disturbances can be vibrations or draughts, organic

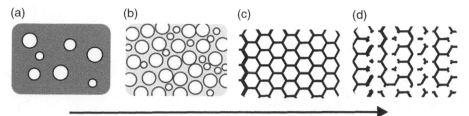

Figure 13.7 *Stages in a foam lifetime. (a) Spherical foam (independent gas bubbles), (b) gravity drainage period, (c) lamella thinning period and (d) film rupture. Berg (2010). Reproduced with permission from World Scientific*

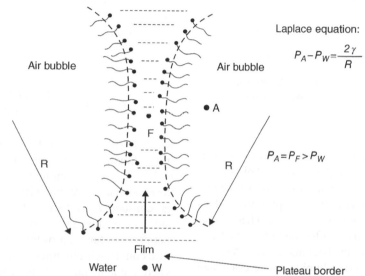

Laplace equation:

$$P_A - P_W = \frac{2\gamma}{R}$$

$$P_A = P_F > P_W$$

Figure 13.8 *Pressure differences inside a foam with monodispersed bubbles. Adapted from Pashley and Karaman (2004), with permission from John Wiley & Sons, Ltd*

contaminants, dust, or particles, temperature gradients, evaporation or heat. Contaminants or dust can work as defoamers by breaking the lamella (see later about defoamers). Temperature gradients will lead to natural convection (slow flow of air).

13.6.1 Gas diffusion

To explain gas diffusion, we start with the pressure difference between points A and W, see Figure 13.8. This pressure difference is given by the Laplace equation, see Chapter 4, which is also shown in the figure.

The pressure inside the bubble is higher than at point W because of the surface tension, which tries to minimize the surface area, and must be counterbalanced

for the bubble not to collapse. Across a flat interface there is no pressure difference, so the pressure in the film (at F) must be greater (by ΔP) than the pressure in the bulk liquid (at W). The pressure increase in the film is due entirely to the repulsive interaction between the two surfaces, which may be of electrostatic origin due to ionization of the surfactant head groups. The reduced pressure in the Plateau border regions acts to "suck" liquid from the lamellar films between bubbles (capillary pump), which can lead to rupture.

The pressure inside the air bubbles is higher than that in the solution or air. This pressure increase can be described by the Laplace equation (Figure 13.9).

Furthermore, the pressure and the solubility of the dispersed air phase are higher for smaller bubbles.

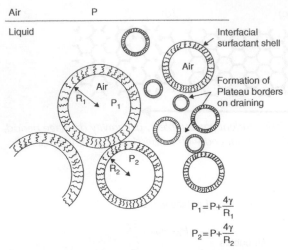

$$P_1 = P + \frac{4\gamma}{R_1}$$

$$P_2 = P + \frac{4\gamma}{R_2}$$

Figure 13.9 *Pressure inside air bubbles in a spherical foam. Adapted from Pugh (1996), with permission from Elsevier*

So there is a driving force for diffusion from small air bubbles to larger ones or to the bulk liquid phase. The rate of diffusion depends on the solubility of the dispersed air phase in the continuous liquid phase. This phenomenon can be seen in beer foam, where larger bubbles are formed rapidly after the initial foam formation upon pouring of the beer into a glass. As mentioned, the pressure inside a bubble can be calculated by the Laplace equation. Note that the pressure inside a bubble is higher than that of the surrounding air and is higher the smaller the bubble. The factor "4" arises because the bubble wall involves two curved interfaces, each of which contributes to the total. However, this is not the case in Figure 13.8 for the Plateau border, where the traditional derivation of the Laplace equation for a capillary tube with only a half-sphere surface can be used. Gas diffusion from small to large bubbles leads to greater polydispersity (broader bubble size distribution) over time. An important parameter is the solubility of the gas in the liquid (quantified by Henry's law). If the solubility is very low, gas diffusion between bubbles will not be important. Gas diffusion, in this context, is also referred to as disproportionation or Ostwald ripening (see Chapter 4).

Gas diffusion can be quantified using the following equation (shrinkage of a bubble over time):

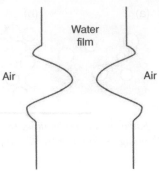

Figure 13.10 *Surface waves travelling through a film lamella. Adapted from Pashley and Karaman, (2004), with permission from John Wiley & Sons, Ltd*

$$R_t(t)^2 = R_o{}^2 - \frac{4R_{ig}TDS_{eq}\gamma_{AL}t}{P_o\delta} \qquad (13.3)$$

where D (in $m^2\ s^{-1}$) is the gas diffusion coefficient in the liquid, S is the solubility of the gas (in mol $m^{-3}\ Pa^{-1}$), R_t is the current radius of a bubble, P_o is the atmospheric pressure (101325 Pa), R_{ig} is the ideal gas constant, R_o is the initial thickness of the bubble, and δ is the lamella thickness between bubbles. Note, that $\delta(t)$ is a function of time, t. As a final note, gas diffusion is seldom important in polyhedral foam, where the lamellae are approximately planar.

13.6.2 Film (lamella) rupture

Film rupture in a foam occurs because of surface waves (originating from disturbances) (Figure 13.10).

In pure water, rupture takes place when the film is 100–400 nm thick. In surfactant solutions, rupture takes place when the film is 5–15 nm thick. The critical rupture thickness of thin films, h_c, can be calculated from the following semi-empirical equation:

$$h_c = 0.207 \left(\frac{A_{121}{}^2 R_B{}^2}{\gamma_{AL}\Delta P} \right)^{\frac{1}{7}} \qquad (13.4)$$

where A_{121} is the effective Hamaker constant (J) for air/liquid/air, ΔP is the excess capillary pressure (Pa) in the film (= $2\gamma_{AL}/R_B$) and R_B is the radius of a bubble (m). Pugh (1996) has reviewed the equation and states that it can be used to give an indication of the effect of various parameters, but also that it does not always match experimental data. For instance, it is well known that many poorly foaming liquids with thick lamella are easily ruptured. The equation was derived based on the capillary wave mechanism by Vrij (Wang and Yoon, 2008). Surfactants can influence both the Hamaker constant and the surface tension and therefore it is not easy to predict the effect of surfactant concentration on the critical rupture thickness.

13.6.3 Drainage of foam by gravity

Drainage from a lamella by gravity can be approximated by drainage from a vertical slab (Figure 13.11).

Hydrodynamic drainage due to gravity is usually the most rapid of the destabilization mechanisms, and if the foam is not stabilized this mechanism leads to total collapse before other mechanisms can become important. In those cases, once the loss of liquid from the lamella produces a critical film thickness of 5–15 nm, the liquid film can no longer support the pressure of the gas in the bubble and film rupture occurs (Myers, 1999). An equation has been derived that can be used to estimate an average drainage velocity, v_{av}, due to gravity (Pugh, 1996):

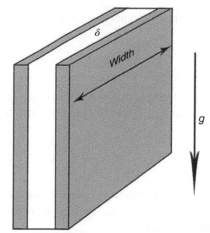

Figure 13.11 Drainage by gravity approximated by drainage from a vertical slab

$$v_{av} = \frac{g\rho_L\delta(t)^2}{8\eta_L} \qquad (13.5)$$

Notice that the film thickness is a function of time and the equation only provides an estimate. It can be used to analyse how we can influence the rate of drainage. High viscosity (η_L) liquids reduce draining and provide a "cushion" effect, which can absorb "shock" energy.

Example 13.2. Draining of beer foam.
The main foam destabilization mechanism in freshly poured beer is vertical draining by gravity. The height of the fresh beer foam is measured to 5 cm and can be considered constant. It can be assumed that 25% of the cross-sectional area of the foam is used for vertical draining, in this initial phase, and that the foam structure remains constant and homogeneous.

If the initial lamella (or film) thickness in the foam is crudely assumed to be 100 μm (and constant), how long does it approximately take to drain the beer foam (vertically) from 10% of its liquid volume when the gas volume fraction of the foam is 75%? Comment on the result.

Solution
A gravity draining time for 10% of liquid volume in the foam must be calculated. The average vertical drainage rate is given by Equation 13.5:

$$v_{av} = \frac{g\rho_L\delta^2}{8\eta_L}$$

The volumetric drainage rate (in m³ s⁻¹) can be taken as constant because the lamella thickness is constant, leading to:

$$Q = v_{av}A_C f_A$$

where A_C is the cross-sectional area of the foam and f_A is the fraction of the cross-sectional area that is available for drainage. Because Q is constant, the drainage time is given by:

$$t_d = \frac{f_d V_L}{Q} = \frac{f_d V_L}{v_{av}A_C f_A} = \frac{f_d l A_C \varphi_L}{v_{av}A_C f_A} = \frac{f_d l \varphi_L}{v_{av}f_A} = \frac{f_d l(1-\varphi_G)}{v_{av}f_A}$$

where f_d is the percentage of liquid that must be drained, l is the height of the beer foam, φ_L is the liquid volume fraction, and φ_G is the gas volume fraction in the foam. Combining the latter equation with Equation 13.5 gives:

$$t_d = \frac{f_d l 8 \eta_L (1-\varphi_G)}{g \rho_L \delta^2 f_A}$$

Inserting numerical numbers gives:

$$t_d = \frac{f_d l 8 \eta_L (1-\varphi_G)}{g \rho_L \delta^2 f_A} = \frac{0.1 \times (0.05 \text{ m}) 8 (2 \times 10^{-3} \text{ kgm}^{-1}\text{ s}^1)(1-0.75)}{(9.81 \text{ ms}^{-2})(1000 \text{ kgm}^{-3})(100 \times 10^{-6} \text{ m})^2 (0.25)} = 0.8 \text{ s}$$

This is a very fast drainage rate and if it continued the foam would collapse within seconds. However, the assumption of a constant lamella thickness is, of course, not correct for more than the initial stage. As liquid drains the lamella thins and the rate of drainage by gravity rapidly decreases (see the lamella thickness dependency in Equation 13.5). Later on the main destabilization mechanism can be gas diffusion.

Drainage can also take place in approximate horizontal directions, where liquid flows from lamella to Plateau borders. An approximation of this situation is shown in Figure 13.12.

The so-called Reynolds equation, valid for $\delta >$ 100 nm, can be used to estimate the horizontal drainage rate (Pugh, 1996; Wang and Yoon, 2009):

$$\frac{d\delta}{dt} = -\frac{2\delta^3 \Delta P}{3\eta_L R_f^2} \qquad (13.6)$$

where ΔP is the pressure difference between lamella and Plateau border. The initial condition is given by:

$$\delta(t=0) = \delta_o \qquad (13.7)$$

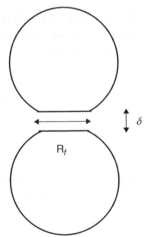

Figure 13.12   *Approximation of horizontal drainage of a foam*

where δ_o is the initial thickness of the lamella. The main assumptions underlying the equation are: (i) the liquid flows between plane parallel disc surfaces; (ii) the film surfaces are tangentially mobile; (iii) the rate of thinning due to evaporation is negligible in comparison with the rate of thinning due to drainage. For very thin films the pressure gradient should also include the so-called disjoining pressure (see Wang and Yoon, 2009, for more information).

Consequently, for a full quantitative description of foam drainage one will need to consider both horizontal and vertical draining. However, such a complete analysis has not yet been undertaken.

13.7　Stabilization of foams

Now that the mechanisms of destabilization are known, we may try to formulate some means to improve foam stabilization. Obviously, changes that may help foam stabilization may have other side effects to a product and this must be kept in mind. Increasing the viscosity, either for the liquid bulk or for the surface, can reduce the rate of draining by gravity or capillary flow. The addition of surfactants and/or polymers to a foaming system can alter some or all system characteristics and therefore enhance the stability of the foam. Electrostatic and steric repulsion between adjacent interfaces reduces drainage through effects on the capillary pressure and thereby opposes bubble coalescence. These

different stabilization mechanisms will now be treated in some more detail.

13.7.1　Changing surface viscosity

Foam drainage is influenced by the surface viscosity and stability is enhanced by increasing the surface viscosity (and thereby also the elasticity of the liquid film). Generally, the surface viscosity can be increased by packing a high concentration of surfactant or particles in the surface, leading to high adhesive or cohesive bonding. The mechanism is illustrated in Figure 13.13. As an example, adding polymers of relatively high molecular mass, such as proteins, polysaccharides or certain types of particles can increase surface viscosity. Many food foams are stabilized by protein polymers adsorbed at the air–water interface (Pugh, 1996).

13.7.2　Surface elasticity

Surface elasticity, sometimes referred to as the "self-healing" effect, is caused by surfactants. The mechanism behind this phenomenon is called the Gibbs–Marangoni effect and is illustrated in Figure 13.14.

As explained by Pugh (1996), when a soap film is stretched, due to an external disturbance, a part of the film will expose a pure water interface before surfactant molecules have time to diffuse from bulk to the surface. The water region will have a much higher

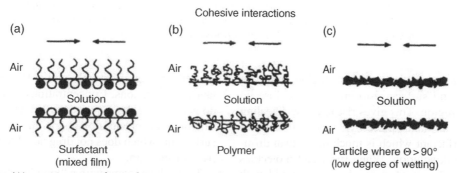

Figure 13.13　*Ways to increase the surface viscosity at the gas–liquid interface of a foam using the presence of a high packing density of certain types of surfactant species which can cause strong cohesive interactions. Adapted from Pugh (1996), with permission from Elsevier*

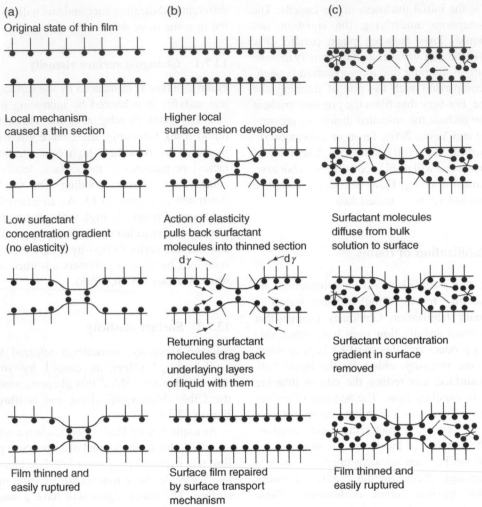

Figure 13.14   *Influence of surfactant concentration on surface elasticity. The concentration must neither be too high nor too low for the Gibbs–Marangoni effect to work. (a) Low surfactant concentration gives only low differential tension in film and low foaming. (b) Intermediate concentration gives a strong effect. (c) High surfactant concentration (>CMC) gives differential concentration that relaxes too rapidly due to diffusion of surfactant resulting in a thin film that easily ruptures. Adapted from Pugh (1996), with permission from Elsevier*

surface tension than the surfactant region and there will be a strong surface-restoring force to close up the exposed region. This gives rise to the self-healing effect. Diffusion takes place by surface diffusion of adsorbed surfactant which is much faster than diffusion from bulk solution. When surfactant molecules diffuse, they drag underlying layers of liquid with them to "close" the thin part of the film. This is the self-healing effect. As can be seen in Figure 13.14,

the self-healing effect only works for certain surfactant concentrations. Concentrations that are too low do not provide the required surface tension gradient. Too high a concentration results in bulk diffusion of surfactants which does not drag new liquid to the thin parts of the film.

To produce foams with a "long" lifetime, it is very important that the surface elasticity is sufficiently large. The elasticity of static foam films, E (unit is

J m^{-2} or mN m^{-1}), is defined by the Gibbs surface elasticity, where A is the film surface area (see, for example, Wang and Yoon, 2008):

$$E = 2\frac{\partial\gamma_{AL}}{\partial\ln A} = 2A\left(\frac{\partial\gamma_{AL}}{\partial A}\right) \qquad (13.8)$$

The larger the value of E, the better the film shock resistance (resiliency). For a pure liquid, γ_{AL} does not change when A changes and $E = 0$ (no foam). The smaller γ_{AL} is, the less work is required for making the foam. Wang and Yoon (2008) have derived a useful expression for film elasticity:

$$E = \frac{4R_{ig}T\Gamma_m{}^2K_L{}^2c}{\delta(1+K_Lc)^2 + 2\Gamma_mK_L} \qquad (13.9)$$

There is also an elasticity for dynamic foams, but that will not be discussed here in more detail. A typical value for E for a static foam is 10 mN m^{-1} (Wang and Yoon, 2008). The Marangoni effect is usually of importance in fairly dilute surfactant solutions and over a relatively narrow concentration range. Using the following equation, it is possible to estimate the time required for the adsorption of a given amount of surfactant at a new interface, n (in numbers per m^2) (which can be compared to the rate of generation of that interface):

$$n(t) = 2\left(\frac{D}{\pi}\right)^{1/2}ct^{1/2}N_A \qquad (13.10)$$

where D is the diffusion coefficient of surfactant (in m^2 s^{-1}), c is concentration of surfactant (in mol m^{-3}), t is time (in s), and N_A is Avogadro's constant (in mol^{-1}).

If the surfactant solution is too dilute, the surface tension of the solution will not differ sufficiently from that of pure solvent for the restoring force to counteract the effects of casual thermal and mechanical agitation. This will lead to a very transient foam. Experimental data have shown that the optimal concentration to use is usually within a factor of two of the critical micelle concentration.

Adsorption of surfactants has been discussed in Chapters 4 and 7. Here is a reminder that the equilibrium adsorption density, Γ, is given by the Gibbs adsorption equation:

$$\Gamma = -\frac{1}{R_{ig}T}\frac{d\gamma}{d\ln c} \qquad (13.11)$$

If the surface tension is known as a function of concentration of surfactant, then Γ can be evaluated. One possibility is to use the Langmuir–Szyszkowski equation (Wang and Yoon, 2008):

$$\gamma = \gamma_0 - R_{ig}T\Gamma_m\ln(1+K_Lc) \qquad (13.12)$$

Differentiation of the equation for surface tension with respect to c provides the differential needed in the general expression (Equation 13.11).

In practice, we often have a mixture of surfactants. Even "pure" surfactants of technical origin will be a mixture because of production impurities. If two surfactants adsorb simultaneously at an interface, the component of the highest bulk concentration will adsorb fastest, as the transport by diffusion is mainly controlled by the concentration of surfactant. A surfactant of lower concentration adsorbs slower, however, and due to a higher surface activity it can displace another surfactant from the interface. The composition of the surface layer at any moment is in a local equilibrium and governed by a respective adsorption isotherm for the surfactant mixture.

13.7.3 Polymers and foam stabilization

Very stable foams can be produced if surface-active polymers are included in the formulation. When polymers, proteins in particular, are adsorbed at the gas–liquid interface they denature, as shown in Figure 13.15, and can perform as a foaming agent.

The relatively dense, somewhat structured adsorbed polymer layer imparts a significant degree of rigidity or mechanical strength to the lamellar walls, producing an increase in the stability of the foam. When the protein is in the bulk solution, hydrophobic parts are "hidden" inside the molecule to get away from water. At the interface, the hydrophobic parts are attracted to the very hydrophobic air leading to unfolding of the protein. Foaming caused by proteins is sometimes a problem in waste water-treatment plants (Myers, 1999).

In summary, it can be stated that the foam ability a given surfactant imparts depends on the surfactants effectiveness in reducing surface tension, its diffusion

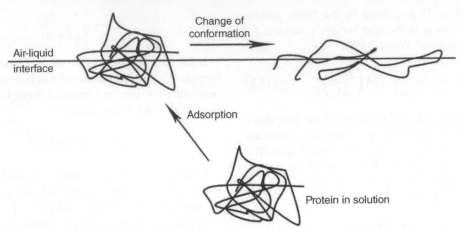

Figure 13.15 *Proteins adsorb and subsequently denaturate at the air–liquid interphase, placing the hydrophobic parts of the molecule into hydrophobic air*

and adsorption rates, effect on electrostatic double layer thickness, elastic properties of the interface, and the effects of surfactants on surface and bulk viscosities.

13.7.4   Additives

The presence of additives can affect the stability of a foam by influencing any of the mechanisms already discussed for foam stabilization. Additives can increase the viscosity of the liquid phase or the interface or alter the interfacial interactions related to the Gibbs–Marangoni effects or the electrostatic repulsions. A proper choice of additive can turn a high-foaming surfactant into one exhibiting little or no foam formation and vice versa. Therefore, the addition of small amounts of additives has become the primary way of adjusting the foaming characteristics of a formulation in many, if not most, practical surfactant applications. There are three classes of additives: (1) inorganic electrolytes, which work well with ionic surfactants (they reduce CMC, but also decrease double layer thickness), (2) polar organic additives (work with all types of surfactants by reducing CMC) and (3) macromolecular materials. The organic polar additives are the most important additives. Examples of such additives are tetradecanol, lauryl glycerol ether, and caprylamide.

13.7.5   Foams and DLVO theory

The DLVO theory can be used to analyse foam stabilization by electrostatic repulsion. When a substance is added to a foam leading to the charge of the surface film this can result in the repulsion of the air bubbles coming near to one another. Soap (surfactant) films are modelled by two air slabs (infinite thickness) with a liquid film in between. The attractive potential is then given by:

$$V_A = -\frac{A_{121}}{12\pi H^2} \qquad (13.13)$$

The electrostatic potential is expressed by (assuming the Debye–Hückel approximation is valid):

$$V_R = 2\varepsilon_0 \varepsilon \kappa \psi_0^2 e^{-\kappa H} \qquad (13.14)$$

Foams are elastic systems because bubbles can be compressed and it may be important to include this effect in a full theoretical DLVO description of foams.

13.8   How to avoid and destroy foams

Foaming is not always desired. There are many processes where foaming can be a major problem. Excessive foaming can lead to production stop or malfunction and loss of valuable materials. Some of the industries that can be affected are: pulp and paper, pharmaceutical production, biochemical engineering (fermentation), wet flue gas desulfurization plants, production of detergent, coatings, certain foods (and beverages), distillation and absorption columns, and crosslinked foams (e.g. polyurethanes).

Antifoamers are usually added to the aqueous phase, prior to foam formation, and prevent or inhibit foam formation from within the aqueous phase. Defoamers or foam breakers are added to eliminate an existing foam and usually act on the outer surface of the foam (a foam is a closed system and the defoamer can only reach the outer surfaces). Frequently, the separation is confusing but often the mechanisms are different; for example, alcohols such as octanol are effective defoamers, but ineffective as antifoamers (Pugh, 1996). There are many commercial formulations available and it is quite a challenge to find the optimal ones to use in a given situation. Some are recommended for special purposes, but usually some practical experience is needed. In addition, we must consider whether the defoamer is compatible with the product or process (environmental issues, cost, availability etc.). Defoamers will often end up in the product being produced and this can sometimes be a problem if the product is an intermediate product to be used for further processing.

13.8.1 Mechanisms of antifoaming/defoaming

The mechanisms of antifoaming and defoaming are not so well understood. However, these compounds interfere in one or more ways with the stabilization methods discussed earlier. Potential mechanisms are to increase surface tension or to decrease elasticity, bulk- or surface viscosity, or electrostatic repulsion in the lamellae. Defoamers are formulations (multicomponent products) and typically contain various oils (e.g. silicone oils) and hydrophobic particles making the possible interactions quite versatile. Some of the basic requirements of a defoamer are: limited solubility in aqueous phase so that it goes to the interface, surface tension should be below the value of foam liquid to enhance entry and spreading at air–liquid interphase, and low interfacial tension with foaming liquid to enhance spreading at air–liquid interface. Formation and use of oil lenses is an often used route to foam destruction.

The oil droplets reduce the surface elasticity whereby rupture of the liquid film is enhanced. Quantification of the defoaming action is performed via entry and spreading coefficients (abbreviated E and S, respectively). When an oil droplet of low surface tension is placed in the interface, the foam liquid, due to a higher surface tension, tends to drag liquid away from the oil droplet whereby the film thins and rupture. A crude way of quantifying this is via the entry coefficient (E) and spreading coefficient (S) both of which must be larger than 0:

$$E = \gamma_{L/A} + \gamma_{L/D} - \gamma_{D/A} \qquad (13.15)$$

$$S = \gamma_{L/A} - \gamma_{L/D} - \gamma_{D/A} \qquad (13.16)$$

where $\gamma_{L/A}$, $\gamma_{D/A}$ and $\gamma_{L/D}$ refer to the interfacial tension of the aqueous phase, the oil (defoamer) phase and interfacial tension of the oil–water interphase, respectively (A = air).

The entry and spreading coefficients basically refer to changes in free energy. However, although these two equations are useful, in that they give a rough guide to antifoaming action, they are often inadequate. This is because they do not consider the geometry of the system or the influence of interfacial interactions within the film. Hence, their application is rather limited according to Pugh (1996), especially because of the potential presence of impurities which may alter the surface tension values needed for using the equations. However, for an initial analysis these equations can be quite useful.

Partially hydrophobic dispersed particles can cause an increase or decrease in foam stability. Provided they are not fully wetted then the small particles may become attached to the interface and give some mechanical stability to the lamella. On one hand, if they are completely dispersed they may cause an increase in the bulk viscosity. On the other hand, larger particles having a higher degree of hydrophobicity (coal dust, sulfur, non-wetting quartz) usually exhibit a finite contact angle when adhering to an aqueous interface, and may cause destabilization of froths (Pugh, 1996). Particle size has a great influence on the foaming ability (Figure 13.16).

As stated by Pugh (1996), the synergistic antifoaming effect of mixtures of insoluble hydrophobic particles and hydrophobic oils (filled antifoams) when dispersed in aqueous medium has been well established in the patent literature from the early 1950s. These mixed type antifoamers are very effective at low concentrations (10–1000 ppm) and are

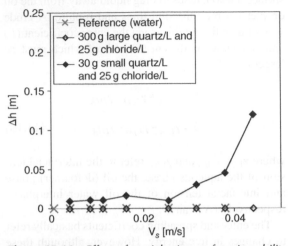

Figure 13.16 *Effect of particle size on foaming ability in an electrolyte solution of CaCl₂. Large, electrolyte-induced hydrophobic quartz particles can cause desta- bilization of froths, whereas small particles can stabilize them. In the figure, "large" particles have an average size of 227 μm and "small" particles an average size of 6 μm; Δh refers to foam height in a Bikerman column. Adapted from Hansen et al. (2008), with permission from Elsevier*

widely used. The hydrophobic particles may be hydrophobized silica (silica that has hydrophobic groups (alkyl or PDMS) chemically bonded to the surface) or glass (and is often referred to as the activator) while the hydrocarbon oil of poly(dimethylsiloxane)s (PDMS) is referred to as the carrier. The effect appears to be quite general and even precipitates of polyvalent metal ions with long-chain alkyl phosphates and carboxylates may be combined with mineral oil of PDMS to produce a synergistic antifoam effect. The size of the particles is within the range 0.001–1 μm and the solid content of the mixture is about 1–20 wt%. One possible explanation of the synergistic effect, is that the spreading coefficient of the PDMS oil is modified by the addition of the hydrophobic particle. It has been suggested that oil–particle mixtures form com- posite entities where the particles can adhere to the oil–water surface. The particles alone give only weak antifoam action. However, the presence of the particle adhering to the oil–water interface may facilitate the movement of oil droplets into the air–water interface to form lenses leading to rupture of the oil–water–air films. This mechanism is similar to that of oil droplets forming mechanically unstable bridging lenses in foam films, where the configuration requires rupture of oil–water–air films by particles. However, the contact angle for rupture is less severe than required for symmetrical air–water–oil films. Ultrasonics can also be used for defoaming action.

Example 13.3.   *Compounds performing defoaming action in beer foam.*
Lipstick and/or remains of potato chips, released from a person during drinking, are known to have a strong destabilizing effect on beer foam.
    Explain, qualitatively, a potential mechanism behind this destabilization.

Solution
Lipstick and remains of potato chips can both destabilize a beer foam upon drinking. Both products contain oils (see Chapter 12) and particles (pigments or small crumbs and salt particles) covered with oil (potentially making them hydrophobic) and they can therefore work as a defoamer. As described earlier in this chapter, commercial defoamers and antifoamers contain oils with hydrophobic particles. The particle-containing oils can go to the air/liquid interphase and cause rupture of the foam lamella (oil lens). The entry and spreading coefficients for the oil should both be positive for effective defoaming. It is not known if the pigment particles in the lipstick emulsion or the crumbs in the chip oil around the mouth of a hungry person have any impor- tance for the defoaming action. This depends on the balance between hydrophobicity and hydrophilicity and the diameter of the particles, but it is certainly a possibility.

13.9   Rheology of foams

Foams have unique rheological properties. The bulk flow of foam is very different from that of either Newtonian (laminar or turbulent) fluid or "conventional" two-phase fluids. The presence of a significant yield stress is required to cause the flow (analogous to semisolids) and there is a strongly shear thinning behaviour. These properties derive largely from the unique microscopic structure of foams.

Finally, one more practical example, illustrating some of the principles of this chapter, is presented.

Example 13.4.   Foaming and defoaming in fermentors.

In fermentation reactors, microorganisms produce enzymes (and other metabolic products) on a large scale. The enzymes are sold to companies producing, for example, detergent powders. Now and then, severe foaming appears in the stirred fermenters, which may cause loss of enzymes upon mechanical removal of the foam.

During a particular foaming incident, a uniform foam with a lamella thickness of 100 µm (assumed here to be independent of time) is formed.

The density and viscosity of the fermentation medium are 1000 kg m^{-3} and 0.03 kg m^{-1} s^{-1}, respectively, and can be assumed constant.

1.   Suggest and discuss a potential mechanism for foam formation in fermenters.
2.   It can be assumed that 30% of the cross-sectional area of the foam is available for vertical draining. How much liquid can be drained vertically (in L m^{-2}) from the newly formed foam in 10 s by gravity?
3.   The foam formed turns out to be rather persistent. Describe in some detail the practical (short term) solution and an underlying mechanism you would use to solve (or at least reduce) the problem?

Solution
1.   The mechanism behind foaming in fermenters is most likely protein-induced foaming. Proteins contain both hydrophobic and hydrophilic groups and can behave as a surfactant. As described earlier in this chapter, when polymers, proteins in particular, are adsorbed at the gas–liquid interface they denaturate. The relatively dense, somewhat structured adsorbed polymer layer imparts a significant degree of rigidity or mechanical strength to the lamellar walls, producing an increase in the stability of the foam. When the protein is in the bulk solution, hydrophobic parts are "hidden" inside the molecule to get away from water. At the interface, the hydrophobic parts are attracted to the very hydrophobic air leading to unfolding of the protein (Figure 13.15). Microorganisms produce many proteins during their metabolism and this is the cause of foaming. The rigorous stirring and air/oxygen injection add gas to the fermentation liquid.
2.   The rate of vertical draining by gravity can be calculated from Equation 13.5:

$$v_{av} = \frac{g\rho_L \delta(t)^2}{8\eta_L} \tag{13.4.1}$$

It is stated in the assignment that the lamella thickness is assumed independent of time. This gives:

$$v_{av} = \frac{g\rho_L \delta^2}{8\eta_L} \tag{13.4.2}$$

Inserting numbers gives:

$$v_{av} = \frac{(9.81\ \text{ms}^{-2})(1000\ \text{kgm}^{-3})\left(100\ \mu\text{m}\dfrac{1\ \text{m}}{10^6\ \mu\text{m}}\right)^2}{8(0.03\ \text{kgm}^{-1}\text{s}^{-1})} = 4.1 \times 10^{-4}\ \text{ms}^{-1} \qquad (13.4.3)$$

The volumetric flow rate of liquid from vertical draining can be calculated from the draining rate:

$$Q = v_{av}Af_A \qquad (13.4.4)$$

where f_A is the cross-sectional area of the foam that is available for vertical draining (provided in assignment). The amount of liquid drained in a given amount of time, t, is given by:

$$q = Qt = v_{av}Af_At \qquad (13.4.5)$$

Upon rearrangement of Equation 13.4.5, we obtain:

$$\frac{q}{A} = v_{av}f_At \qquad (13.4.6)$$

Inserting numbers gives:

$$\frac{q}{A} = v_{av}f_At = (4.1 \times 10^{-4}\ \text{ms}^{-1})0.30(10\ \text{s}) = 0.00123\ \text{m}^3\,\text{m}^{-2} = 1.23\ \text{Lm}^{-2} \qquad (13.4.7)$$

This is a fairly rapid drainage rate. It will slow down dramatically as the lamella thickness is reduced, but this complication is not considered in this assignment.

3. The way to deal with excessive foaming is to use a defoamer. The defoamer is used to break a foam. The mechanisms of defoaming are not so well understood. However, these compounds interfere in one or more ways with the stabilization methods discussed earlier. Potential mechanisms are to increase surface tension or to decrease elasticity, bulk- or surface viscosity, or electrostatic repulsion in the lamellae. Defoamers are formulations (multicomponent products) and typically contain various oils (e.g. silicone oils) and hydrophobic particles making the possible interactions quite versatile. Some of the basic requirements of a defoamer are: limited solubility in aqueous phase so that it goes to the interface, surface tension should be below the value of foam liquid to enhance entry and spreading at air–liquid interphase, and low interfacial tension with foaming liquid to enhance spreading at air–liquid interface. Formation and use of oil lenses is an often used route to foam destruction. The oil droplets reduce the surface elasticity whereby rupture of the liquid film is more likely.

13.10   Concluding remarks

Foams and froths are industrially important in many products and processes. For practical purposes, design of a given foam structure requires laboratory experiments. There are several destabilization mechanisms, which may need to be considered depending on the foam type, application and expected lifetime: gas diffusion, drainage and coalescence. Each of these processes can be strongly coupled, which complicates the analysis. Foaming is a problem in many industrial processes, where pumps and measurements can be seriously affected. Therefore, many commercial antifoam and defoamer formulations are available. The behaviour of foams can be measured and qualitatively described, but reliable mathematical models of the full foam behaviour are not yet available.

Problems

Problem 13.1: Characterization of foams
Two foams contain 50 and 95 vol.% of air, respectively.

1. Calculate the foam number, F, for the two foams.
2. Explain the difference between a spherical foam and a polyhedral foam.
3. Calculate the reduction in surface area when two equal-size bubbles coalesce to one single bubble. Comment on the result.

Problem 13.2: Pressure inside foam bobbles
A foam consists of bubbles with diameters of 2 and 4 mm and a liquid/air surface tension of 40 mN m^{-1}.

1. Calculate the pressure difference between the bubbles and the Plateau borders for (a) a foam consisting of only 2 mm bubbles and (b) a foam consisting of only 4 mm bubbles.
2. Calculate the pressure difference between bubbles of different size.

Problem 13.3: Rupture of foam lamellae
A surfactant foam has a liquid/air surface tension of 25 mN m^{-1}. Bubbles in the foam can be assumed to be spherical and monodisperse with a diameter of 20 μm. The Hamaker constant for the air/liquid/air interface can be taken to 6×10^{-20} J.

Calculate the critical film rupture thickness. Comment on the result.

Problem 13.4: Draining of foams
A newly formed foam with a liquid/air surface tension of 40 mN m^{-1} and bubble diameter of 2 mm is exposed to draining by gravity. It can be assumed that the draining corresponds to that of a vertical water film of uniform thickness. Density and viscosity of the liquid phase can be taken to 1000 kg m^{-3} and 10^{-3} kg m^{-1} s^{-1}, respectively.

1. What is the average rate of vertical draining when the film is 100 and 10 nm thick, respectively?
2. How can the rate of draining by gravity be reduced?

   The foam is also exposed to horizontal draining.
3. What is the rate of horizontal draining for a film of thickness 110 nm and a radius, R_f, equal to 0.1 mm?
4. How can the rate of draining by capillary forces be reduced?

Problem 13.5: Destabilization of foams
1. Explain why a soap bubble formed in air, after a short while, collapses?
2. What are the ideal weather conditions for blowing soap bubbles?
3. List potential destabilization mechanisms for soap bubbles when bubbles exist in a foam and as individual bubbles in air.

Problem 13.6: Elasticity of foam lamellae
The following data are available for a foam made by use of the flotation frother (surfactant) methyl isobutyl carbinol: $c = 20$ mol m^{-3}, $\Gamma_m = 5 \times 10^{-6}$ mol m^{-2}, $K_L = 230$ M^{-1}, $\delta = 20$ nm. The temperature is 20 °C. The Langmuir–Szyszkowski equation is valid for adsorption of the surfactant:

$$\gamma = \gamma_0 - R_{ig}T\Gamma_m \ln(1 + K_L c)$$

1. Calculate the surface elasticity of the foam.
2. Calculate the equilibrium adsorption density of the surfactant.

Problem 13.7: Pressure from van der Waals force
1. The potential energy from a van der Waals' interaction is given by:

$$V_A(\text{energy per unit area}) = -A_{121}/12\pi d^2$$

where A_{121}, the effective Hamaker constant for the interaction of two slabs of air across a thin film of water, can be taken to be approximately 6×10^{-20} J; d is the inter-slab distance. Show that the pressure generated by the van der Waals' force is:

$$P_A = -A_{121}/6\pi d^3$$

and calculate its value when d is 10 nm. A negative pressure implies attraction.

Problem 13.8: Defoaming in waste water plant

A company is struggling with excessive foaming in an industrial waste water plant. It is suggested to solve the problem by using an antifoamer (which in this case is also a defoamer). The following surface tensions are available for the foam liquid (L) and defoamer/antifoamer (D): $\gamma_{L/A} = 40$ mN m^{-1}, $\gamma_{L/D} = 10$ mN m^{-1}, $\gamma_{D/A} = 25$ mN m^{-1}. A = air.

1. In quantitative terms, what must be fulfilled for an antifoamer to work?
2. Calculate the entry and spreading coefficients for the antifoamer chosen. Comment on the results.
3. How reliable are the estimations based on entry and spreading coefficients?

Problem 13.9: Beer foam

1. Discuss, in qualitative terms, which properties are required of a surfactant to obtain good foaming ability and stabilization.

2. What is the stabilization mechanism in beer?
3. What could be the destabilization mechanisms in beer?
4. Can gas diffusion be reduced by using a N_2/CO_2 mixture instead of pure CO_2 for foaming in the beer? Explain why or why not.

References

J.C. Berg, An Introduction to Interfaces and Colloids – the Bridge to Nanoscience, World Scientific, 2010, New Jersey.

B.B. Hansen, S. Kiil, J.E. Johnsson, (2008), Ind. Eng. Chem. Res. 47(9), 3239–3246.

D. Myers, Surfaces, Interfaces, and Colloids – Principles and Applications, 2nd edn, chapter 12 on foams, Wiley-VCH Verlag GmbH, Weinheim, 1999.

R.M. Pashley, M.E. Karaman, Applied Colloid and Surface Chemistry, John Wiley & Sons Ltd, Chichester, 2004.

R.J. Pugh, Adv. Colloid Interface Sci., 64 (1996) 67–142.

S. Qin, B.B. Hansen, S. Kiil, (2013) Foaming in wet flue gas desulphurisation plants: Lab-scale investigation of long-term performance of antifoaming agents, *AIChE. J.*, 59 (10), 3741–3747.

L. Wang, R. Yoon, Langmuir, 20, (2004), 11457–11464.

L. Wang, R. Yoon, Int. J. Miner, Process, 85, (2008), 101–110.

L. Wang, R. Yoon, Langmuir, 25, (2009), 294–297.

14

Multicomponent Adsorption

14.1 Introduction

Chapter 7 focused on binary adsorption, but multi-component adsorption equilibria, e.g. a mixture of gases adsorbing on the same solid, is extremely important in many natural and engineering applications, such as the design of heterogeneous chemical reactors and certain separations for removing low-concentration impurities and pollutants from fluid streams as well as for chromatography. Adsorption is thus important to both reaction engineering and separation technology.

Compared to single gas adsorption, there are relatively few high quality data for multicomponent adsorption, which is possibly one of the reasons why it is difficult to validate the capabilities of existing theories for multicomponent adsorption equilibria. An additional complication is that solids are often so heterogeneous, e.g. data reported on activated carbon or silica by different authors may actually refer to different solids.

The purpose of most theories is to predict multicomponent adsorption based on single (gas) adsorption isotherm data.

Engineering theories for multicomponent adsorption can be roughly divided into three categories: extensions of the Langmuir equation, the thermodynamic approach (ideal and real adsorbed solution theories, IAST and RAST) by Myers and Prausnitz (1965) and finally the potential adsorption theory, especially as extended to multicomponent systems by Shapiro and co-workers (Shapiro and Stenby, 1998; Monsalvo and Shapiro, 2007a, 2009a,b).

Moreover, density functional theory (DFT) has been extensively used for multicomponent adsorption, in combination with equations of state like SAFT see, for example, Shen *et al.* (2013a,b, 2014) for recent related work. Molecular simulation studies for adsorption have also been reported. DFT and molecular simulation will not be covered in this chapter.

Introduction to Applied Colloid and Surface Chemistry, First Edition. Georgios M. Kontogeorgis and Søren Kiil.
© 2016 John Wiley & Sons, Ltd. Published 2016 by John Wiley & Sons, Ltd.
Companion website: www.wiley.com/go/kontogeorgis/colloid

14.2   Langmuir theory for multicomponent adsorption

Of the three theories we will discuss, the Langmuir multicomponent expression is a more or less direct extension of a theory we have previously seen, i.e. of the Langmuir equation for binary systems that was presented in Chapter 7.

Figure 14.1 gives a simplified idea of the physical picture of the Langmuir theory, which we have also presented previously. The Langmuir equation is a, conceptually and computationally, simple model based on a (quasi)chemical picture of adsorption and a large number of rather drastic assumptions:

- "ideal" (clean, homogeneous, smooth, non-porous) surface;
- the surface consists of a fixed amount of localized sites;
- each site may either be free or occupied by an adsorbed molecule;
- there is (almost) no lateral interactions between adsorbed molecules;
- same size of solute and solvent;
- maximum one (mono)layer.

There have been studies of the impact of heterogeneity and improvements have been proposed but they lead to cumbersome equations. Consequently, the Langmuir equation is typically used in the simple form shown here. For a more detailed discussion of the various extensions of Langmuir equation, see the review by Shapiro and Stenby (2002).

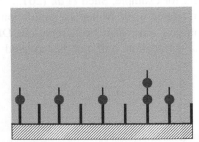

Figure 14.1   *Physical picture of the Langmuir theory. Reproduced with permission from A. Shapiro, 2015*

Despite these drastic assumptions, the Langmuir equation is often applied in the areas where assumptions behind it do not hold (e.g. adsorption of asphaltenes, surfactants or polymers).

The popularity of the Langmuir equation is explained by the fact that it is the simplest dependence predicting linear (Henry-type) behaviour at low partial pressures and the appearance of saturation of the surface at high pressures, which is the expected physical behaviour for most of the systems.

For multicomponent systems, the Langmuir equation is written for the ith component as (using either adsorbed volumes or adsorptions in general):

$$\frac{V_i}{V_{m,i}} = \frac{B_i P_i}{1 + \sum_i B_i P_i} \tag{14.1}$$

$$\frac{\Gamma_i}{\Gamma_{m,i}} = \frac{B_i P_i}{1 + \sum_i B_i P_i} \tag{14.2}$$

The summation is over all gases (or, in general, molecules) adsorbing on a solid. The term P_i is the partial pressure and B_i is the affinity coefficient for component i.

For a single component (gas) adsorption ($i = 1$), Equations 14.1 and 14.2 result to our well-known Langmuir equation from Chapter 7 (Equation 7.1).

Both Equations 14.1 and 14.2 are valid when we have ideal gas mixtures, i.e. obeying the ideal gas law. In this case, the partial pressure is given as:

$$P_i = y_i P \tag{14.3}$$

If the gas is non-ideal, e.g. at very high pressures, the (partial) pressure is replaced with the (pure compound) fugacity, but this is rarely needed unless the pressures are very high.

For every gas component, the affinity coefficient is given as:

$$B_i = B_{0,i} e^{\frac{E_{0,i}}{R_{ig} T}} \tag{14.4}$$

While this equation is obtained based on Langmuir's thermodynamic derivation, it has little value in practice. It may be used for prediction of the thermal behaviour of adsorption isotherms, if the temperature range is not very large.

However, the two parameters of the Langmuir equation should be, in principle, obtained from single gas adsorption data.

The adsorbed amount is related to adsorption as:

$$x_i = \frac{\Gamma_i}{\Gamma} = \frac{\Gamma_i}{\sum_i \Gamma_i} \tag{14.5}$$

The following example shows how the Langmuir equation is used and how it can be simplified for a binary gas system.

Example 14.1. *Langmuir equation for a binary gas mixture.*
Consider an ideal binary gas mixture consisting of gases (1) and (2) which adsorb on the same solid. Assume that the adsorption can be described by the Langmuir multicomponent adsorption equation. Let x indicate the adsorbed mole fractions and y indicate the gas mole fractions.

Show that the mole fractions of the adsorbed and gas phase are linked via the following equation:

$$x_1 = \frac{y_1}{y_1 + a(1 - y_1)} \tag{14.1.1}$$

Express the value of a in terms of the parameters of the multicomponent Langmuir equation.

Solution
From the general expression for Langmuir, Equation 14.2, we write the equations for the adsorption for gas (1) and gas (2) on the solid:

$$\Gamma_1 = \frac{\Gamma_{m,1} B_1 P_1}{1 + B_1 P_1 + B_2 P_2} \tag{14.1.2}$$

$$\Gamma_2 = \frac{\Gamma_{m,2} B_2 P_2}{1 + B_1 P_1 + B_2 P_2} \tag{14.1.3}$$

Since we have ideal solutions these expressions can be written as:

$$\Gamma_1 = \frac{\Gamma_{m,1} B_1 P_1}{1 + B_1 P_1 + B_2 P_2} = \frac{\Gamma_{m,1} B_1 y_1 P}{1 + B_1 y_1 P + B_2 y_2 P} = \frac{\Gamma_{m,1} B_1 y_1 P}{1 + B_1 y_1 P + B_2 (1 - y_1) P} \tag{14.1.4}$$

$$\Gamma_2 = \frac{\Gamma_{m,2} B_2 P_2}{1 + B_1 P_1 + B_2 P_2} = \frac{\Gamma_{m,2} B_2 y_2 P}{1 + B_1 y_1 P + B_2 y_2 P} = \frac{\Gamma_{m,2} B_2 (1 - y_1) P}{1 + B_1 y_1 P + B_2 (1 - y_1) P} \tag{14.1.5}$$

The adsorbed amounts for the two gases of a binary system are given as, based on Equation 14.5:

$$x_1 = \frac{\Gamma_1}{\Gamma} = \frac{\Gamma_1}{\Gamma_1 + \Gamma_2} \tag{14.1.6}$$

$$x_2 = \frac{\Gamma_2}{\Gamma} = \frac{\Gamma_2}{\Gamma_1 + \Gamma_2} \qquad (14.1.7)$$

Combining Equations 14.1.4–14.1.7 we get:

$$x_1 = \frac{\Gamma_{m,1} B_1 y_1 P}{\Gamma_{m,1} B_1 y_1 P + \Gamma_{m,2} B_2 P - \Gamma_{m,2} B_2 y_1 P} \qquad (14.1.8)$$

The pressure is eliminated and by simple division we obtain the given equation:

$$x_1 = \frac{y_1}{y_1 + a(1 - y_1)}$$

where:

$$a = \frac{\Gamma_{m,2} B_2}{\Gamma_{m,1} B_1}$$

The multicomponent Langmuir equation has been successfully applied, e.g. for predicting the adsorption of liquid hydrocarbons in molecular sieves 5A, dry natural gas on activated carbon and simple gases on silica gel (Bartholdy, 2012; Bartholdy *et al.*, 2013). Its performance, also in comparison to the other theories, is discussed in Section 14.5.

Figure 14.2　*Physical picture of the thermodynamic adsorption theory. Reproduced with permission from A. Shapiro, 2015*

14.3　Thermodynamic (ideal and real) adsorbed solution theories (IAST and RAST)

The thermodynamic theory for adsorption has possibly been known in some form since Gibbs time but it really came into use for practical applications when Myers and Prausnitz (1965) proposed IAST, the ideal adsorbed solution theory.

In the thermodynamic adsorption theory (see Figure 14.2), the adsorbed phase is a "black box", i.e. a homogeneous phase obeying common thermodynamic relations. An equation of state for this phase is obtained by analogy with a bulk phase or from experimental data.

Thus, in the adsorbed solution theory we follow a procedure known from thermodynamics in direct analogy to the thermodynamic equations for vapour–liquid equilibria which (at low pressures) are:

$$y_i \hat{\varphi}_i^V P = x_i \gamma_i P_i^{sat} \qquad (14.6)$$

$$y_i P = x_i P_i^{sat} \qquad (14.7)$$

where y and x are the mole fractions in the vapour and liquid phase, P and P^{sat} are the pressure and vapour pressures of each compound and the remaining "correction" parameters in Equation 14.6 (fugacity and activity coefficients) are only used when we have non-ideal vapour and liquid phases, respectively. If both are ideal, then Equation 14.7, the so-called Dalton–Raoult's law, is used.

In the thermodynamic adsorbed solution theory, we assume that there is thermodynamic equilibrium between the adsorbent (solid, a) and the adsorbate (gas) bulk phase:

$$\mu_i = \mu_i^a, or \, Py_i\varphi_i = P_0^i(\pi)\gamma_i x_i \qquad (14.8)$$

In Equation 14.8 we include fugacity and activity coefficients in analogy to Equation 14.6 but in the IAST both are ignored. In IAST we assume that the gas phase is ideal (fugacity coefficients equal to unity) and that the activity coefficient of the gas adsorbate in the solid is also one. Thus, Equation 14.8 solved for the mole fraction of the gas in the solid (which is the property we wish to calculate) is:

$$x_i = \frac{y_i P}{P_0^i(\pi)} \qquad (14.9)$$

Thus, in IAST, the adsorbed pseudo-phase is assumed to be an ideal solution in equilibrium with the bulk phase.

Both P and y have the usual meanings (pressure and gas concentration in the vapour phase), but the last term requires some attention.

The term $P_0^i(\pi)$ is the pressure of pure gas i at the same spreading pressure as the mixture. It must be determined in order to calculate the mole fraction in the solid (adsorption).

Thus, the problem of the adsorption equilibria is, with known values of T, P and y, to estimate x via Equation 14.9.

We need the reference pressures of the gas adsorbate. Here we will use again the Gibbs adsorption equation but written in terms of the spreading pressure at constant temperature as (Appendix 14.1):

$$d\pi = \sum_i \Gamma_i d\mu_i \qquad (14.10a)$$

In the summation we have the surface excesses (molar amounts per unit surface) multiplied by the chemical potentials (equal for bulk and adsorbed phase).

Based on Equation 14.10a, the reference pressures of the gas adsorbate are found from the Gibbs single component adsorption isotherm written in terms of the spreading pressure as:

$$\pi = R_{ig}T \int_0^{P_0^i} \Gamma_0^i d\ln f_0^i(P) \qquad (14.10b)$$

The adsorption depends in principle on both P and T.

Equation 14.10b is valid for both the gas mixture and a pure compound.

For an ideal gas we can use pressures instead of fugacities; hence Equation 14.10b is written:

$$\pi = R_{ig}T \int_0^{P_0^i} \Gamma_0^i d\ln P \qquad (14.11a)$$

$$\pi = R_{ig}T \int_0^{P_0^i} \Gamma_0^i \frac{dP}{P} \qquad (14.11b)$$

The total adsorption is found from the ideal mixing condition, as well as from the adsorptions of the individual gases:

$$\frac{1}{\Gamma} = \sum_i \frac{x_i}{\Gamma_0^i} \qquad (14.12a)$$

$$x_i = \frac{\Gamma_i}{\Gamma} \qquad (14.12b)$$

Due to the form of Equations 14.10 or 14.11, usually an iterative procedure is needed to determine the dependence of the pure component pressure from the spreading pressure which is needed in Equation 14.9.

The pure component pressure is, thus, found by inverting the integral of the single-component adsorption isotherm, Equation 14.11. Thus, we need single gas adsorption data or some expressions for the single component adsorptions to implement in Equations 14.10 or 14.11. We usually employ some well-known adsorption theory, e.g. Langmuir, Freundlich or Toth, and then the integration is carried out numerically. The adsorption isotherms must be known over the pressure range from zero to the value that produces the spreading pressure of the mixed gas adsorbate. This is very demanding and not always possible!

Myers and Prausnitz (1965) mention that the pure component adsorption isotherms should be known very accurately at low surface coverage, because the integration for spreading pressure is sensitive to this section of the pure component adsorption isotherm.

The three-parameter Toth equation is considered particularly successful in representing adsorption data over extensive ranges (from zero to full mono-layer coverage). For using IAST it is thus necessary to have a very good fit of single-component adsorption isotherms for all the gases involved. This in turn requires experimental data at each temperature/gas considered.

In the very simple case where we can use the Henry's adsorption equation, then an analytical solution is obtained, as illustrated in Example 14.2. However, in the general case we will need an iterative (and sometimes numerical) procedure.

For example, let us assume the more realistic case where the adsorption isotherm can be represented by the Langmuir equation (Equation 14.2). Then, by simple integration of Equation 14.11b using (14.2) (for single gas) and solving with respect to the pure component pressure we get:

$$P_0^i = \frac{1}{B}\left[\exp\left(\frac{\pi}{R_{ig}T\Gamma_m}\right) - 1\right] \tag{14.13}$$

In this case the following iterative solution procedure can be used:

Since there are $n + 1$ degrees of freedom, both T and P as well as the gas-phase composition must be specified:

1. We assume a starting spreading pressure value, e.g. from Henry's law, Equation 14.2.3 (see Example 14.2 below) which can be written as:

$$\pi = P\sum_i y_i K_i \tag{14.14}$$

2. With this value of spreading pressure we calculate the pure compound pressure from Equation 14.13.
3. We calculate the adsorbed amount from the Langmuir equation.
4. We check whether the adsorbed amounts satisfy the material balance, $\sum_i x_i = 1$.
5. New value of spreading pressure and iteration with step i until convergence is obtained.

Finally, after convergence is obtained and we have arrived to the final value of the spreading pressure we calculate x from IAST (Equation 14.9) and the adsorbed amounts from Equation 14.12a,b.

The use of the Langmuir equation has made this computational scheme appear rather simple, because we were able to obtain a direct solution for the pure compound pressure (Equation 14.13).

However, most equations for the adsorption isotherm are more complex and an analytical solution like that shown in Equation 14.13 is not possible. Then, the calculation must be performed numerically which increases the computational effort but the procedure is similar. Thus, in the general case, the system of n IAST equations is solved numerically with regard to n variables (spreading pressure and $n - 1$ mole fractions in the adsorbed phase). The best way to do this is to solve numerically the equation for the spreading pressure:

$$\sum_i x_i = 1 \Rightarrow \sum_i \frac{y_i P}{P_0^i(\pi)} = 1$$

Example 14.2. The IAST solution using the Henry's law equation for the single adsorption isotherms.
Assume that the adsorption can be represented by the simple Henry's law equation:

$$\Gamma = (\Gamma_m B)P \tag{14.2.1}$$

Prove that it is then possible to obtain an analytical solution for the adsorbed mole fraction:

$$x_i = \frac{y_i K_i}{\sum_i y_i K_i} \tag{14.2.2}$$

where K is a characteristic constant for each gas adsorbed.

Solution

The starting point is the equation of the spreading pressure as a function of adsorption which can be used both for the gas mixture and for each pure component (i), Equation 14.11b.
Applying this equation twice we have:

$$\pi = R_{ig}T \int_0^{P_0} \Gamma_m BP \frac{dP}{P} = (R_{ig}T\Gamma_m B) = KP \tag{14.2.3}$$

$$\pi = R_{ig}T \int_0^{P_0^i} \Gamma_{m,i} B_i P \frac{dP}{P} = (R_{ig}T\Gamma_{m,i} B_i) = K_i P_0^i \tag{14.2.4}$$

Equating these two equations we find:

$$KP = K_i P_0^i$$

This equation is then substituted to the IAST, Equation 14.9, to give:

$$y_i K_i = x_i K \tag{14.2.5}$$

Taking the sum of this equation we get:

$$K = \sum_i y_i K_i \tag{14.2.6}$$

Combining Equations 14.2.5 and 14.2.6 we obtain an analytical solution for the adsorbed mole fraction:

$$x_i = \frac{y_i K_i}{\sum_i y_i K_i} \tag{14.2.2}$$

(This is essentially the Langmuir equation.)

The adsorbed amounts are:

$$\frac{1}{\Gamma} = \sum_i \frac{x_i}{\Gamma_0^i}$$

$$\Gamma = \frac{1}{\sum_i \left(\frac{x_i}{\Gamma_0^i}\right)} \qquad (14.2.7)$$

Of course, this analytical solution has been obtained at the limit of zero pressure (very low surface coverage) where Henry's law is valid.

IAST is a predictive model for multicomponent adsorption, in the sense that all parameters are obtained from single gas adsorption data. Thus, for a multicomponent gas mixture adsorbing on the solid, IAST can predict the adsorbed concentrations. For a range of relatively simple mixtures containing hydrocarbons and CO_2 and also rather simple adsorbents (e.g. activated carbon) excellent results are obtained at various conditions (temperatures, various solids, etc.). Some illustrative results are shown in Figures 14.3–14.5 from the seminal work of Myers and Prausnitz (1965). In particular, in Figure 14.5 can be seen that the selectivity is also accurately represented by IAST. The selectivity is defined as:

$$S_{1,2} = \frac{y_1/x_1}{y_2/x_2} = \frac{P_1^0}{P_2^0} \qquad (14.15)$$

We can see that selectivity is much higher than one, which means that ethylene is more strongly adsorbed than CO_2 on activated carbon. On the same solid, ethane shows strong preferential adsorption over methane and this is also very well represented by IAST, as shown by Myers and Prausnitz (1965).

Smith, Van Ness and Abbott (2001) have reported that IAST predictions are usually satisfactory when the specific amount adsorbed is less than a third of the saturation value for monolayer coverage.

Despite these very promising results, they are by far not general and there are serious problems

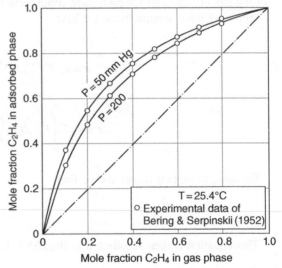

Figure 14.3 *Adsorption of ethylene–carbon dioxide mixture on activated carbon using IAST. Reprinted from Myers and Prausnitz (1965), with permission from John Wiley & Sons, Ltd*

reported by many authors, especially for polar gases and solids. For a review see Bartholdy (2012) and Bartholdy *et al.* (2013).

The origin of some of the problems is that, normally, negative deviations from the molar "volumes" (surfaces) predicted by the IAST are observed in the experiments (contrary to bulk mixtures). This has led many authors to propose an extension of IAST, called

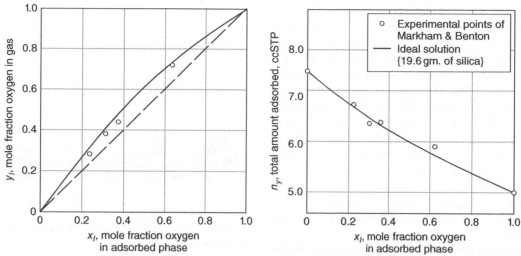

Figure 14.4 Adsorption of carbon monoxide–oxygen on silica using IAST (at 100 °C and 1 atm). Equally good results are obtained at different temperatures. Reprinted from Myers and Prausnitz (1965), with permission from John Wiley & Sons, Ltd

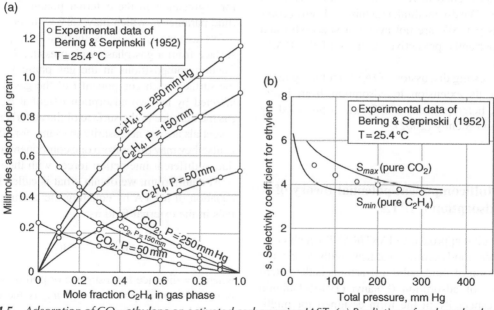

Figure 14.5 Adsorption of CO_2–ethylene on activated carbon using IAST. (a) Prediction of moles adsorbed and (b) selectivity coefficients against total pressure. Reprinted from Myers and Prausnitz (1965), with permission from John Wiley & Sons, Ltd

real adsorbed solution theory (RAST) using non-unity activity coefficients. However, RAST normally gives positive deviations of the molar volume. Activity coefficients from "bulk-phase" thermodynamics may not be appropriate for modelling the behaviour of adsorbed phases.

Smith, Van Ness and Abbott (2001) mentioned that, especially for high adsorbed amounts, appreciable negative deviations from ideality are observed due to differences in size of the adsorbate molecules and because of the adsorbent heterogeneity. This has led to RAST but the activity coefficients are often negative (less than unity), strong functions of both spreading pressure and temperature and sometimes asymmetric with respect to composition (Myers, 1983; Talu, Li and Myers, 1995). These results are in contrast to activity coefficients for liquid phases which are, in most cases, higher than unity (positive deviations) and insensitive to pressure.

RAST is much more complex computationally than IAST. It needs binary adsorption data for correlation of the parameters of the theory. Thus, it is not predictive.

One way to correct this situation is to consider the heterogeneous nature of the surface and asymmetries in gas/solid systems. Myers (2004) postulated that "reliable methods for the predictions of the activity coefficients needed in RAST may be discovered in the future". We do not think that this has been accomplished as yet. We are not aware of successful and most importantly predictive versions of the RAST approach.

An interesting discussion of IAST, including references for its extension to adsorption from liquid solutions, heterogeneous IAST etc., is provided by Shapiro and Stenby (2002).

14.4 Multicomponent potential theory of adsorption (MPTA)

An alternative approach to IAST/RAST which combines a thermodynamic treatment with a specific consideration of fluid–solid interactions is the potential theory of adsorption (Polanyi (1914)–Dubinin (1950)), especially in the form designed for multicomponent mixtures pioneered by Shapiro and Stenby (1998).

In PTA/MPTA (Figure 14.6) the adsorbed phase is considered to be an inhomogeneous phase. Actually, the surface of the adsorbent is supposed to be a source of a potential force field with the molecules (adsorbate) being attracted to the surface by this field.

Figure 14.6 *Physical picture of the potential adsorption theory. Reproduced with permission from A. Shapiro, 2015*

The distribution of the fluid (gas, liquid) near the solid surface is given by the equilibrium conditions for segregation in the external potential field, with fluid parameters which may vary with the distance.

We assume we have isothermal adsorption on a surface from a gas phase with pressure P and mole fraction compositions in the gas phase y. Close to the surface, each component i of the gas mixture is affected by its own adsorption potential $\varepsilon(z)$, where z is the distance to the surface of the adsorbent. These potentials are not necessarily the same for all components; they may differ from one component to another due to different interaction forces with the surface.

At equilibrium, we obtain what is called the basic equation of MPTA by equating the chemical potentials in the external field as:

$$\mu_i(z) - \varepsilon_i(z) = \mu_{gi} \qquad (14.16)$$

where z = distance from surface or porous volume, ε is the adsorption potential and μ_{gi} is the chemical potential in the gas bulk phase.

Equation 14.16 will be written for all components adsorbing. In the specific case of a binary adsorbing mixture, at a given temperature the systems of chemical potentials are given as:

$$\mu_1(z) - \varepsilon_1(z) = \mu_{g1} \qquad (14.17)$$

$$\mu_2(z)-\varepsilon_2(z)=\mu_{g2}$$

where compound (1) is affected by ε_1 and compound (2) is affected by ε_2.

Thus, given P,y in the gas phase, the distributions of all the parameters (pressure and mole fractions) close to the surface are determined from Equation 14.16.

Then, the surface excesses (adsorptions) are found from the distributed profiles of the corresponding extensive variables:

$$\Gamma_i=\int_0^\infty \left(y_i(z)\rho(z)-y_g\rho_g\right)dz \qquad (14.18)$$

i.e. the difference between the actual amount and the amount in absence of adsorption forces. The mole fraction in the adsorbed phase is calculated from Equation 14.5.

The central idea of MPTA is that the chemical potential of a component in the mixture results from two contributions: a bulk-phase contribution, which is represented by a thermodynamic model (equation of state), and a potential energy contribution that depends on the distance between the gas component and the solid adsorbent (fluid–solid interactions).

Thus, to be able to use the MPTA model in practice we need models to describe both fluid–fluid and fluid–solid interactions which are involved in the chemical potential in the adsorbed phase. For the fluid–fluid interactions, a thermodynamic model is used (typically in the form of equations of state like cubic equations of state, e.g. SRK).

For the fluid–solid interactions we need some expression for the potential that reflects the distribution of the components in the adsorbed phase. There are several such potential equations and they typically contain two or three adjustable parameters. In most applications, the adjustable parameters from these potential functions are the only adjustable parameters in MPTA, i.e. there are no additional adjustable parameters from the fluid–fluid interactions (thermodynamic part).

One of the best known potential expressions is the semi-empirical Dubinin/Radushkevich–Astakhov (DA) potential (Shapiro and Stenby, 1998):

$$z=z_0\exp[-(\varepsilon/\varepsilon_0)^n] \qquad (14.19)$$

where the model's two parameters are:

z_0 = parameter related to the porous volume or adsorption capacity;
ε_0 = parameter related to solid–fluid interactions

These two parameters are fitted to experimental adsorption data and can be considered roughly independent of temperature for most applications.

The exponent n is typically set to ($n = 2$) for activated carbons while $n = 3$ for molecular sieves and silica gels. There are, however, exceptions. It is sometimes considered that n is a measure of the heterogeneity of the pores.

If $n = 2$, the DA potential results to the well-known Dubinin–Radushkevich (DR) potential:

$$\varepsilon(z)=\varepsilon_0\sqrt{\log(z_0/z)} \qquad (14.20)$$

DR (and DA) are especially useful for the description of adsorption in micropores, where $z(\varepsilon)$ is interpreted as distribution of the porous volume by the values of the potential. For adsorption of the surface, the Steele or similar potentials are a better choice.

There are many more potential functions, e.g. the so-called Steele potential where the solid is modelled as having slit-like pores ("porous medium as an effective pore").

The choice of the potential may play a very important role in the results, as discussed later.

MPTA is a highly successful model which, in a very effective way, separates fluid–solid and fluid–fluid interactions. It can be used for the prediction of multicomponent adsorption from single-component adsorption isotherms. It has been used (Shapiro and Stenby, 1998; Monsalvo and Shapiro, 2007a,b; 2009a,b) both for the prediction of gas mixture adsorption on solids (including high pressures and supercritical conditions) and for the adsorption from liquid solutions, even for systems exhibiting strongly "non-Langmuir" behaviour. Two typical results, one for gas mixtures and one for liquid solutions are shown in Figures 14.7 and 14.8.

In summary, MPTA yields good results for the prediction of the adsorption of N_2, methane, ethylene and hydrocarbons on activated (and other) carbons as well as a few silica gel samples considered. Successful predictions at various temperatures were obtained as well for a few four- and five-component gas systems for which data are available on activated carbon.

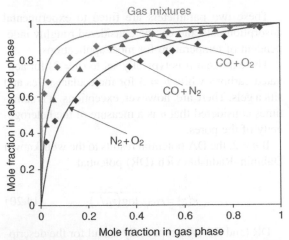

Figure 14.7 *Predictions with MPTA of the adsorption of a gas mixture on zeolites (Mol. Sieve) at T = 144 K and P = 1.01 bar. Reprinted from Monsalvo and Shapiro (2007a), with permission from Elsevier*

In most cases, very good prediction results are obtained using one fitting parameter per substance and one common parameter (the adsorbate capacity). However, there is a need to use different adsorption energies in the DR/DA potentials for the different components. In general, the choice of the potential used for the fluid–solid interactions is very important.

In these investigations very few slightly polar systems were considered and the results were not very successful. It has been stated (but has not as yet fully proved) that very good densities are needed which can be obtained if an advanced thermodynamic model, e.g. PC-SAFT, is used instead of SRK (which is the often used choice).

The result shown in Figure 14.8 requires some discussion. Unlike IAST (whose extension to liquid mixtures is rather complicated), MPTA can be readily extended to liquid mixtures. For liquids, pure-component adsorption does not exist and the binary adsorption isotherms should be correlated.

There are two distinct cases in adsorption of liquids; the so-called *U-shape*, where one of the components prevails in the adsorbed phase (and which is much simpler to correlate with a model) and the much more complex *S-shape* where each component prevails at its high concentration, but may be repelled at small concentration. Figure 14.8 illustrates the

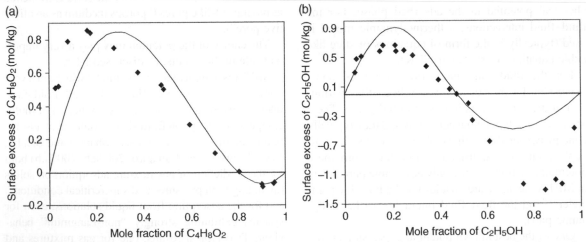

Figure 14.8 *Modelling of the adsorption isotherms (surface excess) of liquid solutions on activated carbon using MPTA. (a) Ethyl acetate + cyclohexane at T = 303.15 K; (b) ethanol + cyclohexane at T = 303.15 K. Reprinted from Monsalvo and Shapiro (2007b), with permission from Elsevier*

applicability of MPTA for two systems exhibiting S-shape curves.

As can be seen in Figure 14.8, rather good representation of the S-shape is obtained, especially for ethyl acetate–cyclohexane. The Steele potential is used here. It is of paramount importance to choose the right potential and for S-curves the Steele performs often much better than DA (Monsalvo and Shapiro, 2009a).

Shapiro and co-workers have shown that MPTA (with the SRK EoS and the Steele potential) performs very well for gas adsorption as well for various gases (hydrocarbons, CO_2, N_2) on various activated carbons and at various temperatures. It must be recalled, however, that the Steele potential has three adjustable parameters. Thus, it is unclear whether the improved performance is due to the better potential function or the additional – compared to the DA potential – adjustable parameter.

14.5 Discussion. Comparison of models

The application of Langmuir and IAST/RAST to adsorption from liquid solutions is limited, thus we concentrate our comparative evaluation on the adsorption of gas mixtures on solids.

14.5.1 IAST – literature studies

From the literature studies, IAST has been applied first of all by Myers and Prausnitz (1965) as mentioned previously with success to the adsorption of various hydrocarbon gases (as well as CO_2, O_2, N_2) on solids like activated carbon and silica gel. Equally good results for the same type of gases on both activated carbon and molecular sieves have been presented by Sievers and Mersmann (1994). Richter, Schotz and Myers (1989) applied IAST to methane–ethane on activated carbon using three different single gas isotherms and concluded that the quality of the predictions depend on the representation of the single component data and, thus, the selection of the single gas isotherm theory is an important parameter in the IAST model. Three-parameter theories like Toth usually perform better than Langmuir when

combined with IAST. Similar results (for the same types of gases and solid adsorbents) with IAST have been presented by other researchers as well (Bartholdy, 2012; Bartholdy *et al.*, 2013) verifying that IAST is a successful predictive model for these relatively simple systems.

Compared to the extensive results of IAST for nonpolar gases, relatively few studies have been presented for polar gases/solids. Li, Xiao and Webley (2009) studied the CO_2–water adsorption on activated alumina, while Perfetti and Wightman (1975) modelled with IAST the adsorption equilibrium of ethanol–benzene, ethanol–cyclohexane and benzene–cyclohexane onto Graphon. Finally Steffan and Akgerman (1998) considered the adsorption of diverse volatile organic compounds (hexane/benzene, hexane/trichloroethylene, chloroform/chlorobenzene) on silica gel. There is a common underlying conclusion from all of these studies and it is that IAST fails for water, ethanol and even the least polar one of the systems studied by Steffan–Akgerman. Thus, IAST does not appear to be a reliable model for polar systems. RAST is an alternative approach but in this way the predictive capability of the approach for modelling multicomponent adsorption is lost.

14.5.2 IAST versus Langmuir

In one of the few comparative studies in the literature, Chen, Ritter and Yang (1990) compared IAST to the multicomponent Langmuir equation for the adsorption of gases at elevated pressures (15–25 bar).

Due to the high pressures, Langmuir is used with fugacities determined from the virial equation of state. It was found that the predictions with the multicomponent Langmuir model were better than with IAST for two binary gas mixtures (H_2–CO and CO–CH_4) at various temperatures, but IAST proved to be superior when modelling the systems CO–CO_2 and CH_4–CO_2 and for all the ternary and quaternary systems. However, overall both models proved to adequately predict the mixed gas data and the predictions from the two models were very similar. From a mathematical and computational point of view, the explicit Langmuir model is simpler, while IAST needs an iterative solution method;

nevertheless with computers the difference is rather small.

Bartholdy (2012) and Bartholdy *et al.* (2013) presented both a review of applications of IAST to multicomponent adsorption and also carried out an extensive comparison of IAST and multicomponent Langmuir over a wide range of adsorbates and adsorbents, including polar and hydrogen bonding ones. These authors considered gas mixtures containing various alcohols (methanol, ethanol butanol, 2-propanol), acetone, water and several hydrocarbons in diverse adsorbents (activated carbon, silica and silica gel, molecular sieves and a zeolite). More than 200 different experimental data points were considered. The multicomponent Langmuir was used in the form of Equation 14.1 as using fugacities predicted by a thermodynamic model made little difference at the conditions of those systems. For IAST the single gas adsorptions were correlated using either the Langmuir or Toth isotherms. The latter always correlates the single adsorption isotherms better as it contains an additional adjustable parameter.

Table 14.1 presents an overview of the main results for those systems for which calculations were possible. None of the models could predict the adsorption of 1-butanol/xylene on the zeolite. None of the models were able either to predict the water–hydrocarbon adsorption on two different activated carbons for which data are available.

First of all, it should be mentioned that several experimental data for multicomponent adsorption show some scatter and inconsistencies which should be accounted for when evaluating the performance of the models.

As can be seen both from Table 14.1 and the detailed results presented by Bartholdy (2012) and Bartholdy *et al.* (2013), none of the approaches perform very satisfactorily in all cases and actually overall IAST and multicomponent Langmuir perform similarly, with IAST being overall a bit better. Even though the differences on percentage deviations are similar in some systems, IAST performs more robustly and actually the multicomponent Langmuir is completely erroneous for some systems, e.g. adsorption on molecular sieves.

Table 14.1 *Percentage absolute average deviations (% AAD) between predicted results and experimental data for the adsorption isotherms (x) and total amounts adsorbed (n) using the multicomponent Langmuir and IAST models. The first value is the % AAD in x and the second is the % AAD in* n

Adsorbates	Adsorbent	NS/NP	IAST-Langmuir	IAST-Toth	Multicomponent Langmuir
Methanol, acetone, benzene, hexane	MS-13X	4/25	11/21	13/16	51/38
Methanol, acetone, benzene, hexane	MS-carbon 5A	4/41	17/13	18/11	21/17
Methanol, acetone, benzene, hexane	AC G-2X	5/32	13/6	16/9	9/9
2-Propanol/water	Silica gel	2/10	10/23	—	19/49
Heptane/water	Silica gel	2/6	14/26	15/26	15/26
Ethanol, benzene, cyclohexane	Cab-o-sil (=293 K)	3/25	6/33	4/26	4/32
Ethanol, benzene, cyclohexane	Cab-o-sil (=303 K)	6/52	5/42	7/41	5/41
Total/average		26/191	11/23	12/22	18/30

NS = number of systems, NP = number of points. MS = molecular sieve, AC = activated carbon.
Data from Bartholdy (2012) and from Bartholdy et al. (2013).

On the other hand, both theories have problems for the most polar gases and solids, especially for water and silica gel systems. For both 2-propanol/water and heptane/water, even though the percentage deviations do not seem that high, especially with IAST, the results are deceptive because the trend of the experimental data with respect to selectivity or the amount adsorbed are not predicted well at all.

On a more detailed level, it is observed that, in agreement to previous studies, the models perform worse for the total amount adsorbed than for the adsorption isotherms (with the error almost doubled in the latter case). As expected, none of the models can predict well those cases with S-shape adsorption curves, e.g. acetone/hexane and methanol/hexane in MS-Carbon-5A.

Finally, somewhat in contradiction to other studies, using a "better" theory for correlating the single adsorption isotherms did not improve the predictions for the multicomponent systems. Actually, the results of IAST using either Langmuir or Toth are overall very similar. There are even some cases where IAST + Langmuir performs better than IAST + Toth. This is surprising since Toth correlates better the single gas adsorption isotherms. We can thus conclude that, despite its positive characteristics, there are several cases where calculations with IAST are successful due to cancellation of errors for multicomponent systems.

14.5.3   MPTA versus IAST versus Langmuir

Monsalvo and Shapiro (2007a) carried out an extensive comparison of MPTA (using two different equations of state for the fluid–fluid interactions) and IAST. They considered the adsorption of 40 binary gas systems (more than 600 points) on activated carbon, molecular sieves, silicates and silica gels as well as eight ternary gas systems with about 120 data points on the same type of solids. They reported results both in terms of adsorption equilibria (mole fractions, x) and, when possible, also for total adsorbed amounts (n). For the binary systems, the average % deviation with MPTA is 5 and 9 (over the whole database, respectively for x,n) and 5 and 7 for IAST (x,n). For the ternary systems, the deviations are higher but nevertheless similar with the two approaches (around 11–12 for both MPTA and IAST; here the differences in the model performance for adsorption equilibria

and total adsorbed amounts is smaller than for binary systems). The authors conclude that both approaches are able to predict binary and ternary gas adsorption with the right trends and good accuracy. There are problems in some cases, e.g. for mixtures exhibiting azeotrope-like behaviour like CO_2–H_2S and CO_2–C_3H_8. It is worth mentioning that (considering also that the Toth equation was used in the IAST case) MPTA employs fewer parameters in the correlation of the single-component adsorption data.

A very extensive investigation on MPTA has been carried out by Bjørner (2012) (see also Bjørner, Shapiro and Kontogeorgis, 2013) and an overview of these results (also compared to the results with IAST and multicomponent Langmuir by Bartholdy, 2012) is presented in Table 14.2. Bjørner used MPTA with two equations of state, but here we show results only for the MPTA+SRK combination. Moreover, to improve the results but also have a fairer (in terms of number of adjustable parameters) comparison to IAST, he considered both common and individual capacities for the adsorbents in the various systems.

As seen from Table 14.2, comparing also the results to those of the other models, overall MPTA performs as accurately as IAST + Langmuir and better than the multicomponent Langmuir. Bjørner's investigations showed that using more advanced equations of state than SRK does not always lead to better adsorption results, thus further investigation on the MPTA model is needed, especially for complex systems (adsorbates and adsorbents).

The similarity in the results between IAST and MPTA is not entirely surprising. As Shapiro and Stenby (1998) have shown, IAST is consistent with the potential theory if the bulk phase is an ideal gas mixture for all the pressures arising under the action of the potential field of the adsorbent. This may be the case for low-pressure adsorption and if the adsorption potentials are rather weak in the adsorbent volume.

14.6   Conclusions

Three theories for multicomponent adsorption have been presented, the extension of Langmuir's theory to multicomponent systems, the ideal and real adsorbed solution theories (IAST, RAST) and the multicomponent potential adsorption theory (MPTA).

Table 14.2 *Percentage absolute average deviations (% AAD) between predicted results and experimental data for the adsorption isotherms (x) and total amounts adsorbed (n) using MPTA with the SRK equation of state. The first value is the % AAD in x and the second is the % AAD in n*

Adsorbates	Adsorbent	NS/NP	MPTA + SRK – common	MPTA + SRK – individual
Methanol, acetone, benzene, hexane	MS-13X	4/25	21/19	22/18
Methanol, acetone, benzene, hexane	MS-carbon 5A	4/41	24/15	16/12
Methanol, acetone, benzene, hexane	AC G-2X	5/32	16/10	7/6
Ethanol, benzene, cyclohexane	Cab-o-sil (=293 K)	3/25	8/35	8/37
Ethanol, benzene, cyclohexane	Cab-o-sil (=303 K)	6/25	4/23	4/21
Total/average			15/20	11/19
Average[a]				
IAST–Langmuir			10/23	
IAST–Toth			12/21	
Multi. Langmuir			18/27	

NS = number of systems, NP = number of points; MS = molecular sieve, AC = activated carbon. "Common" indicates using a common capacity z; "Individual" indicates using an individual capacity z.
[a]Average calculated for the same systems tested with MPTA.
Data from Bjørner (2012), from Bjørner et al. (2013) and from Bartholdy (2012).

All theories have strengths and weaknesses. To our knowledge, there has not been any systematic comparative investigation – comparison of the performance of the various theories over a large number of experimental multicomponent adsorption data. No large databases or adsorption models parameter databases for any of these theories are available.

The advantage of Langmuir's approach for multicomponent systems is its simplicity and often good performance for relatively simple systems. Addition of new physical mechanisms to improve the model for relaxing some of its assumptions has resulted in cumbersome equations which have not found widespread use. An alternative way to use the Langmuir equation is, in its single adsorption form, in combination with the thermodynamic approaches of IAST and RAST.

IAST is a popular approach, especially because of its simplicity and the similarity with the bulk-phase thermodynamics. When IAST works well, for example for relatively simple gas mixtures, then we have an excellent predictive model for multicomponent adsorption. Single gas adsorption isotherms needed in IAST should be modelled with exceptional accuracy. However, in the general case, negative deviations are observed and activity coefficients are needed. In this way RAST can be used but the resulting model is no longer predictive for multicomponent adsorption and, thus, it is of rather limited applicability for engineering calculations.

IAST is from a fundamental point of view a rather narrow model. It attempts to describe adsorption looking only at its thermodynamic aspects, ignoring the molecular interactions in the adsorbed layer (solid).

MPTA has been developed as an alternative to the "Langmuir" and "thermodynamic" type approaches. MPTA with the concept of surface potentials provides more physical information about adsorption equilibria than the purely thermodynamic approach. Like IAST, MPTA is predictive for gas mixtures (as the parameters are obtained from pure gas adsorption data). MPTA and IAST give overall similar results for multicomponent gas adsorption (and are equally predictive). A complete evaluation of the models remains to be carried out over an extensive database.

A positive feature about MPTA is that it has been generalized for several complex cases like liquid solutions, supercritical/high pressure and non-Langmuir adsorption behaviour. In addition, MPTA may probably be considered as a partial step from "macroscopic", "phenomenological" adsorption theories towards the theories based on statistical mechanics and molecular dynamics, like, for example, density functional theory.

All theories have problems for some systems, especially for polar/hydrogen bonding adsorbates and complex solids.

Acknowledgments

The authors wish to thank Associate Professor Alexander Shapiro for many inspiring discussions on all aspects of multicomponent adsorption and for providing material for this chapter. The authors also thank Mr Martin Bjørner for his comments and for many discussions.

Appendix 14.1 Proof of Equations 14.10a,b

The starting point is Equation 4.27 in Chapter 4 (Appendix 4.3), by removing the superscript which indicates the surface phase:

$$\sum_i n_i d\mu_i + A d\gamma = 0 \qquad (14.21)$$

Since $\pi = \gamma_0 - \gamma \Rightarrow d\pi = -d\gamma$ (14.22)

From Equations 14.21 and 14.22 we get:

$$\sum_i n_i d\mu_i - A d\pi = 0 \Rightarrow d\pi = \sum_i \frac{n_i}{A} d\mu_i \qquad (14.23)$$

or using the adsorption definition:

$$d\pi = \sum_i \Gamma_i d\mu_i \qquad (14.24)$$

$$\pi = \int_0^{P_0^i} \Gamma_0^i d\mu_i(P, T) \qquad (14.25)$$

which in terms of fugacity is Equation 14.10b.

Problems

Problem 14.1: Derivation of the Langmuir equation from spreading pressure equations of state

In some cases the following two-dimension equation of state is valid:

$$\pi = \frac{k_B T}{A - b}$$

This model is the two-dimensional analogue of the repulsive term of the van der Waals equation of state.

1. Show that, in the general case, this model is equivalent to the following adsorption isotherm equation:

$$kC = \frac{\Theta}{1 - \Theta} \exp\left(\frac{\Theta}{1 - \Theta}\right)$$

2. Then show that at low surface coverage (much lower than unity), the Langmuir adsorption isotherm is recovered:

$$\Theta = \frac{\Gamma}{\Gamma_{max}} = \frac{kC}{1 + kC}$$

Problem 14.2: Langmuir and IAST models

The Langmuir and IAST models are not entirely different from each other despite their different origins and derivations. Assume a binary ideal gas mixture and using Equations 14.9, 14.13 and 14.1.1, prove that the Langmuir and IAST models lead to mathematically identical equations for the adsorbed mole fractions x under a simplifying assumption. What is the assumption required?

References

S.B. Bartholdy, 2012. Modeling of adsorption equilibrium on molecular sieves and silica gels by AST and Langmuir. M.Sc. Thesis. Technical University of Denmark.

S.B. Bartholdy, M.G. Bjørner, E. Solbraa, A.A. Shapiro, G.M. Kontogeorgis, 2013. Ind. Eng. Chem. Res., 52, 11552.

M.G. Bjørner, A.A. Shapiro, G.M. Kontogeorgis, 2013. Ind. Eng. Chem. Res., 52, 2672.

M.G. Bjørner, 2012. Modeling of adsorption on molecular sieves and silica gel using the Potential Adsorption Theory. M.Sc. Thesis. Technical University of Denmark.

Y.D. Chen, J.A. Ritter, R.T. Yang, Chem. Eng. Sci. 1990, 45, 2877.

G. Li, P. Xiao, P. Webley, 2009, Langmuir 25, 10666–10675.

M.M. Monsalvo, A.A. Shapiro, Fluid Phase Equilibria, 2007a, 261, 292.

M.M. Monsalvo, A.A. Shapiro, Fluid Phase Equilibria, 2007b, 254, 91.

M.M. Monsalvo, A.A. Shapiro, Fluid Phase Equilibria, 2009a, 283, 56.

M.M. Monsalvo, A.A. Shapiro, J. Colloid & Interface Sci., 2009b, 333, 310.

A.L. Myers, 2004. Thermodynamics of adsorption, in Chemical Thermodynamics for Industry, ed. T.M. Letcher, Royal Society of Chemistry: Cambridge, UK, pp. 243–253.

A.L. Myers and J. M. Prausnitz, AIChE J., 1965, 11, 121.

A.L. Myers, AIChE J., 1983, 29, 691.

G.A. Perfetti, J.P. Wightman, Carbon 1975, 13, 473.

E. Richter, W. Schotz, A.L. Myers, Chem. Eng. Sci., 1989, 44, 1609.

A.A. Shapiro and E.H. Stenby, J. Colloid & Interface Sci., 1998, 201, 146.

A.A. Shapiro and E.H. Stenby, 2002. Multicomponent Adsorption. Approaches to modeling adsorption equilibria in Encyclopedia of Surface and Colloid Science. Marcel Dekker, Inc., New York.

G. Shen, X. Ji, X. Lu, J. Chem. Physics, 2013a, 138, 224706.

G. Shen, X. Ji, S. Oberg, X. Lu, J. Chem. Physics, 2013b, 139, 194705.

G. Shen, X. Lu, X. Ji, Fluid Phase Equilibria, 2014, 382, 116.

W. Sievers, A. Mersmann, 1994. Characterization of Porous solids III, Studies in Surface Science and Catalysis, Vol. 87, Elsevier, Amsterdam.

J. M. Smith, H. Van Ness, M. Abbott, 2001. Introduction to Chemical Engineering Thermodynamics; McGraw-Hill Education, New York.

D.G. Steffan, A. Akgerman, Environ. Eng. Sci., 1998, 15, 191.

O. Talu, J. Li and A.L. Myers, 1995. Adsorption, 1: 103.

P. Xiao, P. Webley, 2009. Langmuir, 25, 10666.

15

Sixty Years with Theories for Interfacial Tension – *Quo Vadis?*

15.1 Introduction

For almost 60 years researchers have been actively developing theories for estimating the interfacial tension, for all types of interfaces, with the emphasis on solid–liquid and solid–solid interfaces of importance to wetting and adhesion phenomena. These theories have been presented in Chapter 3 for liquid–liquid interfaces and applied – in combination to Young's contact angle equation – to solid interfaces in Chapter 6. Some advantages and shortcomings have already been discussed in these chapters. But the topic of theories for interfacial tension has been an active research field full of colourful debates for more than 60 years and there are still many questions. Essentially both the oldest and the newest theories are still used today in practical applications and there are researchers that doubt the validity of all or most theories. The purpose of this chapter is to provide a historical and comparative presentation of the theories, with reference to the discussion in Chapters 3, 4 and 6 when necessary. It is the hope that this presentation will serve as a short guide to an immense literature which, together with the references, can be used as a starting point for further studies for interested readers.

15.2 Early theories

The adventures in interfacial theories possibly began with the work of Louis Girifalco and Robert Good in what is known today as the Girifalco–Good (1957) equation (GG). This very simple model relates the interfacial tension solely based upon the surface tensions of individual components and the assumption that the cross-interactions are given by the geometric mean rule, known from thermodynamics (it is strictly valid for non-polar mixtures). Good called these systems "regular surfaces", presumably in analogy with the "regular solutions" which obey the Hildebrand theory with the solubility parameters (see Chapter 3, Section 3.4.2).

Introduction to Applied Colloid and Surface Chemistry, First Edition. Georgios M. Kontogeorgis and Søren Kiil.
© 2016 John Wiley & Sons, Ltd. Published 2016 by John Wiley & Sons, Ltd.
Companion website: www.wiley.com/go/kontogeorgis/colloid

Without any correction this equation has the form:

$$\gamma_{12} = \left(\sqrt{\gamma_1} - \sqrt{\gamma_2}\right)^2 = \gamma_1 + \gamma_2 - 2\sqrt{\gamma_1 \gamma_2} \quad (15.1)$$

This model is of limited practical use, e.g. mixtures of some fluorocarbons with hydrocarbons. Thus, the now well-known GG "ϕ" parameter is often used, (see Equation 3.16). This correction is significant, of the order 0.5–1.15 for water–organic and mercury–organic liquid interfaces. Values in the range 0.51–0.78 are used for liquids not forming strong hydrogen bonds, while values above 0.84 are needed for hydrogen bonding liquids, and for non-metallic liquids with mercury the correction is of the order 0.55–0.75 (Girifalco and Good, 1957).

This equation is also quite successful for many metal–metal systems (see references in Good, 1977).

GG tried to estimate the correction "φ" parameter from molecular properties (dipole moments, polarizabilities and ionization potentials) but this has been only partially successful. Such predictive approaches yield, for example, for water–benzene a correction factor equal to 0.55 (while the experimental correction value is 0.74) and 0.96 for water–nitrobenzene (experimental value is 0.80) (Good, 1992). Thus, there are both under- and overestimations of the experimental correction parameter of the GG equation using these predictive approaches.

As discussed in Chapter 6, the use of the GG equation together with Young's equation yields (if the correction "φ" parameter is equal to one):

$$\cos \vartheta = -1 + 2 \left(\frac{\gamma_s}{\gamma_l}\right)^{1/2} \quad (15.2)$$

Is this equation really in disagreement with the Zisman plot? Equation 15.2 implies that the cosine of the contact angle is proportional to the inverse of the square root of the liquid surface tension instead of the liquid surface tension, indicated by the Zisman plot. The difference is only apparent. Good (1992) expands Equation 15.2 using a Taylor series expansion and he obtains:

$$\cos \vartheta = 1 - \left(\frac{\gamma_l - \gamma_s}{\gamma_s}\right) + \frac{3}{4} \left(\frac{\gamma_l - \gamma_s}{\gamma_s}\right)^2 - \dots \quad (15.3)$$

A limited extrapolation (use of only the first two terms) does indeed verify the Zisman plot and the use of the critical surface tension instead of the solid surface tension:

$$\cos \vartheta = 1 - \left(\frac{\gamma_l - \gamma_s}{\gamma_s}\right) = 2 - \frac{1}{\gamma_s} \gamma_l \Rightarrow \gamma_{crit} = \gamma_s \quad (15.4)$$

Good (1992) mentions that, for better extrapolations, the GG equation should be used instead of the Zisman plot.

Good (1977) shows that the GG plot (Equation 15.2) provides a better representation of data than Zisman plot and it can accommodate a wider range of experimental data, whereas the Zisman plot sometimes is valid for a limited range of fluids and surface tension values. Dann (1970a) also mentions that long extrapolations of the Zisman plot do not always yield straight lines. For example, as Good (1977) shows, for polystyrene (PS) Zisman results in two critical surface tensions (33 and 43 mN m^{-1}) depending on whether hydrogen bonding or non-hydrogen bonding liquids are used. This discrepancy was resolved with the GG plot, which yields a solid surface tension for PS equal to about 43 mN m^{-1}, which is in agreement with other studies. The "ϕ" for PS–water is 0.64, which is close to that of ethylbenzene–water (=0.69). This agreement further supports the hypothesis of the hydroxyl–π electron interaction, which we see in other systems as well (e.g. water with aromatic hydrocarbons).

The GG equation has significant interpretative value. Good (1980) has showed that, using the Zisman experimental data, the Teflon surface tension is calculated to be 24 mN m^{-1}. Using this value and Equation 15.2 we can calculate that cooking oil (with a surface tension 29 mN m^{-1}) has a contact angle of about 35° which is close to the experimental value. This implies that cooking oil will not form a continuous film on a clean Teflon-coated frying pan.

GG does appear in more recent literature as well, e.g. Brady (1999) applied it to relate the surface free energy to bioadhesion (adhesion of marine organisms to various surfaces).

A few years after the first presentation of Girifalco and Good, Frederick Fowkes (1962) presented his modification of the GG equation, where no correction is used but instead the cross term is calculated from the dispersive contribution of the surface tension (see also Section 3.5.2):

$$\gamma_{12} = \gamma_1 + \gamma_2 - 2\sqrt{\gamma_1^d \gamma_2^d} \qquad (15.5)$$

In reality, what Fowkes states is that while all intermolecular forces are important for the surface tension, only the dispersion part of the forces contributes to the free energy (or work) of adhesion.

The success of this equation was shown from the start (Fowkes, 1964) as it was illustrated that experimental liquid–liquid interfacial tension data for ten mercury–hydrocarbon mixtures and for eight water–hydrocarbon mixtures resulted to more or less unique values for the dispersion contribution of mercury and water, respectively (see Example 3.3). Moreover, these values are in agreement with those estimated from theoretical considerations.

In the Fowkes and many subsequent equations, it is assumed that the surface tension can be divided into "imaginary" contributions from the various intermolecular forces, e.g. dispersion and "remaining" specific forces in the Fowkes and subsequently in the Owens–Wendt equation (OW) (Equations 3.19 and 3.22a). These theories have since been known as "surface component" or simply "component theories". It is a very useful concept in the sense that, if these surface components could be calculated reliably, we could obtain information about the contribution of the various intermolecular forces on the

liquid and solid surfaces and interfacial tensions. However, the surface components cannot be estimated directly. They have to be obtained – for liquids – from experimental data, e.g. liquid–liquid interfacial tensions or data for solids (contact angle information, see below). Therefore, the ultimate success of these theories rely on whether they can fit a large amount of experimental data with the same surface component contributions and whether they can be used in some predictive way. The Fowkes equation, in particular, has received much attention and the original article is cited significantly (Table 15.1).

Despite its simplicity, many success stories have been presented. For example, as already mentioned, using liquid–liquid interfacial tension data for several different mercury–hydrocarbons and water–alkanes, we can estimate "almost" unique contributions for the dispersion surface tension of mercury (201 mN m^{-1}) and water (21.9 mN m^{-1}) (Hiemenz and Rajagopalan, 1997; Fowkes, 1964, 1980). These values appear to be reasonable and moreover when used to predict the interfacial tension of water–mercury, quite good agreement is obtained (see Example 3.3 and Problem 3.5). In addition to the success for liquid–liquid interfaces, the Fowkes equation has been used for solid interfaces with good results. As shown in Chapter 6, when combined with the Young equation, we obtain:

$$\cos\vartheta = -1 + \frac{2\sqrt{\gamma_s^d \gamma_l^d}}{\gamma_l} = -1 + 2\sqrt{\gamma_s^d}\frac{\sqrt{\gamma_l^d}}{\gamma_l} \qquad (15.6)$$

As we saw in Figure 6.2 this equation provides an alternative to Zisman's equation co-ordinate for

Table 15.1 *Citations of the most well-known theories for interfacial tensions. Only the citations to the "first" or most cited paper of each method are shown*

Method	2009	2010	2011	2012	2013	Total	Citations per year
Zisman (1950)	15	12	14	17	13	604	9.3
GG (1957)	21	28	25	36	34	871	15
Fowkes (1964)	53	48	58	55	56	1733	34
OW (1969)	235	234	211	274	273	2928	64
Neumann (1974)	8	13	10	10	7	399	10
van Oss–Good (1988)	40	38	64	60	54	964	36
Della Volpe–Siboni (1997)	13	10	11	11	10	189	11

OW = Owens and Wendt (1969); GG = Girifalco and Good (1957).
Based on data extracted from Web of Science, March 2014.

plotting contact angle against surface tensions. The agreement is good in many cases (Figure 6.2) and we can estimate the dispersive contribution to the surface tension of the solid. Reasonable values are obtained which are, moreover, in good agreement with values extracted by other methods, e.g. vapour adsorption data on high energy surfaces obtained via the equation:

$$\pi_{SV} = R_{ig}T \int_{0}^{P^{sat}} \Gamma \, d\ln P \qquad (15.7)$$

A characteristic example is polypropylene, which has been studied by adsorption measurements, resulting in a solid dispersion value of 26–29 mN m^{-1}, which is rather close to 23–35 mN m^{-1} (for paraffin wax) and 31–35 mN m^{-1} (for polyethylene), both obtained via Fowkes equations and contact angle data (using Equation 15.6). The good agreement obtained in this and other cases for solid surface tensions estimated by these two different methods (contact angle and vapour adsorption data) is rather reassuring and certainly talks in favour of the Fowkes approach. In addition, solid surface tensions (dispersion) for many polymers from Fowkes equation are rather close to Zisman's critical surface tensions for apolar polymers, e.g. 19.5 (Teflon), 18.0 (polyhexafluoropropylene) and 35 (PE) compared to Zisman values which are, respectively, 18.5, 16.2 and 31 (all values in mN m^{-1}). For other polymers there are differences (PS: Fowkes value is 44 and Zisman is 33; in mN m^{-1}).

Moreover, dispersion surface tensions from Fowkes are in good agreement with Hamaker constants and experimentally measured van der Waals forces between surfaces (Hiemenz and Rajagopalan, 1997; Fowkes, 1964).

Fowkes (1964) discussed several ways to estimate the dispersion part of solids using diverse types of experimental data (beyond contact angles and Hamaker constants), e.g. heats of adsorption and heats of immersion.

Equation 15.6 can be also used to estimate the dispersion part of polar liquids from solid–liquid interfacial data (contact angles). This is important in many cases as polar liquids are often miscible with water and thus no liquid–liquid interfacial tension data are available. To do this, we need to know the dispersion part of

the solid surface tension which can be determined from non-polar liquids like hydrocarbons and ethers. Fowkes (1964) presented surface tensions for several such polar liquids. The relative magnitude of the dispersion/specific contributions is overall reasonable but we can see that the dispersion contributions of liquids like glycerol (37 mN m^{-1}) and formamide (39.5 mN m^{-1}) are quite high, higher than the corresponding contributions from water (21.8 mN m^{-1}). For these liquids the dispersion interactions are somewhat higher than the "specific" (remaining) interactions which are mostly due to hydrogen bonding.

Table 15.2 presents the Fowkes/Owens–Wendt and van Oss–Good parameters for several liquids.

The above results alone are so impressive that it may be tempting to assume (as some researchers do) that all "cross interactions" should occur via the dispersive contributions and that the Fowkes equation is of widespread applicability. It is, thus, not entirely surprising that many (almost all?) textbooks of relevance to colloid and surface chemistry (even some of the most modern ones), e.g. Shaw (1992), Hunter, Hiemenz (1997), Israelachvili (1985, 1992), Goodwin (2004), etc., limit their discussion about interfacial theories to the Girifalco–Good and Fowkes equations. Even if this is done for illustrative purposes only, it is highly unfortunate. In the aforementioned textbooks, it is not clearly mentioned that, despite their successes, the GG and Fowkes models have serious problems (which both Frederick Fowkes and Robert Good knew well! For example, Fowkes, 1963, 1964). In the case of Fowkes, see Example 3.3 and Problems 3.5 and 3.6, it fails to represent the interfacial tensions of water–aromatic hydrocarbons or fluorocarbons, water–chlorinated compounds and in several other cases. In the case of water–aromatics, the Fowkes equation overestimates the interfacial tension by 40–50%, while in the case of water with aniline, amines, alcohols and other hydrogen-bonding compounds, the overestimation is often more than 100%. This is a typical observation with the Fowkes equation, as Fowkes himself in the original paper (Fowkes, 1964) has stated; the presence of strong hydrogen bonds typically leads to interfacial tension values from the Fowkes equation which are much higher than the experimentally observed values.

Table 15.2 Parameters of the van Oss–Good and Fowkes/Owens–Wendt theories for some liquids (in mN m^{-1}) at room temperature

Liquid	γ	γ^{LW} (van Oss–Good)	γ^{AB} (van Oss–Good) (acid/base)	γ^d (Fowkes-OW)	γ^{spec} (Fowkes-OW)
cis-Decalin	33.3	33.2	0 (0/0)	33.3	0.0
α-Bromonaphthalene	44.4	43.5	0 (0/0)	44.4	0.0
Diidomethane (methylene iodide)	50.8	50.8	0 (0/0)	49.5/50.8	1.3/0
Water	72.8	21.8	51.0 (25.5/25.5)	21.8	51
Glycerol	64.0	34.0	30.0 (3.92/57.4)	37.0	27.0
2-Ethoxyethanol	28.6			23.6	5.0
Formamide	58.0	39.0	19.0 (2.28/39.6)	39.5/32	18.5/26
Ethylene glycol	48.0	29.0	19.0 (1.92/47)	33.8/29	14.2/19
Dimethyl sulfoxide	44.0	36.0	8.0 (0.5/32)		
Aniline	42.9			24.2	18.7

Diverse sources (e.g. Fowkes, 1964, 1972).

Tavana and Neumann (2007) illustrated problems of the Fowkes equation even for the non-polar Teflon AF 1600 surface with various liquids and they state that the geometric mean rule tends to be "too large" when the difference between the solid and liquid surface tension increases, even for rather non-polar systems. Differences in ionization potentials and size of molecules, e.g. fluorocarbons and hydrocarbons, have also been discussed by Fowkes (1964) as one partial explanation of the failure of the theory for specific systems/interfaces.

These early theories have been used by many researchers with great success in many cases. For example, Dann (1970) from Eastman Kodak Company Research Laboratories provided in two articles an extensive evaluation of Fowkes–Good approaches for many systems. In the first of the articles (Dan 1970a, part I), he finds that indeed the Fowkes–Good approaches are comparable to Zisman, at least when a careful choice of liquids is made. Overall good results are obtained unless we have strong hydrogen bonds between solids and liquids or highly polar liquids.

In the second article he proposes the following modification of the Fowkes equation:

$$\gamma_{12} = \gamma_1 + \gamma_2 - 2\sqrt{\gamma_1^d \gamma_2^d} - I_{12}^p \qquad (15.8)$$

The *I*-parameter is obtained in the "usual" way by combining this equation with the Young equation for the contact angle and by subtracting the experimental spreading coefficient from the one estimated from the Fowkes equation.

Considering ten different polymeric solids (PS, PE, PET, PMMA, PVC, etc.) he found that the *I*-parameter is proportional to a "polar" surface tension component of the liquid but he states that it is possibly a complex function of the "polar" contributions to the surface tension from both liquids and solids. There is also some scatter in these plots and he does not propose the simple square root expression adopted by Owens and Wendt (Equation 3.22b). Similar results were presented with the same method by others as well (Schultz *et al.*, 1977a,b) who used it to determine the surface energy components of high energy surfaces like mica. Dann finally found that there is evidence for spreading pressure contributions which cannot be easily assessed.

Actually, Fowkes himself was aware of the limitations of his approach and he added later (Fowkes, 1964; Fowkes *et al.*, 1980, 1990) an acid–base term in his theory, which he suggested that it is estimated from Drago's theory (Drago *et al.*, 1965) for the enthalpy of interaction. The expression for the work of adhesion is:

$$W_{adh} = W_{adh}^d + W_{adh}^{AB} = 2\sqrt{\gamma_1^d \gamma_2^d} - fn^{AB}\Delta H^{AB}$$ (15.9)

$$\Delta H^{AB} = E_A E_B + C_A C_B$$

A and B indicate acid and base while E and C are susceptibilities to undergoing electrostatic and covalent interactions, respectively. The other parameters in Equation 15.9 are a conversion factor (f) and n^{AB} is the number of acid–base interaction sites on the surface. Drago *et al.* (1965) have provided E and C values for about 30–40 acids and bases. In the case of benzene–water, the enthalpy of interaction is −5 kJ mol^{-1}, which incidentally is about ⅓ of heat of vaporization of benzene. This enthalpy of interaction corresponds to 16.5 mN m^{-1} (assuming 50 Å^2 as the area per molecule for benzene) and $f = 1$; both assumptions have been, however, criticized, see Douillard, 1997. Using such a value the "improved" Fowkes equation is, as shown by Fowkes *et al.* (1990), in excellent agreement with the experimental interfacial tension for water–benzene (35 mN m^{-1}):

$$\gamma_{12} = \gamma_1 + \gamma_2 - 2\sqrt{\gamma_1^d \gamma_2^d} + W_{12}^{AB}$$

$$= 72.8 + 28.9 - 2\sqrt{22 \times 28.9} - 16.5 = 34.8$$

Fowkes suggested that the inclusion of acid–base interaction is very useful and certainly a step in the right direction but it was not used very much, possibly because Equation 15.9 was not parameterized for many compounds. Some calculations have been reported, e.g. 5.7 mN m^{-1} for the acid–base interaction between PVC (acid) and DMSO (base), but not for many systems.

Fowkes in his later work (1990) further supported his method for acid–base interactions by stating that many compounds have both acid and base contributions and, thus, are capable of self-association. He states that even acetone has a 31% self-associating degree and tetrahydrofuran a 27% degree. From a thermodynamic viewpoint, we would have characterized both of these compounds as non-self-associating. In this work (Fowkes, 1990), one of the last articles before he passed away, Fowkes again criticized heavily the Owens–Wendt method while he was more positive about the van Oss–Good (acid–base) approach, though he considered that it also has problems.

Actually, the Fowkes equation to account for the acid–base interactions has been seriously criticized by Douillard (1997) who finds that the f-parameter should account for entropic effects and is, in principle, temperature dependent. Even if assumed to be temperature-independent, this parameter has a typical value of 0.3–0.45 for several solid–liquid systems studied by Douillard (1997) and can, thus, not be considered to be unity, as Fowkes assumed.

Partially due to the above limitations and lack of extensive investigations, we have not seen Equation 15.9 mentioned in any general textbook of colloid and surface chemistry which we have come across.

Despite the fact that the Fowkes equation in the form of Equation 15.9 is not widely used, the importance of Lewis acid–base concepts in surface studies has been discussed extensively by many authors. Some of the best reviews have been presented by Jensen (1978, 1982, 1991). The 1991 review was actually written for a presentation in a symposium in honour of Professor Fowkes. Jensen presents various approaches for quantifying the acid–base interactions like the Drago method mentioned above but also the well-known Gutmann *et al.* donor–acceptor number approach and others. According to Jensen, Drago's method is only suitable for acid–base interactions between molecules/surfaces with only acid and only base contributions and is not suitable for self-associating molecules. Jensen (1991) presents several "approximate" Drago equations that can be used in specific cases and require fewer parameters. He also mentions that polar electrostatic and hydrogen bonding contributions cannot be separated and should be treated together. Finally, we should mention that the acid–base interactions are very useful in describing adsorption phenomena especially adsorption of polymers (see Chapter 7 and also Lloyd, 1994; Fowkes, 1991; Jensen, 1981; Fowkes and Mostafa, 1978).

As a consequence of the aforementioned discussion, it is the original Fowkes equation (Equation 15.5) that was and is still used a lot but, as mentioned, the Fowkes equation should be used with care and its limitations should be recognized. This has been indeed done and we saw that the OW and Hansen models (Equations 3.22–3.24) are, more or less, direct extensions of the Fowkes equation to include polar (specific) contributions to the cross interaction term (and thus the work of adhesion), either together or separately as polar and hydrogen bonding effects. While it is certainly a useful addition to include the effect of more intermolecular forces, the use of the geometric mean rule for the polar and especially for the hydrogen bonding interactions is not a good assumption! The geometric mean rule is strictly valid (see Chapter 2) for non-polar fluids and extensions to more complex fluids should be made with caution. Nevertheless, both the OW and Hansen methods can be used successfully in many cases. The OW method has received quite some acceptance in the literature, especially for mixtures containing polymers (Owens and Wendt, 1969; Owens, 1970), and there are many who consider it to be a standard method (even) today for polymer interfaces. Owens and Wendt (1969) and others use water and methylene iodide as reference liquids to calculate the surface tensions of solids. Fowkes and OW use the same values for liquids (see Table 15.1). Alternatively, it is possible to obtain the solid surface tension components from regressing contact angle data from various liquids on solids using a least-squares method (e.g. Fowkes, 1964).

Testing the theories for interfacial tension, either for liquids or solids, is not an easy task. In the case of liquids, small impurities may affect the values of interfacial tensions, as Owens (1970) and others have stated. On the other hand, if accurate data are available for obtaining the surface components and if there is confidence on a specific theory we can get useful information about the solids, e.g. a semiquantitative measure of surface composition (Owens and Wendt,

1969), identifying the sources of contamination (Myers, 1991, 1999) including presence of hydrophilic sites on hydrophobic polymeric surfaces (Fowkes, McCarthy and Mostafa, 1980), etc. This is because the surface component values of a theory must be sensitive to small changes in the nature of the solid substrate. This issue is discussed in some detail by Fowkes, McCarthy and Mostafa (1980) where they present contact angle data for various polyethylene (PE) and Teflon surfaces with and without contamination. In the former case, the contamination consists of hydrophilic impurities, e.g. inorganic oxides like silica, resulting in lower contact angles for both PE and Teflon. Such "impurities" or additives like emulsifiers, plasticizers, stabilizers, lubricants and antioxidants are common in commercial polymers. Some of these compounds can migrate to the interface and it is thus clear that these effects can affect surface and interfacial tensions and they can have a significant influence on the analysis and interpretation of contact angle data in combination with surface tension theories. Of course, it is possible to extract the impurities and make an analysis on a "pure" polymer sample, but this is far too often not done and the polymers are used without further purification. Contamination aspects can create misunderstandings when they are not appropriately accounted for or discussed as illustrated in an old communication in *Nature* (Gray, 1966; Good, 1966). Similarly, the preparation method in general of a solid polymer for use in surface studies can have a large influence on the results, e.g. contact angle measurements (Good, 1977).

Another issue that may complicate the analysis is the importance of spreading (sometimes called film) pressure which is due to adsorption of the vapour from the liquid on the solid surface. If this parameter should be accounted for then all the above equations should be corrected accordingly, e.g. for the Fowkes equation, instead of Equation 15.6, we should use:

$$\cos \vartheta = -1 + \frac{2\sqrt{\gamma_s^d \gamma_l^d}}{\gamma_l} - \frac{\pi_{sv}}{\gamma_l} \qquad (15.10)$$

Similar corrections should apply to Girifalco–Good, Owens–Wendt and all the other theories. But how important is this spreading pressure effect (i.e. the contribution of the last term of Equation 15.10)? Fowkes (1964) stated that this term is often small except for solids with high surface energies. Possible inclusion of the spreading pressure contribution has been mentioned by some researchers as an explanation for the discrepancies between interfacial theories and experimental data.

Owens and Wendt (1969) and many others ignore the contribution of the spreading pressure by assuming that for low energy polymeric surfaces it must be close to zero. Others like Good (1992) and Fowkes, McCarthy and Mostafa (1980) have studied its magnitude based on experimental measurements. Good (1992) has presented experimental spreading pressure values for five alkanes from butane to hexadecane on Teflon and he concludes that their contribution is small on homogeneous, low-energy solids for pure liquids that form non-zero contact angles. The spreading pressure starts becoming of importance for contact angles close to zero (even on polymers). Fowkes, McCarthy and Mostafa (1980) obtained similar results in their analysis of their own spreading pressure measurements for PE and Teflon surfaces. The above studies also show that the complexity of solid surfaces, e.g. the fact that they are often heterogeneous and can become contaminated, certainly complicates their analysis using a few contact angle measurements and theories for estimating the interfacial tension. Good (1977) also states that for low-energy surfaces that have been treated carefully during their preparation in order to approach ideality (molecularly smooth and homogeneous) it can be expected that the spreading pressure is small and it can be neglected.

Table 15.3 shows components of surface tension for some solid polymers estimated by the Owens–Wendt methods; the Zisman critical surface tensions values are also given. There is rather good agreement between the predicted polymer surface tensions from the Owens–Wendt and Zisman methods.

Erbil (1997) writes that most researchers in the field today have abandoned the Owens–Wendt theory (which also predicts measurable interfacial tensions for certain mixtures which are known to be miscible).

Table 15.3   Components of surface tension for various solids (polymers) from the Owens–Wendt (OW) theory and the Zisman plot (critical surface tension). All values are in mN m^{-1}. When two values are mentioned it is because they are based on different experimental values for the contact angle

Polymer	Zisman	γ_s (OW)	γ_s^d (OW)	γ_s^{spec} (OW)
PE	31	33.1	32.0	1.1
		33.2	33.2	0.0
PP		30.2	30.2	0.0
PVC	39	41.5	40.0	1.5
PTFE	18.5	14.0	12.5	1.5
		19.1	18.6	0.5
PET	43	41.3	37.8	3.5
		47.3	43.2	4.1
PMMA	39	40.2	35.9	4.3
Nylon 6,6	46	43.2	34.1	9.1
		47.0	40.8	6.2
PS	43	42.0	41.4	0.6
PDMS	24	22.8	21.7	1.1

Ownes and Wendt (1969). Reproduced with permission from John Wiley & Sons, Ltd.

This is not entirely true. We see many applications of OW in the recent literature, e.g. by Kudela and Liptakova (2006) for adhesion of coatings materials to wood and much more.

Cosgrove (2010) also presents wettability envelopes using the OW model.

The OW method is still the most cited theory of estimation of interfacial tension despite its well-established limitations. This is despite the fact that some of the most important researchers in surface science simply dismiss it, e.g. Robert Good (1992). According to Good (1992), the geometric mean use of the polar term "does not correspond to anything that is at all related to the contact angle" and "a formal solution exists but it has no physical meaning". Actually, in his 1992 article, which contains 67 equations, the OW model remains without an equation number as "numbering it might be taken as giving status comparable to say the van Oss–Good method". Fowkes (1964, 1972, 1990) has been equally critical about the use of the geometric mean rule for the "polar" interactions (including hydrogen bonding) and proposed the Drago-based acid–base theory mentioned earlier (Equation 15.9). In proof of his statements, Fowkes *et al.* (1990) illustrated using his measurements for the liquid–liquid interfacial tension of squalane with a wide range of liquids that not only the OW but also other theories for interfacial tensions fail. Fowkes' statements are as critical as those of Robert Good. He writes (Fowkes *et al.*, 1990): "Owens–Wendt… and the Neumann's equation of state are erroneous and should be abandoned". In addition, also "In light of these results in the first real test ever made to verify the Owens–Wendt equation, it is surprising that

so many investigators have believed that this equation works".

These critical comments may be partially justified but naturally Fowkes *et al.* analysis is not the "first real test ever" to validate the OW model. OW has been proven to work well and provide physically acceptable results for many solid surfaces, especially polymers. However, the OW like many other theories may *indeed* present unacceptable results for *certain* liquid–liquid interfaces. Thus, the quest to find better theories should continue and it has indeed continued, as discussed in the rest of this chapter (and in Chapters 3 and 6).

Example 15.1. *Effect of contact angle measurements on the determination of solid surface tensions using the Owens–Wendt method. Based on data by Owens and Wendt (1969).*

Owens and Wendt (1969) report for PTFE (Teflon) contact angle measurements for water (108°) and for methylene iodide (88° and 77°, the latter with their own measurements). Similarly, they report for PE contact angle measurements for water (94° and 104°, the latter with their own measurements) and for methylene iodide (52°). Calculate the dispersion and specific contribution of the surface tension of the two polymers using the OW method and compare the obtained solid surface tensions with the Zisman values of the critical surface tension which are 18.5 (PTFE) and 31 (PE). All measurements are at room temperature and all surface tensions are in mN m^{-1}. Provide a discussion of the results.

Solution

The Owens–Wendt method for the interfacial tension combined with the Young equation for the contact angle is given by (assuming zero spreading pressure):

$$\cos\vartheta = \frac{2\sqrt{\gamma_s^d \gamma_l^d}}{\gamma_l} + \frac{2\sqrt{\gamma_s^{spec}\gamma_l^{spec}}}{\gamma_l} - 1 \tag{15.1.1}$$

Table 15.1 presents the dispersion and specific contribution of the surface tension of water and methylene iodide (MI) used in conjunction with the OW method. These are also considered to be the "standard" liquids used with this method to estimate the surface tension of solids. When these values are known and are substituted in Equation 15.1.1 we can solve the resulting system of two equations with two unknowns, which are the dispersive and specific surface tensions of the solids.

The final results are provided in the table. All surface tension values are given in mN m^{-1}.

Surface	Contact angle Water (°)	Contact angle MI (°)	γ_s^d	γ_s^{spec}	γ_s
PE	94	52	32.0	1.1	33.1
	104	52	33.2	0.0	33.2
PTFE	108	88	12.5	1.5	14.0
		77	18.6	0.5	19.1

We can see that in the case of PE there is a small effect of the contact angle measurements on the surface tension values (and components) and the final value is in reasonably good agreement with the Zisman critical surface tension (31 mN m^{-1}). For Teflon, however, the effect of contact angle measurements on surface tension determination is significant. Only using the authors' own measurements do we get a solid surface tension that is reasonably close to the Zisman value (18.5 mN m^{-1}). The low value of 14 mN m^{-1} obtained using the other literature contact angle measurements should be considered wrong.

Owens and Wendt (1969) mention also that since the dispersion and specific values of the surface tension are sensitive to the surface composition, then the analysis illustrated in this example can be useful as a semi-quantitative measure of the surface composition.

Example 15.2.   *Spreading of liquids on solids using the Fowkes equation (With data from Fowkes, 1964).*
The table below provides dispersion values of the surface tensions of several solids obtained from spreading pressure measurements of diverse adsorbates (therefore the range of reported values).

Solid	Dispersion value of solid surface tension (mN m^{-1})	Solid	Dispersion value of solid surface tension (mN m^{-1})
Polypropylene	26–29	Copper	60
Graphite	115–130	Silver	74
Silica	78	Lead	99
Ferric oxide	107	Iron	89–108
Barium sulfate	76	TiO$_2$	78–141

1. Show, using the Fowkes equation, that the Harkins spreading coefficient is given by the following equation:

$$S=2\sqrt{\gamma_l^d \gamma_s^d}-2\gamma_l$$

2. Investigate using the Fowkes equation the spreading behaviour of water and formamide on solids (use the liquid surface tension components shown in Table 15.2). On which solids from the ones shown in the table above will water and formamide spread?

Solution
1. The starting point is the definition of the Harkins coefficient (see Chapter 3):

$$S=\gamma_{WA}-(\gamma_{OA}+\gamma_{OW})=W_{adh,OW}-W_{coh,O}$$

Which for a solid–liquid interface can be equivalently written as:

$$S=\gamma_s-(\gamma_l+\gamma_{sl})=W_{adh,sl}-W_{coh,l}$$

By substituting the Fowkes equation into the aforementioned equation we get the requested result.

2. In the case of water we have (using the values of Table 15.2):

$$S = 2\sqrt{\gamma_l^d \gamma_s^d} - 2\gamma_l = 9.338\sqrt{\gamma_s^d} - 145.6 \Rightarrow \gamma_s^d \geq 243$$

As spreading exists only if $S > 0$, this value indicates that water will only fully spread on very high energy solids and thus on none of the solids given in the problem.

On the other hand for formamide we have:

$$S = 2\sqrt{\gamma_l^d \gamma_s^d} - 2\gamma_l = 12.5698\sqrt{\gamma_s^d} - 116.4 \Rightarrow \gamma_s^d \geq 86$$

This result implies that formamide will fully spread ($S > 0$) on at least a few of the aforementioned solids, e.g. PP, copper and possibly also some of the other metals and on silica.

The Hansen method has, somewhat surprisingly, not received much attention, even though the authors have proposed an independent method to estimate the surface tension components using the Hansen solubility parameters (Equations 3.14 and 3.15). Nevertheless, we have shown in Chapters 3 and 6 that the Hansen method can be successfully used for many liquid–liquid and liquid–solid interfaces (e.g. Examples 3.5, 6.2 and 6.4). Both the OW and Hansen methods use for the dispersion part of water's surface tension the same value as the one used by Fowkes. Of course, as already mentioned, Fowkes (e.g. 1972) and Good disagree on the use (in Hansen and related theories) of the geometric mean rule for the polar and hydrogen bonding interactions. The critique is direct. Fowkes (1972) writes for these combining rules: "The literature is filled with contact angle correlations suggesting some physical significance but really more concerned with a simple mathematical form". Later, "If the correlations are not intended to provide information about bonding, we cannot object to their usefulness".

Finally, we should mention that in addition to the geometric mean rules there have been other combining rules recommended for the cross interactions, often known as "harmonic mean approaches" (see Problem 3.8). These have been considered to be variations of the Owens–Wendt approach and have found some applicability especially for mixtures of polymer melts (blends) for which there are a lot of data (see Problem 15.10), also at diverse temperatures; despite often being

shown to perform better than the Fowkes method (see Wu, 1971, 1973) they are not used very much today.

15.3 van Oss–Good and Neumann theories

15.3.1 The two theories in brief

While most (if not all) textbooks on colloids and interfaces limit their discussion about interfacial theories to Girifalco–Good and Fowkes/Owens–Wendt, the discussion is hardly complete without presenting the two most modern, possibly most widely used and certainly most controversial, theories, the acid–base theory of Carel van Oss, Manoj Chaudhury and Robert Good (from now on called here van Oss–Good) and the equation of state approach of A. Wilhelm Neumann. These theories, already presented in Chapter 3 (Equations 3.18 and 3.25 and 3.26), have resulted to extensive discussions – not the least between their developers, often with rather direct and not always entirely polite statements about the capabilities and limitations. Numerous articles have been published about these two theories both by their developers and by others. Thus, the pertinent literature is enormous but we attempt a short review here.

We discuss first the background of the two approaches. Each one presents a different "school" in theories for the interfacial tension. The van Oss–Good is the latest well-known version of the "surface component theories", where the geometric mean rule is maintained only for the van der Waals forces (LW for Lifshitz–van der Waals or London–van der Waals). The hydrogen bonding or in general the acid (+)–base (−) contributions are given by asymmetric combining rules, where only the acid component interacts with the basic one and vice versa:

$$\gamma_{12} = \gamma_1 + \gamma_2 - 2\sqrt{\gamma_1^{LW}\gamma_2^{LW}} - 2\sqrt{\gamma_1^+\gamma_2^-} - 2\sqrt{\gamma_1^-\gamma_2^+}$$

$$(15.11)$$

This last term is what makes the theory fundamentally different from its predecessors. Van Oss *et al.* (1988a,b) and Good (1992) justify the assumption that all van der Waals forces can be "lumped" into a single term by the fact that polar and induction contributions to surface tension should be small, or at least much smaller than the dispersion ones. Fowkes (1972) agrees. He mentions that in the case of acetone, for example, the dipole–dipole interactions contribute to the intermolecular forces only about 14% and that acid–base interactions (hydrogen bonding and the like) are much more important. We have seen a similar trend in modern thermodynamics where some of the most successful models, e.g. the CPA and SAFT theories, are those which explicitly account for hydrogen bonding effects, while polar interactions can often be lumped together with the other van der Waals interactions (Kontogeorgis and Folas, 2010).

Actually, a similar approach to Equation 15.11 was proposed earlier by Small (1953) who also stated that polarity is often far less crucial than the dispersion and hydrogen bonding forces. He also mentioned that the geometric mean rule breaks down if size/shape of the involved compounds differ too much.

The van Oss–Good equation can result in either positive or negative interfacial tensions, the latter simply meaning miscible liquids. Thus, it is possible for the van Oss–Good theory to predict repulsive van der Waals forces which can be present in certain systems (van Oss *et al.*, 1988, 1989). Because van Oss–Good can also predict negative interfacial tensions, it has been shown to predict well the solubility in aqueous polymer solutions (van Oss and Good, 1992) where Owens–Wendt fails. It has also been applied with success to biopolymers (van Oss *et al.*,

1987a,b). Naturally, application of Equation 15.11 to solid interfaces using the contact angle equation requires data for at least three liquids on the same solid. We return later on the choice of liquids.

Neumann and co-workers not only have presented a completely different approach but they do not believe in the "surface tension component theories" at all. They have stated this clearly many times. For example, Kwok and Neumann (2000a, p. 47):

"Surface tension component approaches do not reflect physical reality. Intermolecular forces do not have additional and independent effects on the contact angle."

In addition (Kwok *et al.*, 1998):

"While intermolecular forces determine the interfacial tensions, they do not have additional effects on the contact angles: hence the surface tension component approaches cannot describe physical reality."

Furthermore (Kwok *et al.*, 2000b):

"Specific intermolecular forces which give rise to the surface tensions do not affect the contact angles directly: intermolecular forces determine the surface tensions; the surface tensions then determine the contact angle through the Young equation."

Their own model is an extension of the GG equation with a semi-predictive approach to estimate the "φ" parameter:

$$\gamma_{12} = \gamma_1 + \gamma_2 - 2\sqrt{\gamma_1\gamma_2}\,e^{-\beta(\gamma_1-\gamma_2)^2} \qquad (15.12)$$

The Neumann equation is rarely used for liquid–liquid interfaces. It is most often used in combination with the Young equation for the contact angle and thus its application is to estimate the surface tension of solids. The β parameter in Equation 15.12 can be solid-specific but average values have been estimated based on three solids (FC-721-coated mica, heat pressed Teflon and PET). Thus, this parameter can be considered to be almost universal. The basic argument of Neumann and co-workers is that for a large number of solids which they have tested they find that the liquid surface tension multiplied by the contact angle is a smooth function of the liquid surface tension (like the plots shown in Figure 3.15), for any type of liquid (non-polar, polar, hydrogen bonding, etc.). Hence, having the Young equation in mind it can be stated:

$$\gamma_l \cos \vartheta = f(\gamma_l) \Rightarrow \gamma_{sl} = f(\gamma_s, \gamma_l) \qquad (15.13)$$

This means, according to Neumann, that the solid–liquid interfacial tension can only depend on solid and liquid surface tensions and not on their "surface tension components".

Theoretical interpretations of the Neumann equations and semi-predictive methods to estimate the "beta" parameter have been proposed (Kwok *et al.*, 2000) and validated with good results (Kwok and Neumann, 2000a) but as the results with the original Neumann method are overall similar these modified approaches have not been used in subsequent publications.

We have already mentioned that theories for interfacial tension constitute a controversial topic and the advent of the van Oss–Good and Neumann methods has not made the topic less controversial. Actually, these two are possibly the most widely discussed theories today and for this reason, and to pay justice to all opinions, we will divide the discussion into different sections; the opinions of the developers on their own and on each other's theories and some of the independent studies carried out by others before expressing our own views.

15.3.2 What do van Oss–Good and Neumann say about their own theories?

Good and co-workers have established several "selected" liquids for which they have estimated LW, acid/base components (Table 3.4). These should be used in the analysis of solid surfaces together with contact angle data. However, it is not "a priori" known which liquids should be used for performing contact angle experiments. Van Oss *et al.* (1988) have established that the "optimum" approach is to always use water (essentially all van Oss–Good theory parameters are relative to water), one non-polar liquid (often methylene iodide) and another polar liquid,

e.g. glycol or formamide (see also Good, 1992). They have several times stated that it may be problematic to use, for example, three polar liquids or to exclude water from the liquids used. Other approaches to establish more "universal" ways to obtain the solid parameters have been attempted without success. For example, Good and Hawa (1997) used simultaneously nine liquids on three solids but the result of this (mathematically cumbersome) exercise did not result in physically more meaningful parameters.

Van Oss–Good (and others) have recognized that, even with the recommended approach mentioned above, the basic components are almost always higher than the acidic ones, which is physically not correct. All results of the theory should be interpreted having this model deficiency in mind. Still, neither this nor the fact that different "liquid triplets" yield different results is a serious issue, according to the developers. When the "peculiarities" of the van Oss–Good method mentioned above are taken into consideration, solid surfaces can still be characterized successfully and useful information can be obtained also for practical phenomena such as wetting and adhesion.

Neumann and co-workers have, first of all, questioned the validity of many of the literature experimental data for contact angles on various surfaces, especially if obtained using the standard goniometer method. According to Neumann *et al.* "wrong" contact angle data (due to hysteresis, roughness, contaminations, vapour adsorption, interactions with the solid, etc.) make the interpretation via the "ideal" Young equation for contact angle of the various surface tension theories cumbersome, if not impossible. They have themselves pioneered a rather advanced experimental method for contact angle determination (automated axisymmetric drop shape analysis-profile; ADSA-P), see Kwok and Neumann (2000), and they recommend that only (or mostly) such data should be used together with the Young equation. They have presented excellent universal trends (according to Equations 15.12 and 15.13) which validate their theory for a wide range of solids, ranging from highly non-polar (diverse fluorinated surfaces), non-polar (PS, PE, etc.) and more polar ones (PET, PMMA, a long list of poly-acrylates and poly-methacrylates, including copolymers), see the references in Table 15.5 below. For all of these solids, the final output of the Neumann theory is, via Equation 15.12), a single value of the solid surface tension. Neumann *et al.* observe that the solid surface tensions obtained, e.g. for families of polymers in the same series such as polymethacrylates (Wulf *et al.*, 2000; Kwok *et al.*, 2000b) follow reasonably expected trends (e.g. the very low solid surface for poly(*t*-butyl methacrylate) in Table 15.5 below can be explained by the presence of the large bulky side group).

However, they too have observed several problems with their theory for some solid–liquid combinations, i.e. deviations from the general plots $\gamma_l \cos \vartheta = f(\gamma_l)$, e.g. see Tavana and Neumann (2007) and Tavana *et al.* (2004, 2005).

They believe that these deviations are not due to experimental error or problems of the theory but all have a clear explanation. In most cases, they attribute the problems to either interaction between the solid and the liquid and/or the presence of a non-zero spreading pressure due to vapour adsorption from the liquid. They improve the smoothness of their plots, in these cases, by eliminating several (in some cases many of the) liquids, sometimes even alkanes, used in the analysis in order to maintain the maximum possible inertness of the test liquids used to estimate the solid surface tension using the Neumann method. They mention that with this approach, i.e. careful selection of test liquids (bulky molecules are often useful), some of the experimental contact angle data, even with the goniometer method used in the extensive studies of Zisman, can be used in the context of the Neumann method (Kwok and Neumann, 2000b).

15.3.3   What do van Oss–Good and Neumann say about each other's theories?

Professor Robert J. Good, one of the pioneers of surface science (see biography in Chapter 3) and main contributor in both the Girifalco–Good and van Oss–Good theories is actually co-author of one of the early articles in the Neumann theory (Neumann, Good, Hope, Seipal, 1974). He has now withdrawn his support to the Neumann theory, as he stated explicitly in a review article (Good, 1992). According to Good, the objections to the Neumann theory is a weak thermodynamic foundation, many disagreements for interfacial tensions between water and organics and the fact that hydrogen bonds or acid–base interactions are not considered. He has stated (van Oss *et al.*, 1988) that the validity range of the Neumann theory is rather limited to apolar systems. Professor Good and his co-workers state in this article "We conclude that this equation of state do not raise any serious challenge to the theory of surface tension components".

Good and co-workers have frequently pointed out (e.g. van Oss *et al.*, 1987) that the inability of Neumann theory to predict negative liquid–liquid interfacial tensions results in often poor agreement for liquid–liquid interfaces and for predicting miscibility (miscibility is equivalent to zero or negative interfacial tensions).

It is worth mentioning that Professor Good has not always been so critical. In his 1977 review (Good, 1977) he sees much promise in the – then – recently appeared Neumann theory, while he mentioned several reasons why the surface tension component theory could break down, among others the importance of entropic effects (incidentally discussed by others later, e.g. Douillard, 1997). Nevertheless, Professor Good and Professor Fowkes clearly converged to the notion that the surface tension component theories constitute the most successful approach, with special emphasis on the explicit treatment of the very important acid–base interactions.

Neumann and co-workers have, from their side, illustrated several problems of the van Oss–Good theory. In a 2000 article (Kwok *et al.*, 2000b) they show that use of different liquids gives widely different results for the surface tensions (and their components) for various polymers (PMMA, PBMA, etc.). The apparent inability of the van Oss–Good method (as used by Kwok *et al.*, 2000b) to establish reasonably correct parameters for the acid–base theory led them to rather harsh conclusions. They write: "these results suggest that the acid–base approach is not capable of predicting correct solid surface tensions and solid surface tension components" and "the van Oss and Good approach is shown to be not tenable for solid surface tensions and solid surface tension components, from a nonpolar fluorocarbon surface to polar methacrylate polymers". Neumann concludes in that article that the whole surface component theory is not useful. He regrets that "the widespread view that contact angles contain readily accessible information about intermolecular forces hampers progress immensely, particularly the reluctance to straightforward contact angle interpretations, such as those shown in figures 3 and 4" (of their article – illustrating examples of the Neumann method). Clearly, Neumann and co-workers do not believe that physically meaningful solid surface tension components can be obtained from component theories and contact angle data through the Young equation.

In addition, Kwok (Kwok and Neumann, 1996; Kwok *et al.*, 1998, 1999) showed that the van Oss–Good theory predicts problematic results for several liquid–liquid interfaces for which experimental data are available. While for many aqueous systems the interfacial tensions are predicted rather satisfactorily, the performance of van Oss–Good for several non-aqueous mixtures is not good. In some cases finite values for the interfacial tension are predicted for mixtures which are known to be miscible such as bromonaphthalene with alkanes and squalene–diiodo-methane. The results or at least the main conclusions appear to be independent of the source of "reference liquids" and parameters used for the van Oss–Good method (original ones or those from Lee, see Section 15.3.4). Kwok and Neumann (1996) state, on basis of these calculations, that "it is surprising to see an approach published in well reputed journals when it can be shown to be false by anybody who possesses a simple calculator and a few drops of the liquids used in the approach". Statements like this can be found in several of the articles published by Neumann and co-workers.

15.3.4 What do others say about van Oss–Good and Neumann theories?

As mentioned, both theories have been extensively used by other researchers, with the van Oss–Good theory being more popular in independent studies.

Some applications of the van Oss–Good theory are presented in Chapters 3 and 6. McCafferty (2002) characterized PVC and PMMA surfaces and based on the obtained values he presented a discussion of adhesion phenomena in various systems which are shown to correlate well with the acid–base interactions. As shown in Chapter 6 (Section 6.2), in agreement with other studies, both polymers have basic character, but the "much more basic" PMMA has a much higher basic component. McCafferty showed that different liquid triplets (though water + a non-polar liquid were always used) and other methods to estimate the van Oss–Good parameters do give different surface tension components, but qualitatively this does not change the conclusions. Similar comments (diverse values of the van Oss–Good parameters depending on estimation method but overall similar conclusions) have been made by other researchers as well (e.g. Radelczuk, Holysz and Chibowski, 2002). The acid–base parameters for PMMA and PVC have been also discussed by Morra (1996) who clearly indicates that the topic needs further investigation. He points out that "despite the long history of contact angle data and calculation of interfacial energetics, it may sound frustrating that we are still wondering whether the rather simple PVC surface is acidic or basic".

Osterhold and Armbruster (1998) used the van Oss–Good theory in the study of adhesion changes due to surface modification (see Figure 6.14). Adhesion phenomena and their relation to acid–base interactions and other theories are also summarized by Clint (2001a,b). He showed that adhesion calculated from theories can be linked to experimental values. He presented results with both the van Oss–Good and Owens–Wendt theories but recommends the former for polar systems.

Fowkes himself became later convinced of the importance of acid–base interactions and related them to Drago's acid–base parameters (showing that 23% of the work of adhesion for water–aromatics can be due to acid–base interactions).

We have also successfully used the van Oss–Good theory in various applications in the paint industry (see Case study: Paints, Chapter 6, and Problem 6.1; Svendsen *et al.*, 2007), characterization of polymers (Problem 6.11) and laser-based surface treatment studies (Zhang, Kontogeorgis and Hansen, 2011; Example 6.3). In all cases, we find that the van Oss–Good basic components are systematically higher than the acidic ones but, even with this limitation, the method is useful for a qualitative assessment of the solid–liquid interfaces and characterization of solid substrates.

Several researchers, especially Della Volpe and Siboni (1997, 2000) and Lee (1998) have presented an alternative to van Oss–Good parameters for the "test liquids" (see Table 3.5 in Chapter 3 for Della Volpe–Siboni values). These researchers do not use the assumption of van Oss–Good that the acid and base contributions of water's surface tension should be equal. The Della Volpe and Siboni parameters result in more "balanced" acid–base contributions (see Tables 3.5, 6.3, 15.4 (below) and Example 6.3), but the base contributions are still often higher than the acid ones (although for PVC, see Table 6.3, the acid value of PVC is, on average, higher than the basic one, as should be expected). Lee (1998), like Neumann's group, also discusses the importance of spreading (film) pressure in deriving reliable surface tension components.

First, let us consider the seminal work of Della Volpe and Siboni, which is presented in numerous publications (Della Volpe and Siboni, 1997, 2000, 2007; Della Volpe *et al.*, 2003, 2004a,b). They have concluded, like others, that of course the surfaces should not be overwhelmingly basic as van Oss–Good parameters indicate. However, this happens not because of some general fault of the theory but because of the assumption used in the original work about water (equal acid and base values of water, see Table 3.4). This, they say, is not correct, e.g. it is not in agreement with other independent assessments of the basicity and acidity of water (see also later). An additional problem in the calculations with van Oss–Good is that almost all liquids chosen (except for water) are very basic and there

appears to be rather few acidic liquids which can be included. Equally difficult is to find truly apolar liquids and the "classical" apolar liquids like diiodomethane have problems. They also observed large disagreements in the LW/AB parameters reported in literature for the van Oss–Good theory by various researchers, even when the same reference values for water are used. Della Volpe and Siboni both presented new parameters for liquids to be used in the van Oss–Good method (see Tables 3.5) and guidelines on how the van Oss–Good method should be used. They also applied the method extensively to diverse systems, including liquid–liquid interfaces. Their parameters are based on simultaneously regressing data for eight liquids on three solid polymers (about 40 equations!). Mathematically, this is a formidable task, but it represents a way to accommodate some of the problems they observed with the van Oss–Good approach. Because of the need of extensive database of established and approved contact angle data on well-defined solids, Della Volpe and Siboni have, on several occasions, called for an international collaboration on collecting and validating such data, an effort which as far as we can see has not as yet been undertaken. Their parameters, see Table 3.5, for water result in a much higher acid than basic component with a ratio of 4.33 (2000 values) and 6.5 (1997 values). These acid/base ratios are in better agreement to the acid/base ratio obtained from the Kamlet–Taft solvatochromic parameters (than the original parameters of van Oss–Good, see also Table 15.4). Another difference between their two van Oss–Good sets is that the 2000 values introduce for the first time a higher LW value for water (26.2 mN m^{-1}) i.e. the LW part accounts for 36%

Table 15.4 *Acid/base ratios of the van Oss–Good parameters (+/–) from various sources as compared to the acid/base ratios extracted from Kamlet–Taft parameters α/β (original values and normalized values from Lee)*

Compound	+/– (van Oss–Good)	+/– (Della Volpe, 1997)	+/– (Della Volpe, 2000)	+/– (Lee)	α/β (original)	α/β (normalized)
Water	1.0	6.5	4.33	1.8	2.5	1.8
Glycerol	0.0683	1.31	3.79	0.1247	2.4	1.7
Ethylene glycol	0.0409	0.0372	0.0187	0.0747	1.7	1.3
Formamide	0.0576	0.0297	1.0	0.1065	1.5	1.1
DMSO	0.0156			0.0294	0.0	0.0

of the water surface tension instead of the largely accepted value (21.8 mN m^{-1}), which accounts for only 30% of the surface tension. The new parameters presented by Della Volpe and Siboni result in more balanced parameters for the van Oss–Good theory. In particular, the 2000 set is recommended by the authors as it gives overall better results and is based on physically more correct water parameters.

Della Volpe–Siboni and Lee agree on the importance of choosing parameters based on balanced acid–base contributions, e.g. as indicated by the Kamlet–Taft parameters, but they disagree on the role and importance of the spreading pressure (which is ignored in the work of Della Volpe and Siboni).

Della Volpe and Siboni recommend that the acid–base values in the van Oss–Good theory should not be compared too much within the same solid but comparatively between different solids and it is more correct, for example, to compare acid values and base values between different solids rather than compare acid to base values. They also point out that the liquids to be used in the van Oss–Good theory should be chosen with care, according to the guidelines, e.g. from the proposers.

In addition, Della Volpe and Siboni used the van Oss–Good method for liquid–liquid interfaces and although the predicted interfacial tensions are often larger than the experimental values, the agreement is, in many cases, quite good and much better than the results reported by Kwok and Neumann. Overall, they correctly conclude that it is quite challenging to predict liquid–liquid interfacial tensions from parameters based on solid–liquid interfaces (while the opposite is more manageable, but depends on the availability of liquid–liquid interfacial tension data).

Della Volpe and Siboni are, despite the problems and limitations which they themselves point out, much more confident about the van Oss–Good theory than for the Neumann theory. They have considered the latter theory in several publications and their conclusion is, in agreement with others, that the Neumann theory is of semi-empirical nature and should be applied only to apolar surfaces.

Another pioneering work on the analysis of the van Oss–Good method has been by Lee published in a series of articles (Lee, 1993, 1996, 1997, 1998, 1999, 2000). He shares with Della Volpe

and Siboni the view that the Neumann approach has significant shortcomings and should be used for apolar mixtures. He also shares the view that the van Oss–Good original parameters are not optimum and he proposes new parameter sets which, in some ways, resemble those of Della Volpe and Siboni. Lee related the acid/base parameters of the surface tension to the Kamlet–Taft solvatochromic parameters. His values for the van Oss–Good parameters for liquids are identical to the original ones with respect to the total surface tension value but with lower base components, which is an improvement compared to the "too high" base values presented by van Oss–Good. Lee's values result in physically more correct parameters for polymers, i.e. having a more balanced acid and base contributions (although there are still problems, e.g. PVC appears still much less acidic than it actually is).

Lee mentions also that the acid and base contributions to the surface tension are temperature dependent. For water at 0 °C, the acid and base parts can very well be the same, but at 20 °C the acid/base ratio is different (and should be around 1.8, when the Kamlet–Taft parameters are given in normalized values).

Lee also proposes to use the LSER (linear solvation energy relationship) concept in its full context and he identifies that the "polarity" parameter of LSER is related to the spreading pressure. The remaining parameters of Kamlet–Taft LSER (acid, base and solubility parameter) have already their "analogue" in the van Oss–Good theory (acid, base and LW contributions to the surface tension). Thus, Lee believes that the spreading pressure should be included in new developments and in a proper evaluation of the van Oss–Good approach.

Using his approach and parameters, Lee has presented many promising results with the van Oss–Good method for both solid–liquid and liquid–liquid interfaces, including polymers and biopolymers.

Table 15.4 shows the acid/base ratios of the van Oss–Good parameters from various sources, as compared to the acid/base ratio predicted from the Kamlet–Taft solvatochromic parameters. The latter parameters are used both in the original form and their normalized values from Lee. We notice that the newest sets of parameters are in better agreement with the water trends predicted from the Kamlet–Taft parameters, i.e. a higher acid than base contribution. However, the results are mixed for some of the other liquids especially for ethylene glycol and formamide. It appears that further studies are needed in this respect.

Finally, it is worth mentioning that the van Oss–Good approach is presented as a state-of-the-art approach by several industrial experts, e.g. in industry courses related to colloids and interfaces organized by YKI (Colloid and Surface Science) institute in Sweden (e.g. Notes by Marie Sjoberg on Coating Adhesion).

Despite the many successful applications, several researchers have expressed serious objections about the foundations of the "surface tension component" theories like the one proposed by van Oss *et al.* Douillard (1997) states that it is based on an oversimplified thermodynamic basis because the additivity of the various force contributions can be claimed for the enthalpy but not for the free energy and thus not for the surface tension either. What Douillard (1997) really claims is that the van Oss–Good (and other similar theories) essentially do not account for the entropic effects which can be significant in some cases or imply that enthalpic and entropic effects are proportional and that the effect of temperature is not appropriately accounted for. He suggests that a surface enthalpy components theory is possibly the way ahead but he admits that this may be more complex and – to our knowledge – this has not been as yet pursued. Despite these limitations, Douillard explains why several of these assumptions can be "less important" for several systems and thus the van Oss–Good theory often works quite well.

The "thermodynamic basis" of the equation of state approach has been also criticized by several researchers (e.g. Lee, 1993, 1994; Morrison, 1989, 1991; Drelich and Miller, 1994) and even though there have been replies to some of the arguments by Neumann and co-workers we believe that several criticisms are valid.

As mentioned previously, very few textbooks mention the Neumann and van Oss–Good methods, which we consider to be a rather surprising fact. One book (essentially an edited collection of chapters written by different authors) where these methods are discussed is the *CRC Handbook of Surface and Colloid Chemistry* (1997 and subsequent editions) edited by K.S. Birdi. Many methods for interfacial tension are discussed in Chapters 2 and 9 of the CRC Handbook, both written by H.Y. Erbil (Erbil, 1997). He considers that the Neumann approach is controversial in many respects, especially since it ignores chemical effects and thus its range of applicability should be apolar systems. Erbil does not object to the empirical practical value of the Neumann approach but he believes that his theory is of rather limited applicability range.

On the other hand, only a few researchers, e.g. Balkenende *et al.* (1998), have provided comparative analysis of both theories (van Oss–Good and Neumann) and illustrate that none is accurate enough, both have limitations (especially when chemical interactions play a role), but also that there are areas where they can be applied to. For the Neumann method, Balkenende *et al.* mention that it can fail even for non-polar solids and it has limited predictive power. For the van Oss–Good theory, they mention that the choice of methylene iodide and 1-bromonaphthalene as "inert" compounds is unfortunate and it is actually rather difficult to find a suitable apolar reference fluid. In addition, LW values for apolar liquids estimated in different ways (e.g. from solid data or liquid–liquid interfaces) can be quite different and for 1-bromonaphthalene they range from 33 to 44 mN m^{-1}. They also say that, due to its mathematical form, there is great sensitivity in the parameter estimation to the experimental errors of the used contact angle data.

It appears, among several researchers, to be a consensus that the Neumann theory is best applied to non-polar liquids and solids (Correia *et al.*, 1989; Drelich and Miller, 1994) and that, all things being equal, the van Oss–Good theory performs best overall for the analysis of solid surfaces.

15.3.5   What do we believe about the van Oss–Good and Neumann theories?

The discussion of the van Oss–Good and Neumann theories is complicated by many factors, e.g. the validity range of the Young equation, the role of spreading pressure, the accuracy of experimental contact angle data and others.

The validity of Young equation and the necessity to include the spreading (or film) pressure is still a rather controversial issue and the interpretation of such data is difficult (Birdi, 1997, Ch.9). This is important for high energy surfaces but such adsorption phenomena cannot be fully ignored for low surface energy solids either (e.g. polymers). Moreover, this is one of the major arguments by the Neumann group concerning the problems of using the Young equation and deducing useful information from contact angle theories.

Despite the criticism, the Neumann method is useful in many practical applications. Table 15.5 and other similar results from the literature (see references in the table) provide Neumann values for several polymers obtained from the method. Clearly, the Neumann method has been applied to a wide range of polymers and other solids (over 50 different solid surfaces) with surface tensions from around or below 10 mN m^{-1} up to more than 40 mN m^{-1}. This list includes thus hydrophobic, non-polar and polar polymers. This wide applicability is a positive characteristic of the method.

On the other hand, the criticism of the van Oss–Good theory by Neumann and co-workers is often unjustified. It is important that the theories are compared – treated as the developers would have expected and as they have recommended. In the case of van Oss–Good, we should typically use water, a non-polar and a polar fluid (e.g. formamide or glycerol). When this is done, even by Kwok *et al.* (2000b), the values obtained for PMMA are quite reasonable and in agreement with those obtained from the Neumann theory (see Table 15.5 extracted from Tables 9–11 in Kwok *et al.*, 2000b for PMMA and a few other polymers). Thus, the criticism of the van Oss–Good theory for solid interfaces in the article of Kwok *et al.*, 2000b about the many different values obtained from different triplets is somewhat exaggerated. It is simply not recommended to use so randomly chosen liquids for obtaining the van Oss–Good parameters as Kwok *et al.* do in their 2000b article.

Overall, we believe that the van Oss–Good method is, despite its well-documented limitations, the best among the two recent theories and moreover it can be used for surface analysis as well. The Neumann

method can be applied to a wide range of surfaces, but solely for estimating the surface tension of the solids.

15.4 A new theory for estimating interfacial tension using the partial solvation parameters (Panayiotou)

Panayiotou has recently (2012a, 2014) proposed a new theory for the interfacial tension that – at least at a first level – bears some similarities to the van Oss–Good theory. The surface tension is divided into a van der Waals (essentially a dispersion) part d, a polarity contribution pz and an asymmetric hydrogen bonding contribution hb:

$$\gamma = \gamma_d + \gamma_{pz} + \gamma_{hb} = \gamma_d + \gamma_{pz} + 2\sqrt{\gamma_a \gamma_b} \qquad (15.14)$$

Equation 15.14 resembles indeed the van Oss–Good equation (Equation 3.25), where the dispersion and polar part are, in the van Oss–Good equation, combined in the LW (Lifshitz–van der Waals) term.

Similarly, the equations of the interfacial tension as well as its combination with the Young equation for the contact angle can be written as (for solid–liquid interfaces, s and l indicate solid and liquids, respectively):

$$\gamma_{ij} = \gamma_i + \gamma_j - 2\left\{ \sqrt{\gamma_{d,i}\gamma_{d,j}} + \sqrt{\gamma_{pz,i}\gamma_{pz,j}} \right.$$
$$\left. + \sqrt{\gamma_{a,i}\gamma_{b,j}} + \sqrt{\gamma_{a,j}\gamma_{b,i}} \right\} \qquad (15.15)$$

$$\gamma_l(1 + \cos\vartheta) = 2\left\{ \sqrt{\gamma_{d,l}\gamma_{d,s}} + \sqrt{\gamma_{pz,l}\gamma_{pz,s}} \right.$$
$$\left. + \sqrt{\gamma_{a,i}\gamma_{b,s}} + \sqrt{\gamma_{a,s}\gamma_{b,l}} \right\} \qquad (15.16)$$

Equation 15.16 is the final equation for characterizing solid surfaces.

Indeed, Equations 15.15 and 15.16 resemble very much the corresponding equations of the van Oss–Good theory, e.g. as compared to Equations 15.11 and 6.3.2.

The acid–base (+/–) contributions in the van Oss–Good theory are now called *a/b*; this is largely a notation difference. A more important difference is that two terms are used now for the van der Waals forces

Table 15.5 *Surface tension values for a wide range of polymers estimated from various theories. All surface tension values are given in mN m^{-1}. The original van Oss–Good test liquid parameters are used to generate the van Oss–Good values*

Polymer	Zisman critical surface tension	Surface tension (Neumann)	Surface Tension (van Oss–Good)	LW (van Oss–Good)	Acid component (van Oss–Good)	Basic component (van Oss–Good)
PMMA	39	38.5	39.8	38.44	0.080	18.1
				35	0.04	10.89
				41.2	0.0	12.2
					0.38	7.5
PEMA		33.6				
PBMA		28.8				
P-*tert*-BMA (PtBMA)		18.1				
PMMA-PEMA (30 : 70)		35.1				
PMMA-PBMA		34.4				
PVC	39			43	0.42	5.1
					0.04	3.5
PS	33	28.9–30.2		42–43.7	0	1.1–7.4
PS-PMMA copolymer (70 : 30)		33				
FC-725		12.28				
PET		35.22				
FC-721		11.78	9.0	9.0	0.0	0.81
Teflon-FEP		17.85		16.7	0.07	0.22
Teflon AF 1600		13.61–13.64				
EGC-1700 (fluorinated acryl polymer)		13.84				
Hexatriacontane		20.4				
Siliconized glass		18.2				
Poly(propene-*alt-N*-methylmaleimide)		42.4				
Nylon 6,6		41.5		38.6	0.39	11.3
Poly(vinylidene fluoride)		34.3				
Poly(vinyl fluoride)		35.7				
Poly(vinylidene chloride)		35.7				

References of values: McCafferty (2002); Kwok *et al.* (1998, 2000a,b); Tavana and Neumann (2007); Tavana *et al.* (2004, 2005); Wulf *et al.* (2000); Kwok and Neumann (1998, 2000a,b); Clint and Wicks (2001a,b). For PMMA the first values are from McCafferty (2002) and the second ones from Kwok *et al.* (2000b). The last values are from Lloyd (1994) and Clint and Wicks (2001a,b).

(dispersion and polar interactions) instead of one (in the van Oss–Good theory), but this is also a rather small difference. Both models consider the hydrogen bonding interactions as asymmetric, comprising acid–base interactions between, e.g. solid and liquid surfaces. In this sense, both models represent correctly the acid–base (and other hydrogen bonding) interactions.

However, the most important difference between the two approaches lies on the way the surface tension components are estimated. In the van Oss–Good approach, the surface tension components for liquids and for solids are estimated from a wide range of experimental data (liquid–liquid interfacial tensions, contact angles, etc.) often regressed simultaneously for various solids and liquids. As we discussed, there are no predictive or estimation methods proposed by van Oss–Good for calculating these surface tension components.

On the other hand, Panayiotou has proposed a novel and potential groundbreaking method for estimating the surface tension components. This is a method based on a quantum mechanical approach and, in particular, the COSMO-RS method pioneered by Klamt and co-workers (see Klamt (2005) for a detailed review of the method).

Panayiotou's method is based on the so-called PSPs (partial solvation parameters) which he has recently developed and applied in numerous cases (polymer–polymer miscibility, polymer–solvent interactions, solubility parameters, pharmaceuticals, phase equilibria, etc.) (see Panayiotou, 2012b,c,d, 2013). The PSPs bear similarities with the Hansen solubility parameters presented in Section 3.4.2 but there are four distinct contributions due to dispersion, polarity, acid and base contributions $\sigma_d, \sigma_{pz}, \sigma_{Ga}, \sigma_{Gb}$ and, moreover, there are predictive methods for their estimation.

They are estimated from quantum chemical calculations (COSMO files and the so-called "σ profiles"). Details are presented by Panayiotou (e.g. 2012a, 2014).

Here we only present the final equations (and we refer to Panayiotou articles for details and the derivations):

$$\gamma_d = \frac{V_{cosm}}{A_{cosm}}\sigma_d^2 \qquad (15.17)$$

$$\gamma_{pz} = \frac{\gamma - \gamma_d}{\left(1 + 4\dfrac{\sigma_{Ghb}\sqrt{\sigma_{Ga}\sigma_{Gb}}}{\sigma_{pz}^2}\right)} \qquad (15.18)$$

$$\gamma_{hb} = 2\sqrt{\gamma_a\gamma_b} = 4\gamma_{pz}\frac{\sigma_{Ghb}\sqrt{\sigma_{Ga}\sigma_{Gb}}}{\sigma_{pz}^2} \qquad (15.19)$$

where:

$$\sigma_{Ghb} = \left(\sigma_{Ga}^2 + \sigma_{Gb}^2\right)^{1/2}$$

$$\frac{\gamma_{pz}}{2\gamma_a} = \frac{\sigma_{pz}^2}{2\sigma_{Ga}\sigma_{Ghb}}, \qquad \frac{\gamma_{pz}}{2\gamma_b} = \frac{\sigma_{pz}^2}{2\sigma_{Gb}\sigma_{Ghb}} \qquad (15.20)$$

The four PSPs are obtained in a clear way provided the COSMO file of the molecule is available or it can be calculated from quantum chemical calculations. As is clear from these equations, both the PSPs and thus also the resulting surface tension components rely heavily on the COSMO-RS quantum chemical model.

First of all, the dispersion part should be calculated from Equation 15.17. The reported COSMO volumes and areas should be used together with the dispersion PSP. However, for polymers such values are not available and Panayiotou (2012a, 2014) proposes an alternative method using the polymer monomer COSMO values. Alternatively, the dispersion part of the surface tension can be estimated from contact angle data using the usual approach (Equation 15.16).

This is a very recently developed method and much testing is needed as well as comparison with the established van Oss–Good approach. However, the first results presented by Panayiotou (2012a, 2014) are quite promising. Surface tension components have been presented for many liquids and polymeric solids and good estimations of contact angles are shown using Equation 15.16. In some cases, e.g. for polymers, as mentioned due to difficulties of using Equation 15.17, some contact angle data should be used in estimating the dispersion (and polar) contributions of polymers.

In addition, some interesting conclusions can be mentioned. Panayiotou's (2014) surface tension parameters (acid and base) for water, based on PSPs, are identical, in agreement to the original van Oss–Good parameters, but in disagreement with the most recent van Oss–Good parameters from Della Volpe–Siboni and from Lee. As we mentioned, the latter parameters indicate a much higher acid component for water (see Table 15.4). Panayiotou states that this is not correct and is not in agreement with the symmetric COSMO quantum chemical calculations (symmetric "σ" profiles for water). Moreover, Panayiotou's acid–base parameters are "more balanced" compared to the van Oss–Good parameters and, for example, the reported values for PMMA and PVC clearly indicate that the former is a "basic" polymer, while the latter is an "acidic" polymer. The same can be stated for many solvents, e.g. for glycol, glycerol and formamide the base/acid ratio of the PSP-based surface tensions is in the range 1.1–1.5 versus 15–25 for the same compounds with the van Oss–Good method. The huge base values for the surface tensions with the van Oss–Good method are apparently erroneous but they were considered "necessary" to obtain good results for solid–liquid interfaces (with the van Oss–Good approach).

Finally, it is worth comparing the Panayiotou PSP method with the Hansen method for estimating the surface tension components which we introduced in Chapter 3. Let us consider some examples. For water, see Problem 3.7, the Hansen method yields a dispersion, polar and hydrogen bonding contribution to surface tension equal to 21.85, 23.26 and 27.68 while with PSP we get 21.13 (d+p) and 51.67 for hydrogen bonding. We consider the PSP result more realistic, as it is unlikely that the polar contribution is as high as Hansen's method indicates. For ethyl acetate (Example 3.5), the Hansen method yields 19.57, 1.54 and 2.78 for the dispersion, polar and hydrogen bonding contribution, respectively. The PSP method yields 16.9 + 6.34 (=23.24) for the dispersion + polar contributions and this is the total surface tension. The method correctly predicts a base contribution for ethyl acetate (=7.33) but since the acid contribution is zero (ethyl acetate is

a non-hydrogen bonding) then the total hydrogen bonding contribution is zero. Thus, again the PSP method is physically more correct. The next example is formamide. Hansen's surface tension components are 17.2, 26.8 and 14.6 (dispersion, polar, hydrogen bonding) and PSP's are 17.6, 12.7 and 28.1, which we consider more realistic (more significant hydrogen bonding contribution). The Fowkes/Owens–Wendt and van Oss–Good all give about 40 mN m^{-1} for the "dispersion" or van der Waals part and 19 for the "specific" or hydrogen bonding part. These models do not differentiate between dispersion, polar and hydrogen bonding contributions. However, in the van Oss–Good method the acid and base components are 2.3 and 39.6, respectively, while in the PSP method they are 13.34 and 14.79. Again we consider the more "balanced" acid–base component of formamide with PSP to be more correct than the corresponding value of the van Oss–Good approach (even if the total hydrogen bonding contributions are not all that different).

Finally, for benzene both the Hansen and the PSP methods need improvement. The Hansen method has a dispersion value (25.9) and a hydrogen bonding (3.0) part – the latter to "represent" polar and base effects, but benzene is not a self-associating molecule. The PSP method has a dispersion part (19.95) and a significant polar part (8.25) but no basic contribution, thus assuming that all interactions with water are due to polar effects, which is not correct (benzene is actually a rather non-polar molecule). The result shown in Example 15.3 verifies this problem.

Overall, as based on the above discussion, most results with the PSP method are rather promising and future investigations will show whether this approach will provide a potential substitute of the van Oss–Good method for engineering applications.

Example 15.3 shows an application of the PSP method to some demanding liquid–liquid interfaces.

Example 15.3. *Evaluation of the Panayiotou method for liquid–liquid interfaces.*
The table below presents some surface tension components for a few liquids (from Panayiotou, 2014). They are estimated from quantum chemical calculations using the PSPs as explained by Panayiotou (2012a, 2014).

Compound	γ_W (mN m^{-1})	γ_{pz} (mN m^{-1})	γ_b (mN m^{-1})	γ_α (mN m^{-1})	γ (mN m^{-1})
n-Hexane	17.56	0.35	0	0	17.91
Cyclohexane	18.20	6.45	0	0	24.64
Benzene	19.95	8.25	0	0	28.21
Carbon tetrachloride (CCl$_4$)	26.00	1.00	0	0	27.00
1-octanol	18.96	2.62	2.97	2.22	27.10
Ethylene glycol	18.92	10.82	10.23	8.14	47.99
Water	15.71	5.42	25.84	25.84	72.80

Calculate using the Panayiotou method the liquid–liquid interfacial tension for water–hexane (51.1 mN m^{-1}), water–benzene (35 mN m^{-1}), MEG–hexane (16 mN m^{-1}), MEG–cyclohexane (14 mN m^{-1}), water–CCl$_4$ (45 mN m^{-1}) and water–octanol (8.5 mN m^{-1}).

In parenthesis are the experimental values for the interfacial tensions.

Compare the results both with the experimental data and with those obtained from the Fowkes equation (see Example 3.3 and Problems 3.5, 3.6 and 3.8).

Provide a short discussion.

Solution

For all systems except for water–octanol, Equation 15.15 is written as follows (as there are no acid and base contributions of the hydrocarbons and CCl$_4$):

$$\gamma_{ij} = \gamma_i + \gamma_j - 2\left\{\sqrt{\gamma_{d,i}\gamma_{d,j}} + \sqrt{\gamma_{pz,i}\gamma_{pz,j}}\right\} \tag{15.3.1}$$

We first apply Equation 15.3.1 to water–hexane:

$$\gamma_{ij} = 72.8 + 17.91 - 2\left\{\sqrt{(15.71)\times(17.56)} + \sqrt{(5.42)\times(0.35)}\right\} = 54.74\ \text{mN}\,\text{m}^{-1}$$

This is in very good agreement with the experimental data (7% deviation), considering that this is an entirely predictive result.

The Fowkes equation cannot be used for the prediction of the interfacial tension of this system. As we saw in Example 3.3, we can use the experimental liquid–liquid interfacial tension data to "back calculate" the "dispersion part" of the surface tension of water. The value obtained from the Fowkes equation (21.8 mN m^{-1}) is in very good agreement with the sum of the dispersion + polar contributions from the Panayiotou method (15.71 + 5.42 = 21.13 mN m^{-1}). The advantage of the PSP method is that these values have been calculated "a priori" without use of experimental interfacial tension data.

Unfortunately, the PSP method has the same problem as Fowkes method for water–benzene. It predicts an interfacial tension of 52.23 mN m^{-1} (51.5 mN m^{-1} with Fowkes) as both methods ignore the acid–base interactions between water and benzene.

For MEG–cyclohexane, we obtain, again by using Equation 15.3.1:

$$\gamma_{ij} = 47.99 + 24.64 - 2\left\{ \sqrt{(18.92)\times(18.2)} + \sqrt{(10.82)\times(6.45)} \right\} = 18.81\,mN\,m^{-1}$$

This is again in satisfactory agreement with the experimental value of 14 mN m^{-1}, considering the predictive character of the PSP method.

The deviation is higher for MEG–hexane (PSP prediction is 25.5 mN m^{-1} versus the experimental value of 16 mN m^{-1}).

Nevertheless, the PSP method can qualitatively predict the much lower interfacial tensions of MEG with hexane compared to water–hexane, apparently due to the higher miscibility of the former system.

As we saw in Problem 3.6, again the Fowkes equation can only be applied indirectly to the glycol–hydrocarbon systems. For example, using the data for MEG–cyclohexane and the Fowkes equation, we can calculate a dispersion value for the MEG equal to 34.36 mN m^{-1} (thus the specific value is equal to 13.3 mN m^{-1}). Then, using this value for the dispersion surface tension, we can calculate the interfacial tension for MEG–hexane, which is calculated to be 15.8 mN m^{-1}, very close to the experimental value.

We can conclude that the Fowkes equation can be applied to the MEG–alkane interfaces. However, these results are not predictive but a qualitative validation of the Fowkes equation against the experimental data for MEG–hydrocarbon interfaces.

Similarly, for water–CCl$_4$ we estimate with the PSP method a value equal to 54.7 mN m^{-1}, a difference of about 10 mN m^{-1} from the experimental value. A similar result is obtained from the Fowkes equation (51.2 mN m^{-1}, using the "accepted" value for the dispersion surface tension of water of 21.8 mN m^{-1}).

Finally, for the water–octanol interface we should use the full Equation 15.15. We then arrive at an interfacial tension of about 25.2 mN m^{-1}. The PSP method qualitatively predicts that the interfacial tension is, for this system due to adsorption phenomena, lower than both water's and octanol's surface tension. However, this value is much higher than the experimental value (8.5 mN m^{-1}). The Fowkes equation cannot be used without experimental data for this system as we do not know a priori the dispersion part of the surface tension of octanol.

In conclusion, for these very difficult liquid–liquid interfacial tensions, promising results are obtained with the PSP method but improvements are needed. Of course, it should be mentioned that testing the methods against liquid–liquid interfacial tension data is possibly a very difficult test for all theories. They will usually perform better for liquid–solid interfaces when reliable contact angle data are available.

15.5   Conclusions – *Quo Vadis*?

In Chapters 3 and 6 we have briefly presented the most important theories for estimating interfacial tensions as well as their applications to wetting, adhesion and surface analysis. In this chapter we discuss the historical development of these theories and evaluate their applicability from a wider point of view.

We can conclude that all theories have positive and negative features, advantages and problems or, in other words, areas/systems where they can be reliably used and others where they cannot. Even the classical theories of Fowkes and Owens–Wendt are still much in use. Actually, the latter, despite its criticism by,

among others, the late Frederick Fowkes and Robert Good is still the most cited theory today. It is widely used for polymeric surfaces.

The Fowkes, Owens–Wendt and the Girifalco–Good theories are those most often presented in textbooks on colloids and interfaces but they would not be the ones we recommend using today. The more modern theories of Neumann and the acid–base by van Oss–Good are better choices. Strangely they are not presented in textbooks but they are the ones most widely accepted in engineering practice. The consensus today is that the Neumann theory has a rather weak theoretical justification but can be reliably used to estimate surface tensions of rather non-polar or slightly polar solids. On the other hand, the van Oss–Good theory has a wider applicability and acceptance. It can be used for both non-polar and highly polar/hydrogen bonding surfaces and is also used for surface analysis, i.e. an assessment of the polar or acid–base character of surfaces. This is highly useful, e.g. in adhesion studies.

During the last decade or so, most new developments focused on improving the van Oss–Good theory, especially by introducing new parameters for the "reference" liquids used. These parameters improve the performance and result in relatively more "correct" values for the acid and base components of solids when the theory is used together with contact angle data for surface analysis.

Nevertheless, as also discussed in this chapter, there are several problems even with the van Oss–Good theory and, while useful in many practical applications, it should be used with care and by experienced users who are familiar with the interpretations of the results. Due to these problems, it is most likely that we have not as yet seen the last in the development of theories for the interfacial tension. The recent Panayiotou theory based on the partial solvation parameters may prove to be a promising tool for predictive calculations of interfacial tensions.

Problems

Problem 15.1: Comparison of classical theories

The surface tensions of diethylene glycol (DEG), hexane, heptane, decane and mercury (Hg) are, at 20 °C, equal to 30.9, 18.4, 20.14, 23.83 and 486.50 mN m^{-1}, respectively. The experimental values for the interfacial tensions of DEG–heptane, DEG–decane and Hg–hexane are, at the same temperature, equal to 10.6, 11.6 and 378, in mN m^{-1}, respectively.

Which of the two classical theories, of Fowkes and Owens–Wendt, will perform best in describing or predicting the interfacial tensions for these three mixtures?

Problem 15.2: What is the best model for a wide range of aqueous liquid–liquid interfaces?

The table below presents liquid–liquid interfacial tensions (and liquid surface tensions) in aqueous systems

for several compounds. All values are given at room temperature (20 °C). The surface tension of water has the well-known value (72.8 mN m^{-1}).

System	Interfacial tension (mN m^{-1})	Liquid	Surface tension (mN m^{-1})
Water–benzaldehyde	15.5	Heptanoic acid	28.18
Water–CS$_2$	48.0	Benzaldehyde	38.54
Water–methylene iodide	45.9	CS$_2$	32.32
Water–ethyl acetate	6.8	Methylene iodide	50.8
Water–ethyl ether	10.7	Ethyl acetate	23.9
Water–aniline	5.85	Ethyl ether	17.01
Water–heptanoic acid	7.0	Aniline	42.67

1. Comment on the values of the liquid–liquid interfacial tensions shown in this table and compare them to the surface tensions of the individual liquids. What do you observe?
2. What is the best model which could represent satisfactorily the wide range of liquid–liquid interfacial tensions shown in this table? Justify your answer with some calculations.

Problem 15.3: Interfacial tensions for non-aqueous liquid mixtures

Israelachvili (1985, 1992) presented experimental liquid–liquid interfacial tension data for several liquids against *n*-tetradecane (26). These interfacial tensions are 53 for mixtures with water (72.8), 31–36 for glycerol (64), 29–32 for formamide (58), 18–20 for ethylene glycol (48) and 5 for dimethylformamide (37).

The values in parenthesis are the surface tensions of pure liquids. All surface and interfacial tensions are given at 20 °C in mN m^{-1}.

Compare the Fowkes, Owens–Wendt and van Oss–Good theories for these non-aqueous liquid–liquid interfacial tensions. Which model performs best?

Problem 15.4: Hansen theory against others. How does it compare?

Chapters 3–6 presented (in problems, etc.) dispersion, polar and hydrogen bonding contributions to surface tension according to the Hansen–Beerbower method for several polymers (PMMA, PS, PET, PA-6) and liquids (formamide, ethyl acetate, propylene carbonate).

The Hansen method has not been extensively compared with the classical "surface component" methods. How are the dispersion and polar/hydrogen bonding contributions to surface tension from the Hansen method compared to the corresponding contributions from the Fowkes/Owens–Wendt and van Oss–Good methods? Comment on your findings.

Problem 15.5: Evaluation of the Neumann equation for contact angle (based on the work of Wulf et al., 2000)

Table 15.5 presents surface tension values for several polymers with the Neumann equation. Use the value given by the developers (shown in Table 15.5) for PtBMA and calculate the contact angle of water (72.7), glycerol (65.02) and ethylene glycol (48.66). Compare the results to the experimental values of the contact angle (°) which are, respectively, equal to 108.08, 101.86 and 85.79.

The values in parenthesis are the surface tensions of pure liquids given at 20 °C in mN m^{-1}.

Problem 15.6: The Dann method compared to Owens–Wendt. Based on Dann analysis and data (Dann, 1970).

In his analysis of the wetting properties of several polymers using his model (Equation 15.8) Dann (1970) presented several data for the contact angle of water (72.8 mN m^{-1}) on many polymers. Some of these data together with the dispersive contribution to the surface tension of the solids are given in the table below.

Polymer	γ_s^d(mN m^{-1})	Contact angle (°)
Teflon	18.5	113
Paraffin	25.5	110
PE	38.0	95
PS	40.0	84
PVC	40.0	83
PET	43.0	71
Nylon 6,6	47.0	65

Calculate the specific contributions to the solid surface tension based on the Owens–Wendt theory and then compute the contact angle of water on these polymers using the OW theory. Compare the results to the experimental data.

Problem 15.7: Water and mercury liquid–liquid interfaces with classical theories

The tables below present surface tensions for some liquids and interfacial tensions in mixtures with water (72.8 mN m^{-1}) and mercury (484 mN m^{-1}). All values are at 20 °C.

Hydrocarbon	Surface tension (mN m^{-1})	Interfacial tension with water (mN m^{-1})
Hexane	18.4	51.1
Heptane	18.4	50.2
Octane	21.8	50.8
Decane	23.9	51.2
Tetradecane	25.6	52.2
Cyclohexane	25.5	50.2
Decalin	29.9	51.4

Hydrocarbon	Surface tension (mN m^{-1})	Interfacial tension with mercury (mN m^{-1})
Hexane	18.4	378
Octane	21.8	375
Benzene	28.85	363
o-Xylene	30.1	359

1. Apply the Fowkes and Owens–Wendt equations to these interfaces. What do you conclude on the applicability of these methods for these interfaces?
2. Apply both methods (after having estimated suitable values for the surface tension components) to the interface between mercury and ethanol (22.27 mN m^{-1}) and mercury and methylene iodide (30.1 mN m^{-1}). Compare the results to the experimental values which are equal to 389 and 304 mN m^{-1}, respectively.

Problem 15.8: Prediction of miscibility in liquid mixtures

According to the van Oss–Good theory, zero or negative interfacial tensions indicate miscible liquid mixtures while of course positive interfacial tensions are obtained for immiscible liquids having a distinct interface.

The aqueous mixtures of glycerol, DMSO, EG and formamide are highly miscible. Equally miscible are the mixtures glycerol/DMSO, formamide/DMSO, glycerol/formamide and glycerol/EG.

On the other hand, water–hexadecane is immiscible (interfacial tension 51.3 mN m^{-1}) and the same is the case for water–octanol (with interfacial tension 8.5 mN m^{-1}).

How accurately do the van Oss–Good and Owens–Wendt theories perform for such interfaces? Are these theories able to identify when we get miscible and we get immiscible liquid mixtures? Compare their performance with each other and against the experimental observations.

Compare also your results to the analysis carried out by the opponents of the van Oss–Good theory, the group of Neumann and co-workers (see, for example, Kwok and Neumann, 1996; Kwok *et al.*, 1998, Kwok, 1999). What do you observe?

Problem 15.9: The Neumann equation and other direct methods for estimating solid surface tensions using data from solid–liquid interfaces. Based on the work by Kwok et al. (2000)

For the 70 : 30 PS–PMMA copolymer, Kwok *et al.* (2000) have reported the following contact angles (°) for, respectively, DEG (44.68), formamide (59.08), glycerol (63.13) and water (71.7): 47.10, 67.9, 73.31 and 83.78.

The values in parenthesis are the surface tensions of pure liquids given at 20 °C in mN m^{-1}.

1. Calculate the solid surface tension for the copolymer using the Neumann equation, the Girifalco–Good equation ("φ" = 1) and the Antonov rule:

$$\gamma_{sl} = |\gamma_{lv} - \gamma_{sv}|$$

   What do you observe?

2. Using the average value of the solid surface tension given for the Neumann equation (see

also Table 15.5), estimate with this model the contact angles of the four liquids and compare the predictions to the experimental values given in the problem.

Problem 15.10: Interfacial tensions between molten polymers. Based on Wu (1971)

Wu (1971) presented an extensive investigation of liquid–liquid interfacial tensions for molten polymers at high temperatures. All data and calculations in this problem are at 140 °C. For four PDMS-containing mixtures, he presented the following experimental data for interfacial tensions (in mN m^{-1}):

PVAC–PDMS: 7.4, PCP–PDMS: 6.5, PTHF–PDMS: 6.3 and PBMA–PDMS: 3.8.

The table below presents the surface tensions and Fowkes dispersion parameters for these polymers in mN m^{-1}. Also shown is the polar/total surface tension ratio for the Harmonic method for interfacial tension developed by Wu himself (it is equation HARM-2 in Problems 3.8/Chapter 3).

Polymer	Surface tension	Fowkes (γ_s^d)	Wu Harmonic method γ^p/γ
PCP (polychloropropylene)	33.2	29.5	0.11
PVAC (poly(vinyl acetate))	28.6	18.5	0.33
PDMS (poly (dimethylsiloxane))	14.1	12.0	0.04
PTHF (polytetrahydrofuran)	24.6	20.8	0.14
PBMA (poly(butyl methacrylate))	24.1	19.7	0.16

Compare the Fowkes, Owens–Wendt and Harmonic Wu equations for the four polymer mixtures for which experimental data were provided. Which method performs best? Provide a short discussion.

Problem 15.11: Analysis of surface impurities using the Fowkes equation

Fowkes, McCarthy and Mostafa (1980) have illustrated that the spreading pressure of methylene iodide (50.8 mN m^{-1}) on low energy surfaces like PE and Teflon is zero.

Fowkes and Zisman have also reported contact angle values of this liquid on "clean" PE (where all hydrophilic impurities are extracted) is equal to 52°.

Calculate with the Fowkes method the surface tension of PE and water's contact angle on this polymer.

The experimental contact angle value on "clean PE" is 106°. Other researchers (Zisman, Adamson) have reported contact angles of water on "diverse types and quality" PE in the range 88–98°.

Compare your results to these diverse values from the literature and provide an explanation for the wide range of contact angle values reported. What type of contact angle is calculated with the Fowkes equation and how can the Fowkes equation be used to analyse the PE surfaces in this case?

Problem 15.12: Surface properties of high energy solids

Schultz *et al.* (1977) presented the following data for wetting properties of liquids on mica.

Liquid	Surface tension (mN m^{-1})	Dispersion part of surface tension (mN m^{-1})	Contact angle (°)
Mercury	484	200	134
Tricresyl phosphate	40.9	39.2	38
α-Bromonaphthalene	47	44.6	38.5

Schultz *et al.* (1977). Reproduced with permission from Elsevier

The same authors presented also the following experimental data:

Hydrocarbon	Hydrocarbon surface tension (mN m^{-1})	Water–hydrocarbon interfacial tension (mN m^{-1})	Mica–water contact angle under alkanes (°)
Hexane	18.4	50.1	13
Octane	21.3	49.8	16
Decane	23.4	51.0	17
Hexadecane	27.1	51.3	19.5

Schultz *et al.* (1977). Reproduced with permission from Elsevier

Provide an estimation of the dispersion part of the surface tension of Mica as well as its total surface tension using both series of data. Compare the results to each other. What do you observe?

Problem 15.13: Issues with the van Oss–Good approach. From Balkenende et al. (1998)

Balkenende *et al.* (1998) discussed several issues with the van Oss–Good approach, among others how the dispersive (LW) part of the surface tension of liquids can be estimated from solid–liquid interfacial data and other approaches.

For the solid FC722 (perfluoro polyacrylate), with a solid surface tension equal to 12.1 mN m^{-1}, are reported the following experimental values for the contact angle (°) of diiodomethane (50.8): 102, 1-bromonaphthalene (43.9): 91.6, EG (48.2): 102.2 and formamide (57.5): 109.1.

The values in parenthesis are the surface tensions of pure liquids given at 20 °C in mN m^{-1}.

Calculate from these data the LW contribution of the van Oss–Good theory for these four liquids and compare the results to reported values in literature from other methods, e.g. data from liquid–liquid interfaces or solid–liquid interfaces with other solids (50.8 for diiodomethane, 29–33.6 for EG, 33.5–39 for formamide and 32.8–44.01 for 1-bromonaphthalene).

What do you observe?

Problem 15.14: Solid surface tensions using spreading pressure (adsorption) data

1. For the Fowkes equation, show that the dispersion part of the solid surface tension can be estimated from spreading pressure data (which are obtained from adsorption studies at complete wetting) as follows:

$$\gamma_s^d = \frac{(\pi_{SV} + 2\gamma_l)^2}{4\gamma_l^d}$$

2. Fowkes (1964) reports for the *n*-heptane (18 mN m^{-1})–copper system a spreading pressure equal to 29 mN m^{-1} and for the *n*-heptane–iron system a spreading pressure equal to 53 mN m^{-1}. Estimate the solid surface tension (dispersion part) of copper and iron using these experimental data.

3. What is the expected spreading pressure for the system silica–heptane if the surface tension of silica (dispersion part) is equal to 78 mN $^{-1}$m (Fowkes, 1964)?

4. Formamide has a surface tension equal to 58.2 mN m^{-1} and interfacial tension against *n*-pentane equal to 25.97 mN m^{-1}. The surface tension of *n*-pentane is equal to 16.1 mN m^{-1}. Investigate, using the Fowkes equation, on which high energy solids, of those mentioned in questions 2 and 3 (iron, copper and silica), water and formamide can spread. All values are at 25 °C and book values should be used for the surface tension (total and dispersion) of water.

5. The spreading pressure effect can be neglected for low energy surfaces like polymers. Fowkes, McCarthy and Mostafa (1980) have measured experimental contact angles of water on commercial Teflon (PTFE) equal to 108° and other researchers report even lower values (around 100°).

    Using data on *pure* Teflon, Fowkes *et al.* calculated the dispersion part of Teflon's surface tension to be equal to 19.6 mN m^{-1}.

    Estimate the contact angle of water on *pure* Teflon and compare the results to the experimental values reported. Comment briefly on the results.

Problem 15.15: Validation of various theories using data for squalane liquid–liquid interfacial tensions (data are from Fowkes et al., 1990)

The table below shows Fowkes/OW parameters for some liquids as well as experimental values for the liquid–liquid interfacial tension and experimental observations about the solubility of these compounds in squalane (29.2 mN m^{-1}). All data are at 23 °C and in mN m^{-1}.

Liquid (surface tension)	Dispersion part of surface tension (mN m^{-1})	Specific part of surface tension (mN m^{-1})	Interfacial tension versus squalane (mN m^{-1})	Solubility in squalane (%)
Water (72.4)	21.1	51.3	52.3	0.01
Acetone (23.7)	22.7	1.00	1.4	0.2
Ethanol (22.2)	20.3	1.9	2.7	0.2
Aniline (42.4)	37.3	5.1	5.6	0.01
Pyridine (38.0)	38.0	0.0	0.0	Miscible

Evaluate the GG (without the correction parameter "φ"), the Fowkes and the Owens–Wendt equations for the squalane systems with water, aniline, pyridine, ethanol and acetone as well as for the ethanol–acetone system (which is miscible). Provide a discussion.

Problem 15.16: Characterization of polymer surfaces using the PSP method

Characterize the polyethylene (PE) and Teflon (PTFE) surfaces of Example 15.1 using the PSP method and, when needed, the reported (in Example 15.1) contact angle data for water and methylene iodide (MI).

More specifically:

1. Estimate the contact angle of water and MI on the two polymer surfaces using predicted values of the surface tension components based on PSPs and related information (see also Panayiotou, 2012a, 2014).

2. Use the experimental contact angle data of MI on the two polymer surfaces to estimate the dispersion part of the polymer surface tension. Then, estimate the contact angle of water on both polymers.

3. Compare the results of questions (1) and (2) with each other and with the results of the Owens–Wendt method (Problem 15.1) and the van Oss–Good approach.

References

A.R. Balkenende, H.J.A.P. van de Boogaard, M. Scholten, N.P. Willard, Langmuir, 1998, 14, 5907.

R.F. Brady, Progr. Org. Coatings, 1999, 35, 31.

J.H. Clint, Curr. Opinion in Colloid & Interface Sci., 2001a, 6, 28.

J.H. Clint, A.C. Wicks, Int. J. Adhesion & Adhesives, 2001b, 21, 267.

N.T. Correia, J.M.M. Ramos, B.J.V. Saramago, J.C.G. Calado, J. Colloid Interface Sci., 1989, 361.

T. Cosgrove (ed.), 2010. Colloid Science: Principles, Methods and Applications, John Wiley & Sons, Ltd, Chichester.

J.R. Dann, J. Colloid Interface Sci., 1970a, 32(2), 302.

J.R. Dann, J. Colloid Interface Sci., 1970b, 32(2), 321.

C. Della Volpe, S. Siboni, J. Mathematical Chem., 2007, 43 (3), 1032.

C. Della Volpe, D. Maniglio, S. Siboni, M. Morra, J. Adhesion Sci. Technol., 2003, 17(11), 1477 (part I) & 1425 (part II).

C. Della Volpe, D. Maniglio, M. Brugnara, S. Siboni, M. Morra, J. Colloid Interface Sci., 2004a, 271, 434 (part I) & 454 (part II).

C. Della Volpe, D. Maniglio, S. Siboni, M. Morra, Colloid Interface Sci., 2004b, 271, 434 (part I) & 454 (part II).

C. Della Volpe, S. Siboni, M. Morra, Langmuir, 2002, 18, 1441.

C. Della Volpe, S. Siboni, J. Adhesion Sci. Technol., 2000, 14, 235.

C. Della Volpe, S. Siboni, J. Colloid Interface Sci., 1997, 195, 121.

J.M. Douillard, J. Colloid Interface Sci., 1997, 188, 511.

R.S. Drago, B.B. Wayland, J. Am. Chem. Soc., 1965, 87:16, 3571.

J. Drelich, J.D. Miller, J. Colloid Interface Sci., 1994, 167, 217.

H.Y. Erbil (1997) *CRC Handbook of Surface and Colloid Chemistry*, ed. K.S. Birdi, CRC Press, Boca Raton, FL, Chapters 2 and 9.

H.W. Fox, W.A. Zisman, J. Colloid Sci., 1950, 5, 514.

H.W. Fox, W.A. Zisman, J. Colloid Sci., 1952, 7, 109 & 428.

F.M. Fowkes, M.B. Kaczinski, D.W. Dwight, Langmuir, 1991, 7, 2464.

F.M. Fowkes, F.L. Riddle Jr., W.E.Pastore, A.A. Weber, Colloids and Surfaces, 1990, 43, 367.

F.M. Fowkes, J. Adhesion Sci., Technol., 1990, 4(8), 669.

F.M. Fowkes, J. Phys. Chem., 1980, 84, 510.

F.M. Fowkes, Curr. Content Phys. Chem. Earth Sci., 1980 (issue 18), 12.

F.M. Fowkes, D.C. McCarthy, M.A. Mostafa, J. Colloid Interface Sci., 1980, 78(1), 200.

F.M. Fowkes, M.A. Mostafa, Ind. Eng. Chem. Prod. Res. Dev., 1978, 17(1), 3.

F.M. Fowkes, J. Adhesion, 1972, 4, 155.

F.M. Fowkes, Ind. Eng. Chem., 1964, 56(12), 40.

L.A. Girifalco, R.J. Good, J. Phys. Chem., 1960, 64, 561.

L.A. Girifalco, R.J. Good, J. Phys. Chem., 1957, 61, 904.

R.J. Good, J. Adhesion Sci. Technol., 1992, 6(12), 1269.

R.J. Good, L.K. Shu, H.-C. Chiu, C.K. Yeung, J. Adhesion, 1996, 59, 25.

R.J. Good, A.K. Hawa, J. Adhesion, 1997, 63, 5.

R.J. Good, ChemTech, 1980, 100.

R.J. Good, J. Colloid. Interface Sci., 1977, 59(3), 398.

R.J. Good, Nature, 1966, 212, 276.

J. Goodwin, 2004. Colloids and Interfaces with Surfactants and Polymers. An Introduction. John Wiley & Sons, Ltd, Chichester.

V.R. Gray, Nature, 1966, 209, 608.

C.M. Hansen, 2000. Hansen solubility parameters. A User's Handbook, CRC Press, Boca Raton, FL.

I. Hamley, 2000. Introduction to Soft Matter. Polymers, Colloids, Amphiphiles and Liquid Crystals, John Wiley & Sons, Ltd, Chichester.

P.C. Hiemenz, and R. Rajagopalan, 1997. Principles of Colloid and Surface Science, 3rd edn, Marcel Dekker.

J. Israelachvilli, 1985 and 1992. Intermolecular and Surface Forces, Academic Press.

W.B. Jensen, J. Adhesion Sci. Technol., 1991, 5(1), 1.

W.B. Jensen, ChemTech, 1982, 755.

W.B. Jensen, Chem.Rev., 1978, 78(1), 1.

M.J. Kamlet, J.M. Abboud, M.H. Abraham, R.W. Taft, J. Org. Chem., 1983, 48, 2877.

A. Klamt, *COSMO-RS from Quantum Chemistry to Fluid Phase Thermodynamics and Drug Design*, Elsevier, Amsterdam, 2005.

G.M. Kontogeorgis, G.K. Folas, 2010. Thermodynamic Models for Industrial Applications. From Classical and Advanced Mixing Rules to Association Theories. John Wiley and Sons Ltd, Chichester.

D.Y. Kwok, R. Wu, A. Li, A.W. Neumann, J. Adhesion Sci. Technol., 2000a, 14(5), 519.

D.Y. Kwok, A.W. Neumann, Colloids and Surfaces, A: Physicochemical and Engineering Aspects Sci., 2000a, 161, 31 and 49.

D.Y. Kwok, A.W. Neumann, J. Phys. Chem. B., 2000b, 104, 741.

D.Y. Kwok, C.N.C. Lam, A.W. Neumann, Colloid Journal, 2000b, 62(3), 369.

D.Y. Kwok, Colloids and Surfaces, A: Physicochemical and Engineering Aspects Sci., 1999, 156, 191.

D.Y. Kwok, A.W. Neumann, Adv. Colloid Interface Sci., 1999, 81, 167.

D.Y. Kwok, A.W. Neumann, Prog. Colloid Polym. Sci., 1998, 109, 170.

D.Y. Kwok, Y. Lee, A.W. Neumann, Langmuir, 1998a, 14, 170.

D.Y. Kwok, C.N.C. Lam, A. Li, A. Leung, R. Wu, E. Mok, A.W. Neumann, Colloids Surf., A: Physicochem. Eng. Aspects Sci., 1998b, 142, 219.

D.Y. Kwok, A.W. Neumann, Can. J. Chem. Eng., 1996, 74, 551.

J. Kudela, E. Liptakova, J. Adhesion Sci. Technol., 2006, 20 (8), 875.

L.-H. Lee, J. Colloid Interface Sci., 1999, 214, 64.

L.-H. Lee, J. Adhesion, 1998, 67, 1

L.-H. Lee, J. Adhesion, 1997, 63, 187.

L.-H. Lee, Langmuir, 1996, 12, 1681.

L.-H. Lee, Langmuir, 1993a, 9, 1898.

L.-H. Lee, Langmuir, 1994, 10, 3368.

L.-H. Lee, J. Adhesion Sci. Technol., 1993b, 7(6), 583.

L.-H. Lee, J. Adhesion Sci. Technol., 2000, 14(2), 167.

D. Li, A.W. Neumann, J. Colloid Interface Sci., 1992, 148(1), 190.

T.B. Lloyd, Colloids and Surfaces, A: Physicochemical and Engineering Aspects Sci., 1994, 93, 25.

E. McCafferty, J. Adhesion Sci. Technol., 2002, 16(3), 239.

M. Morra, J. Colloid Interface Sci., 1996, 182, 312.

I.D. Morrison, Langmuir, 1989, 5, 540.

I.D. Morrison, Langmuir, 1991, 7, 1833.

D. Myers, 1991 & 1999. Surfaces, Interfaces, and Colloids. Principles and Applications. VCH, Weinheim.

A.W. Neumann, R.J. Good, C.J. Hope, M. Seipal, J. Colloid Interface Sci., 1974, 49(2), 291.

D.K. Owens, J. Phys. Chem., 1970a, 74(17), 3305.

D.K. Owens, J. Appl. Polymer Sci., 1970b, 14, 1725.

D.K. Owens, R.C. Wendt, J. Appl. Polymer Sci., 1969, 13, 1741.

M. Osterhold and K Armbruster, Prog. Org. Coatings, 1998, 33, 197.

C.G. Panayiotou, *J. Chromatography A*, 1251, 194, 2012a.

C.G. Panayiotou, *Phys. Chem. Chem. Phys.*, 14, 3882, 2012b.

C.G. Panayiotou, *J. Chem. Thermodynamics*, 51, 172, 2012c.

C.G. Panayiotou, *J. Phys. Chem. B*, 116, 7302, 2012d.

C.G. Panayiotou, *Polymer*, 54, 1621, 2013.

C.G. Panayiotou, 2014. Adhesion and wetting: A quantum mechanics based approach. Chapter for the CRC Handbook on Colloid and Interfacial Science, Taylor & Francis (CRC Press), Boca Raton, FL.

H. Radelczuk, L. Holysz, and E. Chibowski (2002) *J. Adhesion Sci.* 16 (12), 1547–1568.

J. Schultz, K. Tsutsumi, J.-B. Donnet, J. Colloid Interface Sci., 1977a, 59(2), 272.

J. Schultz, K. Tsutsumi, J.-B. Donnet, J. Colloid Interface Sci., 1977b, 59(2), 277.

D. Shaw, 1992. Introduction to Colloid & Surface Chemistry. 4th edn, Butterworth-Heinemann, Oxford.

P.A. Small, J. Appl. Chem., 1953, 71.

J.R. Svendsen, G.M. Kontogeorgis, S. Kiil, C. Weinell, C., M. Grønlund, M., J. Colloid Interface Sci., 2007, 316, 678.

H. Tavana, A.W. Neumann, Adv. Colloid Interface Sci., 2007, 132, 1.

H. Tavana, F. Simon, K. Grundke, D.Y. Kwok, M.L. Hair, A.W. Neumann, J. Colloid Interface Sci., 2005, 291, 497.

H. Tavana, C.N.C. Lam, K. Grundke, P. Friedel, D.Y. Kwok, M.L. Hair, A.W. Neumann, J. Colloid Interface Sci., 2004, 279, 493.

C.J. Van Oss, R.J. Good, Langmuir, 1992, 8, 2877.

C.J. Van Oss, R.J. Good, J. Disp. Sci. & Technol., 1990, 11 (1), 75.

C.J. Van Oss, L. Ju, M.K. Chaudhury, R.J. Good, J. Colloid Interface Sci., 1989, 128, 313.

C.J. Van Oss, R.J. Good, M.K. Chaudhury, Langmuir, 1988a, 4, 884.

C.J. Van Oss, M.K. Chaudhury, R.J. Good, Chem. Res., 1988b, 88, 927.

C.J. Van Oss, M.K. Chaudhury, R.J. Good, Adv. Colloid Interface Sci., 1987, 28, 35.

S. Wu, J. Polym. Sci.: Part C, 1971, 34, 19.

S. Wu, J. Adhesion, 1973, 5, 39.

M. Wulf, K. Grundke, D.Y. Kwok, A.W. Neumann, J. Appl. Polymer Sci., 2000, 77, 2493.

Y. Zhang, G.M. Kontogeorgis, H.N. Hansen, 2011. An explanation of the selective plating of laser machined surfaces using surface tension components. J. Adhesion Sci. Technol., 25: 2101.

W.A. Zisman, 1964, in Contact Angle, Wettability and Adhesion, Advances in Chemistry Series, vol. 43, American Chemical Society, Washington D.C., p. 1.

16

Epilogue and Review Problems

Colloidal systems have come a long way since Thomas Graham's narrow definition more than 150 years ago (see Chapter 1). They comprise a much wider range of systems than Graham could have imagined. Suspensions (solid–liquid), emulsions (liquid–liquid), foams (gas–liquid), pastes and gels (solid–liquid, liquid-solid), aerosols (liquids or solids in gas), surfactants and even polymer solutions are all considered to be colloids (in parenthesis are mentioned the phases of the dispersed phase and dispersion media, respectively). Colloidal particles or droplets have a characteristic dimension roughly between the nanometre and micrometre, although in some of these systems the particles or droplets can be larger (several micrometres in several emulsions, for example).

Colloidal systems are notorious for their extensive interfaces whose understanding is very important for their stability, the latter possibly being their most important property. For this reason, colloid and surface phenomena are typically studied together and this has been the approach also followed in this book.

Colloid and surface chemistry is a multidisciplinary field of high importance to numerous products (food, detergents, paints, pharmaceuticals, etc.),

processes (adhesion, lubrication, cleaning, catalysis, etc.) and sciences (chemistry, engineering, biology, medicine, agriculture, etc.). Instead of a process- or product-oriented approach, we have followed a concept-oriented methodology in this book where examples of products, processes and other applications have been presented in all chapters.

We have not studied all types of colloidal systems in detail but limited ourselves to suspensions, surfactants, emulsions and foams. In terms of properties, the stability and associated concepts (double layer, van der Waals forces, steric effects) as well as the DLVO theory have been presented in detail, while kinetic and especially the optical properties have been discussed more briefly.

From surface science, we presented both the "general laws" (Young–Laplace, Kelvin, Gibbs adsorption equation, Young equation for contact angle) and the theories for estimating interfacial tension and how they are used in understanding wetting and adhesion phenomena. Adsorption has been discussed throughout the book but especially in Chapters 7 and 14; the latter focusses on multicomponent adsorption. Another topic that is apparent in essentially the whole

Introduction to Applied Colloid and Surface Chemistry, First Edition. Georgios M. Kontogeorgis and Søren Kiil.
© 2016 John Wiley & Sons, Ltd. Published 2016 by John Wiley & Sons, Ltd.
Companion website: www.wiley.com/go/kontogeorgis/colloid

Table 16.1   *Colloid and surface science – some areas of current and future research*

Theoretical studies	Surface and interfacial energies of solids, adsorption of polymers, steric stabilization, surface energies and bulk properties, density functional theory (DFT), molecular simulation, new theories for interfacial tension based on the partial solvation parameters
Surface chemistry	Adhesion, dynamic wetting, spectroscopic/microscopic analysis of surfaces: AFM, ESCA
Interparticle interactions	Measurement of forces, "special" forces: solvation, etc.
Colloid stability	Emulsions, foams and defoaming, microemulsion phase behaviour, coagulation theory
Complex products	Interplay of adsorption of polymers–surfactants–proteins, rheological properties, electrophoresis
Surfactants	Special structures, theories for micellization
	Complex phase behaviour, prediction of CMC
Lyophilic colloids	(Micro)gels, polymers in solutions and surfaces, liquid crystals
Biocolloids	Membranes, cell adhesion, drug delivery

book concerns the forces between molecules and particles/surfaces. Chapter 2 gives a unified presentation, while these forces are discussed and applied further in the remaining of the book but especially in Chapters 3 and 6 (interfacial theories, wetting, adhesion) and Chapters 10–12 (colloid stability).

Most chapters present exercises for which knowledge from the specific chapter alone is sufficient to provide the solution. This chapter closes with some review problems, several of which require combined knowledge from several chapters.

Finally, a few words are in order on research trends in the field, current and future ones. Colloid and surface science is not just an area of extensive applications but also an active field of research in essentially all its dimensions. Even a brief account of research trends in colloids and interfaces would require a separate book. We have nevertheless attempted to present research examples throughout the book, via case studies and problems, some of them taken from industrially-oriented applications. Several examples have been presented in the fields of paints, wetting and adhesion, surface characterization, surfactants and foams in Chapters 3, 5, 6 and 13.

Chapters 14 and 15 also present recent efforts in the very active fields of theories for interfacial tensions and multicomponent adsorption. Both are areas of not only very active research but also many heated debates. These chapters offer a short account of some research in the field which can serve as an inspiration for the interested reader to search deeper.

There are still many unanswered questions and areas where both data and good models are missing. One example is the multicomponent adsorption of polar compounds like water and glycols in solids such as microporous molecular sieves and silica gels. Such systems find applications in, for example, the drying of natural gas, for which multiple processes are used (glycol absorption and further drying by use of molecular sieves/silica gel). Good data and models for multicomponent adsorption would, in this and many more cases, help in the accurate design of the processes including increasing the lifetime of adsorbents. As recent studies have shown (Bjørner *et al.*, 2013, Bartholdy *et al.*, 2013 – see references in Chapter 14), existing approaches have problems and much more research is needed.

Finally, Table 16.1 presents a few selected ongoing and future research areas in colloid and surface science.

Review Problems in Colloid and Surface Chemistry

Problem R.1: Characterization and wetting of Polystyrene (PS) surfaces

The following data are available for a polystyrene (PS) surface:

repeating unit: -[$CH_2CH(C_6H_5)$]-;
critical surface tension: 33 mN m^{-1};
density: 1.04 g cm^{-3};

solubility parameters (dispersion, polar, hydrogen bonding): (8.9, 2.7, 1.8) (cal cm^{-3})$^{1/2}$;
the surface tension components of glycerol are available in the same order (dispersion polar, hydrogen bonding): 20.6, 6.3, 36.5 mN m^{-1}.

The values of all properties in this problem are at room temperature.

1. The critical surface tension of poly(vinyl chloride) (PVC) is 40 and that of Teflon (PTFE) is 18, both in mN m^{-1}. Which of the three polymers, PS, PVC and Teflon is easiest to wet and which one is the most difficult to wet? Explain briefly your answer.
2. Which of the following liquids can completely wet PS: *n*-hexadecane (27.6), formamide (58.2), acetone (23.3) and 1,1-diphenylethane (37.7)? Explain briefly your answer. The values in parentheses are the surface tensions of the liquids, all in mN m^{-1}.
3. Using the Hansen method for the interfacial tension provide an estimation for the dispersion, the polar and the hydrogen bonding contributions of the surface tension of PS. Discuss briefly the result.
4. Estimate, based on the above, the contact angle of glycerol with PS. Compare your result with the experimental value which is 82°.
5. Provide an estimation of the theoretical work of adhesion of glycerol with PS.

Problem R.2: Adsorption and surface active compounds

The following table shows experimental surface tension data for an aqueous solution of nonylphenyl ethoxylate NPE$_{15}$ at 20 °C (the subscript 15 indicates that we have 15 oxyethylene groups):

Concentration C (10^4 C) mol L^{-1}	Surface tension (mN m^{-1})
0.4	57.5
0.8	42.0
1.2	28.0
1.3	33.0
1.6	33.5
2.0	33.5

1. How are the surfactants organized in the solution at CMC? Explain the minimum in the surface tension versus concentration plot.
2. Provide an estimate of the CMC as well as of the adsorption at a concentration equal to 0.2 × 10^{-4} mol L^{-1}.
3. Estimate the HLB value for this surfactant. When an oil is mixed with water, what type of emulsion is expected to be stabilized with this surfactant and why?
4. We would like to stabilize polystyrene (PS) particles by using either the NPE$_{15}$ surfactant or the classical SDS (sodium dodecyl sulfate) anionic surfactant. Which surfactant would you use and why? (Use the HLB–CPP plot of Problem 7.6.)

Problem R.3: Surfactants, emulsions and steric stabilization

For an aqueous solution of the non-ionic surfactant of the polyoxyethylene type (abbreviated as C$_{10}$EO$_5$) CH$_3$(CH$_2$)$_9$(OCH$_2$CH$_2$)$_5$OH, the critical micelle concentration is CMC= 9.0 × 10^{-4} mol L^{-1} at 25 °C. The adsorption at CMC is $\Gamma = 3.2903 \times 10^{-6}$ mol m^{-2}.

1. How many categories of surfactants do you know according to the type of the hydrophilic group? Mention one example for each surfactant category. How does the CMC of the surfactant (C$_{10}$EO$_5$) change if the number of carbon atoms in the hydrophobic chain is increased to 15 while the hydrophilic group remains the same?
2. Calculate (in nm^2) the area occupied by each adsorbed surfactant molecule at the critical micelle concentration as well as the critical packing parameter, CPP, of the surfactant.
3. Using the value you have estimated in the previous question for the surfactant area, make an estimation for the shape of the micellar structure created by the surfactant under these conditions as well as for the number of surfactant molecules expected to be present in one micelle.

    The hydrophilic–lipophilic balance (HLB) and CPP of surfactants can be related via the plot shown in Problem 7.6.
4. Surfactants are often used for stabilizing emulsions. What type of emulsions (oil-in-water or

water-in-oil) is the surfactant $C_{10}EO_5$ likely to stabilize and why?

5. We are asked to stabilize sterically hydrophobic latex particles using one of the following surfactant choices: (i) SDS (sodium dodecyl sulfate), (ii) $C_{10}EO_5$, (iii) NPE_{15} (i.e. an ethoxylated nonyl phenol with 15 ethylene oxide groups) and (iv) a mixture of 30 wt% $C_{16}EO_4$ and 70 wt% $C_{12}EO_{30}$. Which surfactant would you use and why?

Problem R.4: Adsorption based on the Langmuir and the BET equations

1. The dependency of the surface tension, γ, of a liquid solution with concentration, C, can sometimes be expressed as:

$$\gamma = \gamma_0 - R_{ig}T\Gamma_{max} \ \ln(1+kC) \qquad (R.1)$$

where R_{ig} is the ideal gas constant, T is the temperature, γ_0 is the surface tension of solvent, e.g. water, and k and Γ_{max} are constants.

Starting from Equation R.1, show that the adsorption can be expressed as:

$$\Gamma = \frac{\Gamma_{max}kC}{1+kC} \qquad (R.2)$$

What is the name of the adsorption theory represented by Equation R.2 and what is the physical significance of its parameters? Mention some practical applications of Equation R.2.

How can the surface area occupied by an adsorbed molecule at an interface be estimated using Equation R.2?

2. The following data have been obtained for the adsorption of nitrogen on a sample of microtalk (which is a pigment in paints) at 77 K (P/P_o is the relative pressure):

Show that the adsorption data are likely to be represented by the BET equation.

Estimate using the BET equation:

a. the volume for monolayer coverage V_m (in $cm^3 \ g^{-1}$);
b. the specific surface of this solid in $m^2 \ g^{-1}$; assume that the area occupied by one nitrogen molecule is known to be $16.2 \times 10^{-20} \ m^2$; the molar volume of gases at S.T.P. is 22.414 L.

Problem R.5: Zisman plot and work of adhesion

1. Assume the validity of the Fowkes equation for the interfacial tension of a solid–liquid interface. Show that under certain assumptions:

$$\gamma^{crit} = \gamma_s^d$$

where γ^{crit} is the critical surface tension of the solid and γ_s^d is the dispersion part of the surface tension of the solid. Under which assumptions is this equation correct?

2. According to the Fowkes–Young theory for the interfacial tension and under the assumptions where $\gamma^{crit} = \gamma_s^d$, the Zisman plot is given by the equation:

$$\cos\vartheta = 1 - \frac{\gamma_l - \gamma_s^d}{\gamma_s^d} \quad \text{when } \gamma_l > \gamma_s^d$$

(of course $\cos\vartheta = 1$ when $\gamma_l < \gamma_s^d = \gamma^{crit}$); ϑ is the contact angle.

Derive an expression for the work of adhesion as a function of the surface tension of the liquid, γ_l, and the dispersion part of the surface tension of the solid, γ_s^d. Assume that the spreading pressure is zero. What is the value for the work of adhesion when $\gamma_l = \gamma_s^d$?

3. For which value of the liquid surface tension has the work of adhesion a maximum?

P/P_o	0.02	0.05	0.1	0.2	0.3	0.4	0.6	0.8	0.9
Volume of gas adsorbed per g. of solid ($cm^3 \ g^{-1}$)	0.42	0.53	0.67	0.84	0.95	1.05	1.33	1.61	2.35

Problem R.6: Wetting and adhesion
The following plot shows the cosine of the contact angle of nine liquids on a solid polyester (poly(ethylene terephthalate), PET) surface against the surface tension of the liquids. All measurements have been carried out at 20 °C.

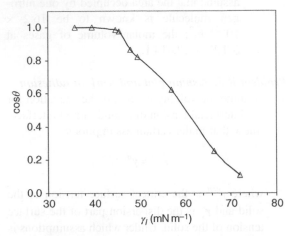

1. What is the name of this plot and how can it be used to estimate the wetting of solid surfaces? Which surfaces are easier to wet, the metals or the polymers and why?
2. What is, based on this plot, the critical surface tension of PET? Is it easier to wet a PET or a polyethylene (31.0 mN m^{-1}) or a Teflon (18.0 mN m^{-1}) surface? Which of the three polymers is easiest to wet and which is the most difficult to wet? Explain briefly your answer. The numbers in parenthesis are the critical surface tension values at 20 °C.
3. Which of following liquids will completely wet PET: water (71.99), benzonitrile (35.79), pyridazine (49.51), glycerol (63.4), adiponitrile (45.5) and acetone (23.3)? Explain briefly your answer. The numbers in parenthesis are the surface tensions in mN m^{-1} at 20 °C.
4. Calculate the theoretical work of adhesion of glycerol and acetone with PET. The surface tension values for the two liquids are given in question 3.
5. Explain briefly how the solid surface tension of a polymer as well as its van der Waals and Lewis acid and Lewis base components can be estimated using the van Oss *et al.* theory. How

many experimental measurements are required (as minimum)?
6. Show, using the Hansen/Beerbower method for the interfacial tension, how the work of adhesion between a liquid and a solid can be estimated.
7. The surface tension of *n*-butanol is 24.6 mN m^{-1} at 20 °C. Estimate, with a method of your own choice, the interfacial tension of a water–*n*-butanol liquid–liquid interface and compare the result to the experimental value (=1.8 mN m^{-1}).

Problem R.7: Wetting of PMMA surfaces
The following data are available for a poly(methyl methacrylate) (PMMA) surface:

repeating unit: -[CH$_2$C(CH$_3$)COOCH$_3$]-;
critical surface tension: 39 mN m^{-1};
density: 1.18 g cm^{-3};
solubility parameters (dispersion, polar, hydrogen bonding): (8.6, 5.6, 2.1) (cal cm^{-3})$^{1/2}$;
the surface tension components of formamide are available in the same order (dispersion polar, hydrogen bonding): 16.5, 27.3, 14.4 mN m^{-1}.

The values of all properties in this problem are at room temperature.
The table below shows the contact angle of several liquids on a PMMA surface.

Liquid	Contact angle (°)	Surface tension (mN m^{-1})
Water	80	72.8
Glycerol	69	63.3
Formamide	57	58.3
Diethylene glycol	30	44.7
Nitromethane	10	36.0
Acetone	0	23.3

1. Explain briefly "the critical surface tension" concept and how its value for PMMA has been estimated. Which of the following liquids will fully wet PMMA: diethylene glycol, formamide and acetone?

2. Using the Hansen method, provide an estimation of the dispersion, the polar and the hydrogen bonding contributions of the surface tension of PMMA ($\gamma_d, \gamma_p, \gamma_h$). Discuss briefly the result.

3. Estimate, based on the above, the contact angle between formamide and the PMMA surface. Compare your results with the experimental value.

4. The Hansen radius of solubility of PMMA is equal to 3.1 $(cal\ cm^{-3})^{1/2}$. Are benzene and methanol good solvents for PMMA? Discuss briefly the results.

Problem R.8: Solid surfaces

1. Explain briefly the Zisman plot and how it can be used for estimating the wetting of solid surfaces.

2. Mention the advantages and the limitations of the Zisman plot approach.

3. The critical surface values of Teflon, polyethylene, poly(methyl methacrylate) and polyacrylonitrile are 18, 31, 39 and 44 mN m^{-1}, respectively. Which of the liquids mentioned in Table 3.1 can fully wet these polymers?

4. Why can objects in outer space be produced using thinner (but still strong) materials?

5. How can solid surfaces be characterized using contact angle data and a theory for the interfacial tension? How many experimental data are needed?

6. Show that in the case of the Girifalco–Good theory the contact angle is related to the solid surface tension via the following equation:

$$\cos\vartheta = 2\phi \left(\frac{\gamma_s}{\gamma_l}\right)^{1/2} - 1 - \frac{\pi_{SV}}{\gamma_l}$$

For which surfaces can the spreading pressure π_{SV} be neglected?

7. Derive, under the assumption of negligible spreading pressure, the relationship between the critical surface tension and the solid surface tension for the Girifalco–Good theory.

Index

Introduction to Applied Colloid and Surface Chemistry, First Edition. Georgios M. Kontogeorgis and Søren Kiil.
© 2016 John Wiley & Sons, Ltd. Published 2016 by John Wiley & Sons, Ltd.
Companion website: www.wiley.com/go/kontogeorgis/colloid

Printed and bound by CPI Group (UK) Ltd, Croydon, CR0 4YY

27/10/2024

14580310-0001